Neuroimaging II

Clinical Applications

HUMAN BRAIN FUNCTION
Assessment and Rehabilitation

SERIES EDITORS:

Antonio Puente, *University of North Carolina at Wilmington North Carolina*

Gerald Goldstein, *Veterans Administration Medical Center Pittsburgh, Pennsylvania*

Erin D. Bigler, *Brigham Young University, Provo, Utah*

NEUROIMAGING I: Basic Science
Edited by Erin D. Bigler

NEUROIMAGING II: Clinical Applications
Edited by Erin D. Bigler

A Continuation Order Plan is available for this series. A continuation order will bring delivery of each new volume immediately upon publication. Volumes are billed only upon actual shipment. For further information please contact the publisher.

Neuroimaging II

Clinical Applications

Edited by

Erin D. Bigler

Brigham Young University
Provo, Utah

PLENUM PRESS • NEW YORK AND LONDON

Library of Congress Cataloging-in-Publication Data

On file

ISBN 0-306-45229-4

A Division of Plenum Publishing Corporation
233 Spring Street, New York, N. Y. 10013

10 9 8 7 6 5 4 3 2 1

Printed in the United States of America

Contributors

ABASS ALAVI, *Division of Nuclear Medicine, Hospital of the University of Pennsylvania, Philadelphia, Pennsylvania 19104*

ERIN D. BIGLER, *Department of Psychology, Brigham Young University, Provo, Utah 84602-5372*

ROBERT B. BURR, *Department of Neurosurgery, University of Utah Medical Center, Salt Lake City, Utah 84112*

B. J. CASEY, *Western Psychiatric Institute and Clinic, University of Pittsburgh Medical Center, Pittsburgh, Pennsylvania 15213*

GORDON J. CHELUNE, *Departments of Psychiatry and Psychology and Neurology, The Cleveland Clinic Foundation, Cleveland, Ohio 44195*

FRANCOIS CHOLLET, *Service de Neurologie B, Hôpital Purpan, Toulouse 31059, France*

JONATHAN D. COHEN, *University of Pittsburgh Medical Center, and Carnegie Mellon University, Pittsburgh, Pennsylvania 15213*

C. MUNRO CULLUM, *Departments of Psychiatry and Neurology, University of Texas Southwestern Medical Center, Dallas, Texas 75235-8898*

ELANA FARACE, *Department of Psychology, University of Virginia, Charlottesville, Virginia 22903-2477*

KATHLEEN A. FLANNERY, *Department of Psychiatry, Dartmouth Medical School, Lebanon, New Hampshire 03756*

LAURA A. FLASHMAN, *New Hampshire Hospital, Concord, New Hampshire, and Department of Psychiatry, Dartmouth Medical School, Lebanon, New Hampshire 03756*

RICHARD S. J. FRACKOWIAK, *Wellcome Department of Cognitive Neurology, Queen Square, London WC1 N3B6, England*

Eric B. Geller, *Department of Neurology, The Cleveland Clinic Foundation, Cleveland, Ohio 44195*

Jay N. Giedd, *National Institute of Mental Health, Bethesda, Maryland 20892*

Josette G. Harris, *Department of Psychiatry, University of Colorado Health Science Center, Denver, Colorado 80262*

John M. Hoffman, *Department of Radiology and Medicine, Division of Neurology, Duke University Medical Center, Durham, North Carolina 27713*

Narinder Kapur, *Wessex Neurological Centre, Southampton General Hospital, Southampton S016 64D, England*

John Lauriello, *Department of Psychiatry and Behavioral Sciences, Stanford University School of Medicine, Stanford, California 94305; and Psychiatry Service, Veterans Affairs Palo Alto Health Care System, Palo Alto, California 94304*

Arthur C. Maerlender, *New Hampshire Hospital, Concord, New Hampshire, and Department of Psychiatry, Dartmouth Medical School, Lebanon, New Hampshire 03756*

Laura Marsh, *Department of Psychiatry and Behavioral Sciences, Stanford University School of Medicine, Stanford, California 94305; and Psychiatry Service, Veterans Affairs Palo Alto Health Care System, Palo Alto, California 94304*

Richard I. Naugle, *Departments of Psychiatry and Psychology and Neurology, The Cleveland Clinic Foundation, Cleveland, Ohio 44195*

Andrew B. Newberg, *Division of Nuclear Medicine, HUP, Philadelphia, Pennsylvania 19104*

Douglas C. Noll, *University of Pittsburgh Medical Center, and Carnegie Mellon University, Pittsburgh, Pennsylvania 15213*

Adolf Pfefferbaum, *Department of Psychiatry and Behavioral Sciences, Stanford University School of Medicine, Stanford, California 94305; and Psychiatry Service, Veterans Affairs Palo Alto Health Care System, Palo Alto, California 94304*

Judith L. Rapoport, *National Institute of Mental Health, Bethesda, Maryland 20892*

Naftali Raz, *Department of Psychology, The University of Memphis, Memphis, Tennessee 38152*

Henry J. Riordan, *Department of Psychiatry, Dartmouth Medical School, Lebanon, New Hampshire 03756*

Andrew J. Saykin, *Departments of Psychiatry and Radiology, Dartmouth Medical School, Lebanon, New Hampshire 03756; and New Hampshire Hospital, Concord, New Hampshire 03301*

Walter Schneider, *University of Pittsburgh, Pittsburgh, Pennsylvania 15213*

Edith V. Sullivan, *Department of Psychiatry and Behavioral Sciences, Stanford University School of Medicine, Stanford, California 94305; and Psychiatry Service, Veterans Affairs Palo Alto Health Care System, Palo Alto, California 94304*

Eric Turkheimer, *Department of Psychology, University of Virginia, Charlottesville, Virginia 22903-2477*

JOHN B. WEAVER, *Department of Radiology, Dartmouth Medical School, Lebanon, New Hampshire 03756*

CORNELIUS WEILLER, *Klinik für Neurologie, Friedrich-Schiller-Universität Jena, 07740 Jena, Germany*

KATHLEEN A. WELSH-BOHMER, *Joseph & Kathleen Bryan Alzheimer's Disease Research Center, and Department of Psychiatry, Duke University Medical Center, Durham, North Carolina 27713*

RONALD A. YEO, *Department of Psychology, University of New Mexico, Albuquerque, New Mexico 87131*

Preface

The focus of Volume I of the *Handbook of Human Brain Function* was on basic scientific principles of brain imaging as it relates to the study of human brain function. Once the scientific bases for a particular discipline are established, clinical applications follow. Such is the status of brain imaging in the study of human brain function. It is of interest to note that the 1952 Nobel Prize for Physics was awarded to Felix Bloch and Edward Purcell, who discovered that nuclei precessing in the radiofrequency range could emit a radiofrquency signal detected by a radio receiver. Their findings initiated a series of very basic research studies on the characteristics of nuclear magnetic resonance. It would take over 25 years of basic research before findings began to point toward truly biomedical applications. However, once realized, clinical applications became standard fare for nuclear magnetic resonance. The example of Bloch and Purcell's work in an area of very basic science expanding to clinical application has been repeated throughout the medical and neurological sciences. This type of progress is what drives science. As a benefit from these scientific advances, research, clinical, and diagnostic imaging from a variety of modalities, not just computerized tomography or magnetic resonance imaging, can be performed. This volume focuses on the clinical applications of various neuroimaging methods.

Chapter 1 introduces the topic of clinical neuroimaging in the study of human brain function. In Chapter 2, Cullum and Harris discuss the integration of neuroimaging with neurophysiological and neurobehavioral methods. With regard to contemporary clinical application, Chapter 3 by Newberg and Alavi reviews common neurological disorders and their typical clinical imaging presentation. As a counterpart, Marsh and colleagues review neuroimaging findings by general psychiatric classification in Chapter 4. Farace and Turkheimer address the issue of gender differences assessed by neuroimaging in Chapter 5, followed by Raz's discussion of aging in Chapter 6. Chapters 1–6 deal with general issues in clinical neuroimaging. Chapters 7–13 deal either with specific clinical syndromes or with the application of a most exciting new technique—

functional magnetic resonance imaging (fMRI). Chapter 7 by Welsh-Bohmer and Hoffman deals with positron emission tomography (PET) imaging and dementia. Memory is addressed in Chapter 8 by Naugle and colleagues and in Chapter 11 by Kapur. Brain injury is reviewed in Chapter 9 by Wheeler and in Chapter 10 by myself. Chapter 12 and 13 deal with fMRI. The volume concludes with Chapter 14, an integration chapter and a look toward the future by Yeo.

Erin D. Bigler

Acknowledgments

The assistance of Tracy Abildskov in preparation of various aspects of both volumes I and II of this series is gratefully acknowledged. Likewise, several of the illustrations in both volumes and in parts of the Appendix in Volume I have relied significantly on the ANALYZE™ software program, Biomedical Imaging Resource, Mayo Foundation.

Contents

CHAPTER 3

Neuroimaging in Neurological Disorders

Andrew B. Newberg and Abass Alavi

CHAPTER 4

Neuroimaging in Psychiatric Disorders

Laura Marsh, John Lauriello, Edith V. Sullivan, and Adolf Pfefferbaum

CHAPTER 5

Gender Differences in Brain Morphometry and Function

Elana Farace and Eric Turkheimer

CHAPTER 6

Neuroanatomy of the Aging Brain Observed *in Vivo*: A Review of Structural MRI Findings

Naftali Raz

PART II CLINICAL SYNDROMES

CHAPTER 7

Positron Emission Tomography Neuroimaging in Dementia

Kathleen A. Welsh-Bohmer and John M. Hoffman

CHAPTER 8

Functional and Structural Measures for Determining Risk Memory Change following Epilepsy Surgery

Richard I. Naugle, Gordon J. Chelune, and Eric B. Geller

1

Introduction

Erin D. Bigler

Of all the major technical revolutions in the clinical neurosciences, imaging methods of visualizing the brain have had the most dramatic impact on diagnostic decision making (Eisenberg, 1994; Oldendorf, 1980). Computerized tomography (CT) and magnetic resonance (MR) imaging have profoundly altered neurodiagnostics. Such techniques have also provided new methods for clinical and basic research in human brain function.

In the early part of this century, with the advent of the x-ray and subsequent development of pneumoencephalography, the skull and a shadow of the ventricular system could be visualized, but the brain could not (see Bigler, Yeo, & Turkheimer, 1989). The brain was still a "black box," inaccessible to direct visualization. During this era, behavioral and neurological scientists studied characteristics of stimulus input and some type of physiological or behavioral response output. However, the *in vivo* workings or structure of the brain could not be directly investigated, short of a neurosurgical procedure. Thus, prior to contemporary imaging methods, only by studying neurosurgical cases, cases with clearly defined lesions (i.e., left hemisphere stroke) or clinically distinct syndromes (i.e., Huntington's disease), and/or postmortem examinations of patients in whom antemortem assessment had been completed did clinical neuroscience progress. For the first seven decades of this century, a direct image of the brain remained elusive. During this period, understanding brain structure-function was basically inferential, due to the inability to study directly the structure or function of the living human brain.

A radical change in the early 1970s began with the work of Godfrey Hounsfield, a physicist, who developed the first working model of what was to

Erin D. Bigler Department of Psychology, Brigham Young University, Provo, Utah 84602-5372.

Neuroimaging II: Clinical Applications, edited by Erin D. Bigler. Plenum Press, New York, 1996.

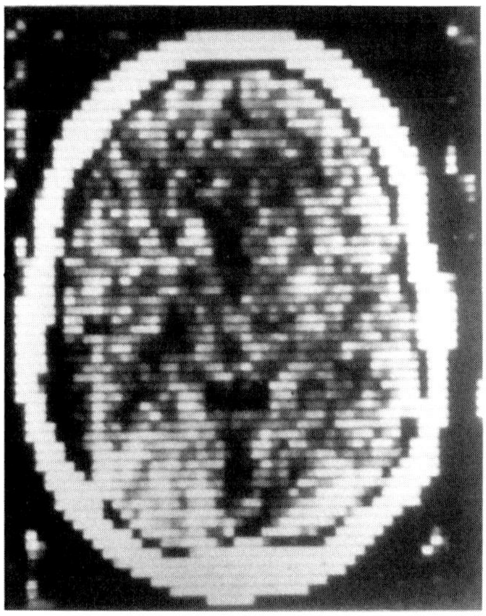

Figure 1. First generation CT scan depicting a very crude image of the brain and skull (outer rim depicted in white). Note that an outline of the ventricular system (dark centrally located) and some other CSF spaces can be observed. Note the large pixel size used to create this image.

become computerized axial tomography (Hounsfield, 1980). Although Hounsfield as well as others (Oldendorf, 1980) had long worked out some of the mathematics and physics of such instrumentation, it took rapid progress in the development of computer technology before the first prototype machines could be developed (Figure 1). However, once developed there was no retreat and the explosion in methods and refinement of brain imaging technology was under way (Figure 2). In a matter of two decades, neuroimaging technology progressed from detection of only gross structural features of the brain (Figure 1) to the refinement of exquisite structural visualization exemplified by MR imaging (Figure 3).

The rapid improvement in imaging technology was driven at an astronomical rate by increasing computer capabilities. The focus during that era was for better and better image quality for clinical detection of pathological states. These improvements in image quality proceeded at a rate far faster than techniques to objectively study brain pathology. In the past two decades, the primitive imaging method of the earliest CT scanner has rapidly improved. Interestingly, because of the speed with which image acquisition and display methods developed, no standard or systematic methods for image analysis were universally established. In these early days of imaging research and clinical application, most systematic research used some type of clinical rating scale or the imaging findings for gross classification (i.e., presence/absence of a tumor, left versus right hemisphere pathology, etc.). The first quantitative approaches to image analysis were with operator controlled planimetric measurements done by hand (see Turkheimer *et al.*, 1984) followed by similar but more automated methods for computer assisted quantitative analysis (see Bigler, Yeo, & Turkheimer, 1989). These early methods of quantification were slow and time-consuming, and a variety of technical problems limited the scope of quantitative image analysis. For example, with CT imaging, problems with clear differentiation between white and gray matter boundaries made it impossible to perform accurate volumetric analyses of whole brain white or gray matter, and bone artifacts made quantification of subtentorial structures difficult. Bone artifact also made differentiation of gyral surface areas and cerebrospinal fluid (CSF) spaces difficult. Thus, the majority of imaging studies addressing clinical questions did not use quantitative, objec-

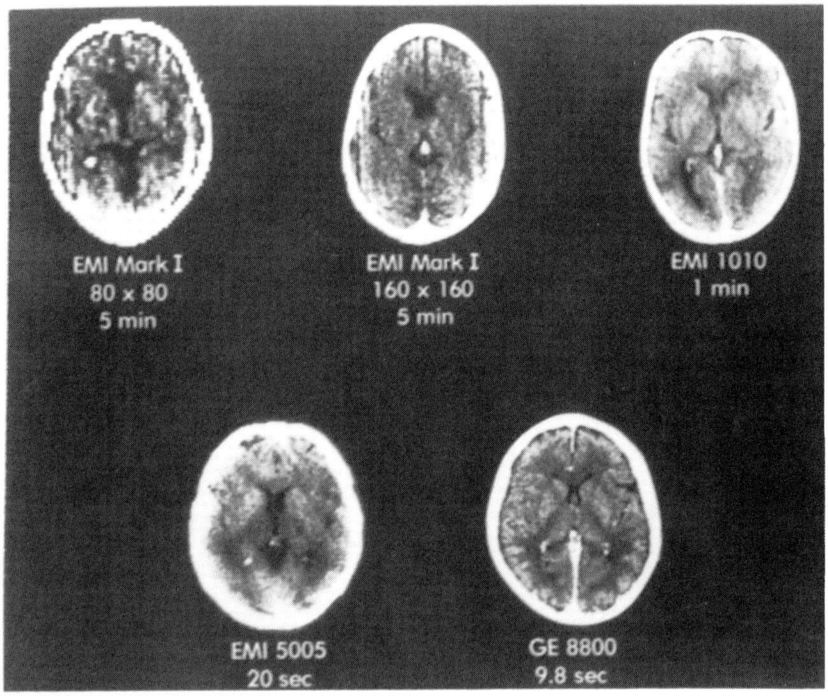

Figure 2. Evolution of CT imaging: top left, 1974; top middle, 1976; top right, 1977; bottom left, 1978; bottom right, 1980. Note the improvement in tissue differentiation with each advancement. The time beneath each scan image represents the time required for each slice acquisition. (Reproduced with permission from Margulis, 1987.)

tive measures for CT analysis but were relegated to clinical rating schemes or simple planimetric measurements. Such measurements permitted only basic quantification of a few brain parameters (i.e., simple single CT section ventricle-to-brain ratio, etc.). It was a case where technology expanded so quickly that a vacuum for image analysis was created. Accordingly, investigations of brain-behavior relationships using CT imaging technology could address questions at only the most basic or gross pathology level.

The same problem confronted the new technology of MR imaging that came in the mid 1980s. Exquisite images of the brain could now be displayed, but how to quantitatively capture that information? Jernigan and colleagues (Jernigan, Press, & Hesselink, 1990; Jernigan, Trauner, Hesselink, & Tallal, 1991), as well as others (see review by Robb, 1995), introduced some computer automated algorithms that permit "segmentation" of the brain into its three major components—white matter, gray matter, and CSF space. Since excellent differentiation between these structures can be achieved with MR imaging, by separating or segmenting each target structure surface area, accurate *in vivo* volumes could be obtained. From such methods, calculation of volumes/areas of any given structure could be achieved, from the whole brain down to a specific structure, such as the hippocampus. These methods are depicted in Figure 3.

Segmentation methods can then be applied to three-dimensional (3-D) image analysis, so that the component parts (either a brain structure or area of pathology) can be displayed from any conceivable angle or orientation. The segmented images obtained from the traumatic brain injury (TBI) patient depicted in Figure 4 were utilized to generate 3-D images of the cortical surface of

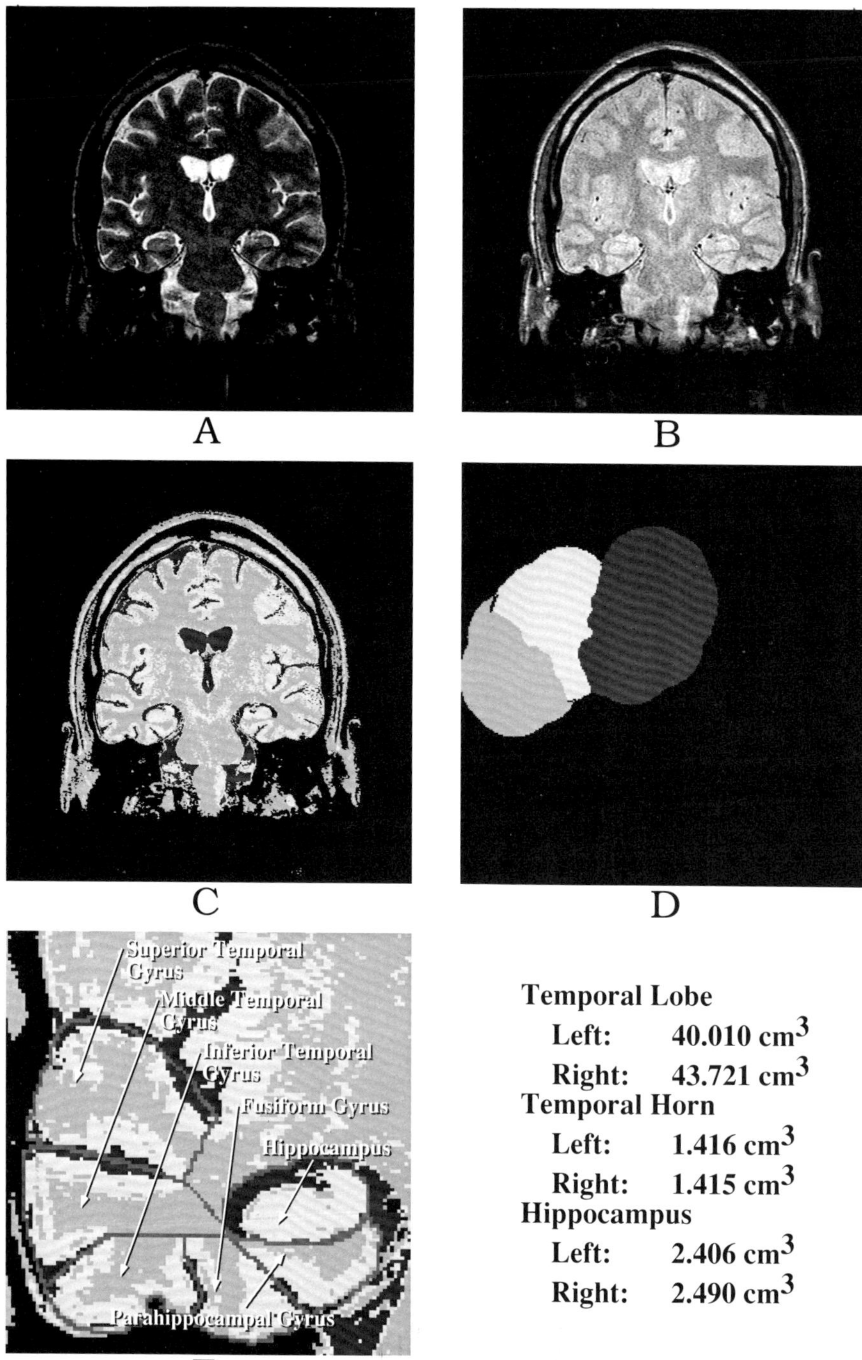

Figure 3. MR brain segmentation. (A) Coronal T2-weighted MR image at the level of the hippocampus. (B) Equally registered proton density–weighted image. (C) Multispectral image based on the combined T2- and proton density–weighted images segmenting the brain into three component parts—white matter (dark gray), gray matter (light gray), and CSF (darkest gray). (D) Two-dimensional feature space map that represents the segmented image pixel classification into the three regions where the proton density image is depicted on the *x*-axis and the T2-weighted image is on the *y*-axis. By developing such a feature space map of all pixel intensities, all aspects of brain parenchyma and CSF are included. (E) Close-up of the temporal lobe depicting the segmented image highlighting the

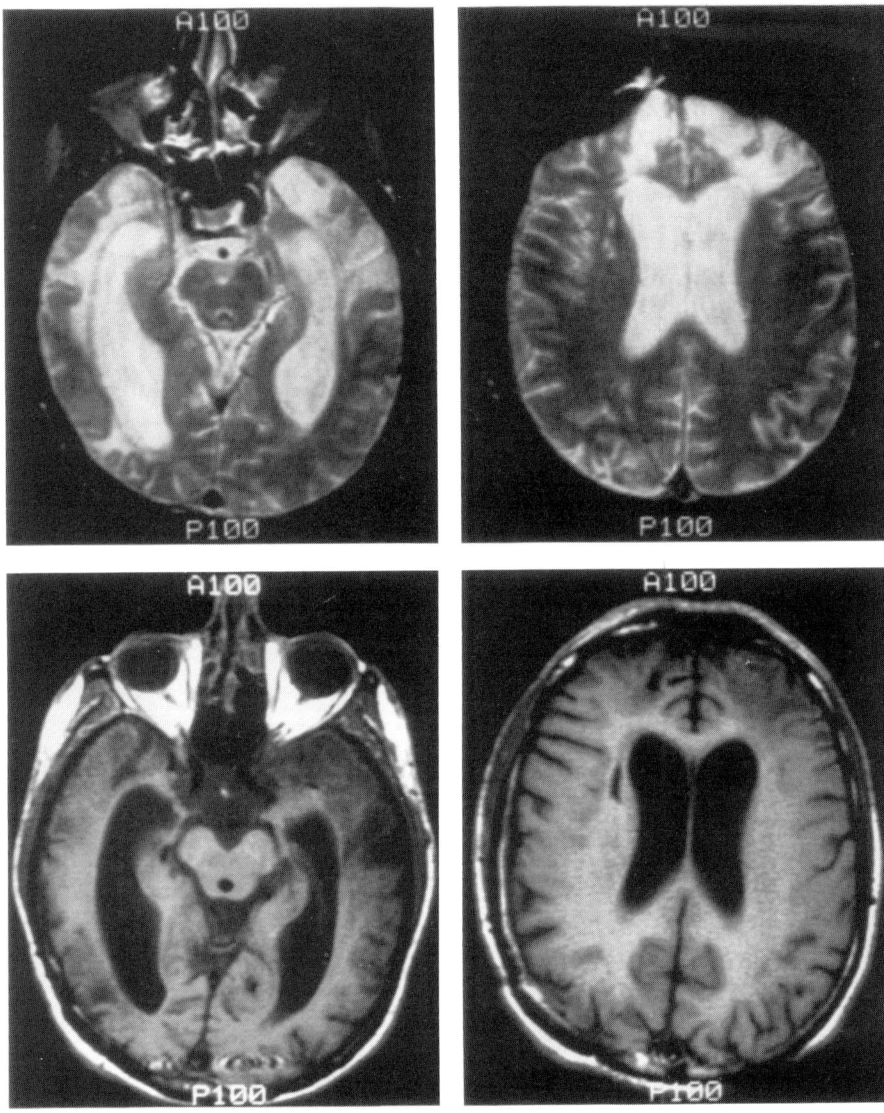

Figure 4. Axial T2 (top) and T1 (bottom) MR images from a patient who sustained a severe TBI. Note the ventricular dilation and various areas of cortical encephalomalacia. Compare these 2-D figures to the 3-D figures presented in Figure 5 from this same patient.

the brain which displays areas of contusion and a dilated ventricular system (Figure 5). These 3-D images are particularly useful for visualization of areas of pathology and many of the subcortical structures, which because of their odd shape, angularity, and irregularity, are difficult to observe at any one level. For example, the ventricular system with its compartments and interconnecting channels is difficult to appreciate in two-dimensional perspective (as seen in Figure 4). However, by using the above methods, 3-D views of the ventricular system can be readily visualized (Figure 5).

←

hippocampus and the five temporal lobe gyri. The numbers presented on the lower right represent volumes of the temporal lobes, temporal horns and hippocampus for this subject. Segmentation based on the ANALYZE™, Biomedical Imaging Resource, Mayo Foundation (see Robb, 1995).

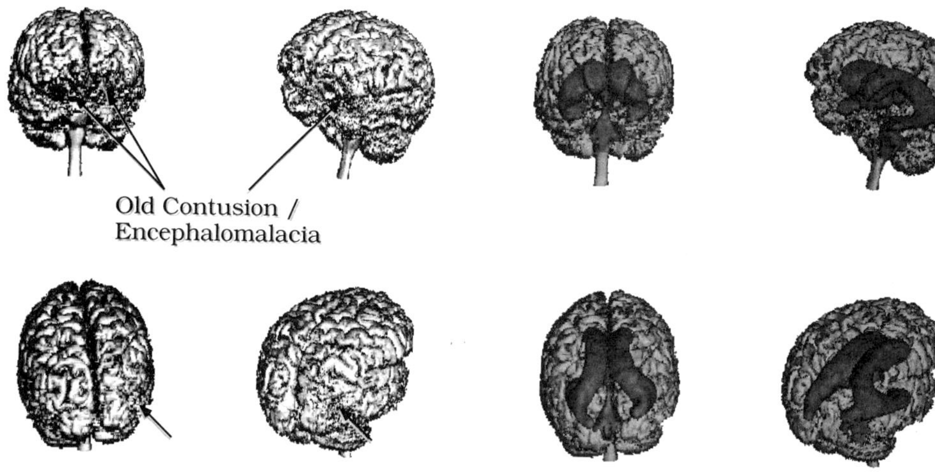

Figure 5. Three-dimensional depiction of the areas of frontal-temporal encephalomalacia as a consequence of old cortical contusion that developed in response to a fall of 30 feet. The patient struck the back of his head and does have an area of focal contusion in the right posterior temporo-occipital cortex, but the major areas of contusion occurred in the frontotemporal region. The ventricular system is depicted at the right, with the ventricle portrayed in dark gray. Note that with these methods the brain can be visualized from any angle, and even structures that are difficult to portray, such as the ventricular system, can be readily segmented from brain parenchyma.

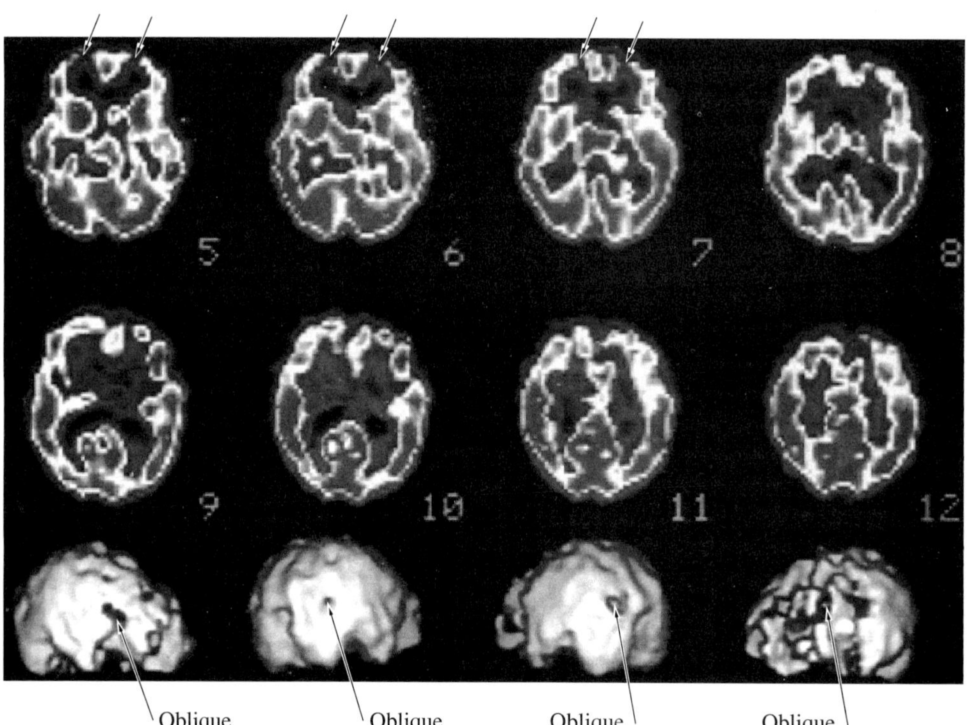

Figure 6. Surface imaging of brain SPECT (bottom row). This is a brain SPECT from a patient with a history of head injury. Notice the disruption in cortical perfusion in the frontal region of the brain as presented in various axial views. This is appreciated more directly in the surface rendered images, particularly in the left frontal oblique view.

Thus, contemporary methods now permit not only single-slice depiction of gross brain morphology using MR technology, but also 3-D representation of any isolated structure. With this excellent method for *in vivo* visualization of brain structure, other methods that relate more to function evolved simultaneously, namely positron emission tomography (PET), single photon emission computed tomography (SPECT), and various computerized methods to assess brain electrophysiology (quantitative electroencephalography) or magnetic field properties (i.e., magnetoencephalography). These methods can be subjected to image analysis and surface rendering so that 3-D images can be portrayed as well (see Figure 6).

Now that the bases of image analysis and quantification have been established, researchers and clinicians have moved on to address questions of brain-behavior relationships using quantitative methods and interfacing different methods of imaging technology. For example, the exquisite gross anatomy depiction of MR can be interfaced with PET so that both structure and function can be portrayed (see Figure 7). In the last five years, a literal explosion of brain imaging research

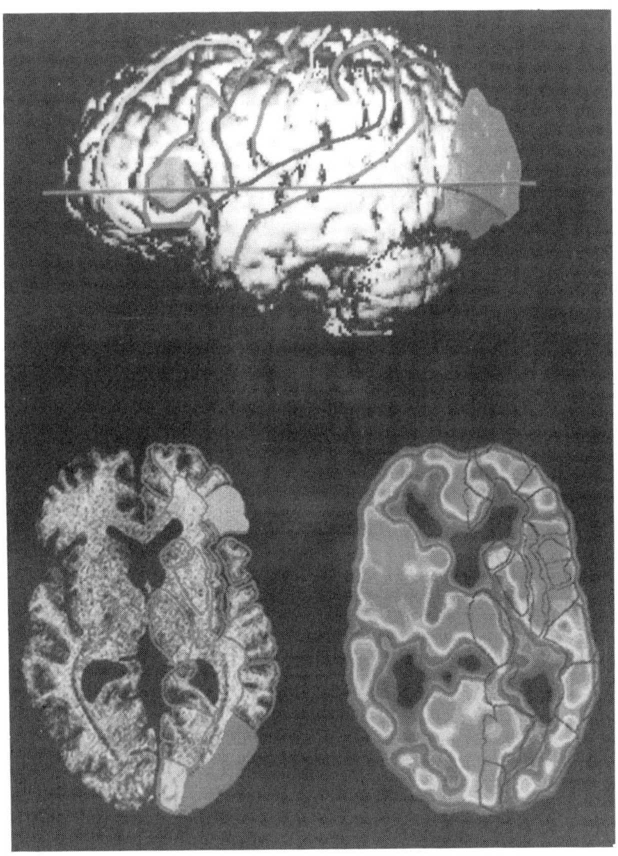

Figure 7. (Top) Three-dimensional brain surface reconstruction based on MR with a horizontal section line depicting the position of both single-slice MR (bottom left) and PET image section (bottom right) so that regions of interest (ROIs) can be overlaid. The axial MR image has several identifiable ROIs and by superimposing the PET image that has been coregistered with the MR, activation patterns based on the PET procedure can be examined by ROI. This image sequence is based on a study by Mellet, Tzourio, Denis, and Mazoyer (1995), who examined PET activation findings in visual and mental exploration. In this study it was determined that the superior occipital cortex (depicted in a dark shade of gray) was involved in the generation and maintenance of visual mental images. (Reproduced with permission from Mellett et al., 1995.)

has occurred. Building on this explosion of information, this text demonstrates various clinical applications of brain imaging technologies in the clinical neurosciences. Applied techniques and quantitative analysis methods are discussed where appropriate. Each chapter deals with a different facet of this most important advancement in the clinical neurosciences, the imaging of the living human brain.

References

Bigler, E. D., Yeo, R. A., & Turkheimer, E. (1989). *Neuropsychological function and brain imaging.* New York: Plenum Press.

Eisenberg, R. L. (1992). *Radiology: An illustrated history.* St. Louis, MO: Mosby Year Book.

Eisenberg, R. L. (1990). *Radiology: An illustrated history.* St. Louis, MO: Mosby Yearbook.

Hounsfield, G. N. (1980). Computed medical imaging: Nobel Lecture. December 8, 1979. *Journal of Computed Tomography, 4,* 665–674.

Jernigan, T. L., Press, G. A., & Hesselink, J. R. (1990). Methods for measuring brain morphologic features on magnetic resonance images: Validation and normal aging. *Archives of Neurology, 47,* 27–32.

Jernigan, T. L., Trauner, D. A., Hesselink, J. R., & Tallal, P. A. (1991). Maturation of human cerebrum observed *in vitro* during adolescence. *Brain, 114,* 2037–2049.

Margulis, A. R. (1987). Radiologic imaging: Changing costs, greater benefits. *American Journal of Roentgenology, 136,* 657–665.

Mellet, E., Tzourio, N., Denis, M., & Mazoyer, B. (1995). A positron emission tomography study of visual and mental spatial exploration. *Journal of Cognitive Neuroscience, 7,* 433–445.

Oldendorf, W. H. (1980). *The quest for an image of the brain.* New York: Raven Press.

Robb, R. A. (1995). Three-dimensional biomedical imaging: Principles and practice. New York: VCH Publishers.

Turkheimer, E., Cullum, C. M., Hubler, D., Paver, S., Yeo, R. A., & Bigler, E. D. (1984). Quantifying cortical atrophy. *Journal of Neurology, Neurosurgery, and Psychiatry, 47,* 1314–1318.

I

General Issues

2

Neuroimaging, Neurophysiological, and Neurobehavioral Techniques
A Beginning Synthesis

C. MUNRO CULLUM AND JOSETTE G. HARRIS

INTRODUCTION

The rapid developments in neuroimaging and neuroscience techniques over the past decade have been remarkable, and a number of procedures now exist for assessing and visualizing various aspects of brain structure and function. As new developments emerge and continue to evolve, we are also likely to see an increased use of *combinations* of techniques. For example, in the early days of computerized tomography (CT), a heretofore unprecedented depiction of gross brain morphology was available, yet it was only with the addition of neuropsychological and neurophysiological techniques to CT information that the important questions of structure-function relationships became more amenable to investigation. This heralded a new era in our understanding of brain structure-function interrelationships. Other important techniques that are discussed throughout this text include magnetic resonance imaging (MRI) for structural imaging and positron emission tomography (PET), single photon emission computed tomography (SPECT), functional MRI, quantitative electroencephalogra-

---

C. MUNRO CULLUM Departments of Psychiatry and Neurology, University of Texas Southwestern Medical Center, Dallas, Texas 75235-8898. JOSETTE G. HARRIS Department of Psychiatry, University of Colorado Health Science Center, Denver, Colorado 80262.

Neuroimaging II: Clinical Applications, edited by Erin D. Bigler. Plenum Press, New York, 1996.

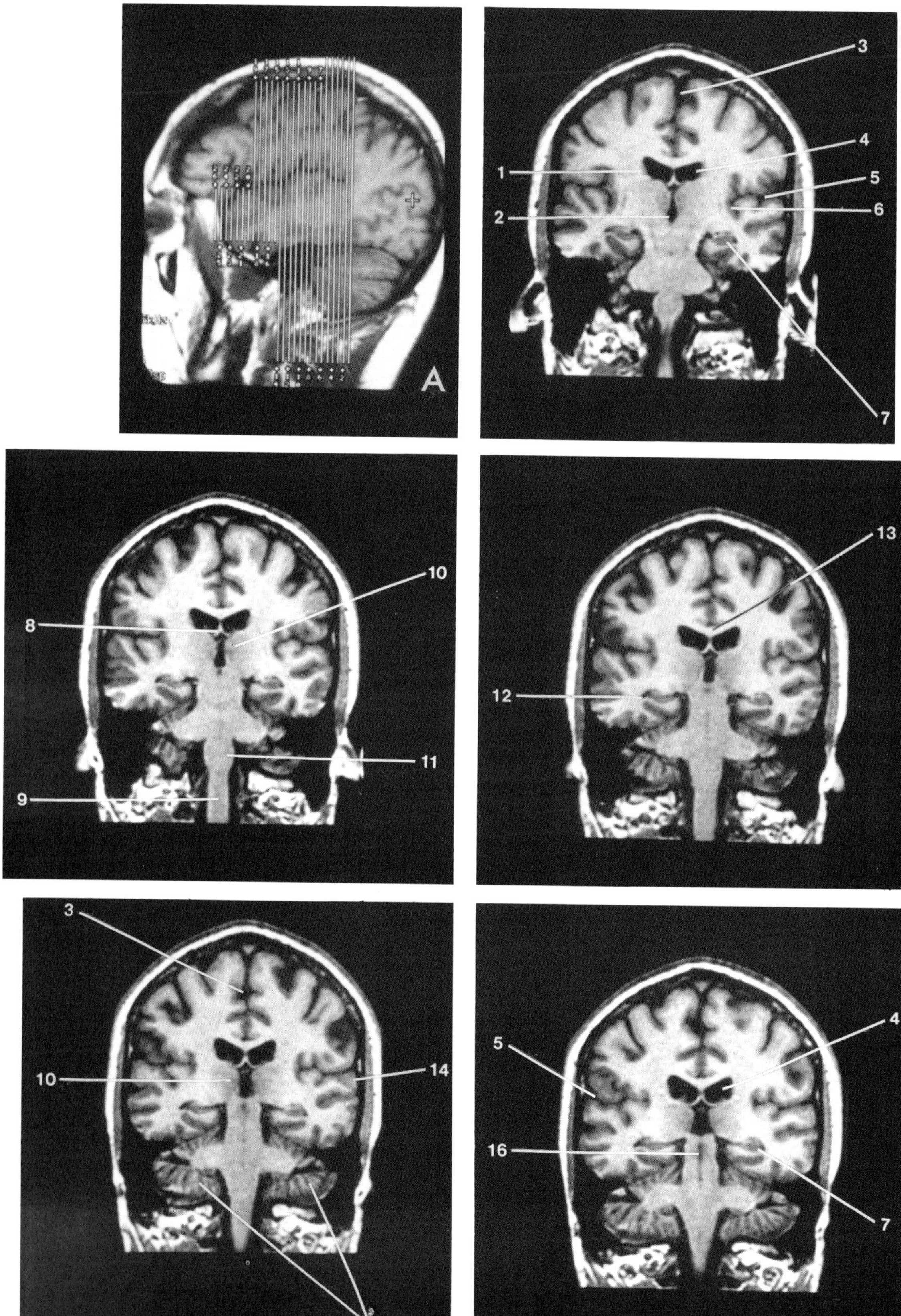

Figure 1.

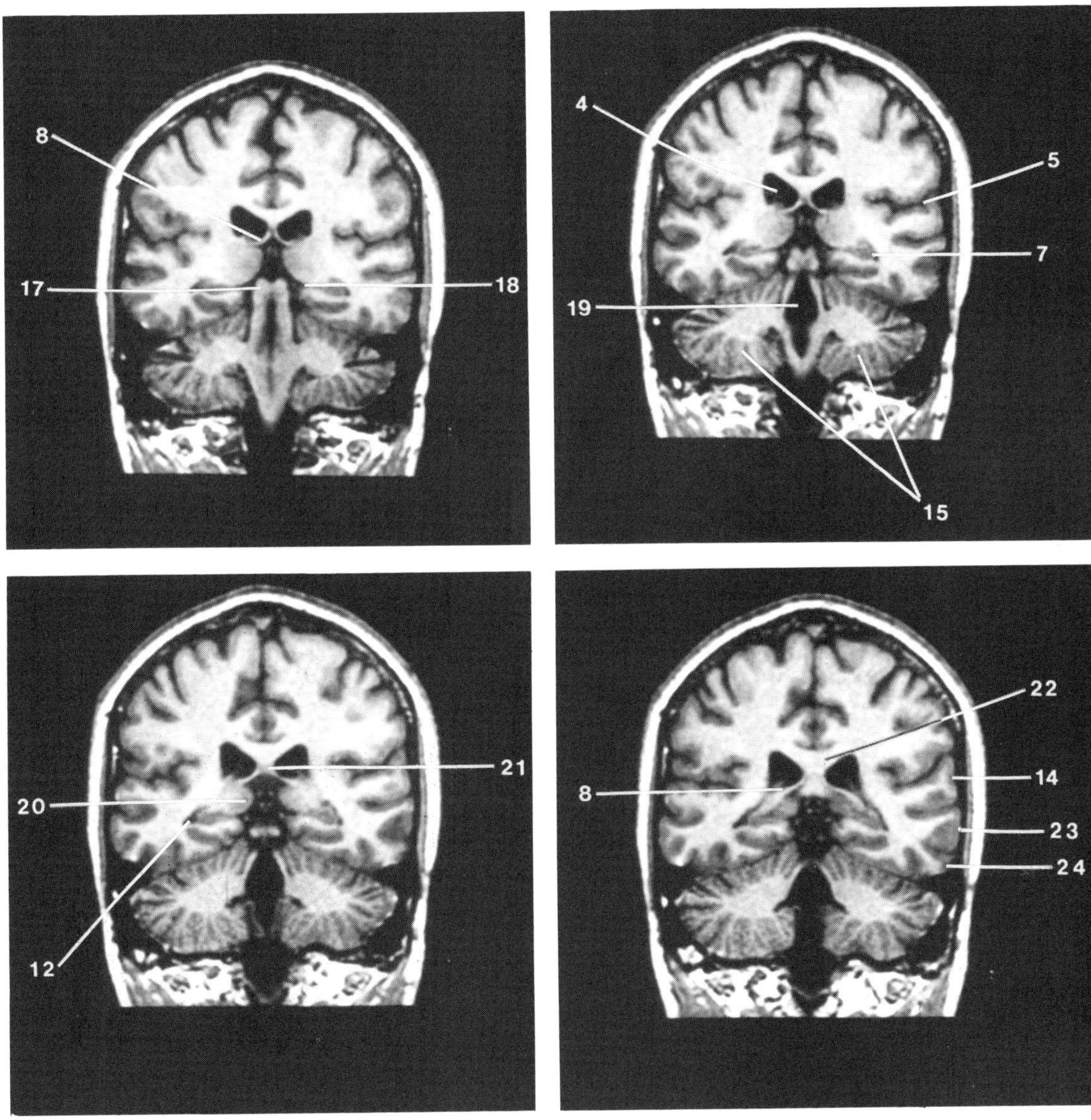

Figure 1. Selected 1.5-mm coronal T 1-weighted spoiled gradient echo MR images acquired in 3-D mode. (A) Orientation of the acquired slices. Some of the major structures are indicated and can be followed in the anterior-posterior plane across serial sections: 1) caudate nucleus, 2) third ventricle, 3) interhemispheric fissure, 4) body of lateral ventricle, 5) sylvian fissure, 6) insula, 7) hippocampus, 8) fornix, 9) medulla, 10) thalamus, 11) pons, 12) posterior horn of lateral ventricle, 13) body of corpus callosum, 14) superior temporal gyrus, 15) cerebellar hemispheres, 16) sylvian aqueduct, 17) tectum, 18) choroidal fissure, 19) fourth ventricle, 20) pulvinar (of thalamus), 21) forniceal commissure, 22) splenium of corpus callosum, 23) middle temporal gyrus, 24) inferior temporal gyrus.

phy (EEG), and magnetoencephalography (MEG) for assessing aspects of brain function. In this chapter, we will provide brief overviews of some of these areas as they relate to indices of neuropsychological functioning. An emphasis will be placed on some of the more recent integrations of such techniques.

Magnetic Resonance Assessment of Brain Structure

Magnetic resonance imaging allows for a detailed inspection of the structural status of the brain. Modern images on high-resolution equipment, for example, provide neurostructural views of the brain that approach gross postmortem inspection of brain sections (e.g., see Hendrick, Russ, & Simon, 1993). While a few years ago we were able to visualize the brain in what were then considered to be striking 10-mm sections, recent technological advances allow for sectioning at 3 mm and even 1 mm, thereby allowing for an exquisitely detailed view of contiguous brain regions. Figure 1 presents a series of coronal MR images through the brain, with some of the major structures highlighted.

To further illustrate the utility of advanced neuroimaging techniques, Figure 2 shows selected sections from the MR image of a patient with a cerebrovascular accident in the left hemisphere. While the lesion is clearly visible in the horizontal plane (Figure 2A), only a rough estimate of its topography can be inferred. With the addition of sagittal and coronal views (Figures 2B and 2C), the true extent of this lesion can be more accurately visualized, suggesting the lesion is roughly the size of a small egg. From a neurobehavioral standpoint, any of a number of cognitive deficits could be hypothesized on the basis of the apparent size and location of this lesion, although the precise neurobehavioral sequelae must be assessed using detailed neuropsychological studies. This patient, for example, demonstrated residual contralateral motor deficits and only mild language abnormalities (particularly verbal learning and word finding difficulties) four months poststroke.

Complementing the advances in MRI acquisition techniques has been the development of software allowing three-dimensional computer reconstruction of regions of interest, including the whole brain, using MRI data. Serial sections can be digitally analyzed and superimposed to create the framework for such reconstructions, and even though this technology is relatively new, impressive results have been reported by a number of laboratories in various populations (e.g., Johnson, Bigler, Burr, & Blatter, 1994; Reite *et al.*, 1994; Shenton *et al.*, 1992). A 3-D MRI reconstruction of an entire human head, exposing a section of cortex, is depicted in Figure 3.

Figure 4 depicts the 3-D reconstruction of the cerebral cortex, with skull removed. Obviously, some of the fine architectural details are lost using 3-mm sections, although remarkable *in vivo* reconstructions of the lateral hemispheric convexities are possible using current technologies.

A further extension of such procedures allows for the selection of specific

Figure 2. (A) Horizontal view of an infarct in a 50-year-old female in the area of the left internal capsule (note that the figures are reversed, i.e., left is depicted on the right). (B) Lesion in the coronal plane, with its superior-inferior extent. (C) A sagittal view of the lesion.

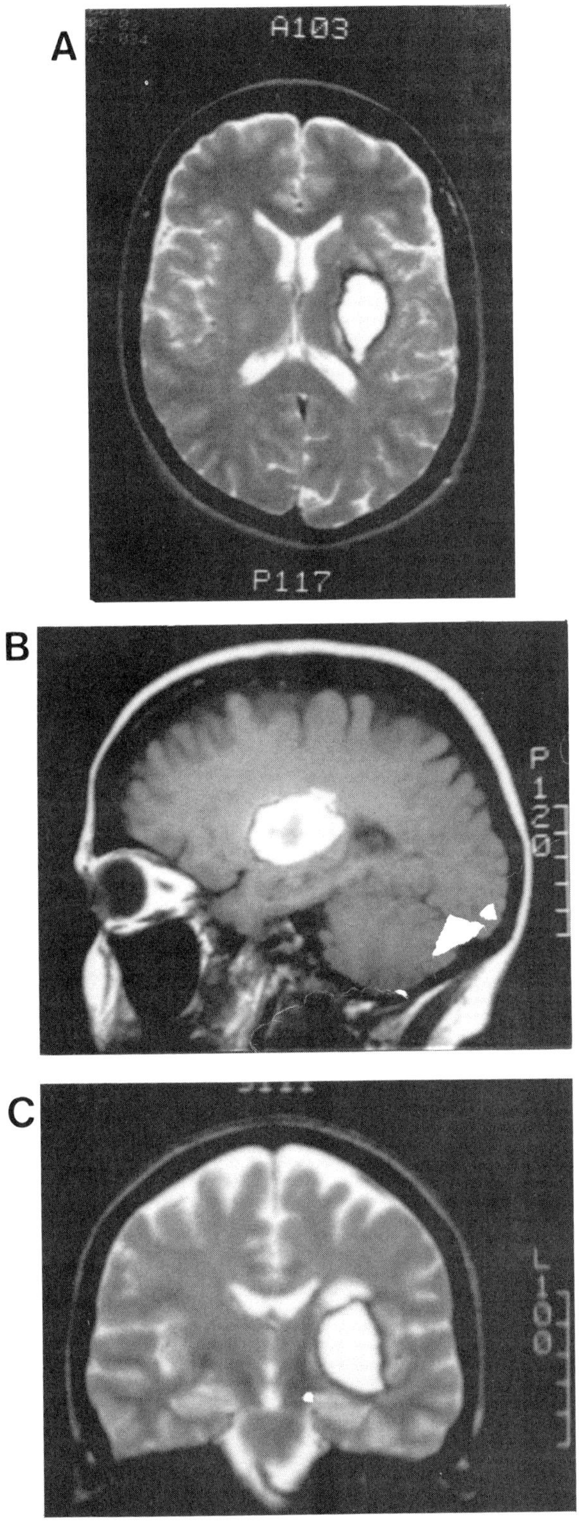
A
A103
P117
B
C

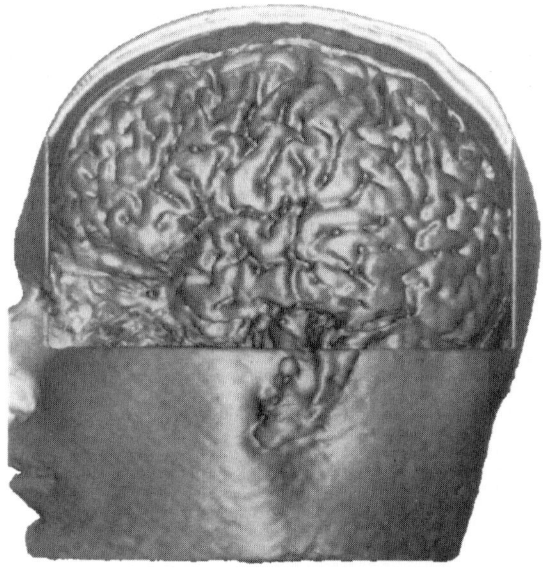

Figure 3. 3-D MRI reconstruction of entire human head, exposing a section of cortex. (Courtesy of Matthew Adams, Neuromagnetism Laboratory, UCHSC.)

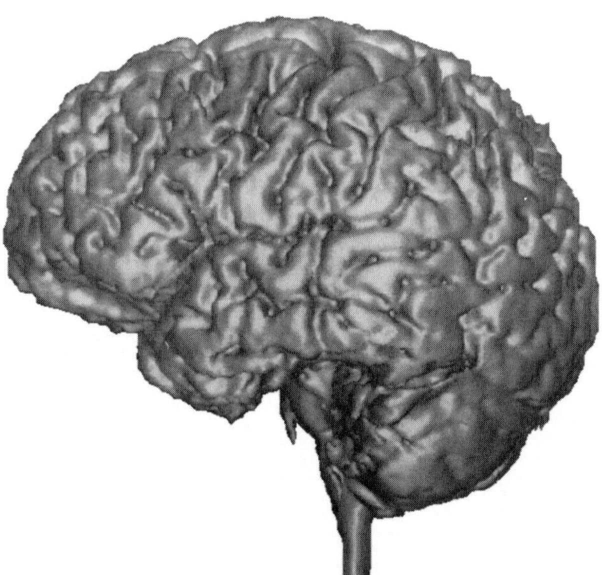

Figure 4. 3-D reconstruction of the cerebral cortex, with skull removed. (Courtesy of Matthew Adams, Neuromagnetism Laboratory, UCHSC.)

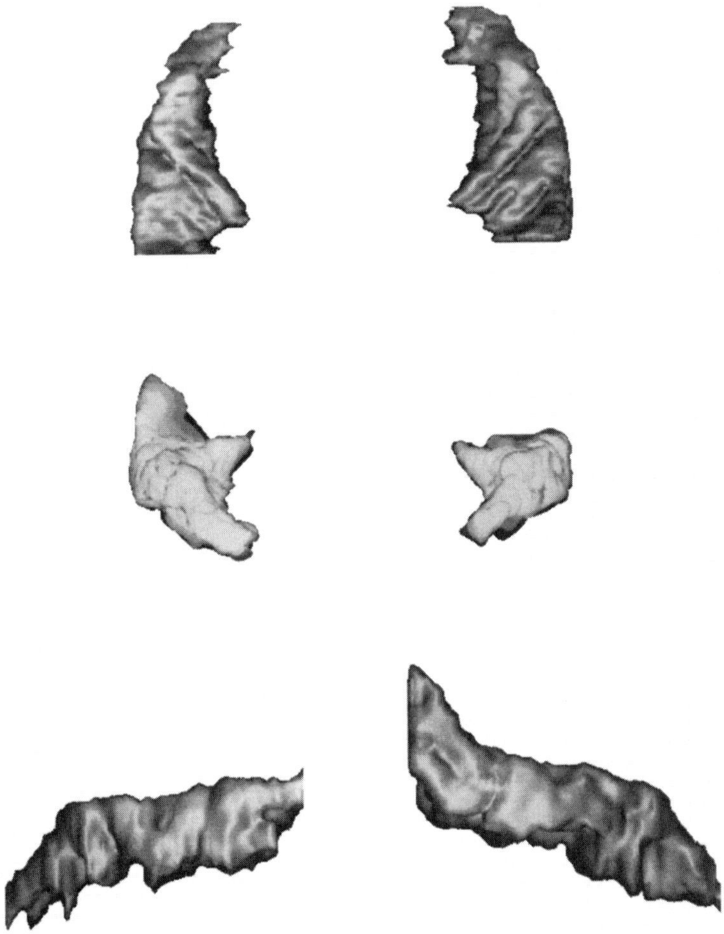

Figure 5. The top figures depict a dorsal view of the left and right superior temporal gyri (STGs), respectively. Center figures show the STGs from an anterior perspective. The lower reconstructions display a lateral view of the left and right STGs. (Courtesy of Matthew Adams, Neuromagnetism Laboratory, UCHSC.)

regions and structures for 3-D visualization. Figure 5 shows 3-D renderings of superior temporal gyri bilaterally. While this technology is in the relatively early stages of development, the implications for normative neuromorphometric databases and neurobehavioral correlations are intriguing.

Volumetric Analysis of Neuroimaging Data: Quantitative Neuromorphometry

Techniques for estimating the size of various brain structures and regions have evolved from linear and planimetric indices to volumetric approaches (e.g., see Turkheimer, 1989; Chapter 2, Volume I of this series). Recent quantitative imaging techniques and software have sought to automate more fully the evaluation of definable regions of interest throughout the brain, although such procedures remain methodologically limited and are undergoing constant refinement. Typically, selected regions are defined manually or in semiautomated fashion, and areas are calculated on each pertinent brain slice. Volumetric reconstructions are then performed, yielding estimates of the 3-D size of various structures and areas.

Such procedures have been successfully utilized to characterize various normal, neurological, and psychiatric populations (Andreasen *et al.*, 1990; Jernigan, Press, & Hesselink, 1990; Lim, Zipursky, Watts, & Pfefferbaum, 1992; Sullivan, Shear, Mathalon, Lim, Yesavage, Tinklenberg, & Pfefferbaum, 1993), although large-scale quantitative normative comparison data are still lacking. Serial quantitative image examinations of individual cases have also proven to be most informative. To illustrate, three-dimensional reconstructions of the brain can be useful in depicting the ventricular system in greater detail, which can be espe-

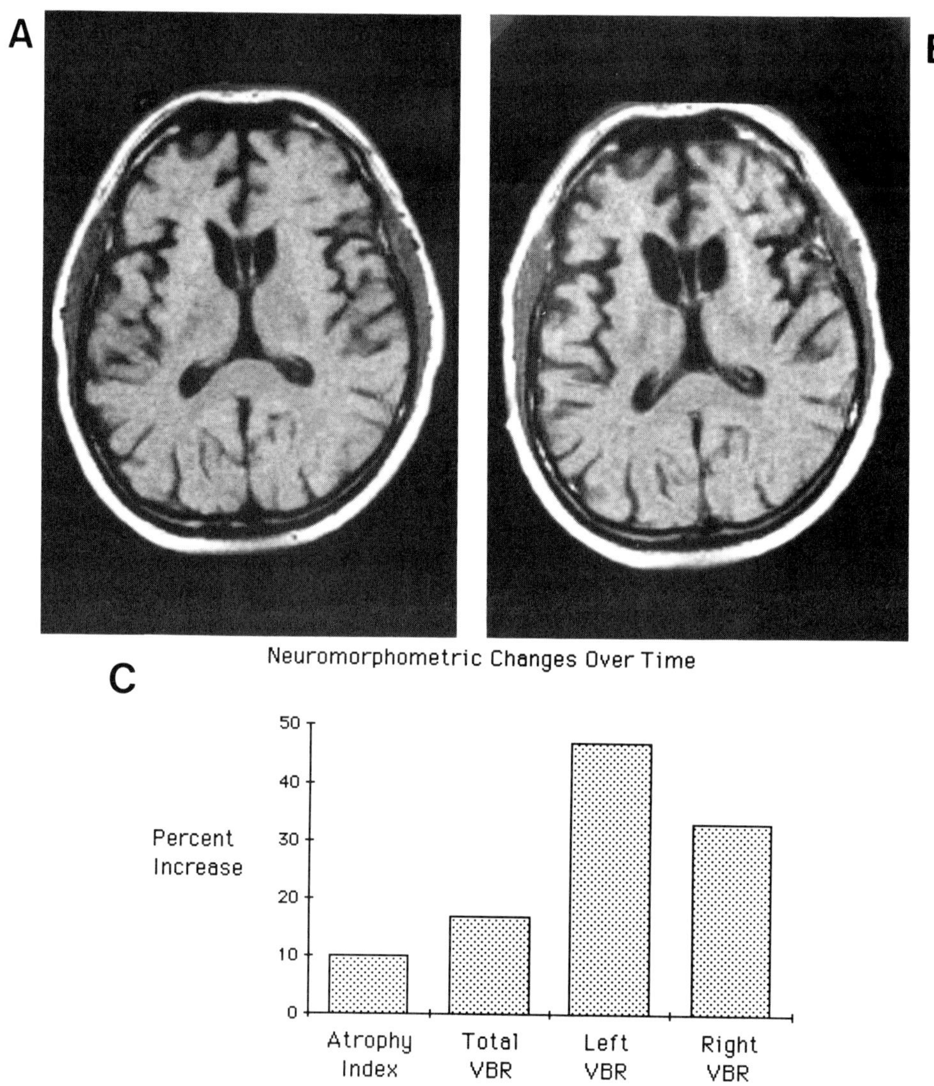

Figure 6. (A) Patient's baseline MR results. (B) MRI findings 30 months later. (C) Quantitative neuromorphometric changes in cortical atrophy and VBRs between baseline and 30 months. (Reproduced with permission from Filley & Cullum, 1993.)

cially useful in the case of moderate to severe head injury (e.g., Bigler, 1992), wherein delayed ventricular dilation is a common sequela (Cullum & Bigler, 1986; Levin, Meyers, Grossman, & Sarwar, 1981). Similarly, the utility of serial neurobehavioral assessments in the prediction of subsequent development/progression of atrophy on MRI was demonstrated by Filley and Cullum (1993). In one such case, the patient's baseline neuropsychological results suggested generalized dementia, with primary involvement of frontal systems. While initial MRI results were not particularly suggestive of frontal pathology *per se,* this became more evident upon follow-up MRI (see Figure 6). Neuromorphometric changes over time were further associated with changes in the patient's neuropsychological status.

Quantitative Neuroimaging: Neuropsychological Relationships

A number of investigations have examined relationships between quantitative neuroimaging findings and neuropsychological indices of brain function. In general, significant, albeit rather modest, relationships between specific neuropsychological abilities and imaging findings have been reported. While no specific relationships have been reported in some populations, a number of studies have shown positive results. For example, using volumetric MRI analyses in healthy college students with low versus high SAT scores, an association between intellectual capacity and brain volume was found by Willerman, Schultz, Rutledge, and Bigler (1991). Furthermore, significant relationships between neuropsychological status and quantitative neuroimaging results have been reported in various neurological groups (Naugle, Cullum, Bigler, & Massman, 1986; Pfefferbaum *et al.,* 1990). More specific relationships between global and regional cortical atrophy and ventricle size and cognitive abilities associated with focal or lateralized brain functions have also been demonstrated (Massman, Bigler, Cullum, & Naugle, 1986). In particular, performance IQ tasks have shown consistent relationships with ventricle-to-brain ratio (VBR) following traumatic brain injury TBI (Johnson *et al.,* 1994).

Magnetoencephalography-Derived Indices of Brain Function: Relationships with MRI and Neuropsychological Measures

Magnetoencephalography involves the extracranial measurement of changes in the weak magnetic fields generated by underlying electrical activity within the brain (see Chapter 8, Volume I of this series). Cortical responses to different stimuli result in measurable alterations in the extracranial magnetic fields, and the estimated physical sources of selected responses to stimuli can be calculated. Differential regional responses to stimuli across the cerebrum can be optimized using multiple channel instrumentation, thereby providing information regarding underlying neuronal activity. Magnetic source imaging techniques can be used to project the source coordinates onto MR images, resulting in estimates of the physical location of sources (e.g., see Williamson, Lu, Karron, & Kaufman, 1991). For example, there is evidence to suggest that the N100 response of evoked magnetic fields (termed the M100 or N100m) may derive from temporal lobe structures, specifically within or just anterior to Heschl's gyrus (Papanicolaou *et al.*, 1990; Reite *et al.*, in press).

Even though MEG is largely experimental at this point, it has shown promise as an investigational technique. For example, schizophrenics have been found to have different magnetic source localizations of the M100 in comparison with controls (Reite, Teale, Goldstein, Whalen, & Linnville, 1989). In addition, gender differences in M100 have been found among healthy young adults (Baumann *et al.*, 1991; Reite *et al.*, 1989; Scheuneman, Teale, Linnville, Goldstein, & Reite, 1991). Specifically, the M100 appears to be localized further anteriorly in the right hemisphere of normal male subjects. This finding may represent a gender difference in functional asymmetry of the auditory cortex or in the interhemispheric asymmetry of the temporal planum area. Addition of quantitative neuromorphometric analyses may further elucidate these findings.

Relationships between MEG and neuropsychological findings have also been reported in the literature, although such endeavors are in their infancy at present. We recently replicated the finding of significant sex differences in M100 source localization in small samples of healthy adult males and females. Specifically, males were found to have more anterior source locations in both hemispheres compared to females, but this difference was particularly pronounced in the right hemisphere ($p < .0001$) (Reite, Cullum, Stocker, Teale, & Kozora, 1993). Figure 7 shows intrahemispheric MEG source locations by gender.

In addition to MEG source location indices in each hemisphere in this sample, a brief battery of neuropsychological measures was administered. Measures were selected to be sensitive to gender differences and included tests of visuoconstructional skill and verbal learning. As predicted, significant gender differences were found on the Block Design subtest of the WAIS-R (Wechsler,

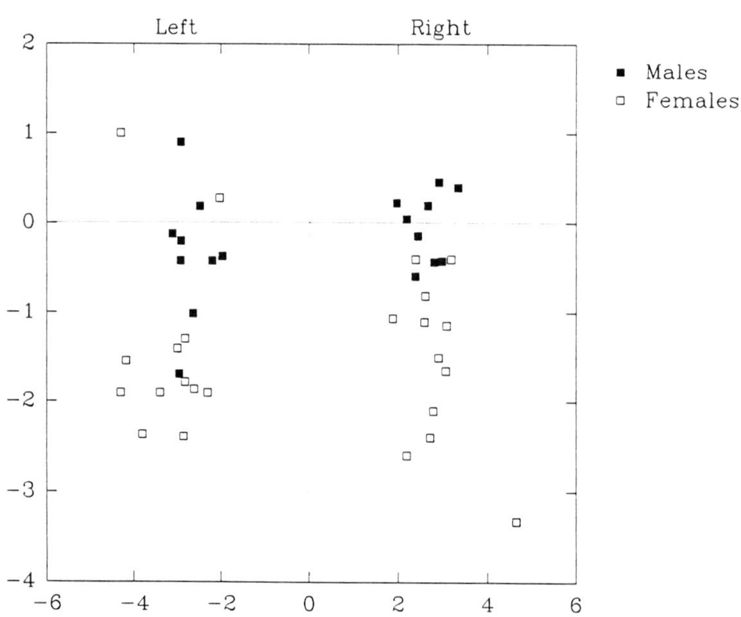

Figure 7. Intrahemispheric MEG source locations by gender. (With permission from Reite *et al.*, 1993.)

1981), wherein males showed a superiority. In contrast, females did better than males on the California Verbal Learning Test (CVLT; Delis, Kramer, Kaplan, & Ober, 1987). Of particular interest, however, was the systematic covariation between MEG source location in the right hemisphere and Block Design performance ($r = .52$, $p = .03$). In contrast, a negative relationship between CVLT performance and right hemisphere source location was observed ($r = -.49$, $p = .04$). These results are presented in Figure 8.

One might speculate that in males, parietotemporal regions associated with

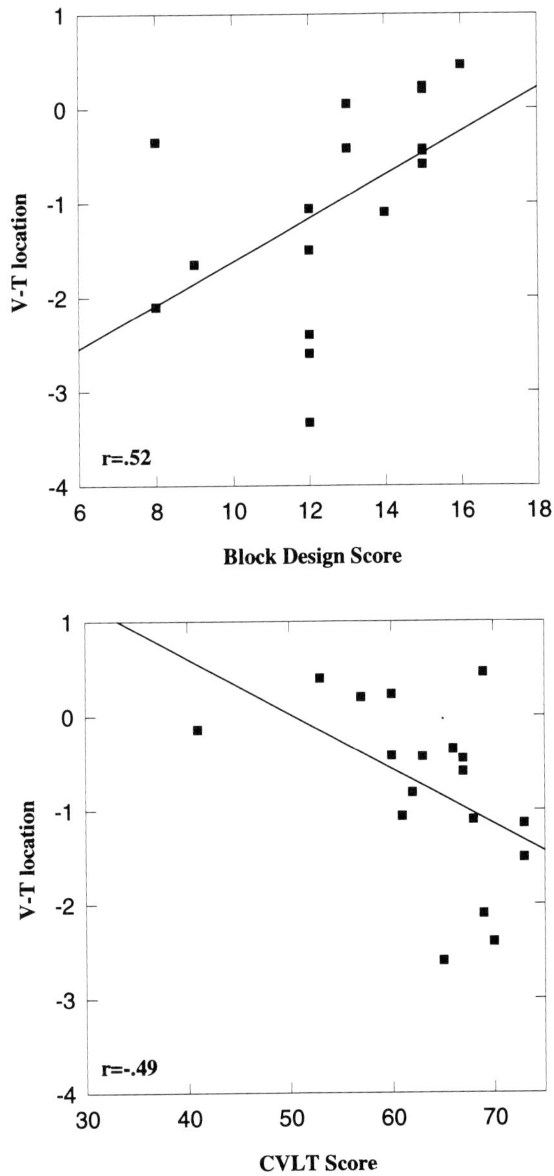

Figure 8. Pearson correlations between right hemisphere MEG source locations and WAIS-R block design (top) and CVLT (bottom) results in a normal cohort. (With permission from Reite *et al.*, 1993).

Block Design performance are more extensively developed. Such development may serve to anteriorly displace the auditory structures responsible for generating the M100. These results, while obviously quite preliminary in nature, suggest the possibility of important interactions between MEG and neuropsychological indices of brain function.

Neurophysiological–Neuropsychological Correlates of Brain Function

Additional neurophysiological-neuropsychological correlations have been reported using other techniques, including EEG, quantitative EEG, and evoked potentials. Our group recently used a conditioning-testing paradigm (Freedman, Waldo, Bickford, & Nagamoto, 1991) to measure P50 auditory evoked responses among groups of healthy controls and patients with schizophrenia (Cullum *et al.*, 1993). This paradigm has been used to demonstrate a specific sensory gating abnormality in patients with schizophrenia which may reflect a CNS-related risk factor for the disease (Adler *et al.*, 1990). It has been postulated furthermore that the P50 component assessed in this manner reflects a neurophysiological correlate of attention. To begin to explore this issue, neuropsychological measures of immediate and sustained attention were examined in conjunction with the P50 conditioning-testing ratio. While no significant relationships were found among controls, several interesting findings were seen in the schizophrenic group. Measures requiring sustained concentration (e.g., digit cancellation) tended to show stronger relationships with P50 ratio, whereas tests of immediate attention showed weaker relationships. In particular, the correlation between digit vigilance (Lewis & Kupke, 1977) total time and the P50 ratio score is depicted in Figure 9.

While these results, too, are preliminary in nature, they nevertheless have interesting implications for additional specific interrelationships between neurophysiological and neuropsychological indices of brain function.

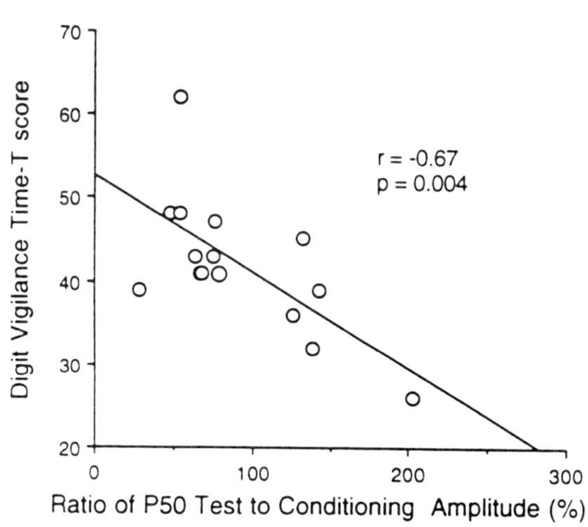

Figure 9. Correlations between P50 conditioning-testing ratio and digit vigilance results in a schizophrenic sample. (With permission from Cullum *et al.*, 1993.)

Future Directions

Relationships between neuropsychological and neuroimaging measures have been demonstrated using a variety of procedures in different populations. Thus, even with the use of disparate measures, some of the procedures clearly are tapping shared aspects of human brain function and underlying anatomy. Clinically, neuroimaging, neurophysiological, and neuropsychological techniques have commonly been used in the context of the comprehensive neurodiagnostic examination. Various combinations of techniques can yield detailed neurodiagnostic information regarding the functional and structural status of the brain. While most of the aforementioned procedures have been well studied in different populations, the *interrelationships* among various neurodiagnostic findings have been less frequent. Such integrative efforts obviously have many benefits, not only in terms of enhancing clinical diagnostic capabilities, but also in increasing our understanding of brain function. It is likely we will see the increased concomitant utilization of neuropsychological, neurophysiological, and neuroimaging techniques as fields interested in brain-behavior relationships show greater collaboration.

What does the future hold in terms of such neurointegrative assessment efforts? We predict that in addition to the more frequent use of multiple neurodiagnostic assessment procedures in the same patients, protocols for the *simultaneous* application of integrated neuropsychological, neurophysiological, and neuroimaging techniques will evolve. For example, some neuropsychological procedures can be performed during MEG, PET, SPECT, quantitative EEG, and functional MRI at present, although more advanced and standardized protocols await development. Neurobehavioral assessment will no doubt evolve to incorporate advances in the fields of neuropsychology, neuroscience, neuroimaging, neurophysiology, cognitive psychology, computer graphics, and virtual reality technology. Such multimodal evaluation procedures likely will lead to a more detailed understanding of brain-behavior relationships as these various fields of inquiry undergo greater integration.

References

Adler, L. A., Gerhardt, G. A., Franks, R., Baker, N., Nagamoto, H., Drebing, C., & Freedman, R. (1990). Sensory physiology and catecholamines in schizophrenia and mania. *Psychiatry Research, 31,* 297–309.

Andreasen, N. C., Ehrhardt, J. C., Swayze, V. W., Alliger, R. J., Yuh, W. T. C., Cohen, G., & Ziebell, S. (1990). Magnetic resonance imaging of the brain in schizophrenia: The pathophysiologic significance of structural abnormalities. *Archives of General Psychiatry, 47,* 35–44.

Baumann, S., Rogers, R., Guinto, F., Saydjari, C., Papanicolaou, A., & Eisenberg, H. (1991). Gender differences in source location for the N100 auditory evoked magnetic field. *Electroencephalography and Clinical Neurophysiology, 80,* 53–59.

Bigler, E. D. (1992). Three-dimensional image analysis of trauma induced degenerative changes: An aid to neuropsychological assessment. *Archives of Clinical Neuropsychology, 7,* 449–456.

Cullum, C. M., & Bigler, E. D. (1986). Ventricle size, cortical atrophy and the relationship with neuropsychological status in closed head injury: A quantitative analysis. *Journal of Clinical and Experimental Neuropsychology, 8,* 437–452.

Cullum, C. M., Harris, J. G., Waldo, M. C., Smernoff, E., Madison, A., Nagamoto, H. T., Griffith, J., Adler, L. E., & Freedman, R. (1993). Neurophysiological and neuropsychological evidence for attentional dysfunction in schizophrenia. *Schizophrenia Research, 10,* 131–141.

Delis, D., Kramer, J. H., Kaplan, E., & Ober, B. A. (1987). *The California Verbal Learning Test-Research edition.* New York: Psychological Corporation.

Filley, C. M., & Cullum, C. M. (1993). Early detection of fronto-temporal degeneration by clinical evaluation. *Archives of Clinical Neuropsychology, 8,* 359–367.

Freedman, R., Waldo, M. C., Bickford, P., & Nagamoto, H. (1991). Elementary neuronal dysfunctions in schizophrenia. *Schizophrenia Research, 4,* 233–243.

Hendrick, R. E., Russ, P. D., & Simon, J. H. (1993). *MRI: Principles and artifacts.* New York: Raven Press.

Jernigan, T. L., Press, G. A., & Hesselink, J. R. (1990). Methods for measuring brain morphologic features on magnetic resonance images: Validation and normal aging. *Archives of Neurology, 47,* 27–32.

Johnson, S. C., Bigler, E. D., Burr, R. B., & Blatter, D. D. (1994). White matter atrophy, ventricular dilation, and intellectual functioning following traumatic brain injury. *Neuropsychology, 8,* 307–315.

Levin, H. S., Meyers, C. A., Grossman, R. G., & Sarwar, M. (1981). Ventricular enlargement after closed head injury. *Archives of Neurology, 38,* 623–629.

Lewis, R., & Kupke, T. (1977). The Lafayette Clinic repeatable neuropsychological test battery: Its development and research applications. Paper presented at the annual meeting of the Southeastern Psychological Association, Hollywood, FL.

Lim, K. O., Zipursky, R. B., Watts, M. C., & Pfefferbaum, A. (1992). Decreased gray matter in normal aging: An in vivo magnetic resonance study. *Journal of Gerontology, 47,* B26–30.

Massman, P. J., Bigler, E. D., Cullum, C. M., & Naugle, R. I. (1986). The relationship between cortical atrophy and ventricular volume. *International Journal of Neuroscience, 30,* 87–89.

Naugle, R. I., Cullum, C. M., Bigler, E. D., & Massman, P. J. (1986). Neuropsychological and computerized axial tomography volume characteristics in senile and presenile dementia. *Archives of Clinical Neuropsychology, 1,* 219–230.

Papanicolaou, A., Baumann, S., Rogers, R., Saydjari, C., Amparo, E., & Eisenberg, H. (1990). Localization of auditory response sources using magnetoencephalography and magnetic resonance imaging. *Archives of Neurology, 47,* 33–37.

Pfefferbaum, A., Sullivan, E. V., Jernigan, T. L., Zipursky, R. B., Rosenbloom, M. J., Yesavage, J. A., & Tinklenberg, J. R. (1990). A quantitative analysis of CT and cognitive measures in normal aging and Alzheimer's disease. *Psychiatry Research, 35,* 115–136.

Reite, M., Adams, M., Simon, J., Teale, P., Sheeder, J., Richardson, D., & Grabbe, R. (1994). Auditory M100 component 1: Relationship to Heschl's gyri. *Cognitive Brain Research, 2,* 13–20.

Reite, M., Cullum, C. M., Stocker, J., Teale, P., & Kozora, E. (1993). Neuropsychological test performance and MEG-based brain lateralization: Sex differences. *Brain Research Bulletin, 32,* 325–328.

Reite, M., Teale, P., Goldstein, L., Whalen, J., & Linnville, S. (1989). Late auditory sources may differ in the left hemisphere of schizophrenic patients: A preliminary report. *Archives of General Psychiatry, 46,* 565–572.

Scheuneman, D., Teale, P., Linnville, S., Goldstein, L., & Reite, M. (1991). Magnetic auditory M100 source localization in normal females. *Brain Research Bulletin, 26,* 747–751.

Shenton, M. E., Kikinis, R., Jolesz, F. A., Pollak, S. D., LeMay, M., Wible, C. G., Hokama, H., Martin, J., Metcalf, D., Coleman, M., & McCarley, R. W. (1992). Abnormalities of the left temporal lobe and thought disorder in schizophrenia: A quantitative magnetic resonance imaging study. *New England Journal of Medicine, 327,* 604–612.

Sullivan, E. V., Shear, P. K., Mathalon, D. H., Lim, K. O., Yesavage, J. A., Tinklenberg, J. R., & Pfefferbaum, A. (1993). Greater abnormalities of brain cerebrospinal fluid volumes in younger than in older patients with Alzheimer's disease. *Archives of Neurology, 50,* 359–373.

Turkheimer, E. (1989). Techniques of quantitative measurement of morphological structures of the central nervous system. In E. D. Bigler, R. A. Yeo, & E. Turkheimer (Eds.), *Neuropsychological function and brain imaging* (pp. 47–64). New York: Plenum Press.

Wechsler, D. (1981). *Wechsler Adult Intelligence Scale-Revised.* San Antonio, TX: Psychological Corporation.

Willerman, L., Schultz, R., Rutledge, J. N., & Bigler, E. D. (1991). In vivo brain size and intelligence. *Intelligence, 15,* 223–228.

Williamson, S. J., Lu, Z-L., Karron, D., & Kaufman, L. (1991). Advantages and limitations of magnetic source imaging. *Brain Topography, 4,* 169–180.

3

Neuroimaging in Neurological Disorders

ANDREW B. NEWBERG AND ABASS ALAVI

INTRODUCTION

Both functional and anatomical imaging have contributed dramatically to our understanding of the causes of various neurological disorders and their diagnosis and management. Anatomical imaging methods such as magnetic resonance imaging (MRI) and x-ray computed tomography (CT) have been used to determine the structural changes associated with various neurological diseases. Functional imaging techniques such as single photon emission computed tomography (SPECT) and positron emission tomography (PET) have improved understanding of the changes in blood flow, metabolism, and neurotransmitter activity in these disease entities. This chapter will review major neurologic disorders and explore the use of neuroimaging in the study of these abnormalities of the central nervous system.

The methods and techniques of MRI, CT, PET, and SPECT have been considered elsewhere in this series. In general, CT is regarded as more available and less expensive than MRI. However, MRI is more sensitive and has better resolution for detecting small lesions than CT. With the advent of better software for quantitative image analysis and volumetric determinations, MRI can offer other advantages. Magnetic resonance is also being developed in order to measure cerebral blood flow and possibly other physiologic parameters.

The relationship of SPECT and PET is similar to that of CT and MRI.

ANDREW B. NEWBERG AND ABASS ALAVI Division of Nuclear Medicine, Hospital of the University of Pennsylvania, Philadelphia, Pennsylvania 19104.

Neuroimaging II: Clinical Applications, edited by Erin D. Bigler. Plenum Press, New York, 1996.

Table 1. A Partial Listing of Radioligands Used in Neurological PET Imaging

Compound	Application
[^{15}O]-H_2O	Blood flow
[^{18}F]-fluorodeoxyglucose	Glucose metabolism
$^{15}O_2$	Oxygen metabolism
[^{11}C]-L-methionine	Amino acid metabolism
[^{11}C]-raclopride, [^{11}C]-methylspiperone, 6-^{18}F]-fluorodopamine, [^{18}F]-spiperone, [^{18}F]-fluoroethylspiperone	Dopamine receptor activity
[^{11}C]-carfentanil, [^{11}C]-etorphine	Opiate receptor activity
[^{11}C]-flunitrazepam	Benzodiazepine receptor activity
[^{11}C]-scopolamine, [^{11}C]-quinuclidinyl benzilate	Muscarinic cholinergic receptors
6-[^{18}F]-fluoro-L-dopa, 4-[^{18}F]-floruo-*m*-tyrosine	Presynaptic dopaminergic system
[^{11}C]-ephedrine, [^{18}F]-fluorometaraminol	Adrenergic terminals

SPECT is more available and less expensive than PET. However, PET has better resolution and provides absolute quantification of cerebral blood flow and metabolism. The recent introduction of technetium-99m-labeled compounds has greatly expanded the ability of SPECT to determine various aspects of brain function. Technetium-99m has a 6-hour half-life, can be labeled on site to the appropriate compound, and has optimal photon characteristics for imaging with modern SPECT instruments (Holman & Devous, 1992). Further, with newer, dedicated SPECT cameras, such as three-headed instruments, high-resolution images comparable to those from PET can be obtained. PET instruments, with their associated cyclotron facilities, are expensive and require highly skilled personnel to operate. The most commonly used radiopharmaceuticals for neurologic PET imaging (see Table 1) are ^{18}F-fluoro-deoxyglucose (FDG) (Kung *et al.*, 1990; Diksic & Reba, 1991), which measures the cerebral metabolic rate for glucose (CMRGlc), and $H_2{}^{15}O$ (Diksic & Reba, 1991; Hoffman & Coleman, 1992), which measures regional cerebral blood flow (rCBF). Both SPECT and PET are used to measure neurotransmitter concentrations and receptor activity in various parts of the brain. A partial list of the most common neurotransmitter analogs for PET and SPECT are shown in Tables 1 and 2 (Holman *et*

Table 2. A Partial Listing of Radioligands Used in Neurological SPECT Imaging

Compound	Application
HMPAO,[a] IMP,[b] HIPDM[c]	Cerebral blood flow
3-quinuclidinyl benzilate (IQNB)	Muscarinic cholinergic receptor
Iodopride, IBZM,[d] iodospiperone	Dopamine receptor activity
AMIK,[e] DOI[f]	Serotonin receptor activity
Iomazenil	Benzodiazepine activity
2-iodomorphine	Opioid receptor activity
I-d$(CH_2)_5$[Tyr$(Me)_2$, Tyr$(NH_2)_9$]AV	Vasopressin activity

[a]HMPAO = Technetium-99m hexamethyl propylene amine oxime
[b]IMP = Iodine-123-N,N′,N,-trimethyl-N′-[2-hydroxyl-3-methyl-5-iodo-benzyl]-1,3-propane diamine
[c]HIPDM = iodine 123-N-N′,N,-trimethyl-N′-[2-hydroxyl-3-methyl-5-iodo-benzyl]-1,3 propane diamine
[d]IBZM = 3-iodo-N-[(1-ethyl-2-pyrrolidinyl)]-methyl-2-hydroxy-6-methoxybenzamide
[e]AMIK = 7-amino-8-iodo-ketanserin
[f]DOI = 1-(2,5-dimethoxy-4-iodophenyl)-2-aminopropane

al., 1985; Wouters, VanDun, & Laduron, 1986; Mertens *et al.*, 1989; Gatley *et al.*, 1990; Sedvall, 1990; Sadzot, Mayberg, & Frost, 1990; Kung, 1990; Kung *et al.*, 1990; Diksic & Reba, 1991; Phelps, 1991; Alavi & Hirsch, 1991; Schubiger *et al.*, 1991; Maziere & Maziere, 1991; Maziere *et al.*, 1992; Weinberger *et al.*, 1991, 1992; Frost, 1992; Abadie *et al.*, 1992; Varastet *et al.*, 1992; Baron, 1992).

Aging and Dementia

The topics regarding the use of neuroimaging techniques in the study of aging and dementia are covered elsewhere in this textbook (see Chapter 7). We will only mention here that all of the imaging techniques have a role to play in the study of aging and dementia. Anatomical imaging with MRI and CT helps determine the degree of morphological changes in the brain during aging and dementia. These techniques can also rule out other causes of dementia symptoms such as tumor or infarcts. Functional imaging is useful in determining changes that occur in the blood flow, metabolism, and neurotransmitter activity in the brain before structural alterations are noted.

Primary Progressive Aphasia

Primary progressive aphasia (PPA) was first described by Mesulam (1982) as a slow process without signs of a generalized dementia. Qualitative CT and MRI studies on these patients (see Figure 1) have revealed both sylvian fissure and temporal horn enlargement on the left, consistent with the location of the language center in the brain (Sapin, Anderson, & Pulaski, 1989; Tyrell *et al.*, 1990). It is believed that the atrophic changes in this area reflected the deficit in lan-

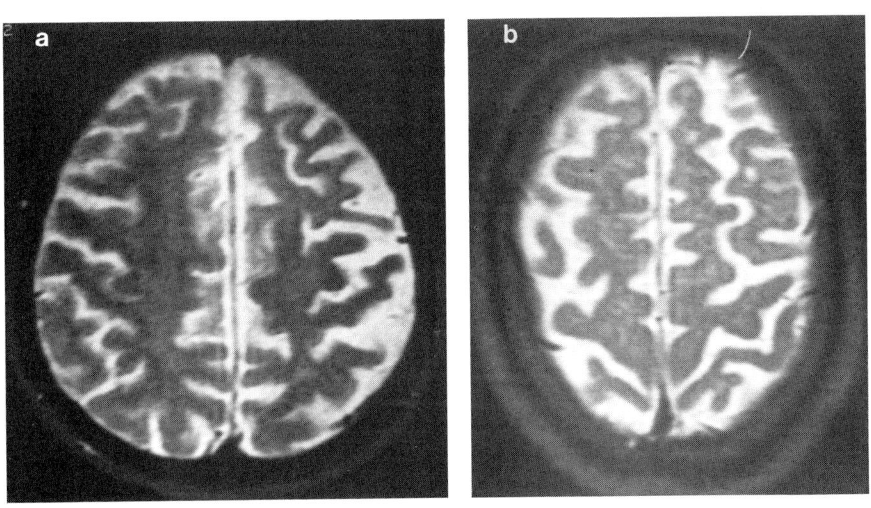

Figure 1. MRI of a patient with primary progressive aphasia (a) showing increased CSF in the left hemisphere specifically in the frontoparietal lobe. This is in contrast to the symmetrical atrophy seen in patients with diffuse cortical atrophy (b).

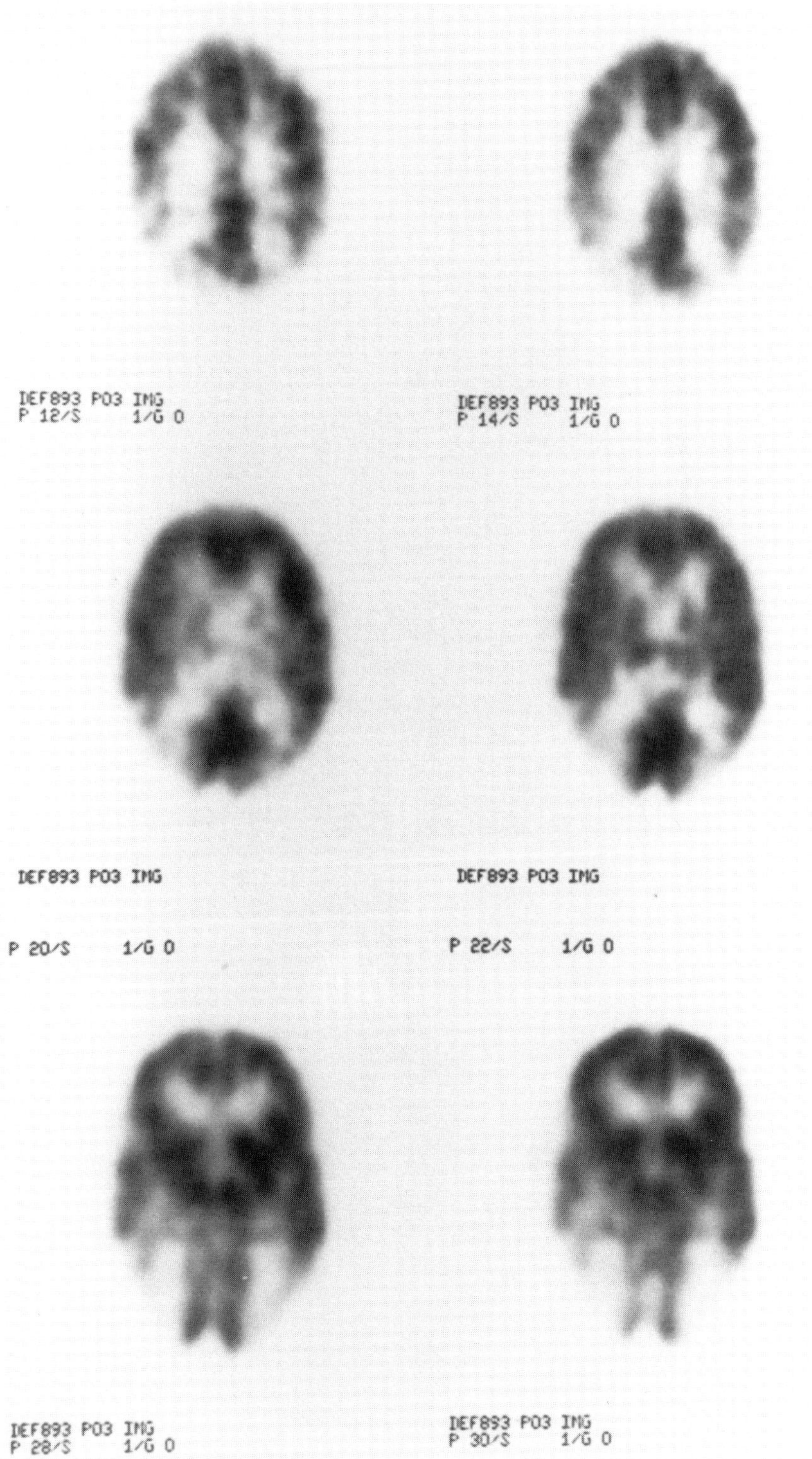

Figure 2. PET image of a patient with AD showing symmetric decreases primarily in the parietal lobe.

guage abilities since the left temporal lobe is generally the home of the language center. Our group reported significant decreases in the brain volume of the left hemisphere compared to the right, which distinguished PPA patients from both Alzheimer's disease (AD) patients and controls (Newberg, Alavi, Souder, Reivich, & Grossman, 1992). These changes were not primarily confined to the left temporal lobe.

There have been only a few studies using SPECT in PPA. In one study of a patient with normal MRI and CT, SPECT imaging found a left temporal lobe abnormality (Lee & Kramer, 1992). This indicates that perfusion changes might be more sensitive for detecting PPA and might occur earlier in the disease process.

Chawluk *et al.* (1986) showed that PPA patients had a markedly decreased glucose metabolism in the left parietal lobe and a less marked decrease in the left temporal lobe. However, it was noted that PPA patients had no contralateral or global abnormalities on PET as is usually the case in AD. Tyrrell *et al.* (1990) reported six cases of PPA in which PET studies revealed hypometabolism localized to the left hemisphere, specifically the left temporal lobe. Kempler *et al.* (1990) analyzed PET and CT in three patients with the presentation of PPA. They concluded that PPA does not have a uniform syndrome complex, that there was only mild atrophy in the left temporal lobe, and that left temporal lobe metabolism was markedly reduced.

In a recent study by our group using FDG-PET (see Figure 2), we found that PPA patients have markedly decreased glucose metabolism in the left temporal lobe compared to controls; the rate of metabolism is similar in absolute value to that of AD patients (Newberg *et al.*, 1992). The left/right ratio of glucose metabolic rates was markedly lower not only in the temporal lobes, but in all other lobes as well compared to controls and AD patients. This finding suggested that PPA is a hemispheric process and that the metabolic derangement extends beyond the focus of the language center. However, PPA still was well distinguished from the typical AD pattern.

Brain Tumors

Primary intracranial tumors comprise approximately 5% to 9% of all cancers. Together, brain tumors carry a median survival of approximately one year. Anatomical imaging allows for the best determination of the location, size, and extent of the tumor. Metabolic imaging is very helpful in the management of brain tumors and allows the grading of tumors, determination of prognosis, and the differentiation of recurrent tumor from radiation necrosis (DiChiro, 1986; Wilson, 1989).

The anatomical imaging techniques with CT and MRI have aided in making early diagnoses (see Figure 3), in planning radiation therapy, and predicting adverse reactions to treatment (Jaeckle, 1991). MRI has the advantage in evaluating the cerebellum and brain stem since CT may have severe beam hardening artifacts in these areas. CT is better than MRI in detecting foci of calcification which can be seen in meningiomas and oligodendrogliomas (Bradley *et al.*, 1984; Kucharczyk *et al.*, 1985; Felix *et al.*, 1985; Kelly & Brant-Zawadski, 1988). The use of contrast CT and gadolinium MRI have clearly made these techniques highly sensitive in the evaluation of brain tumors (Lewin, Estes, & Ross, 1992).

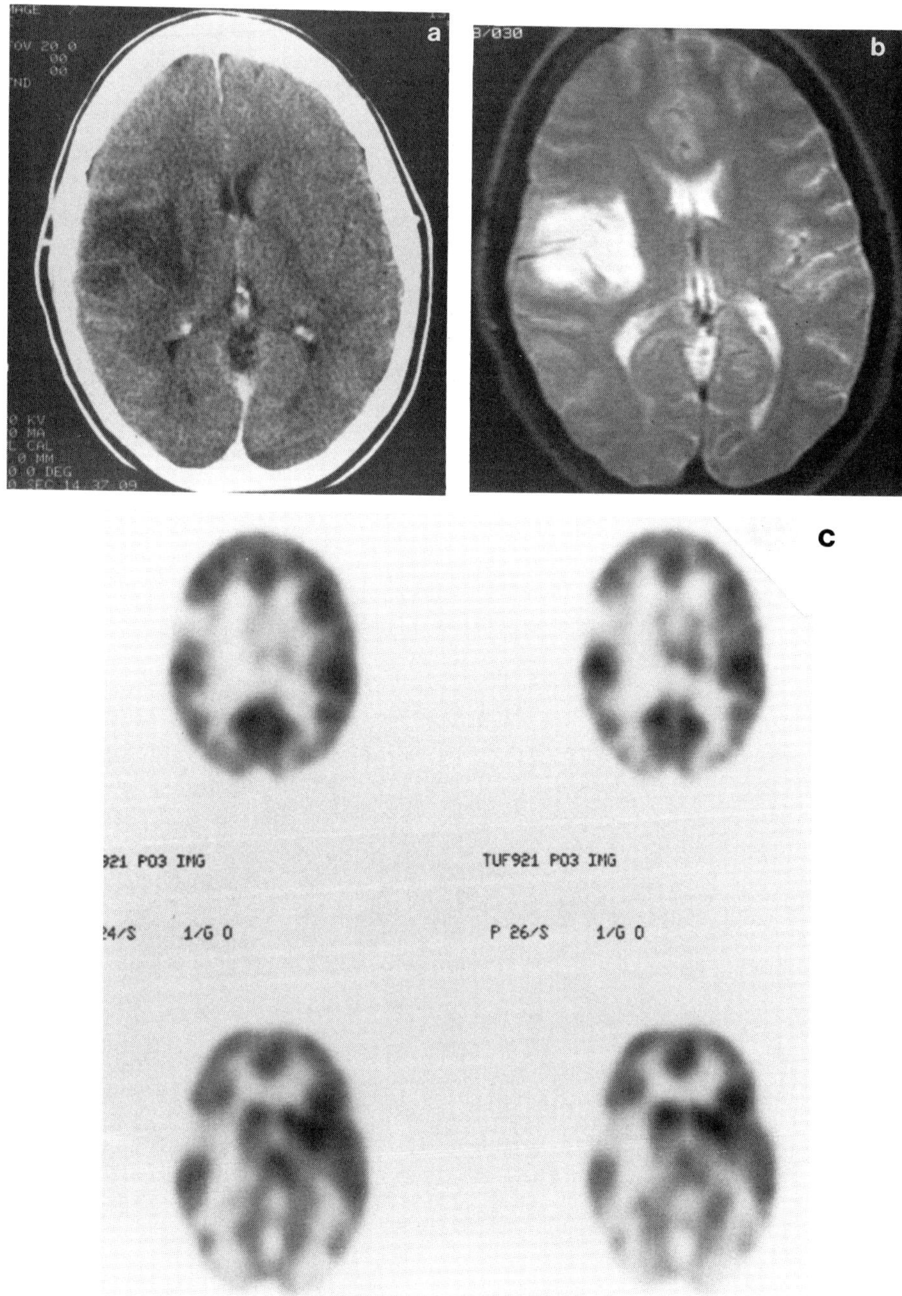

Figure 3. This CT scan (a) shows a low-grade tumor in the right temporoparietal area characterized as a hypodense area with no edema and little mass effect. A T2-weighted MR image (b) shows the same lesion as a hyperintense area. Findings on CT or MRI are nonspecific for disease activity. An FDG-PET image (c) of the same lesion shows a hypometabolic area in the right temporoparietal area indicating an inactive and therefore low-grade tumor.

The newer technique of MR spectroscopy allows for determining the metabolism of tumors by measuring the phosphorus spectra of these tumors (Barker, Glickson, & Bryan, 1993). While still in its early development, MR spectroscopy might help determine the tumor aggressiveness and response to therapy.

Astrocytomas are well characterized by MRI and CT (Jolles, Chapman, & Alavi, 1989). CT imaging shows low-grade astrocytomas as hypodense lesions with little mass effect of edema. They may be associated with a well-defined border and may have central calcifications. These lesions appear as high signal intensity lesions on T 2-weighted MR images and low signal intensity lesions on T 1-weighted images. MRI will detect large calcifications but will miss punctate ones (Brant-Zawadski *et al.*, 1984). High-grade astrocytomas appear as irregular, inhomogeneous density lesions on CT and do have associated mass effects with surrounding edema. Similarly, MR images of these tumors show high-intensity, inhomogeneous signals on T 2-weighted images (Kelly & Brant-Zawadski, 1988). While both MRI and CT are excellent at detecting tumors, they are not useful in distinguishing radiation necrosis from recurrent tumor (Dooms *et al.*, 1986; Wallner *et al.*, 1989). Another type of tumor, meningiomas, appear as sharply delineated, oval hyperdense masses that are adjacent to a dural surface (New, Aronow, & Hesselink, 1980), but they are best detected with gadolinium enhanced MRI which can aid in determining the extent and aggressiveness of the tumor (Zee *et al.*, 1992).

SPECT is imprecise in localizing tumors compared to the anatomical imaging modalities because of its relatively low resolution. However, agents such as thallium-201 localize in both primary and metastatic brain tumors (Kaplan *et al.*, 1987). Further, thallium-201 studies have been used to distinguish between high-grade and low-grade gliomas (Black *et al.*, 1989; Kim *et al.*, 1990). Generally, high-grade tumors have increased uptake, and low-grade tumors have decreased uptake of this radiotracer.

The use of thallium-201 in conjunction with HMPAO (see Table 2) has been useful in distinguishing between radiation necrosis and recurrent tumor, although studies have shown that thallium-201 by itself is not sensitive enough to be useful (Schwartz *et al.*, 1991). The combination of the two may show three possible patterns: high thallium uptake is associated with recurrent tumor, moderate thallium uptake with low HMPAO uptake is not associated with tumor recurrence, and moderate thallium uptake with high uptake of HMPAO is not predictive of either tumor recurrence or reactive tissue (Holman & Devous, 1992).

Most FDG-PET studies (see Figures 3 and 4) have indicated that high-grade tumors are hypermetabolic while low-grade tumors are hypometabolic (DiChiro *et al.*, 1982; Patronas *et al.*, 1983; Alavi *et al.*, 1985, 1988). These differences in metabolism make it easy to distinguish between high- and low-grade tumors and thus can help determine prognosis even better than the other imaging modalities (DiChiro *et al.*, 1982; DiChiro, 1986; Patronas *et al.*, 1983, 1985; Coleman *et al.*, 1991; Alavi *et al.*, 1985, 1988; Herholz *et al.*, 1988). Hypometabolism is also associated with local edema, cystic changes, and necrosis near the tumor and areas neuronally connected to the tumor (DeLaPaz *et al.*, 1983; Lammertsma *et al.*, 1983; Lammertsma, Wise, & Jones, 1984; Patronas *et al.*, 1984). In patients suffering from increased intracranial pressure due to a tumor, the contralateral hemisphere also has been shown to have decreased blood flow and oxygen metabolism compared to normal controls (Beaney *et al.*, 1985). Further, these de-

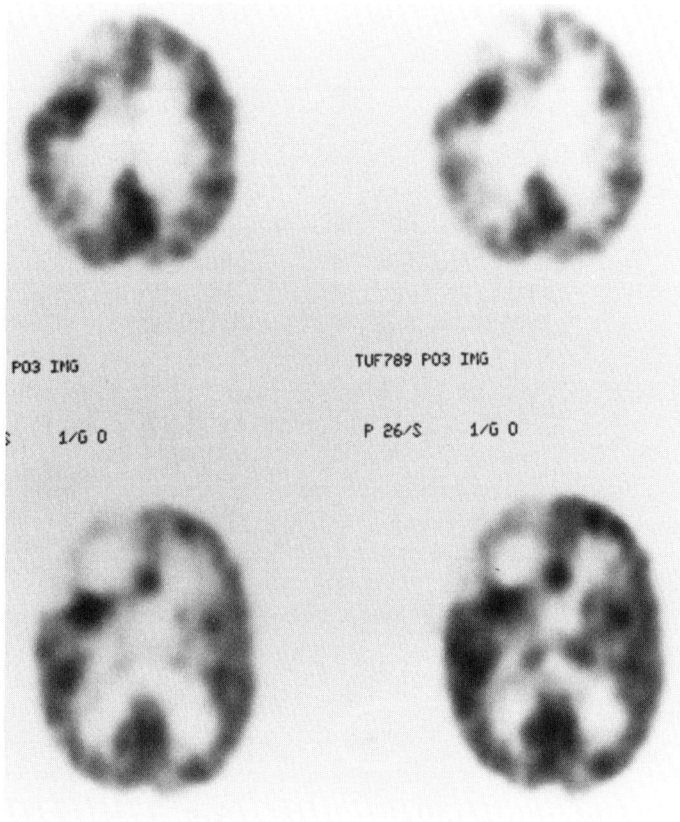

Figure 4. This FDG-PET image shows a high-grade tumor in the right frontal area. The high-grade tumors are characterized by hypermetabolism in the tumor site, distinguishing them from low-grade tumors that are hypometabolic. Adjacent cortex appears hypometabolic while there is some evidence of central necrosis in this rapidly growing tumor.

creases are reversed after decompressive surgery such as craniotomy or resection of the tumor.

In addition to its sensitivity in evaluating brain tumors and their prognosis, PET also has another advantage over anatomical imaging. Unlike CT or MRI, PET can distinguish radiation necrosis from tumor recurrence (Doyle *et al.*, 1987; Valk *et al.*, 1988; Rozental *et al.*, 1991). In general, radiation necrosis areas are hypometabolic and tumor recurrence areas are hypermetabolic on FDG-PET (see Figure 5). Because PET can distinguish areas of tumor activity from those of normal brain tissue and radiation necrosis, PET may be useful in titrating the amount of radiation therapy necessary to damage the tumor tissue without destroying the surrounding normal brain tissue (Ogawa *et al.*, 1988).

PET has also been used to determine tumor response to radiation and chemotherapy and eventual patient survival (Ogawa *et al.*, 1988; Rozental *et al.*, 1991). Further, those patients with greater responses as determined by PET to either chemotherapy or radiation tended to have a better prognosis.

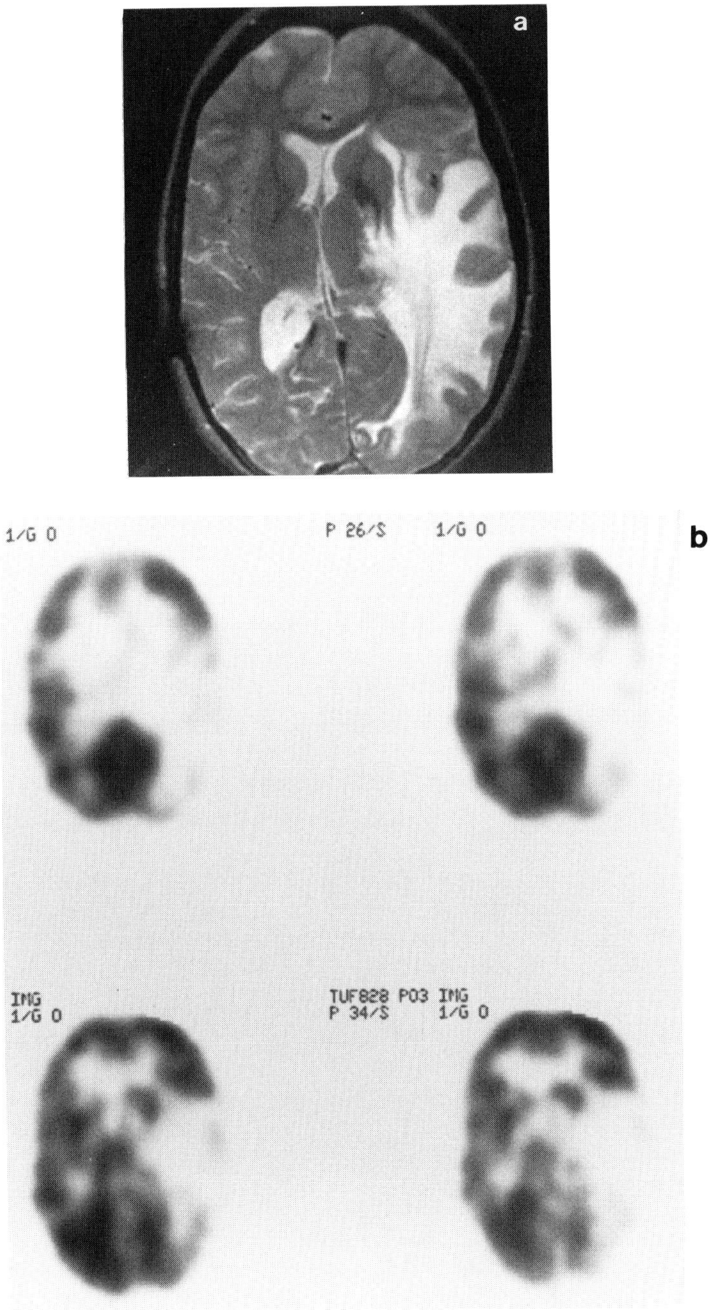

Figure 5. Radiation necrosis appears as a hyperintense area on MRI (a). However, it is difficult to distinguish the radiation effects from the tumor recurrence. The FDG-PET image of radiation necrosis (b) is characterized by the hypometabolism in the area of necrosis and distinguishes radiation effects from hypermetabolic appearance of tumor recurrence.

PARKINSON'S DISEASE

Parkinson's disease (PD) is believed to be caused by loss of the pigmented neurons in the substantia nigra and locus ceruleus and is characterized by the triad of bradykinesia, tremor, and rigidity. Eventually, PD can lead to dementia in 20% to 30% of the patients. The loss of pigmented neurons is associated with decreased production of dopamine, decreased storage of dopamine, and nigrostriatal system dysfunction. It is believed that the mechanism of disease is an initial "upregulation" of dopamine receptors (Marsden, 1982) followed by a downregulation that occurs as the disease progresses.

The decrease in size of the substantia nigra due to degeneration of the neurons can be observed on MR images (Braffman *et al.*, 1989; Olanow, 1992). However, CT and MRI are limited to their analysis of structural changes in PD. It is possible that as MR imaging techniques improve, we will be able to distinguish PD from atypical parkinsonism as well as help determine patients at risk for the development of PD (Olanow, 1992). Conversely, Huber *et al.* (1989) found that MRI was not useful in differentiating dementia of PD and AD.

SPECT studies have shown that patients with PD show mild diffusely decreased blood flow throughout the cortex except for the cerebellum (Smith *et al.*, 1988b; Davidson, 1989; Nagel, Ichise, & Holman, 1991). It is still controversial whether there is hypoperfusion in the basal ganglia in PD. Some studies have shown decreases in HMPAO uptake in the basal ganglia (Podreka *et al.*, 1987; Pizzolato *et al.*, 1988). However, Smith *et al.* (1988b), as well as other investigators (Song & Lin, 1990; Lin *et al.*, 1990), have found no characteristic decreases in perfusion in the basal ganglia while reduced flow is noted in the cerebral cortex. One study (Costa *et al.*, 1988) noted decreased basal ganglia perfusion following therapy with L-dopa.

Dementia can occur in patients with Parkinson's disease, particularly in the later stages of this disorder. Dementia in PD is likely the result of functional alterations in both the subcortical and cortical structures, often resulting in diffuse hypoperfusion seen on SPECT. Sawada *et al.* (1992), utilizing IMP SPECT (see Table 2), found that the rCBF in nondemented PD patients showed no significant differences compared to controls but that PD patients with dementia had frontal or frontoparietal hypoperfusion. This result concurs with the hypothesis that frontal lobe dysfunction occurs in patients with dementia and PD. However, other studies have shown that PD patients with dementia tended to have hypoperfusion in a temporoparietal pattern similar to that found in AD (Pizzolato *et al.*, 1988; Gemmell *et al.*, 1989; Kuwabara *et al.*, 1990; Spampinato *et al.*, 1991; Nagel, Ichise, & Holman, 1991; Holman, Johnson, *et al.*, 1992). Still another group found both parietal flow defects and decreased perfusion in the posterior frontal lobe in PD patients with dementia (Smith *et al.*, 1988b).

PET has also been used extensively in the study of PD. Playford *et al.* (1992, 1993) compared $^{15}O_2$ PET images between PD patients and controls during movement tasks. They found that PD patients had decreased activation of the putamen, anterior cingulate gyrus, supplementary motor area, and dorsolateral prefrontal cortex compared to controls. Several groups have reported hypermetabolism in the basal ganglia in early, untreated PD (Rougemont *et al.*, 1984;

Eidelberg *et al.*, 1990). Similarly, hemiparkinsonism is associated with hypermetabolism in the contralateral basal ganglia (Martin *et al.*, 1984, 1986). However, one group found no significant striatal changes in patients with PD (Kuhl *et al.*, 1984), and another group (Przedborski *et al.*, 1993) reported decreases in glucose metabolism in the basal ganglia contralateral to the side of the symptoms in patients with hemiparkinsonism-hemiatrophy syndrome. Parkinson's disease patients have been shown to have mild diffuse cortical hypometabolism that worsens with the severity of bradykinesia (Rougemont *et al.*, 1984).

Several groups have indicated that levodopa therapy does not correct or change the local or global metabolic rates (Rougemont *et al.*, 1984; Kuhl *et al.*, 1984). However, Blesa *et al.* (1991) reported a reversal of pallidal hypermetabolism with levodopa therapy. A study by Jenkins *et al.* (1992) indicated that PD patients had improved activation in the supplementary motor cortex during a motor function task when akinesia was reversed with apomorphine infusion (a dopamine agonist).

Peppard *et al.* (1990) showed that PD patients with dementia differed from PD patients without dementia in that the former had hypometabolic perirolandic and angular gyrus regions. However, PD patients with dementia did not have significantly different CMRGlc values than AD patients (Kuhl *et al.*, 1985). Severe dementia in PD may be indistinguishable from AD on PET images, both showing significant parietal hypometabolism (Rougemont *et al.*, 1984).

Fluorodopa (^{18}F-dopa) has been used for PET imaging in PD to evaluate presynaptic dopaminergic function and has defective uptake in the nigrostriatal dopaminergic projection (Nahmias *et al.*, 1985; Martin *et al.*, 1986; Brooks *et al.*, 1990, 1992) as well as reduced basal ganglia activity, especially in patients with the "on/off" phenomena (Leenders *et al.*, 1986, 1990). Garnett, Nahmias, and Firnau (1984) showed that in hemiparkinsonism, there is a marked decrease in activity in the contralateral basal ganglia. However, there is also decreased activity, although to a lesser extent, in the ipsilateral basal ganglia. Fluorodopa studies have been used to investigate clinical course and the effects of therapy in patients with PD. These patients have been followed longitudinally and were found to have a continual decrease in the accumulation of fluorodopa, which correlates with worsening bradykinesia scores (Leenders *et al.*, 1990; Eidelberg *et al.*, 1990). Other studies have shown an increase in fluorodopa uptake in PD patients 14 months after receiving fetal mesencephalon transplantation (Lindvall *et al.*, 1990; Widner *et al.*, 1991). Fluorodopa PET studies have also found abnormalities in the nigrostriatal projection in asymptomatic patients with increased risk of developing Parkinson's disease such as co-twins of patients with PD (Bhatt *et al.*, 1991; Burn *et al.*, 1991; Brooks, 1991a, 1991b; Sawle, 1992) and first-degree relatives in families with inherited PD (Bhatt *et al.*, 1991; Sawle *et al.*, 1992).

Several groups (Teddroff *et al.*, 1988, Leenders *et al.*, 1990) have shown that in PD patients less than 65 years of age, there is a marked reduction in striatal uptake of ^{11}C-nomifensine, most pronounced in the putamen with the caudate slightly less affected. However, in older PD patients, there is similar binding activity to age-matched controls. Studies comparing ^{11}C-nomifensine to fluorodopa PET in PD patients have found similar decreases, both qualitatively and quantitatively (Leenders *et al.*, 1990; Teddroff *et al.*, 1990). Wijnand *et al.* (1987) found a bilateral increase in receptor binding following L-dopa therapy. It was suggested that this unexpected increase might be the result of either persistent postsynaptic D2 receptor hypersensitivity or a change in the local pharmacokinetics of the tracer.

Huntington's Disease

Huntington's disease (HD) is an autosomal dominant disorder that presents usually in middle age and is characterized by progressive motor abnormalities such as involuntary choreiform movements and akinetic rigidity as well as behavioral disturbances and progressive cognitive deterioration (Mazziotta, 1989; Nagel, Ichise, & Holman, 1991).

CT and MRI scans have often shown no changes early in the course of the disease, but have shown atrophy of the head of the caudate and frontal cortex in the later stages of the disease (Terrence *et al.*, 1977, Simmons *et al.*, 1986; Starkstein *et al.*, 1989). Jernigan *et al.* (1991) studied HD using quantitative MR analysis to measure brain volumes and found the greatest reductions in the striatal structures, although there were also significant abnormalities in the inferior cortical structures. Further, they suggested that the cause of the dementia symptoms in HD was likely the result of degeneration of the thalamus and temporal lobe in contrast to the hippocampal area, which is believed to be involved in AD symptoms.

In general, patients with HD tend to have bilaterally decreased uptake of both IMP or HMPAO in the basal ganglia regions compared to controls, with no significant decrease in the cortical rCBF (Leblhuber *et al.*, 1987; Reid *et al.*, 1988; Nagel *et al.*, 1988, 1991; Gemmell *et al.*, 1989). As HD progresses, decreased rCBF is noted in the cortex only in the most severe cases. Since the basal ganglia are believed to control and regulate movement, it seems that the decreased flow to these areas indicates neuronal dysfunction which likely results in the movement symptoms (Nagel *et al.*, 1991). Nagel *et al.* (1991) also noted that the functional deficits precede morphological changes which can be seen on CT or MRI. Further, in severe, progressive HD, cortical defects eventually can be observed on SPECT. Thus, it has been suggested that SPECT can be useful in determining patients at risk for the development of HD when used in conjunction with other clinical data (Nagel *et al.*, 1991).

PET findings in HD have generally corroborated SPECT findings. HD patients have consistently been found to have hypometabolism in the caudate nuclei on PET images (Phelps & Mazziotta, 1985; Phelps *et al.*, 1985; Hawkins & Phelps, 1986; Young *et al.*, 1986; Kuwert *et al.*, 1989; Pahl, 1989) which often precedes the atrophy as determined by CT (Kuhl *et al.*, 1982; Hayden *et al.*, 1986; Mazziotta, 1989; Brooks, 1989). Further studies have found correlations between caudate hypometabolism and neuropsychological deficits in verbal learning and memory (Berent *et al.*, 1988). Most studies have not reported cortical changes in glucose metabolism except in those patients with dementia. The duration of the chorea correlated significantly with the rCMRGlc in the frontoparietal and temporo-occipital areas. Further, the frontoparietal and temporo-occipital rCMRGlc correlated with the severity of dementia in the HD patients.

Several PET studies have been conducted to aid in the diagnosis of HD in asymptomatic patients who are at risk for developing the disease. The results from these studies have been mixed. Mazziotta *et al.* (1987) found hypometabolism in the caudate of 31% of at-risk patients. When the caudate hypometabolism was compared to DNA polymorphism studies that identified the HD gene in at-risk patients, there was a concordance rate of 91%. While this success has been corroborated by other investigators (Kuhl *et al.*, 1982; Hayden, Hewitt, Martin *et al.*, 1987; Hayden, Hewitt, Stoessl *et al.*, 1987), Young *et al.* (1987) did not find any significant difference in caudate metabolism between patients at risk for developing HD and control subjects.

PET has also been used to study the changes in the dopaminergic pathways in HD patients. Leenders *et al.* (1986) used ^{18}F-dopa and found normal striatal uptake, but a decreased D2 receptor density. This finding is believed to be consistent with the neuropathology of HD, which consists of neuronal loss in the neostriatum while the nigrostriatal pathway remains intact. Other investigators (Wong *et al.*, 1985; Hagglund *et al.*, 1987) have reported similar decreases in dopamine receptor densities in the basal ganglia of HD patients.

Progressive Supranuclear Palsy

Progressive supranuclear palsy (PSP) is a disease in which patients suffer from cognitive dysfunction, particularly in tasks sensitive to frontal lobe dysfunction (Maher, Smith, & Lees, 1985). PSP is characterized by a paralyzed gaze, dystonia, axial rigidity, and, eventually, dementia. CT imaging can show some of the atrophic changes in PSP, although MRI is particularly useful in showing focal atrophic changes in the tectal region (Ambrosetto, 1987; Savoiardo *et al.*, 1989). HMPAO SPECT studies have shown hypoperfusion in the temporal and frontal lobes bilaterally without accompanying defects in the parietal lobes, thus differentiating it from the AD pattern (Testa *et al.*, 1988). Neary *et al.* (1987) indicated that there were anterior cortical or subcortical decreases in HMPAO uptake in seven of eight PSP patients. An IMP SPECT study by Johnson *et al.* (1989) reported asymmetrically decreased uptake in the basal ganglia and diffuse defects in the superior frontal lobe.

On PET studies, PSP is associated with hypometabolism of the basal ganglia, thalamus, pons, and cerebral cortex particularly in the frontal lobe (Blin *et al.*, 1990), but not the cerebellum (Foster *et al.*, 1988). Studies have shown that the superior frontal cortex has the most marked hypometabolism particularly in the motor and premotor areas (D'Antona *et al.*, 1985; Foster *et al.*, 1988; Goffinet *et al.*, 1989). Significantly decreased dopamine formation and storage in the striatum has been reported in PSP patients (Brooks *et al.*, 1992; van Royen *et al.*, 1993). This decrease also correlated with the degree of reduced frontal lobe blood flow.

Atrophic Disorders

Olivopontocerebellar atrophy (OPCA) is a disorder in which there is neuronal degeneration of the cerebellar cortex, pons, and inferior olives. This atrophy is most easily determined with MR imaging and can be distinguished from other disorders (Giuliani *et al.*, 1992; Testa *et al.*, 1993). Several investigators have shown that olivopontocerebellar atrophy patients had significant hypometabolism in the cerebellar hemispheres, vermis, and brain stem (Kluin *et al.*, 1988; Rosenthal *et al.*, 1988). Further, there was a strong correlation between the degree of ataxic dysarthria and the decrease in metabolism in the cerebellum and brain stem.

Striatonigral degeneration is an akinetic-rigid syndrome that resembles PD without the tremor and has no signs of PSP. MR images reveal atrophy in the putamen which corresponds to histopathological findings (O'Brien *et al.,* 1990). PET studies have found striatal and prefrontal hypometabolism in patients with striatonigral degeneration (De Volder *et al.,* 1989). This finding corroborates MR images in these patients as well as histopathologic reports.

Corticobasal ganglionic degeneration (CBGD) is a disorder that causes akinetic-rigid asymmetrical parkinsonism with lateralized cortical dysfunction. Symptoms such as apraxia, reflex myoclonus, sensory disturbances, and pyramidal dysfunction are caused by this cortical dysfunction. PET studies have shown metabolic asymmetries in the motor cortex, inferior parietal lobe, and thalamus (Eidelberg *et al.,* 1989; 1991; Blin *et al.,* 1992) as well as the superior temporal lobe (Sawle *et al.,* 1989). These findings correspond to the abnormal regions on MRI and histopathologic studies. Fluorodopa (^{18}F-dopa) PET imaging in these patients showed asymmetric dysfunction of the nigrostriatal dopaminergic pathways (Eidelberg *et al.,* 1991). The findings from these studies indicate that CBGD may be distinguishable from PD using functional imaging.

Multiple system atrophy (MSA), also called Shy-Drager syndrome, can include syndromes of olivopontocerebellar atrophy, striatonigral degeneration, pallidopyramidal degeneration, and pure autonomic failure. MR images may show the atrophic changes and allow for the distinction between MSA and OPCA (Kato *et al.,* 1992; Testa *et al.,* 1993). The results of PET imaging using several different ligands (Brooks, 1989) have shown that patients with MSA have decreased striatal, brain stem, and cerebellar glucose metabolism. These patients also have decreased ^{18}F-dopa uptake in the putamen and caudate and normal or decreased D2 binding compared to controls. Kata *et al.* (1992) indicated that there was no significant correlation between the FDG-PET and MRI findings.

Wilson's disease is an autosomal recessive disorder of copper metabolism that can eventually result in movement disorders and psychiatric disorders due to a buildup of copper in the brain. MR images usually show hyperintense lesions in the lenticular, thalamic, and caudate nuclei as well as in the white matter (Prayer *et al.,* 1990; Thuomas *et al.,* 1993). Patients also can have diffuse atrophy; however, there has been no correlation between the MR findings and clinical symptoms. Also, MR imaging might be helpful in detecting smaller lesions that CT cannot (Abdollah, Tampieri, & Melanson, 1993). MR imaging can be used to follow the changes in the brain before and after treatment (Thuomas *et al.,* 1993). An MR study of a Wilson's disease patient with dystonia showed atrophy and low signal intensity changes in the basal ganglia (Schwarz *et al.,* 1992). In the same study, IBZM SPECT scans (see Table 2) revealed decreased uptake in the basal ganglia compared to patients without dystonia and also to controls. Several PET studies have found that rCMRGlc was significantly decreased in the cerebellum, striatum, cortex, and thalamus of these patients compared to controls (Hawkins, Mazziotta, & Phelps, 1987; Kuwert *et al.,* 1992), although one study did not report any differences (Pappata *et al.,* 1989). In addition, decreases in the cerebellum, thalamus, and cortex correlated with the degree of pyramidal signs. Patients recently started on decoppering therapy had significantly reduced rCMRGlc values compared to patients on therapy for a longer duration.

Cerebrovascular Disorders

Stroke

Cerebrovascular disease is the third leading cause of death in the United States. However, stroke is often associated with a poor outcome in part due to the lack of understanding of the mechanisms that underlie stroke and how recovery might occur. PET and SPECT imaging have been of great benefit in advancing the understanding of the pathophysiology of stroke (Alavi *et al.*, 1982; Frackowiak & Wise, 1983; Baron *et al.*, 1989; Herholz & Heiss, 1991). Furthermore, both imaging techniques allow for the detection of stroke earlier than anatomical imaging modalities such as MRI or CT. Functional imaging has also been useful in evaluating the extent of the damage since areas not immediately affected by the infarct may show hypometabolism or decreased blood flow.

CT is most useful in the early evaluation of an acute stroke. Specifically, CT is helpful in distinguishing hemorrhagic versus nonhemorrhagic stroke (Weisberg, 1990; Lassen, 1990). This is particularly important when evaluating a patient for thromobolytic therapy or for surgical revascularization (Lassen, 1990). However, if there is no hemorrhage, CT images are usually negative until 12–24 hours postinfarct (see Figure 6). At that time, the infarct may be detectable as an ill-defined hypodense region and may or may not be associated with edema (Wang *et al.*, 1988). An ischemic stroke may be detectable earlier as an

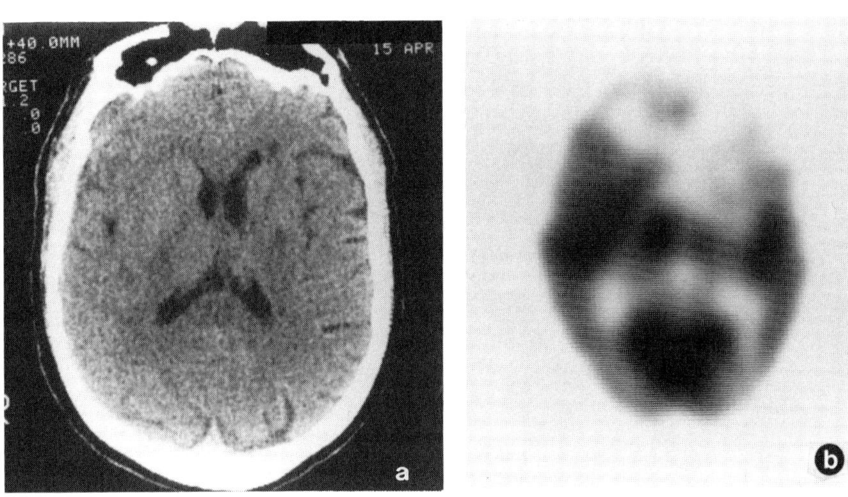

Figure 6. Findings on CT images of cerebral infarcts are often subtle in the early days of infarct. This CT image (a) does show some evidence of stroke anterior to the left caudate nucleus with some atrophy in the left hemisphere as well as enlargement of the frontal horn of the left lateral ventricle. A PET image (b) of the same patient shows an area of decreased glucose metabolism in the entire left frontal lobe and basal ganglia. Also, there is decreased metabolism in the ipsilateral thalamus. This demonstrates the functional consequence of lesions located in anatomically important sites.

area of increased density in the lumen of a cerebral vessel due to embolism or thrombus (Pressman, Tourje, & Thompson, 1987). In the early phase, there may also be a decreased peak amplitude in dynamic CT scans of the affected hemisphere (Shih & Huang, 1988) or a partial disappearance of the lenticular nucleus in patients with a middle cerebral artery infarct (Tomura *et al.*, 1988). Later in the course of the infarct, the area may become isodense followed by a stage of contrast enhancement secondary to "luxury perfusion" (see below). The end stage of stroke is associated with a well-defined hypodense region often with enlargement of the local CSF spaces (Jolles, Chapman, & Alavi, 1989).

MRI may be more sensitive than CT in the evaluation of stroke patients. However, CT is typically more available than MRI and has a lower cost. Animal studies have reported that MRI can detect regions of infarct as soon as 1 to 2 hours postinfarct (Spetzler, 1983). Infarcts typically appear as low signal intensity lesions on T 1-weighted images and as high signal lesions on T 2-weighted images. Further, MRI can detect hemorrhagic infarcts which can sometimes be missed by CT in the early stages of stroke (Jolles, Chapman, & Alavi, 1989). When the stroke is fully evolved, the lesions appear as cystlike CSF intensity areas.

SPECT is very useful in the detection of stroke since decreases in rCBF occur immediately following the infarct, while anatomical changes occur after several hours to days. It has been shown that by 8 hours postinfarct, almost 90% of patients will have changes on SPECT (Podreka *et al.*, 1987; DeRoo *et al.*, 1989) while only 20% will have changes on CT (Fieschi, 1989; Bose *et al.*, 1990). Other studies have also indicated that SPECT can detect acute cerebral infarct when CT shows no abnormality (Buell *et al.*, 1984; Hill *et al.*, 1984; Killen, Oster, & Golditz, 1989; Rango *et al.*, 1989). Usually by 72 hours postinfarct, the differences between the ability of SPECT and CT to detect areas of stroke disappear. SPECT studies have revealed a "luxury perfusion," shown by increased blood flow to the affected regions which may be due to local metabolic acidosis (Buell *et al.*, 1984; Bushnell *et al.*, 1989). The problem with SPECT imaging in stroke is that between approximately 5 and 20 days after the insult, CBF and cerebral metabolism become uncoupled. Thus, during this time, blood flow measurements with SPECT cannot be correlated with cerebral metabolism (Moretti *et al.*, 1990).

SPECT may be useful in determining the potential success of medical intervention in patients with acute cerebral ischemia. SPECT has been used to detect ischemic tissue before irreversible damage occurs (Alavi & Hirsch, 1991), but is at risk for infarction (Creutzig *et al.*, 1986); to determine prognosis based upon areas of redistribution (Moretti *et al.*, 1987; Defer *et al.*, 1987); and to predict which patients might benefit from revascularization (Maurer *et al.*, 1988). It should be noted that some investigators have found a direct correlation between rCBF and clinical outcome (Lee *et al.*, 1984; Nagata *et al.*, 1986; Defer *et al.*, 1987; Kushner *et al.*, 1987; Heiss *et al.*, 1988) while others have not (Demeurisse *et al.*, 1983; Vallar *et al.*, 1988; Hayman *et al.*, 1989). However, other studies have found a very strong positive predictive value between the volume of the flow defect and clinical outcome (Bushnell *et al.*, 1989; Giubilei *et al.*, 1990; Limburg *et al.*, 1991; Fayad & Brass, 1991).

It has also been suggested that SPECT might be useful in determining patients who are at risk for stroke before cerebral ischemia occurs. Some studies have shown that patients at risk for stroke will have decreased responses to CO_2

or acetazolamide (Norrving, Nilsson, & Risberg, 1985; Sullivan *et al.*, 1987; Chollet *et al.*, 1989). SPECT might also be useful in evaluating cerebrovascular reserve and perfusion pressure by determining regional cerebral blood volume (rCBV) and comparing it to rCBF. The measure of cerebrovascular reserve might help quantify the risk for developing stroke (Gibbs, Frackowiak, & Legg, 1986; Knapp, von Kummer, & Kubler, 1986; Buell *et al.*, 1988).

Another finding in SPECT is crossed cerebellar diaschisis in patients with cortical strokes, and sometimes in patients with transient ischemic attacks (TIAs) (Kushner *et al.*, 1984; Meneghetti, Vorstupt, & Mickey, 1984; Pantano *et al.*, 1986, 1987; Perani, Gerundini, & Lenzi, 1987; Perani *et al.*, 1988). Crossed cerebellar diaschisis, first described by Baron *et al.* (1980), refers to hypometabolism and hypoperfusion in the cerebellar cortex contralateral to the site of the infarct and usually occurs during the first 2 months after infarction. It is believed that this is due to an interruption of the cerebropontocerebellar pathways as a result of the stroke (Pantano *et al.*, 1986).

PET has also been used extensively in the investigation of stroke and the results are similar to those reported from SPECT studies (see Figure 6). However, PET imaging has superior resolution and sensitivity for detecting stroke and ischemia when compared to SPECT. Further, SPECT images tend to show less contrast between the normal brain tissue and areas of infarct than PET. HMPAO SPECT tends to underestimate the rCBF in areas of high flow in the cerebral cortex and overestimate the rCBF in the cerebellum compared to blood flow studies using PET (Heiss *et al.*, 1990).

In patients who have suffered a stroke, there is a characteristic uncoupling between CBF and metabolism in the infarcted area (Kuhl 1980; Ackerman *et al.*, 1981; Baron *et al.*, 1981, 1983; Lenzi, 1982; Alavi *et al.*, 1982). Several studies using $H_2{}^{15}O$ have described "misery perfusion" in and near areas of infarct within the first hours to days after a stroke. This misery perfusion is described as a relative decrease in rCBF compared to the regional glucose metabolism or oxygen metabolism ($rCMRO_2$). Further, studies have shown that there is a marked increase in the oxygen extraction fraction (OEF) in response to the diminished blood flow (Baron *et al.*, 1981, 1983, 1984; Wise *et al.*, 1983). Approximately one week after infarct, "luxury perfusion" occurs, which is a relative increase of rCBF compared to cerebral metabolism (Lassen, 1966). Wise *et al.* (1983) found that rCBF increased compared to $rCMRO_2$ with a subsequent decrease in rOEF several days postinfarct.

In addition to the infarcted area, there exists a "penumbral" zone, a hypometabolic and presumably ischemic area which surrounds the infarcted area (Marchal *et al.*, 1987). This area also has increased OEF, suggesting that this area has decreased perfusion for the necessary oxygen requirements. Further, if blood flow to this ischemic area is restored before irreversible damage occurs, then the tissue will likely recover and resume normal function (Baron *et al.*, 1981). Studies have also indicated that there is a period of misery perfusion with decreased perfusion and increased rOEF as much as 2 days from the onset of stroke (Syrota *et al.*, 1983, 1985; Hakim *et al.*, 1987; Senda *et al.*, 1989).

PET studies, like SPECT, have found regions distant from the stroke which also have alterations in metabolism despite being normal on anatomical imaging such as CT or MRI (Baron *et al.*, 1980; Kushner *et al.*, 1984; Gibbs *et al.*, 1984; Powers, Grubb, & Raichle, 1984; Pantano *et al.*, 1986; Feeney & Baron, 1986; Yao *et al.*, 1990; Herholz *et al.*, 1990; Meyer *et al.*, 1991). The most common of these is crossed cerebellar diaschisis mentioned above, which has been observed as cerebellar hypometabolism using PET studies. There are also other areas that are hypometabolic after a cortical infarct (Broich, Alavi, & Kushner, 1992). These areas include the ipsilateral thalamus, the ipsilateral caudate nucleus, and the ipsilateral primary visual cortex (if the infarct is in the anterior visual pathways).

PET studies have also been performed to determine the pathophysiology of chronic ischemia in an attempt to determine the possibility of stroke in high-risk patients. Gibbs *et al.* (1984) found increased rCBV with normal rCBF and $rCMRO_2$ ipsilateral to occlusion of the internal carotid artery. The increase in rCBV is likely due to the vasodilation that occurs in response to the decreased perfusion pressure. It has been postulated that when the compensatory vasodilator response is at a maximum, any further decrease in perfusion pressure would result in a decreased rCBF with the high likelihood of ischemia and eventually stroke (Brooks, 1991b). Since rCBF remains relatively constant until the maximum rCBV is attained, the rCBF/rCBV ratio (which correlates well with rOEF) decreases via autoregulation before ischemia develops (Baron *et al.*, 1989; Sette *et al.*, 1989). SPECT has similarly been used to determine the rCBF/rCBV ratio as an index that measures "perfusion reserve," which can thereby aid in quantifying a patient's risk for stroke (Knapp *et al.*, 1986; Buell *et al.*, 1988). Unfortunately, a number of PET studies have not been successful in using cerebral hemodynamic parameters to assess risk of stroke or the potential efficacy of surgical intervention in patients with carotid artery disease (Powers *et al.*, 1984, 1987; Powers, Grubb, & Raichle, 1989; Leblanc, 1991). These studies found no correlation between the hemodynamic state and the risk of stroke or the degree of carotid stenosis. However, studies of patients before and after successful extraintracranial bypass surgery showed decreases in rCBV and normalization of the rCBF/rCBV ratio following surgery (Gibbs *et al.*, 1984; Samson, Baron, & Boussen, 1985; Leblanc *et al.*, 1987; Powers, Grubb, & Raichle, 1989). Unfortunately, there was no evidence to indicate that the hemodynamic status of patients as determined by PET with carotid artery disease could adequately predict which patients would actually benefit from bypass surgery as determined by the incidence of stroke.

PET studies have been used to determine prognosis and recovery of function after stroke. DiPiero *et al.* (1992) showed that patients with increased oxygen metabolism in the motor cortices after acute stroke were found to have good motor function recovery. Even patients with increased metabolism in the contralateral hemisphere, without increases in the ipsilateral hemisphere, showed better motor recovery than patients who had no increased oxygen metabolism in either hemisphere. This and other studies suggest that there is bilateral recruitment of additional sensorimotor areas in motor function recovery from stroke (Weiller *et al.*, 1992). Kushner *et al.* (1987) found that the pattern of metabolic derangement correlated significantly with the degree of recovery in stroke patients. In a study of patients with aphasia due to stroke, patients with increased rCMRGlc in the left cerebral cortex during active speech production tests had a better recovery than those patients who had metabolic activation restricted to the right hemisphere (Heiss *et al.*, 1989). PET studies have also been used to monitor the success of various treatment regimens for stroke. Treatment with nimodipine has been found to result in an increased rCBF and increased glucose metabolism in stroke areas (Hakim *et al.*, 1989; Heiss *et al.*, 1990).

Multi-infarct dementia (MID) has been extensively studied with SPECT, PET, CT and MRI. MID is associated with multiple foci of decreased flow which represent the areas of infarct. These regions are usually randomly distributed throughout the cortex without a predilection for any particular area. Cortical infarcts are readily observed on CT or MR images (see Figure 7) and are distinguished from the pattern of diffuse atrophy seen in AD (R. D. Zimmerman *et al.*, 1986; Duara *et al.*, 1989; Erkinjuntti *et al.*, 1987; Bonte *et al.*, 1990; Kertesz, Polk, & Carr, 1990; Lechner, Schmidt, & Goetz, 1990; Jagust & Eberling, 1991). Thus, MRI has been successfully used to distinguish MID from AD (R. D. Zimmerman *et al.*, 1986; Harrell, Callaway, & Sekar, 1987; Erkinjuntti *et al.*, 1987; Duara *et al.*, 1989; Kertesz, Polk, & Carr, 1990; Jagust & Eberling, 1991).

Since SPECT can detect the areas of decreased blood flow in MID (see Figure 8), often before they manifest in anatomical changes, SPECT may be an important clinical tool in distinguishing MID from AD even in the face of normal or minimally abnormal MR images (Cohen *et al.*, 1986; Sharp *et al.*, 1986; Smith *et al.*, 1988; Deutsch & Tweedy, 1987; Gemmell *et al.*, 1987; Weinstein *et al.*, 1989, 1991; Kuwabara *et al.*, 1990; Bonte *et al.*, 1990; Launes *et al.*, 1991). In fact, the sensitivity and specificity of SPECT in differentiating MID and AD may be as high as 80% (Dewan & Gupta, 1992). However, Weinstein *et al.* (1991) found that statistically, HMPAO SPECT was unable to differentiate AD from MID based on regional tracer uptake, even in the temporoparietal lobes. SPECT may be useful in determining whether there is an underlying AD pattern in patients who also have MID (Toghi *et al.*, 1991). Another study has questioned the use of SPECT in distinguishing true MID from various psychiatric disorders (Launes *et al.*, 1991).

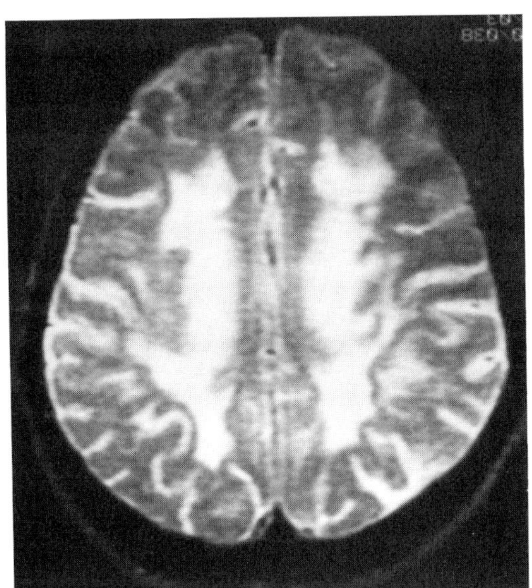

Figure 7. Multi-infarct dementia is similar in appearance to stroke in that areas of infarct appear as hyperintense lesions on T2-weighted MR images. However, MID is characterized by multiple areas of infarct predominantly seen in the white matter especially in the periventricular area.

There have been a number of FDG-PET studies which have investigated the changes seen in MID and how they compared to patients with AD. In patients with MID, PET images show scattered focal areas of hypometabolism (Benson *et al.*, 1983; Duara *et al.*, 1989; Duara, 1990; Heiss *et al.*, 1989) corresponding to the multiple lesions seen on MRI. Thus, qualitative PET can be used to successfully distinguish MID from AD (Pawlik *et al.*, 1989). PET has also been used to quantitatively measure glucose metabolism in patients with MID. In a study by Alavi *et al.* (1992), patients with MID were found to have whole brain CMRGlc values comparable to patients with AD and markedly decreased compared to controls. However, quantitative PET cannot be used to distinguish MID from AD. This study did indicate that MID patients had significantly reduced glucose metabolism even after brain atrophy was taken into account. This differed from AD suggesting that in patients with MID, there is hypometabolism that is disproportionate to the degree of atrophy. The implication is that the pathophysiology of MID differs from AD in that AD patients have relatively normal brain metabolism with decreased tissue volume while patients with MID have decreased metabolism in the existing tissue in addition to decreased brain tissue.

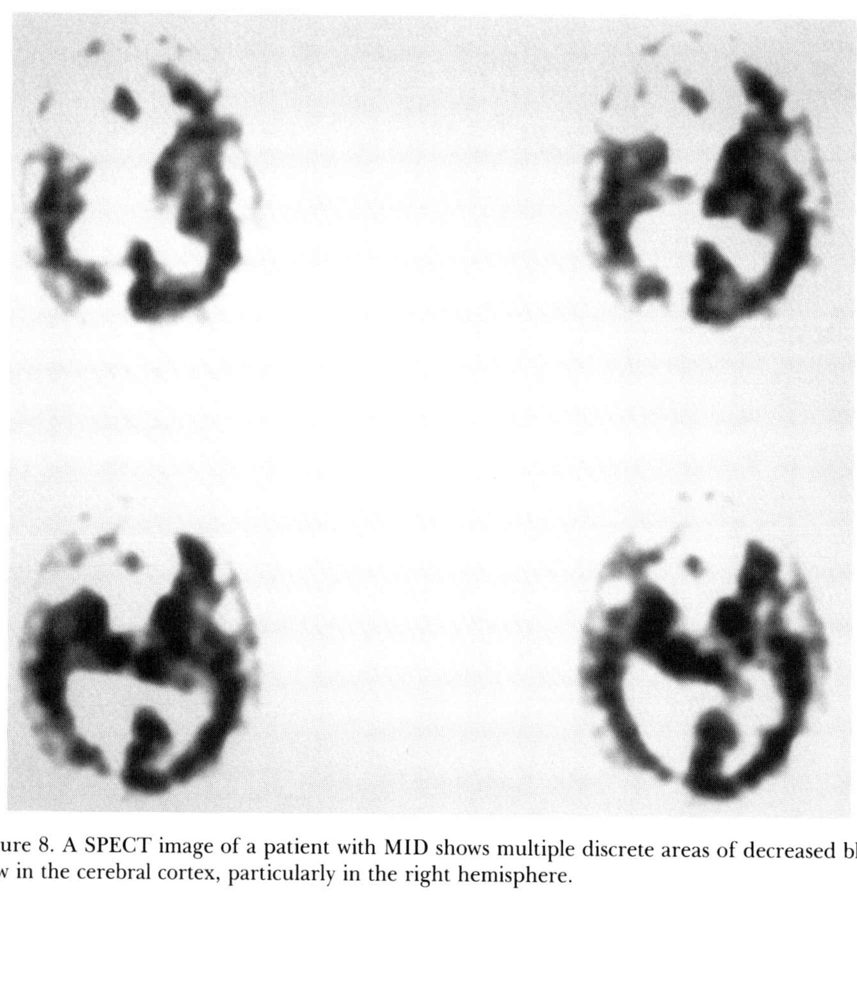

Figure 8. A SPECT image of a patient with MID shows multiple discrete areas of decreased blood flow in the cerebral cortex, particularly in the right hemisphere.

Binswanger's Disease

Binswanger's disease is a disease process involving multiple infarcts eventually resulting in a microangiopathic leukoencephalopathy and is clinically characterized by the progressive onset of dementia. On CT, it is difficult to distinguish Binswanger's disease from MID. On CT, Binswanger's disease appears as multiple hypodense lesions in the periventricular white matter with lacunar infarcts in the basal ganglia and enlarged ventricles (Kinkel *et al.*, 1985; Lotz, Ballinger, & Quisling, 1986). On MR images, these changes appear as confluent regions of hyperintense foci on T 2-weighted images. There is relative sparing of the peripheral white matter, although there can be hyperintense lesions in the basal ganglia (Drayer, 1988).

In Binswanger's disease, periventricular and supraventricular white matter hypoperfusion is a characteristic finding (De Cristofaro *et al.*, 1990) along with clinical symptoms of subcortical dementia, urinary incontinence, motor deficits, pseudobulbar palsy, and subacute progression (Takahashi *et al.*, 1992). The cerebral/cerebellar ratio (CCR) has been found to be significantly decreased in moderately affected (involvement of the frontal lobe and the basal ganglia) and severely affected (in all regions) patients with Binswanger's disease compared to controls (Toghi *et al.*, 1991). Further, the decrease in the CCR in patients with Binswanger's disease correlated with the severity of the disease as determined by psychometric testing. The investigators suggested that the difference in perfusion patterns in AD patients and those with Binswanger's disease reflects the characteristic clinical features of the two disorders.

PET studies of patients with Binswanger's disease have shown decreased rCBF and $rCMRO_2$ in white matter and the frontal, temporal, and parietal cortices (Yao *et al.*, 1990). These changes were accompanied by normal CT and MRI scans. It was suggested that the white matter decrease in rCBF and $rCMRO_2$ reflects damaged tissue from ischemia while decreases in the cortex likely reflect disconnection between cortical and subcortical structures. However, there has been no misery perfusion detected in patients with Binswanger's disease (Yao *et al.*, 1990; Meyer *et al.*, 1991). However, Meguro *et al.* (1990) found that some patients with Binswanger's disease had increased rCBV, increased rOEF, and decreased $rCMRO_2$, indicating misery perfusion.

Head Trauma

The development of anatomical imaging has had a dramatic impact on the diagnosis and management of head trauma. Even though MRI is more sensitive and accurate in diagnosing cerebral pathology (Ogawa *et al.*, 1992; Levin *et al.*, 1992), CT still remains the most beneficial anatomical imaging modality for head trauma in the acute setting (Kelly *et al.*, 1988; Sklar *et al.*, 1992). Yealy and Hogan (1991) indicated that CT is the procedure of choice for moderate- to high-risk head injury patients. CT is also superior to MRI in detecting bony fractures of the skull as well as subarachnoid hemorrhage (Yealy & Hogan, 1991; Jennett, 1991). However, the extent of the lesions detected with CT does not correlate well with the neurological deficit, the level of consciousness, or the clinical prognosis (Ruff *et al.*, 1989; Levin *et al.*, 1992).

In recent years it has become easier to provide optimal patient monitoring during imaging and acquire images in short times. This has allowed MRI to become an acceptable tool in the investigation of acute head injury (Crow, 1991). Yealy and Hogan (1991) suggested that MRI is useful after CT is obtained if subtle acute nonhemorrhagic and subacute hemorrhagic lesions are suspected. MRI is very sensitive for detecting intracranial lesions (Kelly *et al.,* 1988; Gentry *et al.,* 1988a; R. D. Zimmerman *et al.,* 1986; Gentry, Godersky, & Thompson, 1988b; Ogawa *et al.,* 1992; Levin *et al.,* 1987, 1992; Sklar *et al.,* 1992; Yokota *et al.,* 1991) as well as for characterizing lesions as diffuse axonal injury (DAI), cortical contusion, subcortical gray matter injury, or brain stem injury (see Figure 9). However, no study has shown any clinical relevance of the large number of lesions found on MRI in head trauma patients. Thus, despite the high sensitivity of MRI there has been no demonstrable relationship between MRI results and neuropsychological testing (Ruff *et al.,* 1989). Gadolinium enhanced MRI does not offer any advantage to nonenhanced MRI in detecting lesions caused by head injury (Lang *et al.,* 1991). A newer MRI technique allows the acquisition of images for cerebral angiography. When applied to head injury patients, the use of magnetic resonance angiography can be a valuable tool for detecting vascular abnormalities (Sklar *et al.,* 1992). These may include arterial occlusions, arteriovenous fistulae, dissecting aneurysms, and venous sinus occlusions.

SPECT has been found to be more sensitive than CT for detecting lesions in patients with head injury. Specifically, with HMPAO SPECT imaging, investigators have found a greater number of lesions earlier in time than with CT (Abdel-Dayem, Sadek, & Kouris, 1987; Reid *et al.,* 1990; Newtown *et al.,* 1992; Gray *et al.,* 1992). Roper *et al.* (1991) found that some types of contusions are associated with decreased blood flow. Fumeya *et al.* (1990) found that lesions detected with CT ranged from hyperperfused to hypoperfused on SPECT; however, lesions not detected with CT were found to be only hypoperfused on SPECT.

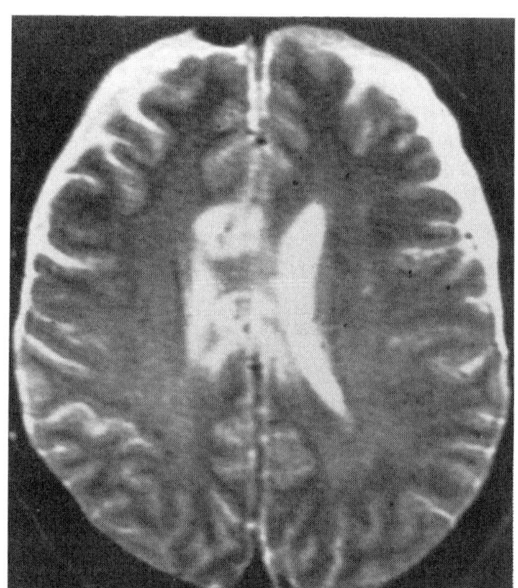

Figure 9. An MR image of a patient with acute head trauma secondary to an automobile accident shows high signal intensity lesions located primarily in the corpus callosum. This type of lesion is considered typical for diffuse axonal injury.

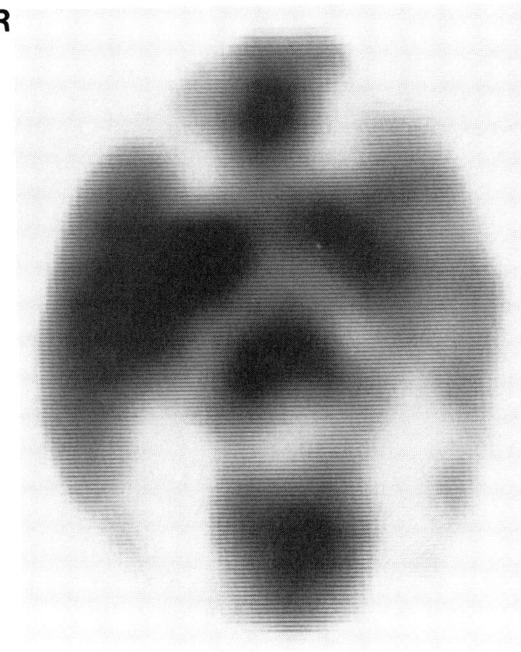

Figure 10. An FDG-PET image of a patient with a subdural hematoma secondary to head injury taken four days after successful evacuation of the hematoma shows diffuse decreased glucose metabolism on the side of the lesion (left side).

It has been found that SPECT is better than CT in distinguishing lesions with a poor prognosis from those with a favorable prognosis (Reid *et al.*, 1990). Patients with larger lesions, multiple defects, and lesions involving the brain stem tend to have a poor prognosis compared to patients with smaller, nonfocal lesions. In a similar study, patients with the greatest disability were found to have the greatest number of lesions and the lowest cerebral blood flow as determined by SPECT (Newtown *et al.*, 1992).

There has been a limited number of studies using PET in the evaluation of patients with head trauma (see Figure 11). One of the problems with the use of PET in these cases is that PET cannot distinguish between structural damage and cerebral dysfunction since these may all result in areas of decreased metabolism (Langfitt *et al.*, 1986). Thus, it is helpful to compare PET to anatomical images such as those obtained by MRI or CT since cerebral dysfunction can extend beyond the boundary of anatomical lesions (Langfitt *et al.*, 1986; Alavi *et al.*, 1987). Like SPECT, abnormalities seen with PET are often not associated with any anatomical lesions.

Lesions such as cortical contusions, intracranial hematoma, and resultant encephalomalacia have metabolic effects that are confined primarily to the site of injury. However, subdural and epidural hematomas often cause widespread hypometabolism (see Figure 10) and may even affect the contralateral hemisphere (George *et al.*, 1989). Crossed cerebellar hypometabolism or diaschisis, as well as ipsilateral cerebellar hypometabolism, has been found in head injury patients with supratentorial lesions (Alavi *et al.*, 1990). Another entity, diffuse axonal injury, has been found to cause diffuse cortical hypometabolism, with a predilection for the parieto-occipital cortex (Alavi, 1989).

Alavi (1989) found a good correlation between the severity of head trauma as measured by the Glasgow Coma Scale (GCS) and the extent of whole brain

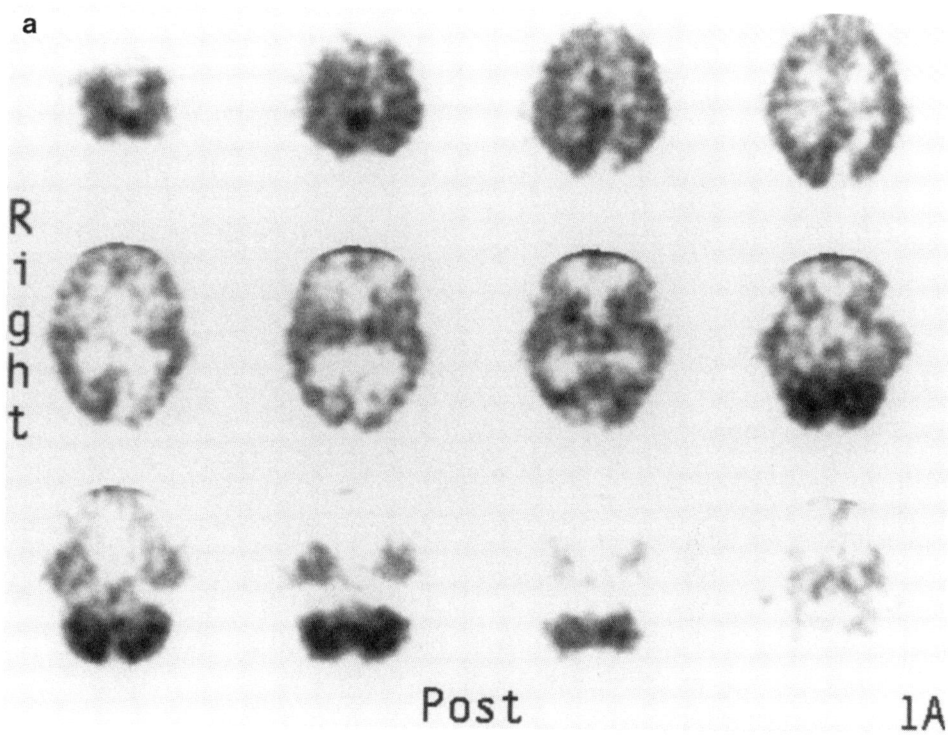
a
Right
Post
1A

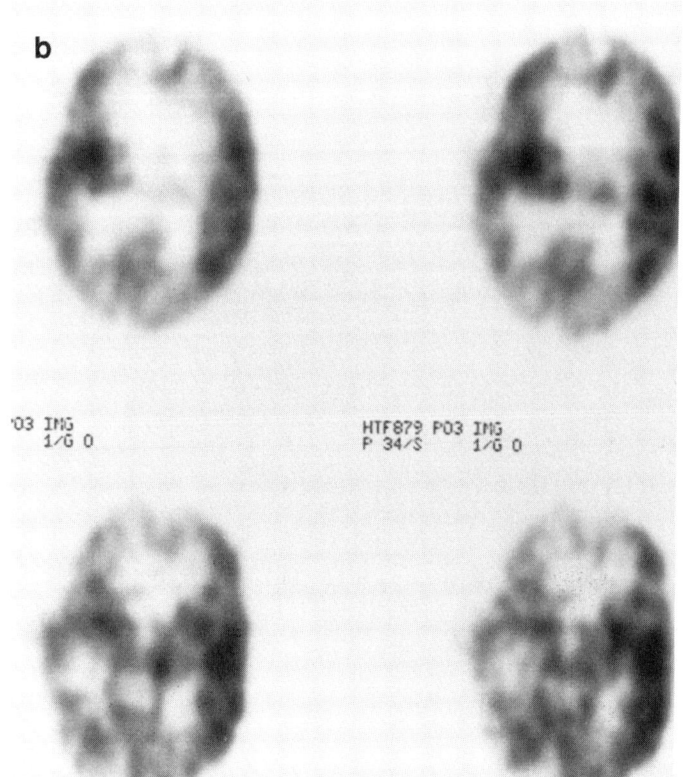
b
'03 IMG
1/G 0
HTF879 P03 IMG
P 34/S 1/G 0

hypometabolism as measured by FDG-PET (the GCS scores range from 1 to 15 with 15 being normal consciousness and 1 being total unresponsiveness). Cerebral metabolism also is associated with neuropsychological deficits (Ruff *et al.*, 1989). Thus, cerebral metabolism is believed to be a good indicator of functional activity in patients with head trauma and also correlates with neuropsychological and language testing, which has implications for rehabilitation potential (Rao *et al.*, 1984; Souder *et al.*,1990). Further, global and regional metabolic rates have been found to improve as patients recover from head trauma (Alavi *et al.*, 1987; George *et al.*, 1989).

Seizure Disorders

Epilepsy, which affects 0.5% to 1% of the population, can cause focal or generalized seizures in patients and usually begins in childhood. In general, during an epileptic seizure, cerebral metabolism and blood flow are markedly increased. SPECT and PET are useful in determining seizure foci (Bonte *et al.*, 1983; Biersack *et al.*, 1985; Reichman *et al.*, 1986; Lee *et al.*, 1986, 1987; Stefan *et al.*, 1987; Ryding *et al.*, 1988; Lang *et al.*, 1988; Homan *et al.*, 1989; Rowe *et al.*, 1989; LaManna *et al.*, 1989; Fisher & Frost, 1991; Tyler & Starr, 1992).

CT and MRI have limited use in the diagnosis and evaluation of seizure patients except when the seizures are caused by an intracerebral lesion such as a tumor or mesial temporal sclerosis (MTS), which is due to a focal neuronal loss and gliosis in the temporal lobe (a common cause of partial seizures) (Engel, Brown, *et al.*, 1982; Jolles, Chapman, & Alavi, 1989). One study comparing MRI and CT in the evaluation of temporal lobe epilepsy found that CT detected abnormalities 32.5% of the time with an accuracy of 19% while MRI detected abnormalities in 81% of the cases with an accuracy of 67% (Swartz *et al.*, 1992). CT images are limited because of bone artifact and therefore can only detect large tumors in the temporal lobe (Blom *et al.*, 1984; Latack *et al.*, 1986; Theodore *et al.*, 1986; Sperling *et al.*, 1986). MRI is very sensitive in detecting nonsclerosing lesions and more recently has been found to be highly sensitive in detecting MTS based on volumetric analysis (Cascino *et al.*, 1992; Sperling *et al.*, 1986; Fobben *et al.*, 1988; Gates & Cruz-Rodriguez, 1990). MRI is also useful in detecting intracerebral lesions that may be causing seizures such as those in Wegener's granulomatosis (Miller & Miller, 1993) and cerebral aneurysm (Putty, 1990–1991). Other studies (Lenoz *et al.*, 1992; Cascino *et al.*, 1992) used MR images to quantitatively determine the brain volume of various regions in patients with medically refractory temporal lobe epilepsy. They found smaller hippocampal areas and temporal lobes on the side of the seizure focus compared to controls, particularly in patients with MTS (Cascino *et al.*, 1992).

SPECT studies are most useful in the detection of ictal foci in patients with partial seizures (see Figure 12). Interictal SPECT imaging (see Figure 13) has been found to have a detection rate of 50–64% for temporal lobe foci (Bonte *et al.*, 1983; Stefan *et al.*, 1987), which is slightly worse than MRI and slightly better than CT. Interictal SPECT generally does not have the sensitivity of PET due to

←

Figure 11. A SPECT image of (a) chronic head trauma shows decreased perfusion in the occipital-visual cortices. A PET image of (b) a patient with acute head trauma shows decreased metabolism especially in the right hemisphere and the occipital-visual cortices. (With permission from publisher.)

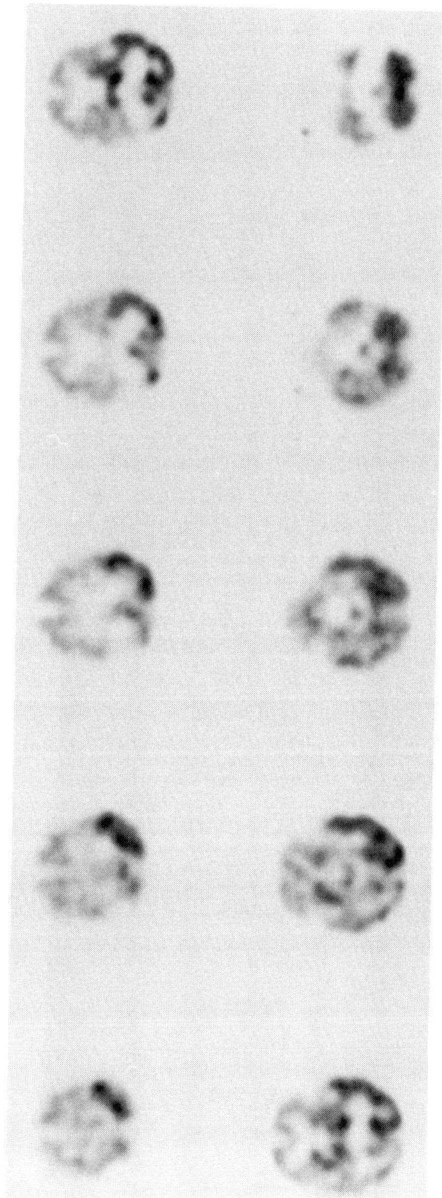

Figure 12. HMPAO SPECT of a patient in status epilepticus showing ictal uptake of the tracer in the left parieto-occipital area.

Figure 13. A coronal SPECT image of a patient with a seizure focus in the right temporal lobe characterized by areas of decreased perfusion, compared to the right, during the interictal phase.

its poorer resolution. However, the lower cost and greater availability of SPECT might make it a better tool for evaluating patients with partial seizures (Tyler & Starr, 1992). With the introduction of high-resolution SPECT instruments, this technique may be used as the primary method for the detection of seizure foci.

As described above, the focus of partial seizures can be identified using FDG-PET (see Figure 14) since these areas have increased metabolism during the seizure and decreased metabolism in the interictal period (Brown, Engel, *et al.*, 1982, 1983; Theodore *et al.*, 1983, 1984; Abou-Khalil *et al.*, 1987). It has been shown that single hypometabolic regions can be identified in 55% to 80% of patients with focal EEG abnormalities (Henry *et al.*, 1991; Engel, 1991; Duncan, 1992). The temporal lobe has been the most common site for partial epilepsy, and studies have shown that FDG-PET has good sensitivity in detecting temporal lobe foci (Engel, Brown *et al.*, 1982, Engel, Kuhl *et al.*, 1982b; Theodore *et al.*, 1986; Theodore, Fishbein, & Dubinsky, 1988) as does $C^{15}O_2$ (Bernardi, Trimble, Frackowiak, Wise, Jones, 1983; Och, Yamamoto, & Gloor, 1984). A study by Sperling *et al.* (1986) has shown a positive finding on PET in 44% of patients with temporal lobe epilepsy (TLE) who had normal CT scans. FDG-PET has been similarly successful in identifying seizure foci in the frontal lobe (Franck *et al.*, 1992; Robitaille *et al.*, 1992). However, performing ictal studies with PET is clinically impractical because of the logistics of preforming scans under these circumstances (Alavi & Hirsch, 1991). Therefore, PET may be replaced by modern SPECT imaging techniques.

PET may be useful for measuring neurotransmitter activity in patients with seizures. Initial studies have found that in TLE, decreased benzodiazepine receptor activity is noted in the medial temporal lobe (Savic *et al.*, 1988). Further,

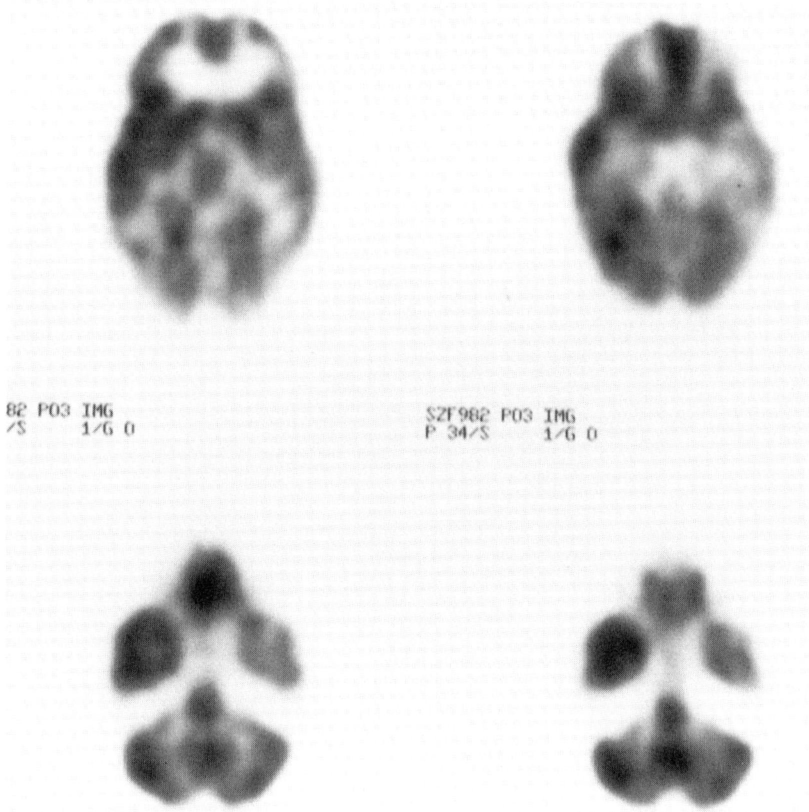

Figure 14. A PET image in a seizure patient shows decreased metabolism in the left temporal lobe (identifying this area as the seizure focus) in the interictal period.

some patients have been shown to have increased opiate receptor activity (using ^{11}C-carfentanil) in the temporal lobe neocortex (Frost *et al.*, 1988; Fisher & Frost, 1991).

AIDS Dementia

Neuroimaging techniques have been widely used in the clinical investigation of AIDS-related neurological disorders. Among the most common of these disorders are lymphoma, opportunistic infections, and AIDS-related dementia (Pedersen *et al.*, 1991). AIDS patients without specific neurological diseases are found to have infratentorial atrophy in the early stages (Raininko *et al.*, 1992). This is followed by supratentorial atrophy in patients with AIDS-related complex and generalized atrophy in patients with AIDS. Lymphoma usually appear as large contrast enhancing lesions which are often multifocal and are associated with mass effect and edema (Peretti-Viton *et al.*, 1991). However, it may be difficult to distinguish lymphoma from toxoplasmosis on the basis of MR imaging (Kupfer *et al.*, 1990). Tuberculosis of the central nervous system appears as hydrocephalus and meningeal enhancement. However, CT studies may appear normal in 25% of patients, and MR imaging is generally more sensitive for detecting lesions (Villoria *et al.*, 1992). Mundinger *et al.* (1992), utilizing CT and MRI, found that patients presenting with neurological symptoms and focal lesions or diffuse atrophy on imaging studies had markedly decreased survival times compared to normal patients. AIDS dementia complex has been shown to be significantly associated with the appearance of white matter lesions on MRI when the CD4 count is less than 200 (Pedersen *et al.*, 1991). Furthermore, brain atrophy clearly occurs in AIDS patients, but its clinical significance is yet to be determined.

Studies have shown that SPECT is highly sensitive in the diagnosis of AIDS dementia complex (Ell, Costa, & Harrison, 1987; Pohl *et al.*, 1988; Tran Dink *et al.*, 1990). Further, perfusion changes can even be seen in early stages of the disease and is seen in as many as half of patients with normal CT and MR images (Schielke *et al.*, 1990; Tatsch *et al.*, 1990; Masdeu *et al.*, 1991). One study (Holman & Devous, 1992) showed that there was a high incidence of cortical defects in the frontal, temporal, and parietal lobes using HMPAO SPECT. There was also significant basal ganglia involvement in 40% of AIDS dementia patients. A study by Kuni *et al.* (1991), using IMP SPECT, showed a significant decrease in pixel intensity (as measured by the fraction of pixels in the lowest quartile of the intensity range) in AIDS patients with mild dementia compared to controls without AIDS. Further, this fraction correlated highly with the severity of the disease as measured by neuropsychological testing. However, a study by Holman, Johnson, *et al.* (1992) also examined a population of HIV-negative chronic cocaine and polydrug abusers and found similar (and statistically undistinguishable) cortical and basal ganglia defects compared to those patients with AIDS-related dementia. Thus, the diagnosis of AIDS-related dementia must be made carefully as there may also be effects on brain perfusion from drug abuse in these patients.

Alcoholism

It is well established that anatomic studies of alcoholic patients have revealed cortical atrophy in the early stages of the disorder with progression to ventricular enlargement in the later stages. CT may also have value in determining whether these patients have suffered from head trauma. MR images have detected brain lesions associated with multiple etiologies in alcoholic patients, such as Wernicke's encephalopathy, ethanol neurotoxicity, central pontine myelinolysis, and head trauma (Charnes, 1993). Studies of alcoholic patients with PET have generally found decreased whole brain metabolic activity (Samson *et al.*, 1986; Wik *et al.*, 1988; Sachs *et al.*, 1987) with a predilection for the left hemisphere (Volkow & Fowler, 1992). Chronic alcoholics with alcoholic cerebellar degeneration have been found to have significantly reduced rCMRGlc in the superior vermis compared to controls (Gilman *et al.*, 1988). Compared to controls, alcoholics had a more marked metabolic deficit after ethanol ingestion, but had less clinical symptoms, suggesting a tolerance to alcohol (Volkow *et al.*, 1990). Preliminary receptor studies using benzodiazepine radioligands have reported decreases in benzodiazepine receptors in chronic alcoholics (Litton *et al.*, 1991).

Cocaine Abuse

Brain metabolism studies have shown that acute administration of cocaine in chronic cocaine abusers results in decreased metabolism in the cortical and subcortical structures (London *et al.*, 1990). Further, the extent of metabolic decreases correlated with the subjective strength of the euphoria. Also, subjects with enlarged ventricles (as determined by CT) were found to be less sensitive to the metabolic effects of cocaine than those with less ventricular enlargement. While it is known that chronic cocaine abusers tend to have enlarged ventricles, how this is related to the effects of cocaine is not yet understood.

One study (Volkow *et al.*, 1991) of the acute changes after withdrawal of the drug showed that cocaine abusers had hypermetabolism in the orbitofrontal cortex and the basal ganglia compared to normal controls. Further, hypermetabolism in these regions correlated with the subjective craving for cocaine. A $H_2{}^{15}O$ PET study of cerebral blood flow showed that chronic cocaine abusers had markedly decreased perfusion in the frontal cortex and the left hemisphere (Volkow *et al.*, 1988). These findings have been correlated by SPECT blood flow studies (Mena, Miller, & Garrett, 1989; Tumeh *et al.*, 1991).

References

Abadie, P., Baron, J. C., Bisserbe, J. C., Boulenger, J. P., Rioux, P., Travere, J. M., Barre, L., Petit-Taboue, M. C., & Zarifian, E. (1992). Central benzodiazepine receptors in human brain, estimation of regional Bmax and KD values with positron emission tomography. *European Journal of Pharmacology, 213,* 107–115.

Abdel-Dayem, H. M., Sadek, S. A., Kouris, K., Bahar, R. H., Higazi, I., Erikson, S., *et al.* (1987). Changes in cerebral perfusion after acute head injury, Comparison of CT with Tc-99m HM-PAO SPECT. *Radiology, 165,* 221–226.

Abdollah, A., Tampieri, D., & Melanson, D. (1993). Wilson's disease, computed tomography and magnetic resonance imaging findings. *Canadian Association of Radiology Journal, 42,* 130–134.

Abou-Khalil, B. W., Siegel, G. J., Sackellares, J. C., Gilman, S., Hichwa, R., & Marshall, R. (1987). Positron emission tomography studies of cerebral glucose metabolism in chronic partial epilepsy. *Annals of Neurology, 22,* 480–486.

Ackerman, R., Correia, J., Alpert, N., Baron, J. C., Gouliamos, A., Grotta, J. C., Brownell, G. L., & Taveras, J. M. (1981). Positron imaging in ischemic stroke disease using compound labeled with oxygen-15. *Archives of Neurology, 38,* 537–543.

Alavi, A. (1989). Functional and anatomical studies of head injury. *Journal of Neuropsychiatry, 1,* S45–50.

Alavi, J. B., Alavi, A., Chawluk, J., Kushner, M., Powe, J., Hickey, W., & Reivich, M. (1988). Positron emission tomography in patients with glioma. A predictor of prognosis. *Cancer, 62,* 1074–1078.

Alavi, J., Alavi, A., Dann, R., Kushner, M., Chawluk, J., Powlis, W., & Reivich, M. (1985). Metabolic brain imaging correlated with clinical features and brain tumors (Abstract). *Journal of Nuclear Medicine, 18,* P64.

Alavi, A., Fazekas, T., Alves, W., *et al.* (1987). Positron emission tomography in the evaluation of head injury. *Journal of Cerebral Blood Flow and Metabolism, 7,* S646.

Alavi, A., Gosfield, T., Cho, W., Alves, W., Kushner, M., Reivich, M., & Gennarelli, T. (1990). Unusual patterns of cerebellar hypometabolism in head trauma (HT) (Abstract). *Journal of Nuclear Medicine, 31,* 741.

Alavi, A., & Hirsch, L. J. (1991). Studies of central nervous system disorders with single photon emission computed tomography and positron emission tomography, evolution over the past two decades. *Seminars in Nuclear Medicine, 21,* 58–81.

Alavi, A., Newberg, A., Souder, E., & Reivich, M. (1992). Comparison of quantitative MRI and PET between patients with dementia of the Alzheimer type and multi-infarct dementia. *Journal of Stroke and Cerebrovascular Disease, 2,* 218–221.

Alavi, A., Ferris, S., Wolf, A., Christman, D., Fowler, J., MacGregor, R., Farkas, T., Greenberg, J., & Reivich, M. (1982). Regional cerebral glucose metabolism in aging and senile dementia as determined by 18F-deoxyglucose and positron emission tomography. In S. Hoyer (Ed.), *The Aging Brain–Physiological and Pathophysiological Aspects.* (pp. 687–693). Berlin: Springer-Verlag.

Ambrosetto, P. (1987). CT in progressive supranuclear palsy. *American Journal of Neuroradiology, 8,* 849–851.

Barker, P. B., Glickson, J. D., & Bryan, R. N. (1993). In vivo magnetic resonance spectroscopy of human brain tumors. *Topics in Magnetic Resonance Imaging, 5,* 32–45.

Baron, J. (1992). PET studies of neuroreceptors in the human frontal cortex. *Advances in Neurology, 57,* 465–470.

Baron, J., Bousser, M., Comar, D. *et al.* (1980). Crossed cerebellar diaschisis, in human supratentorial brain infarction. *Transactions of the American Neurological Association, 105,* 459–461.

Baron, J., Bousser, M., Rey, A., Guillard, A., Comar, D., & Castaigne, P. (1981). Reversal of focal "misery-perfusion syndrome" by extra-intracranial arterial bypass in hemodynamic cerebral ischemia. *Stroke 12,* 454–459.

Baron, J., Frackowiak, R., Herholz, K., Jones, T., Lammertsma, A. A., Mazoyer, B., & Weinhard, K. (1989). Use of PET methods for measurement of cerebral energy metabolism and hemodynamics in cerebrovascular disease. *Journal of Cerebral Blood Flow and Metabolism 9,* 723–742.

Baron, J., Rougemont, D., Bousser, M., *et al.* (1983). Local CBF, oxygen extraction fraction (OEF), and $CMRO_2$. Prognostic value in recent supratentorial infarction in humans. *Journal of Cerebral Blood Flow and Metabolism, 3,* A-2.

Baron, J., Rougemont, D., Bousser, M., Soussaline, F., Bustany, P., Krouzel, C., Bousser, M. G., & Comar, D. (1984). Local interrelationships of cerebral oxygen consumption and glucose utilization in normal subjects and in ischemic stroke patients. *Journal of Cerebral Blood Flow and Metabolism, 4,* 140–149.

Beaney, R. P., Brooks, D. J., Leenders K. L., Thomas, D. G., Jones, T., & Halnan, K. E. (1985). Blood flow and oxygen utilization in the contralateral cerebral cortex of patients with untreated intracranial tumours as studied by positron emission tomography, with observations on the effect of decompressive surgery, *Journal of Neurology, Neurosurgery, and Psychiatry, 48,* 310–319.

Benson, D. F., Kuhl, D. E., Hawkins, R. A., Phelps, M. E., Cummings, J. L., & Tsai, S. Y. (1983). The fluorodeoxyglucose 18F scan in Alzheimer's disease and multi-infarct dementia. *Archives of Neurology, 40,* 711–714.

Berent, S., Giordani, B., Lehtinen, S., Markel, D., Penney, J. B., Buchtel, H. A., *et al.* (1988). Positron emission tomographic scan investigations of Huntington's disease. Cerebral metabolic correlates of cognitive function. *Annals of Neurology, 23,* 541–546.

Bernardi, S., Trimble, M. R., Frackowiak, R. S., Wise, R. J., & Jones, T. (1983). An interictal study of partial epilepsy using positron emission tomography and the oxygen-15 inhalation technique. *Journal of Neurology, Neurosurgery & Psychiatry, 46*(6), 473–477.

Bhatt, M. H., Snow, B. J., Calne, D. R., *et al.* (1991). Fluorodopa PET studies of twins in idiopathic parkinsonism (Abstract). *10th International Symposium on Parkinson's Disease.*

Biersack, H. J., Reichmann, K., Winkler, C., Stefan, H., Kuhnen, K., Bulau, P., Penin, H., Tyrrell, D. A., Neirinckx, R. D., & Gruner, K. R. (1985). 99mTc-labeled hexamethylpropyleneamine oxime photon emission scans in epilepsy. *Lancet, 2,* 1436–1437.

Black, K. L., Hawkins, R. A., Kim, K. T., Becker, D. P., Lerner, C., & Marciano, D. (1989). Use of thallium-201 SPECT to quantitate malignancy grade of gliomas. *Journal of Neurosurgery, 71,* 342–346.

Blesa, R., Blin, J., *et al.* (1991). Levodopa-reduced glucose metabolism in striatopallidothalamocortico circuit in Parkinson's disease. *Neurology, 41,* 359.

Blin, J., Baron, J. C., Dubois, B., Pillon, B., Cambon, H., Cambier, J., & Agid, Y. (1990). Positron emission tomography study in progressive supranuclear palsy. Brain hypometabolic pattern and clinicometabolic correlations. *Archives of Neurology, 47,* 747–752.

Blin, J., Vidailhet, M. J., Pillon, B., Dubois, B., Feve, J. R., & Agid, Y. (1992). Corticobasal degeneration, decreased and asymmetric glucose consumption as studied with PET. *Movement Disorders, 7,* 348–354.

Blom, R. J., Vinuela, F., Fox, A. J., Blume, W. T., Girvin, J., Kaufmann, J. C. (1984). Computed tomography in temporal lobe epilepsy. *Journal of Computer-Assisted Tomography, 8,* 401–405.

Bonte, F. J., Devous, Sr., M. D., Stokely, E. M., & Homan, R. W. (1983). Single-photon tomographic determination of regional cerebral blood flow in epilepsy, *American Journal of Neuroradiology, 4,* 544–546.

Bonte, F. J., Hom, J., Tintner, R., & Weiner, M. F. (1990). Single photon tomography in Alzheimer's disease and the dementias. *Seminars in Nuclear Medicine, 20,* 342–352.
Bose, A., Pacia, S. B., Fayad, P., *et al.* (1990). Cerebral blood flow imaging compared to CT during the initial 24 hours of cerebral infarction. *Neurology, 40,* 190.
Bradley, Jr., W. G., Waluch, V., Yadley, R. A., & Wycoff, R. R. (1984). Comparison of CT and MR in 400 patients with suspected disease of the brain and cervical spinal cord. *Radiology, 152,* 695–702.
Braffman, B. H., Grossman, R. I., Goldberg, H. I., Stern, M. B., Hurtig, H. I., Hackney, D. B., Bilaniuk, L. T., Zimmerman, R. A. (1989). MR imaging of Parkinson's disease with spin-echo and gradient sequences. *American Journal of Roentgenology, 152,* 159–165.
Brant-Zawadzki, M., Badami, J. P., Mills, C. M., Norman, D., & Newton, T. H. (1984). Primary intracranial tumor imaging, a comparison of magnetic resonance and CT. *Radiology, 150,* 435–440.
Broich, K., Alavi, A., & Kushner, M. (1992). Positron emission tomography in cerebrovascular disorders, *Seminars in Nuclear Medicine, 22,* 224–232.
Brooks, D. J. (1989). Positron emission tomographic studies of the subcortical degenerations and dystonias, *Seminars in Neurology, 9,* 351–359.
Brooks, D. J. (1991a). Detection of preclinical Parkinson's disease with PET, *Neurology, 4,*(suppl. 2), 24–27.
Brooks, D. J. (1991b). The clinical role of PET in cerebrovascular disease, *Neurosurgical Review, 14,* 91–96.
Brooks, D. J., Ibanez, V., Sawle, G. V., Quinn, N., Lees, A. J., Mathias, C. J., Bannister, R., Marsden, C. D., Frackowiak, R. S. (1990). Differing patterns of striatal (18F)-dopa uptake in Parkinson's disease, multiple system atrophy, and progressive supranuclear palsy. *Annals of Neurology, 28,* 547–555.
Brooks, D. J., Ibanez, V., Sawle, G. V., Playford, E. D., Quinn, N., Mathias, C. J., Lees, A. J., Marsden, C. D., Bannister, R., Frackowiak, R. S. (1992). Striatal D2 receptor status in patients with Parkinson's disease, striatonigral degeneration, and progressive supranuclear palsy, measured with 11C raclopride and positron emission tomography, *Annals of Neurology, 31,* 184–192.
Buell, U., Braun, H., Ferbert, A., Stirner, H., Weiller, C., Ringelstein, E. G. (1988). Combined SPECT imaging of regional cerebral blood flow (99mTc-hexamethyl-propyleneamine oxime, HMPAO) and blood volume (99mTc-RBC) to assess regional cerebral perfusion reserve in patients with cerebrovascular disease. *Nuklearmedzin, 27,* 51–56.
Buell, U., Moser, E. A., Schmiedek, P., Leinsinger, G., Kreisig, T., Kirsch, C. M., & Einhaupl, K. (1984). Dynamic SPECT with 133 Xe. Regional cerebral blood flow in patients with unilateral cerebrovascular disease. Concise communication. *Journal of Nuclear Medicine, 25,* 441–446.
Burn, D. J., Mark, M. H., Playford, E. D., *et al.* (1991). Parkinson's disease in twins, an (18F)-dopa study (Abstract). *10th International Symposium on Parkinson's Disease.*
Bushnell, D. L., Gupta, S., Mlcoch, A. G., & Barnes, W. E. (1989). Prediction of language and neurological recovery after cerebral infarction with SPECT imaging using N-isopropyl-(I-123)-p-iodoamphetamine. *Archives of Neurology, 46,* 665–669.
Cascino, G. D., Jack, Jr., C. R., Parisi, J. E., Marsh, W. R., Kelly, P. J., Sharbrough, F. W., Hirschorn, K. A., Trenerry, M. R. (1992). MRI in the presurgical evaluation of patients with frontal lobe epilepsy and children with temporal lobe epilepsy, pathological correlation and prognostic importance. *Epilepsy Research, 11,* 51–59.
Charnes, M. E. (1993). Brain lesions in alcoholics. *Alcohol Clinical Experimental Research 17,* 2–11.
Chawluk, J. B., Mesulam, M. M., Hurtig, H., Kushner, M., Weintraub, S., Saykin, A., Rubin, N., Alavi, A., & Reivich, M. (1986). Slowly progressive aphasia without generalized dementia, studies with positron emission tomography. *Annals of Neurology 19,* 68–74.
Chollet, F., Celsis, P., Clanet, M., Guiraud-Chaumeil, B., Rascol, A., & Marc-Vergnes, J. P. (1989). SPECT study of cerebral blood flow reactivity after acetazolamide in patients with transient ischemic attacks. *Stroke, 20,* 458–464.
Cohen, M. B., Graham, L. S., Lake, R., Metter, E. J., Fitten, J., Kulkarni, M. K., Sevrin, R., Yamada, L., Chang, C. C., Woodruff, N., *et al.* (1986). Diagnosis of Alzheimer's disease and multiple infarct dementia by tomographic imaging of iodine-123 IMP. *Journal of Nuclear Medicine, 27,* 769–774.
Coleman, R. E., Hoffman, J. M., Hanson, M. W., Sostman, H. D., & Schold, S. C. (1991). Clinical application of PET for the evaluation of brain tumors, *Journal of Nuclear Medicine, 32,* 616–622.
Costa, D. C., Ell, P. J., Burns, A., Philpot, M., & Levy, R. (1988). CBF tomograms with Tc-99m-HM-

PAO in patients with dementia (Alzheimer type and HIV) and Parkinson's disease–Initial results. *Journal of Cerebral Blood Flow and Metabolism, 8,* S109–S115 (suppl 1).
Creutzig, H., Schober, O., Gielow, P., Friedrich, R., Becker, H., Dietz, H., & Hundeshagen, H. (1986). Cerebral dynamics of N-isopropyl p-iodoamphetamine. *Journal of Nuclear Medicine, 27,* 178–183.
Crow, W. (1991). Aspects of neuroradiology of head injury. *Neurosurgical Clinics of North America, 2,* 321–339.
D'Antona, R., Baron, J. C., Samson, Y., Serdaru, M., Viader, F., Agid, Y., & Cambier, J. (1985). Subcortical dementia. Frontal cortex hypometabolism detected by positron tomography in patients with progressive supranuclear palsy. *Brain, 108,* 785–799.
Davidson, J. (1989). A new approach to brain imaging. *Radiography Today, 55,* 19.
De Cristofaro, M. T., Mascalchi, M., Pupi, A., Nencini, P., Formiconi, A. R. Inzitari, D., Dal Pozzo, G., Meldolesi, U. (1990). Subcortical arteriosclerotic encephalopathy. Single photon emission computed tomography-magnetic resonance imaging correlation. *American Journal of Physiological Imaging, 5,* 68–74.
Defer, G., Moretti, J. L., Cesaro, P., Sergent, A., Raynaud, C., & Degos, J. D. (1987). Early and delayed SPECT using N-isopropyl p-iodoamphetamine iodine 123 in cerebral ischemia. A prognostic index for clinical recovery. *Archives of Neurology, 44,* 715–718.
DeLaPaz, R. L., Patronas, N. J., Brooks, R. A., Smith, B. H., Kornblith, P. L., Milam, H., & Di Chiro, G. (1983). Positron emission tomographic study of suppression of gray matter glucose utilization by brain tumors. *American Journal of Neuroradiology, 4,* 826–829.
Demeurisse, G., Verhas, M., Capon, A., & Paternot, J. (1983). Lack of evolution of the cerebral blood flow during clinical recovery of a stroke. *Stroke, 14,* 77–81.
De Roo, M., Mortelmans, L., Devos, P., Verbruggen, A., Wilms, G., Carton, H., Wils, V., Van den Bergh, R. (1989). Clinical experience with Tc-99m HM-PAO high resolution SPECT of the brain in patients with cerebrovascular accidents. *European Journal of Nuclear Medicine, 15,* 9–15.
Deutsch, G., & Tweedy, J. R. (1987). Cerebral blood flow in severity matched Alzheimer and multi-infarct patients. *Neurology, 37,* 431–438.
De Volder, A. G., Francart, J., Laterre, C., Dooms, G., Bol, A., Michel, C., & Goffinet, A. M. (1989). Decreased glucose utilization in the striatum and frontal lobe in probably striatonigral degeneration. *Annals of Neurology, 26,* 239–247.
Dewan, M. J., & Gupta, S. (1992). Toward a definite diagnosis of Alzheimer's disease. *Comparative Psychiatry, 33,* 282–290.
DiChiro, G. (1986). Positron emission tomography using (^{18}F) fluorodeoxyglucose in brain tumors. A powerful diagnostic and prognostic tool. *Investigative Radiology, 22,* 360–371.
DiChiro, G., DeLaPaz, R. L., Brooks, R. A., Sokoloff, L., Kornblith, P. L., Smith, B. H., Patronas, N. J., Kufta, C. V., Kessler, R. M., Johnston, G. S., Manning, R. G., Wolf, A. P. (1982). Glucose utilization of cerebral gliomas measured by ^{18}F fluorodeoxy-glucose and positron emission tomography. *Neurology, 32,* 1323–1329.
Diksic, M., & Reba, R. C. (Eds.) (1991). *Radiopharmaceuticals and brain pathology studied with PET and SPECT.* Boca Raton: CRC Press.
Di Piero, V., Chollet, F. M., MacCarthy, P., Lenzi, G. L., & Frackowiak, R. S. (1992). Motor recovery after acute ischaemic stroke, a metabolic study. *Journal of Neurology, Neurosurgery, and Psychiatry, 55,* 990–996.
Dooms, G. C., Hecht, S., Brant-Zawadzki, M., Berthiaume, Y., Norman, D., & Newton, T. H. (1986). Brain radiation lesions, MR imaging. *Radiology, 158,* 149–155.
Doyle, W. K., Budinger, T. F., Valk, P. E., Levin, V. A., & Gutin, P. H. (1987). Differentiation of cerebral radiation necrosis from tumor recurrence by (^{18}F)FDG and ^{82}Rb positron emission tomography. *Journal of Computerized Axial Tomography, 11,* 563–570.
Drayer, B. P. (1988). Imaging of the aging brain, II. Pathologic conditions. *Radiology, 166,* 797–806.
Duara, R. (1990). Utilization of positron emission tomography in research and clinical applications. In R. Duara (Ed.), *Frontiers of Clinical Neuroscience: Vol. 10* (pp. 1–12). New York: Wiley Liss.
Duara, R., Barker, W., Loewenstein, D., Pascal, S., & Bowen, B. (1989). Sensitivity and specificity of positron emission tomography and magnetic resonance imaging studies in Alzheimer's disease and multi-infarct dementia. *European Neurology, 29,* 9–15.
Duncan, R. (1992). Epilepsy, cerebral blood flow, and cerebral metabolic rate, *Cerebrovascular Brain Metabolism Review, 4,* 105–121.
Eidelberg, D., Dhawan, V., Moeller, J. R., Sidtis, J. J., Ginos, J. Z., Strother, S. C., Cederbaum, J., Greene, P., Fahn, S., Powers, J. M., *et al.* (1991). The metabolic landscape of cortico-basal

ganglionic degeneration. Regional asymmetries studies with positron emission tomography. *Journal of Neurology, Neurosurgery, and Psychiatry, 54,* 856–862.

Eidelberg, D., Moeller, J. R., Dhawan, V., *et al.* (1990). The metabolic anatomy of Parkinson's disease. Complementary (18F) fluorodeoxyglucose and (18F) fluorodopa positron emission tomographic studies. *Movement Disorders, 5,* 203–213.

Eidelberg, D., Moeller, J. R., Sidtis, J. J., *et al.* (1989). Corticodentatonigral degeneration, metabolic asymmetries studied with 18F-fluorodeoxyglucose and positron emission tomography, *Neurology, 39*(suppl. 1), 164.

Ell, P. J., Costa, D. C., & Harrison, M. (1987). Imaging cerebral damage in HIV infection. *Lancet, 2,* 569–570.

Engel, Jr., J. (1991). PET scanning in partial epilepsy. *Canadian Journal of Neurological Science, 18*(suppl), 588–592.

Engel, Jr., J., Brown, W. J., Kuhl, D. E., Phelps, M. E., Mazziotta, J. C., Crandall, P. H. (1982). Pathological findings underlying focal temporal lobe hypometabolism in partial epilepsy. *Annals of Neurology, 12,* 518–528.

Engel, Jr., J., Kuhl, D. E., Phelps, M. E., & Crandall, P. H. (1982). Comparative localization of the epileptic foci in partial epilepsy by PCT and EEG. *Annals of Neurology, 12,* 529–537.

Engel, Jr., J., Kuhl, D. E., Phelps, M. E., Rausch, R., & Nuwer, M. (1983). Local cerebral metabolism during partial seizures. *Neurology, 33,* 400–413.

Erkinjuntti, T., Ketonen, L., Sulkava, R., Sipponen, J., Vuorialho, M., & Iivanainen, M. (1987). Do white matter changes in MRI and CT differentiate vascular dementia from Alzheimer's disease? *Journal of Neurology, Neurosurgery, and Psychiatry, 50,* 37–42.

Fayad, P. B., & Brass, L. M. (1991). Single photon emission computed tomography in cerebrovascular disease. *Stroke, 22,* 950–954.

Feeney, D., & Baron, J. (1986). Diaschisis. *Stroke, 17,* 817–830.

Felix, R., Schorner, W., Laniado, M., Niendorf, H. P., Claussen, C., Fiegler, W., & Speck, U. (1985). Brain tumors: MR imaging with gadolinium-DTPA. *Radiology, 156,* 681–688.

Fieschi, C., Argentino, C., Lenzi, G. L., Sacchetti, M. L., Toni, D., & Bozzao, L. (1989). Clinical and instrumental evaluation of patients with ischemic stroke within the first six hours. *Journal of the Neurological Sciences, 91*(3), 311–321.

Fisher, R. S., & Frost, J. J. (1991). Epilepsy. *Journal of Nuclear Medicine, 32,* 651–659.

Fobben, E. S., Zimmerman, R. A., Sperling, M. R., *et al.* (1988). MR imaging in temporal lobe epilepsy (abstract). *Radiology, 169* (Suppl.), 142.

Foster, N. L., Gilman, S., Berent, S., Morin, E. M., Brown, M. B., & Koeppe, R. A. (1988). Cerebral hypometabolism in progressive supranuclear palsy studied with positron emission tomography. *Annals of Neurology, 24,* 399–406.

Frackowiak, R., & Wise, R. (1983). Positron tomography in ischemia cerebrovascular disease. *Neurology Clinics, 1,* 183–201.

Franck, G., Maquet, P., Sadzot, B., Salmon, E., Debets, R., Dive, D., Grisar, T., Guillaume, D., Van Veelen, C., Van Huffelen, A., *et al.* (1992). Contribution of positron emission tomography to the investigation of epilepsies of frontal lobe origin, *Advances in Neurology, 57,* 471–485.

Frost, J. J. (1992). Receptor imaging by positron emission tomography and single-photon emission computed tomography, *Investigative Radiology, 27* (Suppl. 2), S54–S58.

Frost, J. J., Mayberg, H. S., Fisher, R. S., Douglass, K. H., Dannals, R. F., Links, J. M., Wilson, A. A., Ravert, H. T., Rosenbaum, A. E., Snyder, S. H., *et al.* (1988). Mu-opiate receptors measured by positron emission tomography are increased in temporal lobe epilepsy. *Annals of Neurology, 23,* 231–237.

Fumeya, H., Ito, K., Yamagiwa, O., Funatsu, N., Okada, T., Asahi, S., Ogura, H., Kubo, M., Oba, T. (1990). Analysis of MRI and SPECT in patients with acute head injury. *Acta Neurochirurgica Supplement, 51,* 283–285.

Garnett, E. S., Nahmias, C., & Firnau, G. (1984). Central dopaminergic pathways in hemiparkinsonism examined by positron emission tomography. *Canadian Journal of Neurological Science, 11,* 174–179.

Gates, J. R., & Cruz-Rodriguez, R. (1990). Mesial temporal sclerosis, pathogenesis, diagnosis, and management. *Epilepsia 31*(Suppl.), S55–66.

Gatley, S. J., DeGrado, T. R., Kornguth, M. L., Holden, J. E. (1990). Radiopharmaceuticals for positron emission tomography. Development of new, innovative tracers for measuring the rates of physiologic and biochemical processes, *Acta Radiology Supplement* (Stockholm), *374,* 7–11.

Gemmell, H. G., Sharp, P. F., Besson, J. A., Crawford, J. R., Ebmeier, K. P ., Davidson, J., & Smith,

F. W. (1987). Differential diagnosis in dementia using the cerebral blood flow agent 99mT HM-PAO. A SPECT study. *Journal of Computer Assisted Tomography, 11,* 398–402.

Gemmell, H. G., Sharp, P. F., Smith, F. W., Besson, J. A., Ebmeier, K. P., Davidson, J., Evans, N. T., Roeda, D., Newton, R., Mallard, J. R. (1989). Cerebral blood flow measured by SPECT as a diagnostic tool in the study of dementia. *Psychiatry Research, 29,* 327–329.

Gentry, L. R., Godersky, J. C., Thompson, B., & Dunn, V. D. (1988a). Prospective comparative study of intermediate-field MR and CT in the evaluation of closed head trauma. *American Journal of Roentgenology, 150,* 673–682.

Gentry, L. R., Godersky, J. C., & Thompson, B. (1988b). MR imaging of head trauma, review of the distribution and radiopathologic features of traumatic lesions. *American Journal of Roentgenology, 150,* 663–672.

George, J. K., Alavi, A., & Zimmerman, R. A., *et al.* (1989). Metabolic (PET) correlates of anatomic lesions (CT/MRI) produced by head trauma. *Journal of Nuclear Medicine, 30,* 802.

Gibbs, J., Frackowiak, R., & Legg, N. (1986). Regional cerebral blood flow and oxygen metabolism in dementia due to vascular disease. *Gerontology, 32,* 84–89.

Gibbs, J., Wise, R., Leenders, K., & Jones, T. (1984). Evaluation of cerebral perfusion reserve in patients with carotid-artery occlusion. *Lancet, 1,* 310–314.

Gilman, S., Adams, K., Koeppe, R. A., Berent, S., Kluin, K. J., Modell, J. G., Kroll, P., & Brunberg, J. A. (1988). Cerebellar and frontal hypometabolism in alcoholic cerebellar degeneration studied with FDG and PET, *Neurology, 38,* 365.

Giubilei, F., Lenzi, G. L., Di Piero, V., Pozzilli, C., Pantano, P., Bastianello, S., Argentino, C., & Fieschi, C. (1990). Predictive value of brain perfusion single-photon emission computed tomography in acute cerebral ischemia. *Stroke, 21,* 895–900.

Giuliani, G., Chiaramoni, L., Foschi, N., & Terziani, S. (1992). The role of MRI in the diagnosis of olivopontocerebellar atrophy. *Italian Journal of Neurological Science, 13,* 151–156.

Goffinet, A. M., De Volder, A. G., Gillian, C., Rectem, D., Bol, A., Michel, C., Cogneau, M., Labar, D., & Laterre, C. (1989). Positron tomography demonstrates frontal lobe hypometabolism in progressive supranuclear palsy. *Annals of Neurology, 25,* 131–139.

Gray, B. G., Ichise, M., Chung, D. G., Kirsh, J. C., & Franks, W. (1992). Technetium-99m-HMPAO SPECT in the evaluation of patients with a remote history of traumatic brain injury, a comparison with x-ray computed tomography. *Journal of Nuclear Medicine, 33,* 52–58.

Hagglund, J., Aquilonius, S. M., Eckernas, S. A., Hartvig, P., Lundquist, H., Gullberg, P., & Langstrom, B. (1987). Dopamine receptor properties in Parkinson's disease and Huntington's chorea evaluated by positron emission tomography using 11C-methyl-spiperone. *Acta Neurologia Scandinavia, 75,* 87–94.

Hakim, A. M., Evans, A. C., Berger, L., Kuwabara, H., Worsley, K., Marchal, G., Biel, C., Pokrupa, R., Diksic, M., Meyer E., *et al.* (1989). The effect of nimodipine on the evolution of human cerebral infarction studies by PET. *Journal of Cerebral Blood Flow and Metabolism, 9,* 523–534.

Hakim, A. M., Pokrupa, R. P., Villanueva, J., Diksic, M., Evans, A. C., Thompson, C. J., Meyer, E., Yamamoto, Y. L., Feindel, W. H. (1987). The effect of spontaneous reperfusion on metabolic function in early human cerebral infarcts. *Annals of Neurology, 21,* 279–289.

Harrell, L. E., Callaway, R., & Sekar, B. C. (1987). Magnetic resonance imaging and the diagnosis of dementia. *Neurology, 37,* 540–543.

Hawkins, R. A., Mazziotta, J. C., & Phelps, M. E. (1987). Wilson's disease studied with FDG and positron emission tomography, *Neurology, 37,* 1707–1711.

Hawkins, R. A., & Phelps, M. E. (1986). Positron emission tomography for evaluation of cerebral function. *Current Concepts in Diagnostic Nuclear Medicine, 3,* 4–13.

Hayden, M. R., Hewitt, J., Martin, W. R. W., *et al.* (1987). Studies in persons at risk for Huntington's disease. *New England Journal of Medicine, 317,* 382–383.

Hayden, M. R., Hewitt, J., Stoessl, A. J., Clark, C., Ammann, W., & Martin, W. R. (1987). The combined use of positron emission tomography and DNA polymorphisms for preclinical detection of Huntington's disease. *Neurology, 37,* 1441–1447.

Hayden, M. R., Martin, W. R., Stoessl, A. J., Clark, C., Hollenberg, S., Adam, M. J., Ammann, W., Harrop, R., Rogers, J., Ruth, T., *et al.* (1986). Positron emission tomography in the early diagnosis of Huntington's disease. *Neurology, 36,* 888–894.

Hayman, L. A., Taber, K. H., Jhingran, S. G., Killian, J. M., & Carroll, R. G. (1989). Cerebral infarction: Diagnosis and assessment of prognosis by using 123 IMP SPECT and CT. *American Journal of Neuroradiology, 10,* 557–562.

Heiss, W., Pawlik, G., Hebold, I., *et al.* (1989). Can positron emission tomography be used to gauge the brain's capacity for functional recovery following stroke? In M. Ginsberg & W. Dietrich (Eds.), *Cerebrovascular diseases (sixteenth research Princeton conference)* pp. 345–352. New York: Raven Press.

Heiss, W. D., Zeiler, K., Havelec, L., Reisner, T., & Bruck, J. (1988). Long-term prognosis in stroke related to cerebral blood flow. *Archives of Neurology 34,* 671–676.

Heiss, W. D., Herholz, K., Podreka, I., Neubauer, I., & Pietrzyk, U. (1990). Comparison of [99mTc]HMPAO SPECT with [18F]fluoromethane (FM) PET in cerebrovascular disease. *Journal of Cerebral Blood Flow and Metabolism, 10,* 687–697.

Henry, T. R., Sutherling, W. W., Engel, Jr., J., Risinger, M. W., Levesque, M. F., Mazziotta, J. C., Phelps, M. E. (1991). Interictal cerebral metabolism in partial epilepsies of neocortical origin, *Epilepsy Research, 10,* 174–182.

Herholz, K., Heindel, W., Rackl, A., Neubauer, I., Steinbrich, W., Pietrzyk, U., Erasmi-Korber, H., Heiss, W. D. (1990). Regional cerebral blood flow in leukoaraiosis and artherosclerotic carotid disease. *Archives of Neurology, 47,* 392–397.

Herholz, K., & Heiss, W. (1991). Use of PET to evaluate acute stroke and other cerebrovascular disorders. In M. Diksic & R. Reba (Eds.), *Radiopharmaceuticals and brain pathology studies with PET and SPECT.* (pp. 217–239). Boca Raton, FL: CRC Press.

Herholz, K., Ziffling, P., Staffen, W., Pawlik, G., Wagner, R., Wienhard, K., & Heiss, W. D. (1988). Uncoupling of hexose transport and phosphorylation in human gliomas demonstrated by PET. *European Journal of Cancer and Clinical Oncology 24,* 1139–1150.

Hill, T. C., Magistretti, P. L., Holman, B. L., Lee, R. G., O'Leary, D. H., Uren, R. F., Royal, H. D., Mayman, C. I., Kolodny, G. M., Clouse, M. E. (1984). Assessment of regional cerebral blood flow (rCBF) in stroke using SPECT and N-isopropyl-(I-123)-p-iodoamphetamine (IMP). *Stroke, 15,* 40–45.

Hoffman, J. M., & Coleman, R. E. (1992). Perfusion quantitation using positron emission tomography, *Investigative Radiology, 27*(Suppl. 2), S22–S26.

Holman, B. (1986). Perfusion and receptor SPECT in the dementias—George Taplin memorial lecture. *Journal of Nuclear Medicine, 27,* 855.

Holman, B. L., & Devous, Sr., M. D. (1992). Functional brain SPECT. The emergence of a powerful clinical method. *Journal of Nuclear Medicine, 33,* 1888–1904.

Holman, B. L., Garada, B., Johnson, K. A., Mendelson, J., Hallgring, E., Teoh, S. K., Worth, J., Navia, B. (1992). A comparison of brain perfusion SPECT in cocaine abuse and AIDS dementia complex. *Journal of Nuclear Medicine, 33,* 1312–1315.

Holman, B. L., Gibson, R. E., Hill, T. C., Eckelman, W. C., Albert, M., & Reba, R. C. (1985). Muscarinic acetylcholine receptors in Alzheimer's disease in vivo imaging with iodine-1,2,3-labeled 3-quinuclidinyl-4-iodobenzilate and emission tomography. *Journal of the American Medical Association 254,* 3063.

Holman, B. L., Johnson, K. A., Gerada, B., Carvalho, P. A., Satlin, A. (1992). The scintigraphic appearance of Alzheimer's disease, a prospective study using technetium-99m-HMPAO SPECT. *Journal of Nuclear Medicine, 33,* 181–185.

Homan, R. W., Paulman, R. G., Devous, Sr., M. D., Walker, P., Jennings, L. W., & Bonte, F. J. (1989). Cognitive function and regional cerebral blood flow in partial seizures. *Archives of Neurology, 46,* 964–970.

Huber, S. J., Shuttleworth, E. C., Christy, J. A., Chakeres, D. W., Curtin, A., & Paulson, G. W. (1989). Magnetic resonance imaging in dementia of Parkinson's disease. *Journal of Neurology, Neurosurgery, and Psychiatry, 52,* 1221–1227.

Jaeckle, K. A. (1991). Neuroimaging for central nervous system tumors. *Seminars in Oncology, 10,* 150–157.

Jagust, W. J., & Eberling, J. L. (1991). MRI, CT, SPECT, PET, their use in diagnosing dementia. *Geriatrics, 46,* 28–35.

Jenkins, I. H., Fernandez, W., Playford, E. D., Lees, A. J., Frackowiak, R. S., Passingham, R. E., Brooks, D. J. (1992). Impaired activation of the supplementary motor area in Parkinson's disease is reversed when akinesia is treated with apomorphine. *Annals of Neurology, 32,* 749–757.

Jennett, B. (1991). Diagnosis and management of head trauma. *Journal of Neurotrauma, 8,* S15–19.

Jernigan, T. L., Salmon, D. P., Butters, N., & Hesselink, J. R. (1991). Cerebral structure on MRI. Part II. Specific changes in Alzheimer's and Huntington's disease. *Biological Psychiatry, 29,* 68–81.

Johnson, K. A., Sperling, R. A., Holman, B. L., *et al.* (1989). IMP-SPECT in supra nuclear palsy. *Neurology, 39,* (Suppl. 1) 204.

Jolles, P. R., Chapman, P. R., & Alavi, A. (1989). PET, CT, and MRI in the evaluation of neuropsychiatric disorders, current applications. *Journal of Nuclear Medicine, 30,* 1589–1606.

Kaplan, W. D., Takvorian, T., Morris, J. H., Rumbaugh, C. L., Connolly, B. T., & Atkins, H. L. (1987). Thallium-201 brain tumor imaging, a comparative study with pathological correlation. *Journal of Nuclear Medicine, 28,* 47–52.

Kato, T., Kume, A., Ito, K., Tadokoro, M., Takahashi, A., & Sakuma, S. (1992). Asymmetrical FDG-PET and MRI findings of striatonigral system in multiple system atrophy with hemiparkinsonism. *Radiation Medicine, 10,* 87–93.

Kelly, A. B., Zimmerman, R. D., Snow, R. B., Gandy, S. E., Heier, L. A., & Deck, M. D. (1988). Head trauma. Comparison of MR and CT experience in 100 patients. *American Journal of Neuroradiology, 9,* 699–708.

Kelly, W. M., & Brant-Zawadski, M. (1988). Magnetic resonance imaging and computed tomography of supratentorial tumors. In J. M. Taveras & J. T. Ferrucci (Eds.), *Radiology, diagnosis-imaging-intervention.* Philadelphia: J. P. Lippincott.

Kempler, D., Metter, E. J., Riege, W. H., Jackson, C. A., Benson, D. F., & Hanson, W. R. (1990). Slowly progressive aphasia: Three cases with language, memory, CT and PET data. *Journal of Neurology, Neurosurgery, and Psychiatry, 53,* 987–993.

Kertesz, A., Polk, M., & Carr, T. (1990). Cognitive and white matter changes on magnetic resonance imaging in dementia. *Archives of Neurology, 47,* 1135–1140.

Killen, A. R., Oster, G., & Golditz, G. A. (1989). An assessment of the role of I-123-N-isopropyl-p-iodoamphetamine with single-photon emission computed tomography in the diagnosis of stroke and Alzheimer's disease. *Nuclear Medicine Communications, 10,* 217–284.

Kim, K. T., Black, K. L., Marciano, D., Mazziotta, J. C., Guze, B. H., Grafton, S., Hawkins, R. A., Becker, D. P. (1990). Thallium-201 SPECT imaging of brain tumors. Methods and results. *Journal of Nuclear Medicine, 31,* 965–969.

Kinkel, W. R., Jacobs, L., Polachini, I., Bates, V., & Heffner, Jr., R. R. (1985). Subcortical arteriosclerotic encephalopathy (Binswanger's disease). Computed tomographic nuclear magnetic resonance and clinical correlation. *Archives of Neurology, 42,* 951–959.

Kluin, K. J., Gilman, S., Markel, D. S., Koeppe, R. A., Rosenthal, G., & Junck, L. (1988). Speech disorders in olivopontocerebellar atrophy correlate with positron emission tomography findings. *Annals of Neurology, 23,* 547–554.

Knapp, W. H., von Kummer, R., & Kubler, W. (1986). Imaging of cerebral blood flow-to-volume distribution using SPECT. *Journal of Nuclear Medicine, 27,* 465–470.

Kucharczyk, W., Brant-Zawadzki, M., Sobel, D., Edwards, M. B., Kelly, W. M., Norman, D., & Newton, T. H. (1985). Central nervous system tumors in children: Detection by magnetic resonance imaging. *Radiology 155,* 131–136.

Kuhl, D. E., Metter, E. J., Benson, D. F., *et al.* (1985). Similarities of cerebral glucose metabolism in Alzheimer's and parkinsonian dementia. *Journal of Nuclear Medicine 26,* P69.

Kuhl, D. E., Metter, E. J., Riege, W. H., & Markham, C. H. (1984). Patterns of cerebral glucose utilization in Parkinson's disease and Huntington's disease. *Annals of Neurology, 15*(Suppl.), S119–S125.

Kuhl, D. E., Phelps, M. E., Kowell, A. P., Metter, E. J., Selin, C., Winter, J. (1980). Effects of stroke on local cerebral metabolism and perfusion: Mapping by emission computed tomography of 18FDG and 13NH3. *Annals of Neurology, 8*(1); 47–60.

Kuhl, D. E., Phelps, M. E., Markham, C. H. (1982). Cerebral metabolism and atrophy in Huntington's disease determined by 18-FFDG and computed tomographic scan. *Annals of Neurology, 12,* 425–434.

Kung, H. F. (1990). Radiopharmaceuticals for CNS receptor imaging with SPECT. *Nuclear Medicine Biology, 17,* 85–92.

Kung, H. F., Alavi, A., Chang, W., Kung, M. P., Keyes, Jr., J. W., Velchik, M. G., Billings, J., Pan, S., Noto, S., Rausch, *et al.* (1990). In vivo SPECT imaging of CNS D-2 dopamine receptors: Initial studies with iodine-123-IBZM in humans. *Journal of Nuclear Medicine, 31,* 573–579.

Kuni, C. C., Rhame, F. S., Meier, M. J., Foehse, M. C., Loewenson, R. B., Lee, B. C., Boudreau, R. J., duCret, R. P. (1991). Quantitative I-123-IMP brain SPECT and neuropsychological testing in AIDS dementia. *Clinical Nuclear Medicine, 16,* 174–177.

Kupfer, M. C., Zee, C. S., Colletti, P. M., Boswell, W. D., & Rhodes, R. (1990). MRI evaluation of AIDS-related encephalopathy: Toxoplasmosis vs. lymphoma. *Magnetic Resonance Imaging, 8,* 51–57.

Kushner, M., Alavi, A., Reivich, M., Dann, R., Burke, A., & Robinson, G. (1984). Contralateral cerebellar hypometabolism following cerebral insult: A positron emission tomographic study. *Annals of Neurology, 15,* 425–434.

Kushner, M., Reivich, M., Fieschi, C., Silver, F., Chawluk, J., Rosen, M., Greenberg, J., Burke, A., & Alavi, A. (1987). Metabolic and clinical correlates of acute ischemic infarction. *Neurology, 37,* 1103–1110.

Kuwabara, Y., Ichiya, Y., Otsuka, M., Tahara, T., Fukumura, T., Gunasekera, R., Ichimiya, A., & Masuda, K. (1990). Comparison of I-123 and Tc-99m HMPAO SPECT studies with PET in dementia. *Annals of Nuclear Medicine, 4,* 75–82.

Kuwert, T., Hefter, H., Scholz, D., Milz, M., Weiss, P., Arendt, G., Herzog, H., Loken, M., Hennerici, M., & Feinendegen, L. E. (1992). Regional cerebral glucose consumption measured by positron emission tomography in patients with Wilson's disease. *European Journal of Nuclear Medicine, 19,* 96–101.

Kuwert, T., Lange, H. W., Langen, K. J., Herzog, H., Aulich, A., & Feinendegen, L. E. (1989). Cerebral glucose consumption measured by PET in patients with and without psychiatric symptoms of Huntington's disease. *Psychiatry Research, 29,* 361–362.

Kuwert, T., Lange, H. W., Langen, K. J., Herzog, H., Aulich, A., & Feinendegen, L. E. (1990). Cortical and subcortical glucose consumption measured by PET in patients with Huntington's disease, *Brain, 113,* 1405–1423.

LaManna, M. M., Sussman, N. M., Harner, R. N., Kaplan, L. R., Hershey, B. L., Bernstein, D. R. Goldstein, P., Parker, J. A., Wolodzko, J. G., Popky, G. L. (1989). Initial experience with SPECT imaging of the brain using I-123 p-iodoamphetamine in focal epilepsy. *Clinical Nuclear Medicine, 14,* 428–430.

Lammertsma, A., Wise, R., Gibbs, J., *et al.* (1983). The pathophysiology of human cerebral tumors and surrounding white matter and remote cortex. *Journal of Cerebral Blood Flow and Metabolism, 3*(Suppl. 1), S9–S10.

Lammertsma, A. A., Wise, R. J. S., & Jones, T. (1984). Regional cerebral blood flow and oxygen utilization in edema associated with cerebral tumours. In K. G. Go & A. Baethmann (Eds.), *Recent progress in the study and therapy of brain edema.* New York: Plenum Press.

Lang, D. A., Hadley, D. M., Teasdale, G. M., Macpherson, P. & Teasdale, E. (1991). Gadolinium DTPA enhanced magnetic resonance imaging in acute head injury. *Acta Neurochirurgica, 109,* 5–11.

Lang, W., Podreka, I., Suess, E., Muller, C., Zeitlhofer, J., & Deecke, L. (1988). Single photon emission computed tomography during and between seizures. *Journal of Neurology, 235,* 277–284.

Langfitt, T. W., Obrist, W. D., Alavi, A., Grossman, R. I., Zimmerman, R., Jaggi, J., Uzzell, B., Reivich, M., & Patton, D. R. (1986). Computerized tomography, magnetic resonance imaging, and positron emission tomography in the study of brain trauma. Preliminary observations. *Journal of Neurosurgery, 64,* 760–767.

Lassen, N. (1966). The luxury perfusion syndrome and its possible relation to acute metabolic acidosis localized within the brain. Lancet, 2, 1113–1115.

Lassen, N. (1990). Pathophysiology of brain ischemia as it relates to the therapy of acute ischemic stroke. *Clinical Neuropharmacology 13*(Suppl.), S1–8.

Latack, J. T., Abou-Khalil, B. W., Siegel, G. J., Sackellares, J. C., Gabrielsen, T. O., & Aisen, A. M. (1986). Patients with partial seizure: Evaluation by MR, CT, and PET imaging. *Radiology, 159,* 159–163.

Launes, J., Sulkava, R., Erkinjuntti, T., Nikkinen, P., Lindroth, L., Liewendahl, K., & Iivanainen, M. (1991). 99Tcm-HMPAO SPECT in suspected dementia. *Nuclear Medicine Communications, 12,* 757–765.

Leblanc, R. (1991). Physiologic studies of cerebral ischemia. *Clinical Neurosurgery, 37,* 289–311.

Leblanc, R., Yamamoto, Y. L., Tyler, J. L., Diksic, M., Hakim, A. (1987). Borderzone ischemia. *Annals of Neurology, 22,* 707–713.

Leblhuber, F., Holl, K., Gebetsberger, B., *et al.* (1987). Long latency responses (LLRs) and IMP-SPECT findings in early detection of Huntington's chorea. *Electroencephalography Clinics in Neurophysiology, 66,* S59.

Lechner, H., Schmidt, R., & Goetz, B. (1990). Long term experience of a trial in multi-infarct dementia. *Neuroepidemiology, 9,* 228–232.

Lee, B. I., Markand, O. N., Siddiqui, A. R., Park, H. M., Mock, B., Wellman, H. H., Worth, R. M., Edwards, M. K. (1986). Single photon emission computed tomography (SPECT) brain imaging using N,N,N′-trimethyl-N′-(2 hydroxy-3-methyl-5-123I-iodobenzyl)-1,3-propanediamine 2 HCl (HIPDM): Intractable complex partial seizures. *Neurology, 36,* 1471–1477.

Lee, B. I., Markand, O. N., Wellman, H. N., Siddiqui, A. R., Park, H. M., Mock, B., Worth, R. M., Edwards, M. K., Krepshaw, J. (1987). HIPDM-SPECT in patients with medically intractable complex partial seizures. Ictal study. *Archives of Neurology, 45,* 397–402.

Lee, H., & Kramer, J. H. (1992). Brain SPECT imaging in progressive aphasia. *Clinical Nuclear Medicine, 17,* 800–802.

Lee, R. G., Hill, T. C., Holman, B. L., Royal, H. D., O'Leary, D. H., & Clouse, M. E. (1984). Predictive value of perfusion defect size using N-isopropyl-(I123)-p-iodoamphetamine emission tomography in acute stroke. *Journal of Neurosurgery, 61,* 449–452.

Leenders, K. L., Palmer, A. J., Quinn, N., Clark, J. C., Firnau, G., Garnett, E. S., Nahmias, C., Jones, T., & Marsden, C. D. (1986). Brain dopamine metabolism in patients with Parkinson's disease measured with positron emission tomography. *Journal of Neurology, Neurosurgery, and Psychiatry, 49,* 853–860.

Leenders, K. L., Salmon, E. P., Tyrrell, P., Perani, D., Brooks, D. J., Sager, H., Jones, T., Marsden, C. D., Frackowiak, R. S. (1990). The nigrostriatal dopaminergic system assessed in vivo by positron emission tomography in healthy volunteer subjects and patients with Parkinson's disease. *Archives of Neurology, 47,* 1290–1298.

Lencz, T., McCarthy, G., Bronen, R. A., Scott, T. M., Inserni, J. A., Sass, K. J., Novelly, R. A., Kim, J. H., Spencer, D. D. (1992). Quantitative magnetic resonance imaging in temporal lobe epilepsy. Relationship to neuropathology and neuropsychological function. *Annals of Neurology, 31,* 629–637.

Lenzi, G., Frackowiak, R., & Jones, T. (1982). Cerebral oxygen metabolism and blood flow in human cerebral ischemic infarction. *Journal of Cerebral Blood Flow and Metabolism, 2,* 321–335.

Levin, H. S., Amparo, E., Eisenberg, H. M., Williams, D. H., High, Jr., W. M., McArdle, C. B., & Weiner, R. L. (1987). Magnetic resonance imaging and computerized tomography in relation to the neurobehavioral sequelae of mild and moderate head injuries. *Journal of Neurosurgery, 66,* 706–713.

Levin, H. S., Williams, D. H., Eisenberg, H. M., *et al.* (1992). Serial MRI and neurobehavioral findings after mild to moderate closed head injury. *Journal of Neurology, Neurosurgery, and Psychiatry, 55,* 255–267.

Lewin, J. S., Estes, M. L., & Ross, J. S. (1992). Magnetic resonance imaging of adult intra-axial brain tumors. *Topics in Magnetic Resonance Imaging, 4,* 17–40.

Limburg, M., van Royen, E. A., Hijdra, A., & Verbeeten, Jr., B. (1991). rCBF-SPECT in brain infarction: When does it predict outcome? *Journal of Nuclear Medicine, 32,* 382–387.

Lin, X. T., Song, W. Z., Liu, Y. C., *et al.* (1990). Clinical applications of perfusion brain imaging. *European Journal of Nuclear Medicine, 16,* S22.

Lindvall, O., Brundin, P., Widner, H., Rehncrona, S., Gustavii, B., Frackowiak, R., Leenders, K. L., Sawle, G., Rothwell, J. C., Marsden, C. D., *et al.* (1990). Grafts of fetal dopamine neurons survive and improve motor function in Parkinson's disease, *Science, 247,* 574–577.

Litton, J. E., Farde, L., Neiman, J., *et al.* (1991). Positron emission tomography of benzodiazepine receptors in alcoholics-equilibrium quantification. *Journal of Cerebral Blood Flow and Metabolism 11,* 5617.

London, E. D., Cascella, N. G., Wong, D. F., Phillips, R. L., Dannals, R. F., Links, J. M., Herning, R., Grayson, R., Jaffe, J. H., & Wagner, Jr., H. N. (1990). Cocaine-induced reduction of glucose utilization in human brain. A study using positron emission tomography and (fluorine-18)-fluorodeoxyglucose. *Archives of General Psychiatry, 47,* 567–574.

Lotz, P. R., Ballinger, Jr., W. E., & Quisling, R. G. (1986). Subcortical arteriosclerotic leukoencephalopathy, CT spectrum and pathologic correlation. *American Journal of Neuroradiology, 7,* 817–822.

Maher, E. R., Smith, E. M., & Lees, A. J. (1985). Cognitive deficits in the Steele-Richardson-Olszewski syndrome (progressive supranuclear palsy). *Journal of Neurology, Neurosurgery, and Psychiatry, 48,* 1234–1239.

Marchal, G., Evans, A., Dagher, A., *et al.* (1987). The evolution of cerebral infarction with time. A PET study of the ischemia penumbra. *Journal of Cerebral Blood Flow and Metabolism, 7,* S99.

Mardsen, C. D. (1982). Basal ganglia disease. *Lancet, ii,* 1141–1147.

Martin, W. R., Beckman, J. H., Calne, D. B., Adam, M. J., Harrop, R., Rogers, J. G., Ruth, T. J., Sayre, C. I., Pate, B. D. (1984). Cerebral glucose metabolism in Parkinson's disease. *Canadian Journal of Neurological Science, 11,* 169–173.

Martin, W. R., Stoessl, A. J., Adam, M. J., Ammann, W., Bergstrom, M., Harrop, R., Laihinen, A., Rogers, J. G., Ruth, T. J., Sayre, C. I. *et al.* (1986). Positron emission tomography in Parkinson's disease. Glucose and dopa metabolism. *Advances in Neurology, 45,* 95–98.

Masdeu, J. C., Yudd, A., Van Heertum, R. L., Grundman, M., Hriso, E., O'Connell, R. A., Luck, D., Camli, U., King, L. N. (1991). Single photon emission computed tomography in human immunodeficiency virus encephalopathy. A preliminary report. *Journal of Nuclear Medicine, 32,* 1471–1475.

Maurer, A. H., Vitti, R. A., Brooks, A., *et al.* (1988). Significance of patterns of I-123 iodoamphetamine redistribution for predicting successful revascularization following carotid endarterectomy. *Journal of Nuclear Medicine, 29,* 844.

Maziere, B., Coenen, H. H., Halldin, C., Nagren, K., & Pike, V. W. (1992). PET radioligands for dopamine receptors and re-uptake sites. Chemistry and biochemistry. *International Journal of Radiologically Applied Instruments, 19,* 4997–512.

Maziere, B., & Maziere, M. (1991). Positron emission tomography studies of brain receptors. *Fundamentals in Clinical Pharmacology, 5,* 61–91.

Mazziotta, J. C. (1989). Huntington's disease, studies with structural imaging techniques and positron emission tomogrpahy, *Seminars in Neurology, 9,* 360–369.

Mazziotta, J. C., Phelps, M. E., Pahl, J. J., Huang, S. C., Baxter, L. R., Riege, W. H., Hoffman, J. M., Kuhl, D. E., Lanto, A. B., Wapenski, J. A., *et al.* (1987). Reduced cerebral glucose metabolism in asymptomatic patients at risk for Huntington's disease. *New England Journal of Medicine, 316,* 357–362.

Meguro, K., Hatazawa, J., Yamaguchi, T., Itoh, M., Matsuzawa, T., Ono, S., Miyazawa, H., Hishinuma, T., Yanai, K., Sekita, Y., *et al.* (1990). Cerebral circulation and oxygen metabolism associated with subclinical periventricular hyperintensity as shown by magnetic resonance imaging. *Annals of Neurology, 28,* 378–383.

Mena, I., Miller, B., & Garrett, K. (1989). Neurospect in cocaine abuse, rCBF and HMPAO findings. *Clinical Nuclear Medicine 50,* 1412.

Meneghetti, G., Vorstupt, S., & Mickey, B. (1984). Crossed cerebellar diaschisis in ischemic stroke. A study of regional cerebral blood flow by 133 Xe inhalation and single-photon emission computerized tomography. *Journal of Cerebral Blood Flow and Metabolism, 4,* 235–240.

Mertens, J., Bossuyt-Piron, C., Guns, M., *et al.* (1989). High selective serotonin S2 receptor mapping with SPECT in baboon brain. *Journal of Nuclear Medicine, 30,* 741.

Mesulam, M. M. (1982). Slowly progressive aphasia without generalized dementia. *Annals of Neurology, 11,* 592–598.

Meyer, E., Delpia, P., Petrides, M., *et al.* (1991). PET metabolic and neuropsychological correlates of periventricular lucencies. In A. Hartmann, W. Kuschinsky, & S. Hoyer (Eds.), *Cerebral ischemia and dementia.* Berlin: Springer-Verlag.

Miller, K. S., & Miller, J. M. (1993). Wegener's granulomatosis presenting as a primary seizure disorder with brain lesions demonstrated by magnetic resonance imaging. *Chest, 103*(1), 316–318.

Moretti, J. L., Defer, G., Cesaro, P., *et al.* (1987). Early and delayed IMP I-123 SPECT as a prognostic index for clinical recovery in cerebral ischemia. *Journal of Nuclear Medicine, 28,* 623.

Moretti, J. L., Defer, G., Cinotti, L., Cesaro, P., Degos, J. D., Vigneron, N., Ducassou, D., Holman, B. L. (1990). "Luxury-perfusion" with Tc-99m-d, 1 HMPAO and 123I-IMP SPECT imaging during the subacute phase of stroke. *European Journal of Nuclear Medicine, 16,* 17–22.

Mundinger, A., Adam, T., Ott, D., Dinkel, E., Beck, A., Peter, H. H., Volk, B., & Schumacher, M. (1992). CT and MRI, prognostic tools in patients with AIDS and neurological deficits. *Neuroradiology, 35,* 75–78.

Nagata, K., Yunoki, K., Kabe, S., Suzuki, A., & Araki, G. (1986). Regional cerebral blood flow correlates of aphasia outcome in cerebral hemorrhage and cerebral infarction. *Stroke, 17,* 417–423.

Nagel, J. S., Ichise, M., & Holman, B. L. (1991). The scintigraphic evaluation of Huntington's disease and other movement disorders using single photon emission computed tomography perfusion brain scans. *Seminars in Nuclear Medicine, 21,* 11–23.

Nagel, J. S., Johnson, K. A., Ichise, M., English, R. J., Walshe, T. M., Morris, J. H., & Holman, B. L. (1988). Decreased iodine-123 IMP caudate nucleus uptake in patients with Huntington's disease. *Clinical Nuclear Medicine, 13,* 486–490.

Nahmias, C., Garnett, E. S., Firnau, G., & Lang, A. (1985). Striatal dopamine distribution in Parkinsonian patients during life. *Journal of Neurological Sciences, 69,* 223–230.

Neary, D., Snowden, J. S., Shields, R. A., Burjan, A. W., Northen, B., MacDermott, N., Prescott, M. C., Testa, H. J. (1987). Single photon emission tomography using Tc-99m-HM-PAO in the investigation of dementia. *Journal of Neurology, Neurosurgery, and Psychiatry, 50,* 1101–1109.

New, P. F. J., Aronow, S., & Hesselink, J. R. (1980). National Cancer Institute Study. Evaluation of computed tomography in the diagnosis of intracranial neoplasms. IV. Meningiomas. Radiology, 136, 665–675.

Newberg, A., Alavi, A., Souder, E., Reivich, M., & Grossman, M. (1992). Quantitative PET and MRI analysis in patients with primary progressive aphasia (PPA). *Journal of Nuclear Medicine, 33,* 857.

Newton, M. R., Greenwood, R. J., Britton, K. E., Charlesworth, M., Nimmon, C. C., Carroll, M. J., Dolke, G. (1992). A study comparing SPECT with CT and MRI after closed head injury. *Journal of Neurology, Neurosurgery, and Psychiatry 55,* 92–94.

Norrving, B., Nilsson, B., & Risberg, J. (1985). rCBF in patients with carotid occlusion. Resting and hypercapnic flow related to collateral pattern. *Stroke, 13,* 155–162.

O'Brien, C., Sung, J. H., McGeachie, R. E., & Lee, M. C. (1990). Striatonigral degeneration, clinical MRI, and pathologic correlation. *Neurology, 40,* 710–711.

Och, R. F., Yamamoto, Y., & Gloor, P. (1984). Correlations between the positron emission tomography measurement of glucose metabolism and oxygen utilization in focal epilepsy. *Neurology, 34*(Suppl. 1), 125.

Ogawa, T., Sekino, H., Uzura, M., Sakamomto, T., Taguchi, Y., Yamaguchi, Y., Hayashi, T., Yamanaka, I., Oohama, N., & Imaki, S. (1992). Comparative study of magnetic resonance and CT scan imaging in cases of severe head injury. *Acta Neurochirurgica Supplement 55,* 8–10.

Ogawa, T., Uemura, K., Shishido, F., Yamaguchi, T., Murakami, M., Inugami, A., Kanno, I., Sasaki, H., Kato, T., Hirata, K., *et al.* (1988). Changes of cerebral blood flow and oxygen and glucose metabolism following radiochemotherapy of gliomas. A PET study. *Journal of Computer Assisted Tomography, 12,* 290–297.

Olanow, C. W. (1992). Magnetic resonance imaging in parkinsonism. *Neurology Clinics, 10,* 405–420.

Pahl, J. J. (1989). Cerebral metabolic decline. A biochemical marker of genetic expression in Huntington's disease. *Advances in Functional Neuroimaging, 2,* 3–8.

Pantano, P., Baron, J. C., Samson, Y., Bousser, M. G., Derouesne, C., & Comar, D. (1986). Crossed cerebellar diaschisis. Further studies. *Brain, 109,* 677–694.

Pantano, P., Lenzi, G. L., Guidetti, B., Di Piero, V., Gerundini, P., Savi, A. R., Fazio, F., & Fieschi, C. (1987). Crossed cerebellar diaschisis in patients with cerebral ischemia assessed by SPECT and I-123-HIPDM. *European Neurology, 27,* 142–148.

Pappata, S., Levasseur, M., Tran-Dink, S., *et al.* (1989). PET study of cerebral glucose utilization in 10 patients with Wilson's disease. *Journal of Cerebral Blood Flow and Metabolism, 9,* S23.

Patronas, N. J., Brooks, R. A., DeLaPaz, R. L., Smith, B. H., Kornblith, P. L., Di Chiro, G. (1983). Glycolytic rate (PET) and contrast enhancement (CT) in human cerebral gliomas. *American Journal of Neuroradiology, 4,* 533–535.

Patronas, N. J., DiChiro, G., Kufta, C., Bairamian, D., Kornblith, P. L., Simon, R., & Larson, S. M., (1985). Prediction of survival in glioma patients by means of positron emission tomography. *Journal of Nuclear Medicine, 18,* P64.

Patronas, N. J., Di Chiro, G., Smith, B. H., DeLaPaz, R., Brooks, R. A., Milam, H. L., Kornblith, P. L., Bairamian, D., & Mansi, L. (1984). Depressed cerebellar glucose metabolism in supratentorial tumors. *Brain Research, 291,* 93–101.

Pawlik, G., Holthoff, V., Rudolf, J., *et al.* (1989). Vasculitic dementia. Characteristic changes of regional brain function demonstrated by PET. *Journal of Cerebral Blood Flow and Metabolism, 9,* S534.

Pedersen, C., Thomsen, C., Arlien-Soborg, P., Praestholm, J., Kjaer, L., Boesen, F., Hansen, H. S., Nielsen, J. O. (1991). Central nervous system involvement in human immunodeficiency virus disease. A prospective study including neurological examination, computerized tomography, and magnetic resonance imaging. *Danish Medical Bulletin, 30,* 374–379.

Peppard, R. F., Martin, W. R., Clark, C. M., Carr, G. D., McGeer, P. L., & Calne, D. B. (1990). Cortical glucose metabolism in Parkinson's and Alzheimer's disease. *Journal of Neuroscience Research, 27,* 561–568.

Perani, D., DiPiero, V., Lucignani, G., Gilardi, M. C., Pantano, P., Rossetti, C., Pozzilli, C., Gerundini, P., Fazio, F., & Lenzi, G. L. (1988). Remote effects of subcortical cerebrovascular lesions. A SPECT cerebral perfusion study. *Journal of Cerebral Blood Flow and Metabolism, 8,* 560–567.

Perani, D., Gerundini, P., & Lenzi, G. L. (1987). Cerebral hemispheric and contralateral cerebellar hypoperfusion during a transient ischemic attach. *Journal of Cerebral Blood Flow and Metabolism, 7,* 507–509.

Peretti-Viton, P., Margain, D., Arnaud, O., Perez-Castillo, A. M., & Graziani, N. (1991). Primary and secondary lymphomas of the brain. An MRI study. *Journal of Neuroradiology, 18,* 173–188.

Phelps, M. E. (1991). PET, a biological imaging technique. *Neurochemical Research, 16,* 929–940.

Phelps, M. E., & Mazziotta, J. C. (1985). Positron emission tomography. Human brain function and biochemistry. *Science, 228,* 799–809.

Phelps, M. E., Mazziotta, J. C., Wapenski, J., *et al.* (1985). Cerebral glucose utilization and blood flow in Huntington's disease. *Journal of Nuclear Medicine, 26,* 47.

Pizzolato, G., Dam, M., Borsato, N., Saitta, B., Da Col, C., Perlotto, N., Zanco, P., Ferlin, G., Battistin, L. (1988). Tc-99m-HM-PAO SPECT in Parkinson's disease. *Journal of Cerebral Blood Flow and Metabolism, 8,* S101–S108.

Playford, E. D., Jenkins, I. H., Passingham, R. E., Nutt, J., Frackowiak, R. S., & Brooks, D. J. (1992). Impaired mesial frontal and putamen activation in Parkinson's disease. A positron emission tomography study. *Annals of Neurology, 32,* 151–161.

Playford, E. D., Jenkins, I. H., Passingham,, R. E., *et al.* (1993). Impaired activation of frontal areas during movement in Parkinson's disease. A PET study. *Advances in Neurology, 60,* 506–510.

Podreka, I., Suess, E., Goldenberg, G., Steiner, M., Brucke, T., Muller, C., Lang, W., Neirinckx, R. D., & Deecke, L. (1987). Initial experience with technetium-99m HM-PAO brain SPECT. *Journal of Nuclear Medicine, 28,* 1657–1666.

Pohl, P., Vogl, G., Fill, H., Rossler, H., Zangerle, R., & Gerstenbrand, F. (1988). Single photon emission computed tomography in AIDS dementia complex. *Journal of Nuclear Medicine, 29,* 1382–1386.

Powers, W., Grubb, R., & Raichle, M. (1984). Physiological response to focal cerebral ischemia in humans. *Annals of Neurology, 16,* 546–552.

Powers, W., Grubb, R., & Raichle, M. (1989). Clinical results of extracranial-intracranial bypass surgery in patients with hemodynamic cerebrovascular disease. *Journal of Neurosurgery, 70,* 61–67.

Powers, W. J., Martin, W. R., Herscovitch, P., Raichle, M. E., & Grubb, Jr., J. L. (1984). Extracranial-intracranial bypass surgery; hemodynamic and metabolism effects. *Neurology, 34,* 1168–1174.

Powers, W. J., Press, G. A., Grubb, Jr., R. L., Gado, M., & Raichle, M. E. (1987). The effect of hemodynamically significant carotid artery disease on the hemodynamic status of the cerebral circulation. *Annals of Internal Medicine, 106,* 27–35.

Prayer, L., Wimberger, D., Kramer, J., Grimm, G., Oder, W., & Imhof, H. (1990). Cranial MRI in Wilson's disease. *Neuroradiology, 32,* 211–214.

Pressman, B. D., Tourje, E. J., & Thompson, J. R. (1987). An early sign of ischemic infarction, increased density in a cerebral artery. *American Journal of Neuroradiology, 8,* 645–648.

Przedborski, S., Goldman, S., Giladi, N., Dhawan, V., Takikawa, S., Hildebrand, J., Fahn, S., Eidelberg, D. (1993). Positron emission tomography in hemiparkinsonism-hemiatrophy syndrome. *Advances in Neurology, 60,* 501–505.

Putty, T. K., Luerssen, T. G., Campbell, R. L., Boaz, J. C., & Edwards, M. K. (1990–1991). Magnetic resonance imaging diagnosis of cerebral aneurysm in an infant. Case report and review of the literature. *Pediatric Neurosurgery, 16,* 48–51.

Raininko, R., Elovaara, I., Virta, A., Valanne, L., Haltia, M., & Valle, S. L. (1992). Radiological study of the brain at various stages of human immunodeficiency virus (HIV-1) infection. Early development of brain atrophy. Neuroradiology, 34, 180–188.

Rango, M., Candelise, L., Perani, D., Messa, C., Scarlato, G., Canal, N., Franceschi, M., & Fazio, F. (1989). Cortical pathophysiology and clinical neurological abnormalities in acute cerebral ischemia. A serial study with single photon emission computed tomography. *Archives of Neurology, 46,* 1318–1322.

Rao, N., Turski, P. A., Polcyn, R. E., Nickels, R. J., Matthews, C. G., & Flynn, M. M. (1984). 18F positron emission computed tomography in closed head injury. *Archives of Physical and Medical Rehabilitation, 65,* 780–785.

Reichmann, K., Biersack, H. J., Basso, L., *et al.* (1986). A comparative study of brain uptake and early kinetics of 99mTc-dl HM-PAO and other PNAO derivatives in baboons. *Nuclear Medicine, 25,* 134.

Reid, I. C., Besson, J. A., Best, P. V., Sharp, P. F., Gemmell, H. G., & Smith, F. W. (1988). Imaging of cerebral blood flow markers in Huntington's disease using single photon emission computed tomography. *Journal of Neurology, Neurosurgery, and Psychiatry, 51,* 1264–1268.

Reid, R. H., Gulenchyn, K. Y., Ballinger, J. R., & Ventureyra, E. C. (1990). Cerebral perfusion imaging with technetium-99m HMPAO following cerebral trauma. Initial experience. *Clinical Nuclear Medicine, 15,* 383–388.

Robitaille, Y., Rasmussen, T., Dubeau, F., Tampieri, D., & Kemball, K. (1992). Histopathology of nonneoplastic lesions in frontal lobe epilepsy. Review of 180 cases with recent MRI and PET correlations. *Advances in Neurology, 57,* 499–511.

Roper, S. N., Mena, I., King,, W. A., Schweitzer, J., Garrett, K., Mehringer, C. M., & McBride, D. (1991). An analysis of cerebral blood flow in acute closed-head injury using Tc-99m-HMPAO SPECT in computed tomography. *Journal of Nuclear Medicine, 32,* 684–687.

Rosenthal, G., Gilman, S., Koeppe, R. A., Kluin, K. J., Markel, D. S., Junck, L., & Gebarski, S. S. (1988). Motor dysfunction in olivopontocerebellar atrophy is related to cerebral metabolic rate studied with positron emission tomography. *Annals of Neurology, 24,* 414–419.

Rougemont, D., Baron, J. C., Collard, P., Bustany, P., Comar, D. & Agid, Y. (1984). Local cerebral glucose utilization in treated and untreated patients with Parkinson's disease. *Journal of Neurology, Neurosurgery, and Psychiatry, 47,* 824–830.

Rowe, C. C., Berkovic, S. F., Sia, S. T., Austin, M., McKay, W. J., Kalnins, R. M., & Bladin, P. F. (1989). Localization of epileptic foci with postictal single photon emission computed tomography. *Annals of Neurology, 26,* 660–668.

Rozental, J. M., Levine, R. L., & Nickles, R. J. (1991). Changes in glucose uptake by malignant gliomas. Preliminary study of prognostic significance. *Journal of Neurooncology, 10,* 75–83.

Ruff, R. M., Buchsbaum, M. S., Troster, A. I., *et al.* (1989). Computerized tomography, neuropsychology, and positron emission tomography in the evaluation of head injury. *Neuropsychiatry, Neuropsychology, and Behavioral Neurology, 2,* 103–123.

Ryding, E., Rosen, I., Elmqvist, D., *et al.* (1988). SPECT measurements with 99mTc-HM-PAO in focal epilepsy. *Journal of Cerebral Blood Flow and Metabolism, 8,* S95.

Sachs, H., Russel, J. A. G., Christman, D. R., & Cook, B. (1987). Alterations in regional cerebral glucose metabolic rate in non-Korsakoff chronic alcoholism. *Archives of Neurology 44,* 1242–1251.

Sadzot, B., Mayberg, H. S., & Frost, J. J. (1990). Detection and quantification of opiate receptors in man by positron emission tomography. Potential applications to the study of pain. *Neurophysiology Clinics, 20,* 323–334.

Samson, Y., Baron, J., & Bousser, M. (1985). Effects of extra-intracranial arterial bypass on cerebral blood flow and oxygen metabolism in humans. *Stroke, 16,* 609–616.

Samson, Y., Baron, J. C., Feline, A., Bories, J., & Crouzel, C. (1986). Local cerebral glucose utilization in chronic alcoholics. A positron tomography study. *Journal of Neurology, Neurosurgery, and Psychiatry, 49,* 1165–1170.

Sapin, L. R., Anderson, F. H., & Pulaski, P. D. (1989). Progressive aphasia without dementia. Further documentation. *Annals of Neurology, 25,* 411–413.

Savic, I., Persson, A., Roland, P., Pauli, S., Sedvall, G., & Widen, L. (1988). In-vivo demonstration of reduced benzodiazepine receptor binding in human epileptic foci. *Lancet, 2*(8616), 863–866.

Savoiardo, M., Girotti, F., Strada, L., & Ciceri, E. (1989). MR imaging in progressive supranuclear palsy and Shy-Drager syndrome. *Journal of Computer Assisted Tomography, 13,* 555–560.

Sawada, H., Udaka, F., Kameyama, M., Seriu, N., Nishinaka, K., Shindou, K., Kodama, M., Nishitani, N., & Okumiya, K. (1992). SPECT findings in Parkinson's disease associated with dementia. *Journal of Neurology, Neurosurgery, and Psychiatry 55,* 960–963.

Sawle, G. V. (1992). The detection of preclinical Parkinson's disease. What is the role of positron emission tomography? *Movement Disorders, 8,* 271–277.

Sawle, G. V., Brooks, D. J., Thompson, P. D. *et al.* (1989). Corticobasal degeneration, metabolic and 18F-dopa studies with positron emission tomography. *Journal of Cerebral Blood Flow and Metabolism, 9,* S338.

Sawle, G. V., Wroe, S. J., Lees, A. J., Brooks, D. J., & Frackowiak, R. S. (1992). The identification of presymptomatic parkinsonism, clinical and [18F] dopa positron emission tomography studies in an Irish kindred. *Annals of Neurology, 32,* 609–617.

Schielke, E., Tatsch, K., Pfister, H. W., Trenkwalder, C., Leinsinger, G., Kirsch, C. M., Matuschke, A., & Einhaupl, K. M. (1990). Reduced cerebral blood flow in early stages of human immunodeficiency virus infection. *Archives of Neurology, 47,* 1342–1345.

Schubiger, P. A., Hasler, P. H., Beer-Wohlfahrt, H., Bekier, A., Oettli, R., Cordes, M., Ferstl, F., Deisenhammer, E., De Roo, M., Moser, E., *et al.* 1991). Evaluation of multicentre study with iomazenil—a benzodiazepine receptor ligand. *Nuclear Medicine Communications, 12,* 569–582.

Schwartz, R. B., Carvalho, P. A., Alexander, E., Loeffler, J. S., Folkerth, R. & Holman, B. L. (1991).

Radiation necrosis vs. high-grade recurrent glioma. Differentiation by using dual-isotope SPECT with 201T1 and 99mTc-HMPAO. *American Journal of Neuroradiology, 12,* 1187–1192.

Schwartz, J. Tatsch, K., Vogl, T., Kirsch, C. M., Trenkwalder, C., Arnold, G., Gasser, T., & Oertel, W. H. (1992). Marked reduction of striatal dopamine D2 receptors as detected by 123IBZM-SPECT in a Wilson's disease patient with generalized dystonia. *Movement Disorders, 7,* 58–61.

Sedvall, G. (1990). PET imaging of dopamine receptors in human basal ganglia. Relevance to mental illness. *Trends in Neuroscience, 13,* 302–308.

Senda, M., Alpert, N. M., Mackay, B. C., Buxton, R. B., Correia, J. A., Weise, S. B., Ackerman, R. H., Dorer, D., Buonanno, F. S. (1989). Evaluation of the $^{11}CO_2$ positron emission tomographic method for measuring brain pH. II. Quantitative pH mapping in patients with ischemic cerebrovascular diseases, *Journal of Cerebral Blood Flow and Metabolism, 9,* 859–873.

Sette, G., Baron, J. C., Mazoyer, B., Levasseur, M., Pappata, S., & Crouzel, C. (1989). Local brain haemodynamics and oxygen metabolism in cerebrovascular disease, *Brain, 112,* 931–951.

Sharp, P., Gemmell, H., Cherryman, G., Besson, J., Crawford, J., & Smith, F. (1986). Application of iodine-123-labeled isopropylamphetamine imaging to the study of dementia. *Journal of Nuclear Medicine, 27,* 761–768.

Shih, T. T. F., & Huang, K.-M. (1988). Acute stroke, detection of changes in cerebral perfusion with dynamic CT scanning. *Radiology, 169,* 469–474.

Simmons, J. T., Pastakea, B., Chase, T. N., & Shults, C. W. (1986). Magnetic resonance imaging in Huntington's disease. *American Journal of Neuroradiology, 7,* 25–28.

Sklar, E. M., Quencer, R. M., Bowen, B. C., Altman, U., & Villaneuva, P. A. (1992). Magnetic resonance applications in cerebral injury. *Radiological Clincs of North America, 30,* 353–366.

Smith, F. W., Besson, J. A., Gemmell, H. G., & Sharp, P. F. (1988a). The use of technetium-99m-HM-PAO in the assessment of patients with dementia and other neuropsychiatric conditions. *Journal of Cerebral Blood Flow and Metabolism 8,* S116–122.

Smith, F. W., Gemmell, H. G., Sharp, P. F., & Besson, J. A. (1988b). Technetium-99m HMPAO imaging in patients with basal ganglia disease. *British Journal of Radiology, 61,* 914–920.

Song, W. Z., & Lin, X. T. (1990). Observing of image of brain blood perfusion in Parkinson's disease. *European Journal of Nuclear Medicine, 16,* S23.

Souder, E., Alavi, A., Uzzell, B., *et al.* (1990). Correlation of FDG-PET and neuropsychological findings in head injured patients. Preliminary data. *Journal of Nuclear Medicine, 31,* 876.

Spampinato, U., Habert, M. O., Mas, J. L., Bourdel, M. C., Ziegler, M., de Recondo, J., Askienazy, S., & Rondot, P. (1991). (99mTc)-HM-PAO SPECT and cognitive impairment in Parkinson's disease, a comparison with dementia of the Alzheimer type. *Journal of Neurology, Neurosurgery, and Psychiatry, 54,* 787–792.

Sperling, M. R., Wilson, G., Engel, Jr., J., Babb, T. L. Phelps, M., & Bradley, W. (1986). Magnetic resonance imaging in intractable partial epilepsy. Correlative studies. *Annals of Neurology, 20,* 57–62.

Spetzler, R. F., Zabramski, J. M., Kaufman, B., & Yeung, H. N. (1983). Acute NMR changes during MCA occlusion. A Preliminary study of primates. *Stroke, 14,* 185–190.

Starkstein, S. E., Folstein, S. E., Brandt, J., Pearlson, G. D., McDonnell, A., & Folstein, M. (1989). Brain atrophy in Huntington's disease. A CT-scan study. *Neuroradioloogy, 31,* 156–159.

Stefan, H., Pawlik, G., Bocher-Schwarz, H. G., Biersack, H. J., Burr, W., Penin, H., & Heiss, W. D. (1987). Functional and morphologic abnormalities in temporal lobe epilepsy. A comparison of interictal and ictal EEG, CT, MRI, SPECT and PET. *Journal of Neurology, 234,* 377–384.

Sullivan, H. G., Kingsbury, T. B., Morgan, M. E., Jeffcoat, R. D., Allison, J. D., Goode, J. J., & McDonnell, D. E. (1987). The rCBF response to Diamox in normal subjects and cerebrovascular disease patients. *Journal of Neurosurgery, 67,* 525–534.

Swartz, B. E., Tomiyasu, U., Delgado-Escueta, A. V., Mandelkern, M., & Khonsari, A. (1992). Neuroimaging in temporal lobe epilepsy. Test sensitivity and relationships to pathology and postoperative outcome. *Epilepsia, 33,* 624–634.

Syrota, A., Castaing, M., Rougemont, D., Berridge, M., Baron, J. C., Bousser, M. G., & Pocidalo, J. J. (1983). Tissue acid-base balance and oxygen metabolism in human cerebral infarction studies with positron emission tomography. *Annals of Neurology, 14,* 419–428.

Syrota, A., Samson, Y., Boullais, C., Wajnberg, P., Loc'h, C., Crouzel, C., Maziere, B., Soussaline, F., & Baron, J. C. (1985). Tomographic mapping of brain intracellular pH and extracellular water space in stroke patients. *Journal of Cerebral Blood Flow and Metabolism, 5,* 358–368.

Takahashi, N., Nishihara, M., Odano, I., Sakai, K., Koyama, A., & Atsumi, T. (1992). Regional cerebral blood flow measured with I-123 IMP SPECT in a case of subcortical arteriosclerotic encephalopathy (Binswanger's disease). *Clinical Nuclear Medicine, 17,* 882–884.

Tatsch, K., Schielke, E., Bauer, W. M., Markl, A., Einhaupl, K. M., & Kirsch, C. M. (1990). Functional and morphological findings in early and advanced stages of HIV infection. A correlation of 99mTc HMPAO SPECT with CT And MRI studies. *Nuklearmedizin, 29,* 252–258.

Tedroff, J., Aquilonius, S. M., Laihinen, A., Rinne, U., Hartvig, P., Anderson, J., Lundqvist, H., Haaparanta, M., Solin, O., Antoni, G., *et al.* (1990). Striatal kinetics of (11C)-(+)-nomifensine and 6-(18F) fluoro-L-dopa in Parkinson's disease measured with positron emission tomography. *Acta Neurologia Scandinavia, 81,* 24–30.

Terrence, C. F., Delaney, J. F., Alberts, M. C., *et al.* (1977). Computed tomography for Huntington's disease. *Neuroradiology, 13,* 173–175.

Testa, D., Savoiardo, M., Fetoni, V., Strada, L., Palazzini, E., Bertulezzi, G., & Girotti, F. (1993). Multiple system atrophy. Clinical and MR observations on 42 cases. *Italian Journal of Neurological Science, 14,* 211–216.

Testa, H. J., Snowden, J. S., Neary, D., Shields, R. A., Burjan, A. W., Prescott, M. C., Northen, B., & Goulding, P. (1988). The use of Tc-99m HM-PAO in the diagnosis of primary degenerative dementia. *Journal of Cerebral Blood Flow and Metabolism, 8,* S123–S126.

Theodore, W. H., Brooks, R., Sato, S., Patronas, N., Margolin, R., DiChiro, G., & Porter, R. J. (1984). The role of positron emission tomography in the evaluation of seizure disorders. *Annals of Neurology, 15,* S1176–S1179.

Theodore, W. H., Dorwart, R., Holmes, M., Porter, R. J., & DiChiro, G. (1986). Neuroimaging in refracatory partial seizures. Comparison of PET, CT, and MRI. *Neurology, 36,* 750–759.

Theodore, W. H., Fishbein, D., & Dubinsky, R. (1988). Patterns of cerebral glucose metabolism in patients with partial seizures. *Neurology, 38,* 1201–1206.

Theodore, W. H., Newmark, M. E., Sato, S., Brooks, R., Patronas, N., DeLaPaz, R., DiChiro, G., Kessler, R. M., Margolin, R., Manning, R. G., *et al.* (1983). 18F fluorodeoxyglucose positron emission tomography in refractory complex partial seizures. *Annals of Neurology, 14,* 537.

Thuomas, K. A., Aquilonius, S. M., Bergstrom, K., *et al.* (1993). Magnetic resonance imaging of the brain in Wilson's disease. *Neuroradiology, 35,* 134–141.

Tohgi, H., Chiba, K., Sasaki, K., Hiroi, S., & Ishibashi, Y. (1991). Cerebral perfusion patterns in vascular dementia of Binswanger type compared with senile dementia of the Alzheimer type. A SPECT study. *Journal of Neurology, 238,* 365–370.

Tomura, N., Uemura, K., Inugammi, A., Fujita, H., Higano, S., & Shishido, F. (1988). Early CT finding in cerebral infarction. Obscuration of the lentiform nucleus. *Radiology, 168,* 463–467.

Tran Dinh, Y. R., Mamo, H., Cervoni, J., Caulin, C., & Saimot, A. C. (1990). Disturbances in the cerebral perfusion of human immune deficiency virus-1 seropositive asymptomatic subjects. A quantitative tomography study of 18 cases. *Journal of Nuclear Medicine, 31,* 1601–1607.

Tumeh, S. S., Nagel, J. S., English, R. J., *et al.* (1991). Use of SPECT perfusion brain scintigraphy to investigate effects of cocaine in the brain. In G. G. Nahas & C. Latur (Eds.), *Physiopathology of illicit drugs, cannabis, cocaine, opiates.* Oxford: Pergamon Press.

Tyler, J. L., & Starr, C. W. (1992). Functional imaging of epileptic foci with PET And SPECT. *JSC Medical Association, 88,* 258–262.

Tyrrell, P. J., Warrington, E. K., Frackowiak, R. S. J., & Rossor, M. N. (1990). Heterogeneity in progressive aphasia due to focal cortical atrophy. *Brain, 113,* 1321–1336.

Valk, P. E., Budinger, T. F., Levin, V. A., Silver, P., Gutin, P. H., & Doyle, W. K. (1988). PET of malignant cerebral tumors after interstitial brachytherapy. Demonstration of metabolic activity and correlation with clinical outcome. *Journal of Neurosurgery, 69,* 830–838.

Vallar, G., Perani, D., Cappa, S. F., Messa, C., Lenzi, G. L. & Fazio, F. (1988). Recovery from aphasia and neglect after subcortical stroke. Neuropsychological and cerebral perfusion study. *Journal of Neurology, Neurosurgery, and Psychiatry, 51,* 1269–1276.

van Royen, E., Verhoeff, N. F., Speelman, J. D., Wolters, E. C., Kuiper, M. A. & Janssen, A. G. (1993). Multiple system atrophy and progressive supranuclear palsy. Diminished striatal D2 dopamine receptor activity demonstrated by 123IBZM single photon emission computed tomography. *Archives of Neurology, 50,* 513–516.

Varastet, M., Brouillet, E., Chavoix, C., Prenant, C., Crouzel, C., Stulzaft, O., Bottlaender, M., Cayla, J., Maziere, B., & Maziere, M. (1992). In vivo visualization of central muscarinic receptors using [11C] quinuclidinyl benzilate and positron emission tomography in baboons. *European Journal of Pharmacology, 213,* 275–284.

Villoria, M. F., de la Torre, J., Fortea, F., Munoz, L., Hernandez, T., & Alarcon, J. J. (1992). Intracranial tuberculosis in AIDS. CT and MRI findings. *Neuroradiology, 34,* 11-14.

Volkow, N. D., & Fowler, J. S. (1992). Neuropsychiatric disorders. Investigation of schizophrenia and substance abuse. *Seminars in Nuclear Medicine, 22,* 254–267.

Volkow, N. D., Fowler, J. S., Wolf, A. P., Hitzemann, R., Dewey, S., Bendriem, B., Alpert, R., & Hoff, A. (1991). Changes in brain glucose metabolism in cocaine dependence and withdrawal. *American Journal of Psychiatry, 148*(5), 621–626.

Volkow, N. D., Hitzemann, R., Wolf, A. P., Logan, J., Fowler, J. S., Christman, D., Dewey, S. L., Schlyer, D., Burr, G., Vitkum, S., *et al.* (1990). Acute effects of ethanol on regional brain glucose metabolism and transport. *Psychiatry Research, 35,* 39–48.

Volkow, N. D., Mullani, N., Gould, K. L., Adler, S. & Krajewski, K. (1988). Cerebral blood flow in chronic cocaine users. *British Journal of Psychiatry, 152,* 641–648.

Wallner, K. E., Galicich, J. H., Malkin, M. G., Arbit, E., Krol, G. & Rosenblum, M. K. (1989). Inability of computed tomography appearance of recurrent malignant astrocytoma to predict survival following reoperation. *Journal of Clinical Oncology, 7,* 1492–1496.

Wang, A.-M., Lin, J.C.-T., & Rumbaugh, C. L. (1988). What is expected of CT in the evaluation of stroke? *Neuroradiology, 30,* 54–58.

Weiller, C., Chollet, F., Friston, K. J., Wise, R. J., & Frackowiak, R. S. (1992). Functional reorganization of the brain in recovery from striatocapsular infarction in man. *Annals of Neurology, 31,* 463–472.

Weinberger, D. R., Gibson, R., Coppola, R., Jones, D. W., Molchan, S., Sunderland, T., Berman, K. F., & Reba, R. C. (1991). The distribution of cerebral muscarinic acetylcholine receptors in vivo in patients with dementia. A controlled study with 123IQNB and single photon emission computed tomography. *Archives of Neurology, 48,* 169–176.

Weinberger, D. R., Jones, D., Reba, R. C., Mann, U., Coppola, R., Gibson, R., Gorey, J., Braun, A., & Chase, T. N. (1992). A comparison of FDG PET and IQNB SPECT in normal subjects and in patients with dementia. *Journal of Neuropsychiatry and Clinical Neuroscience, 4,* 239–248.

Weinstein, H. C., Haan, J., van Royen, E. O., Derix, M. M., Lanser, J. B., van der Zant, F., Dunnewold, R. J., van Kroonenburgh, M. J., Pauwels, E. K., van der Velde, E. A. *et al.* (1991). SPECT in the diagnosis of Alzheimer's disease and multi-infarct dementia. *Clinical Neurology and Neurosurgery, 93,* 39–43.

Weinstein, H. C., Hijdra, A., van Royen, E. A., & Derix, M. M. (1989). Determination of cerebral blood flow by SPECT, a valuable tool in the investigation of dementia? *Clinical Neurology and Neurosurgery 91,* 13–19.

Weisberg, L. A. (1991). How to identify and manage brain hemorrhage. *Postgraduate Medicine, 88,* 169–175.

Widner, H., Ttetrud, J. W., Snow, B. J., *et al.* (1991). Fifteen months follow-up on immature neural tissue grafts in two cases of severe MPTP-induced parkinsonism (abstract). *10th International Symposium on Parkinson's Disease.*

Wijnand, A., Rutgers, F., Lakke, J. P. W., *et al.* (1987). Tracing of dopamine receptors in hemiparkinsonism with positron emission tomography (PET). *Journal of the Neurological Sciences, 80,* 237–248.

Wik, G., Borg, S., Sjogren, I., Wiesel, F. A., Blomqvist, G., Borg, J., Greitz, T., Nyback, H., Sedvall, G., Stone-Elander, S., *et al.* (1988). PET determination of regional cerebral glucose metabolism in alcohol-dependent men and healthy controls using ^{11}C-glucose. *Acta Psychiatry Scandinavia, 78,* 234–241.

Wilson, C. B. (1989). Metabolic imaging of human brain tumors. *Seminars in Neurology, 9,* 388–393.

Wise, R. J., Bernardi, S., Frackowiak, R. S., Legg, N. J., & Jones, T. (1983). Serial observations on the pathophysiology of acute stroke. The transition from ischemia to infarction as reflected in regional oxygen extraction. *Brain, 106,* 197–222.

Wong, D. F., Links, J. M., Wagner, H. N., *et al.* (1985). Dopamine and serotonin receptors measured in vivo in Huntington's disease with C-11-N-methylspiperone PET imaging. *Journal of Nuclear Medicine, 26,* P107.

Wouters, W., Van Dun, J., & Laduron, P. M. (1986). 7-amino-8-([125I]AMIK), a highly sensitive serotonin-S2 receptor ligand. *Biochemical Pharmacology, 35,,* 3199–3202.

Yao, H., Sadoshima, S., Kuwabara, Y., Ichiya, Y., & Fujishima, M. (1990). Cerebral blood flow and oxygen metabolism in patients with vascular dementia of the Binswanger type. *Stroke, 21,* 1694–1699.

Yealy, D. M., & Hogan, D. E. (1991). Imaging after head trauma. Who needs what? *Emergency Medicine Clinics of North America, 9,* 707–717.

Yokota, H., Kurokawa, A., Otsuka, T., Kobayashi, S., & Nakazawa, S. (1991). Significance of magnetic resonance imaging in acute head injury. *Journal of Trauma, 31,* 351–357.

Young, A. B., Penney, J. B., Starosta-Rubinstein, S., Markel, D. S., Berent, S., Giordani, B., Ehrenkaufer, R., Jewett, D., & Hichwa, R. (1986). PET scan investigations of Huntington's disease. Cerebral metabolic correlates of neurological features and functional decline. *Annals of Neurology, 29,* 296–303.

Young, A. B., Penny, J. B., Starosta-Rubinstein, S., Markel, D., Berent, S., Rothley, J., Betley, A., & Hichwa, R. (1987). Normal caudate glucose metabolism in persons at risk for Huntington's disease. *Archives of Neurology, 44,* 254–257.

Zee, C. S., Chin, T., Segall, H. D., Destian, S., & Ahmadi, J. (1992). Magnetic resonance imaging of meningiomas. *Seminars in Ultrasound, CT, and MR, 13,* 154–169.

Zimmerman, R. A., Bilaniuk, L. T., Hackney, D. B., Goldberg, H. I., & Grossman, R. I. (1986). Head injury, early results comparing CT and high-field MR. *American Journal of Roentgenology, 147,* 1215–1222.

Zimmerman, R. D., Fleming, C. A., Lee, B. C., Saint-Louis, L. A., & Deck, M. D. (1986). Periventricular hyperintensity as seen by magnetic resonance. Prevalence and significance. *American Journal of Neuroradiology, 7,* 13–20.

4

Neuroimaging in Psychiatric Disorders

LAURA MARSH, JOHN LAURIELLO, EDITH V. SULLIVAN, AND ADOLF PFEFFERBAUM

INTRODUCTION

One of the most exciting advances in psychiatric research has been the application of *in vivo* structural and functional neuroimaging to explore brain dysmorphology and metabolic dysfunction in psychiatric disorders. A central research issue in psychiatric neuroimaging is identification of a salient pattern of pathology which also reflects the clinical pathology of the disorder. The ability to visualize the brain in the living subject is especially important in investigations of psychiatric disturbances, because they tend to be nonfatal and manifest when individuals are relatively young. This chapter will review progress to date on three psychiatric conditions—schizophrenia, affective disorders, and obsessive-compulsive disorder. It will focus primarily on *in vivo* structural neuroimaging techniques of computed tomography (CT) and magnetic resonance imaging (MRI) and will relate these findings to functional imaging of glucose metabolism and blood flow with single positron emission computed tomography (SPECT) and positron emission tomography (PET).

LAURA MARSH, JOHN LAURIELLO, EDITH V. SULLIVAN, AND ADOLF PFEFFERBAUM Department of Psychiatry and Behavioral Sciences, Stanford University School of Medicine, Stanford, California, and Psychiatry Service, Veterans Affairs Palo Alto Health Care System, Palo Alto, California 94304.

Neuroimaging II: Clinical Applications, edited by Erin D. Bigler. Plenum Press, New York, 1996.

Neuroimaging abnormalities associated with idiopathic psychiatric disorders tend to be subtle and therefore especially susceptible to errors in the neuroimaging techniques themselves. Thus, interpretations of the results must consider the advantages and limitations of the neuroimaging technique as well as the analysis methods. Neuroimaging technology and the significance of different acquisition parameters in data collection are discussed in Volume I of this series. To provide a framework for presenting the neuroimaging findings from studies on psychiatric disorders, we will first review the commonly used image analysis approaches and address certain methodological issues in psychiatric neuroimaging research. We will then focus on the neuroimaging findings for each condition and relate these to clinical manifestations of the illness and variables such as illness course, gender, or aging. Since pathognomic abnormalities have not been detected for schizophrenia, affective disorders, and obsessive-compulsive disorder, neuropsychiatric conditions with similar clinical manifestations to the idiopathic disorders will be related to the neuroimaging findings as appropriate.

Image Analysis of Structural and Functional Neuroimages in Psychiatric Research

In vivo structural neuroimaging techniques are complex, and the technology for both image acquisition and image analysis has advanced considerably since the initial CT studies on schizophrenia reported by Johnstone, Crow, Frith, Husband, and Kreel (1976). The majority of structural imaging studies have attempted to measure the size of various brain structures with CT or MR images using either qualitative or quantitative approaches. Qualitative scales typically assess the extent to which specific features (e.g., sulcal width or ventricle size) are observed to deviate from normal. Quantitative morphometry involves calculating linear (e.g., length, width) or area (in pixels or cm^2) measures. When the values of specific areas are summed over multiple images, volumes can be calculated. CT images can be segmented into cerebrospinal fluid (CSF) and tissue regions, but the method is hampered by poor anatomical resolution, two-dimensional (2-D) visualization of brain structures in a fixed plane, and beam hardening artifacts which limit visualization of the temporal regions and cortical sulci. With MRI, structures can be viewed in all three orthogonal planes (i.e., axial, coronal, and sagittal), and there is good anatomical resolution of gross brain morphology. Recent methods enable 3-D visualization of brain structures.

Analysis of regional image intensity values from MRI and use of different echo times provide a basis for distinguishing the brain image into CSF, gray matter, and white matter compartments and determining measures of the area or volume of these different compartments. In addition, morphological parameters which are not specifically dependent on structural size can be examined. Subcortical and white matter MRI signal hyperintensities, a nonspecific sign of pathology, can be characterized in terms of number, severity, and volume. Their occurrence is associated with a number of conditions such as aging, vascular disease, demyelination, edema, flow enhancement, dilated perivascular spaces, and ischemic injury (Coffey & Figiel, 1991; Englund, Brun, & Alling, 1988; Jernigan, Archibald, *et al.*, 1991). MRI time constants, T1 and T2, represent aspects of the chemical environment of the imaged tissue (e.g., the ratio of free to bound water in the case of T1), although the pathological basis for deviations from normal is not always clear. Structural shape analyses provide methods for examining focal dysmorphology which is not necessarily apparent in gross volumetric estimates of a given structure, such as the configuration or contour of sulcal-gyral boundaries. Despite these capacities, structural neuroimaging cannot provide data on the cytoarchitecture of brain structures, nor is it currently capable of reliably detecting functional processes. Differences in image acquisition parameters and morphometric image analysis techniques obviously influence the detection of structural dysmorphology and may account for inconsistencies in the psychiatric neuroimaging literature.

Dynamic imaging studies using PET and SPECT use a wide range of isotopes to assess brain function in terms of metabolic rate [e.g., 2-[^{18}F]-fluoro-2-deoxy-D-glucose (^{18}FDG); [^{11}C]-glucose] or regional cerebral blood flow (rCBF) [e.g., ^{15}O; ^{99m}Technetium-hexamethylpropyleneamineoxime (^{99m}Tc-HMPAO)], or quantify neuroreceptor density and distribution using radioactive ligands (e.g., ^{11}C-N-methylspiperone) (Bachneff, 1991; Bench, Dolan, Friston, & Frackowiak, 1990; Farde, Wiesel, Halldin, & Sedvall, 1988; Sedvall, Farde, Persson, & Wiesel, 1986; Wong, Wagner, Tune, & Dannals, 1986). ^{18}F-fluorodeoxyglucose allows examination of both global activity and regional cerebral metabolic rate for glucose (rCMRGlc) (Cohen & Nordahl, 1988). While not strictly an imaging technique, the ^{133}Xenon (^{133}Xe) inhalation technique is also used to measure rCBF. Issues of clinical state versus individual traits confound interpretations of functional imaging data, although most studies attempt to control for medication status or cognitive activity during the scan. Phelps, Kuhl, & Mazziotta (1981) first demonstrated the importance of sensory state (e.g., eyes open or closed, etc.) upon the regional cerebral metabolic rate. Physiological or cognitive probes which activate central nervous system (CNS) function during the imaging procedure are used to assess putative structure-function relationships. Cognitive tasks during PET studies have been shown to reduce within-group variance (Duara *et al.*, 1987; Kessler, Herholz, Grond, & Heiss, 1991) and accentuate group differences (Cohen *et al.*, 1987; Volkow *et al.*, 1987; Buchsbaum *et al.*, 1990) between control and psychiatric patient populations. For example, healthy controls demonstrate increased rCBF to the prefrontal cortex when performing tasks which usually tap this area, whereas schizophrenic patients fail to show the expected increase in rCBF in the prefrontal cortex when performing the same cognitive tasks (Berman, Doran, Pickar, & Weinberger, 1993).

A particular shortcoming of functional imaging is that the boundaries of the metabolic signal are undefined. Consequently, in order to establish anatomical boundaries of activity, such studies require quantitative structural imaging, preferably MRI, which can be superimposed onto the metabolic maps. A further problem is that most PET and SPECT investigations are performed on small numbers of patients and only on patients who can comply with the procedure, thus affecting how well results from one study can be generalized. This bias in subject selection is less of a problem for structural imaging studies, although still relevant.

Methodological Considerations in Psychiatric Neuroimaging Research

Investigators using neuroimaging methods are increasingly aware of the need for research methods that minimize the influence of extraneous variables on the designated outcome measures, namely brain morphology and regional function. In particular, interpretations of the data must take into account the range and nature of variability seen in healthy control comparison populations. Such issues involve accounting for head size and age-related variation in CSF and tissue volumes and the choice of comparison groups. These issues will be discussed in terms of structural imaging results, although the principles also apply to functional imaging.

Normal Variation in Head Size

Humans vary widely in head size even when final somatic size is achieved (Dekaban & Sadowsky, 1978). In general, larger people have larger heads, and men, on average being larger than women, have commensurately larger heads than women (Dekaban & Sadowsky, 1978). Thus, overall head size, which is driven by growth in brain size and highly correlated with adult brain size (Dorst, 1971; Davis & Wright, 1977), may be a source of unwanted variance to be taken into account in estimates of regional brain or tissue compartment size. The brain/head size correction is complicated by the different relationships head or brain size has with CSF, gray matter, and white matter volumes (Arndt, Cohen, Alliger, Swayze, & Andreasen, 1991; Mathalon, Sullivan, Rawles, & Pfefferbaum, 1993). In healthy community volunteers who served as control subjects in our studies (Pfefferbaum *et al.*, 1992; Pfefferbaum, Mathalon, *et al.*, 1994; Pfefferbaum, Sullivan, *et al.*, 1993), head size (i.e., intracranial volume) was particularly related to white matter volume, less so to gray matter volume, and much less so to lateral ventricular volume. To remove the effect of head size, the implication is that each relationship must be adjusted independently for effects of intracranial volume. The approach we have taken to account statistically for these relationships between head size and brain structures is regression analysis. Other statistical approaches, such as proportions and ratios, have been used, but they incompletely remove the correlation head size normally has with measures of brain volume (Mathalon *et al.*, 1993). The estimated volume for each of the three tissue types (CSF, gray matter, and white matter) for each subject is regressed on that subject's intracranial volume and divided by the standard deviation in head size for the control group. The residual values for each of the three regressions represent the volume expected for a given head size and these values become the unit of further analysis.

Normal Variation with Aging

Cross-sectional neuroimaging studies spanning five decades in healthy community volunteers have consistently demonstrated that aging exerts a dramatic and systematic effect on brain volumes (Zatz, Jernigan, & Ahumada, 1982; Pfefferbaum, Zatz, & Jernigan, 1986; Pfefferbaum, Mathalon, *et al.*, 1994; Jernigan, Archibald, *et al.*, 1991; Lim, Zipursky, Watts, & Pfefferbaum, 1992; Murphy, Decarli, Schapiro, Rapoport, & Horwitz, 1992; Coffey *et al.*, 1992). CSF volume increases complementary to a decrease in cortical gray matter volume, while white matter volume appears fairly constant at least up to the seventh decade of life (Pfefferbaum *et al.*, 1992; Pfefferbaum, Mathalon, *et al.*, 1994) (Figure 1, left panel). In addition, MRI signal hyperintensities, typically in the periventricular regions and white matter, occur with greater frequency in old age (Jernigan, Archibald, *et al.*, 1991). A further consequence of aging is increased variability of volumes, or heteroscedacity, which provides an additional

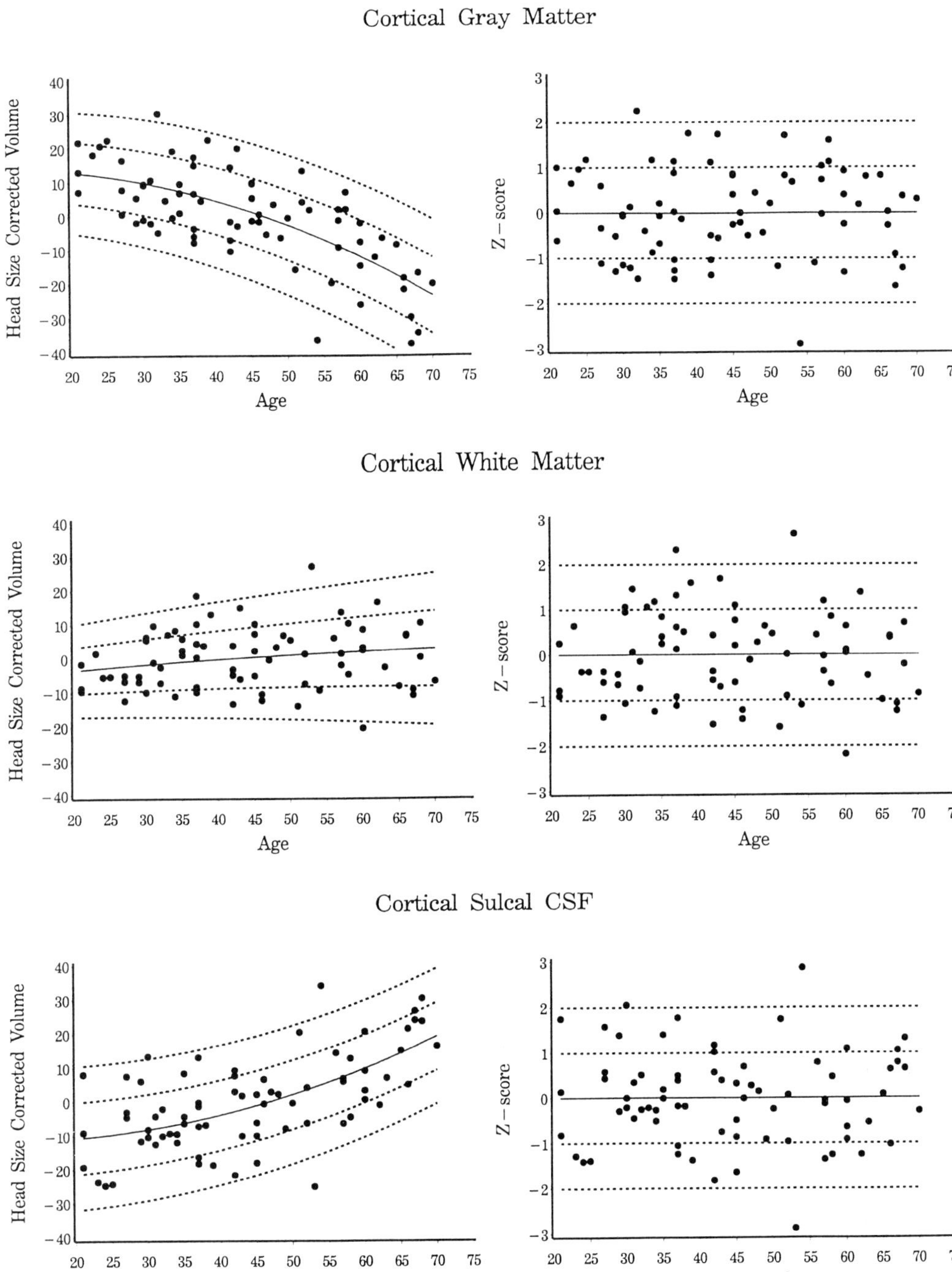

Figure 1. Left: Plots of MRI measures of cortical CSF, gray matter, and white matter versus age in 75 healthy controls spanning the adult age range (21–70 years) after adjusting for head size (head size residual scores) (Pfefferbaum *et al.*, 1992; Pfefferbaum, Mathalm, *et al.*, 1994). Note that cortical CSF volumes increase with age, presumably at the expense of gray matter volumes which decrease with aging; by contrast, white matter volumes do not vary with age over this age range. Right: The same MRI measures, plotted as *z*-scores against age, after adjusting for both head size and age. The variance associated with age for the cortical CSF, gray matter, and white matter has been removed, as suggested by the flat regression lines. By definition, the mean of these *z*-scores is zero, and the standard deviation is 1, and about 95% of the observations fall within 2 SD.

challenge when statistically accounting for age-related effects on brain volumes. Thus, age correction approaches are especially relevant to studies on affective disorders, which tend to have their initial onset when patients are older.

To address these effects of aging, we have again applied regression analysis and a weighted quadratic solution, which deals with the heteroscedacity by permitting values more distant from the mean to exert less influence than those closer to the mean (Glejser, 1969). The head size residualized scores, as derived above, are used in the second regression involving age. The result of this two-step regression, performed on the control data, is a *z*-score for each individual subject for each tissue compartment measured. By definition, the expected mean of the control subjects as a group, regardless of age, is 0, the standard deviation is 1, and about 95% of all control subjects fall within 2 standard deviations (Figure 1, right panel). These *z*-scores then provide normative values for performing head size and age corrections on neuroimaging values obtained from individual patients. The resulting values are estimates of a patient's brain or regional brain volume *for a given head size and a given age.* Thus, the *z*-scores are the final unit of analysis used to determine whether there are quantitative morphological differences in the brains of neuropsychiatric patients, after the expected normal variation in head size and age is removed. Since metabolic rate and receptor density in functional neuroimaging studies are presumably based on the volumes of functioning gray matter, the effects of head size and age might be similarly accounted for in functional imaging data analysis. Alternatively, the absolute gray matter volume can be determined and the functional data can be expressed in terms of cm^3 of tissue.

Comparison Groups

A common approach to studying psychiatric disorders is to compare the patient group with a medically and psychiatrically healthy control group representing the nonaffected population. This method is costly because it requires recruiting volunteers to be screened for the presence of medical and psychiatric illnesses and to undergo brain scanning. An alternative but controversial approach has been to use image data collected from groups of "medical controls," that is, patients who have undergone neuroimaging scans but who had negative clinical readings and no evidence of clinical CNS dysfunction. According to some investigators, the use of "medical controls" method may be susceptible to producing false group differences (Smith & Iacono, 1986). Others contend that similar results are evident regardless of control group deviation (Raz, Raz, & Bigler, 1988). Clearly, the use of healthy nonpatient groups as comparison groups in psychiatric research is a conservative practice.

Neurological conditions are another source of comparison for psychiatric disorders. In particular, psychiatric illnesses occurring in the context of known CNS disturbances serve as models for understanding relationships between psychiatric phenomena and identified brain abnormalities. In these conditions, structure-function relationships may be more apparent because the neuroimaging and clinical abnormalities are clearly linked (e.g., psychomotor automatisms, medial temporal epileptiform discharges, medial temporal sclerosis on MRI, and interictal medial temporal hypometabolism on PET in complex partial epilepsy) (Ryvlin *et al.,* 1992). In contrast to idiopathic psychiatric conditions in which the underlying pathophysiology is unknown, the occurrence of similar psychiatric syndromes, in terms of clinical phenomenology, in patients with neurological disorders suggests that the underlying neuropathology is related to the psychiatric disturbance. For example, depressive illnesses commonly occur in basal ganglia disorders such as Parkinson's and Huntington's disease (Cummings, 1992; Folstein, Folstein, & McHugh, 1979), and temporal lobe pathology in epilepsy can be associated with schizophrenia-like syndromes (Stevens, 1991). In this light, the circumscribed structural or functional brain abnormalities in neuropsychiatric conditions may provide clues as to the nature, locus, and extent of pathophysiology in idiopathic psychiatric disorders.

Schizophrenia

Neuroimaging studies of schizophrenia have generally been guided by speculations regarding the neural substrates of the clinical and neuropsychological manifestations of the disease as well as the putative locus of action of neuroleptic therapies. For example, deficits in complex problem solving, social functioning, and motivation possibly reflect frontal lobe dysfunction (Weinberger, 1987; Gold, Goldberg, & Braff, 1992; Sullivan, Mathalon, *et al.*, 1993). Dysfunction of the basal ganglia and associated cortical projections is suggested by certain motor, memory, and attentional difficulties (Manschreck, 1986). Hallucinations, delusions, and thought disorder implicate temporal-limbic structures (Torrey & Peterson, 1974; Trimble, 1991). To a lesser degree, dysfunction in other brain regions is also apparent. Impairments in selective attention suggest abnormalities in the parietal lobes (Mesulam & Geschwind, 1983), as well as the frontal lobes and pontine nuclei such as the locus ceruleus (Posner, Early, Reiman, Pardo, & Dhawan, 1988). The occurrence of cerebellar-like speech and motor disturbances in schizophrenia (Taylor, 1991) and the role of neural connections from the cerebellar vermis to limbic areas in modulation of emotional and autonomic activity further implicate cerebellar pathology. To the extent that disruption of selective brain regions or networks underlie these clinical phenomena, these brain regions are likely targets of *in vivo* imaging investigations.

Neuroimaging Findings in Schizophrenia

Ventricular and Cortical Sulcal Dilatation. Since the first published CT study on schizophrenia (Johnstone *et al.*, 1976), the most consistent *in vivo* evidence for neuropathology in schizophrenia has been enlargement of the third and lateral ventricles and cerebral sulci observed with CT and MRI (reviewed by Shelton & Weinberger, 1986; Pfefferbaum & Zipursky, 1991; Cleghorn, Zipursky, & List, 1991; Pearlson & Marsh, 1993). Differences in subject sampling, control group selection, and morphometric methods (Raz *et al.*, 1988; Lewis, 1990; Iacono *et al.*, 1988) may obscure group differences and appear to account for lack of evidence for ventricular enlargement in schizophrenia in the relatively few negative studies (Jernigan, Zatz, Moses, & Berger, 1982; Smith *et al.*, 1984; Smith, Baumgartner, & Calderon, 1987; Rossi *et al.*, 1990; Johnstone *et al.*, 1989; Young *et al.*, 1991; Shenton *et al.*, 1992; Colombo *et al.*, 1993). In addition, although ventricular enlargement was not evident, some studies reported tissue volume deficits (Shenton *et al.*, 1992; Colombo *et al.*, 1993). The presence of ventriculomegaly in schizophrenic patients as a group provides evidence for wide-reaching, nonselective brain abnormalities and a context in which to place the observation of more focal dysmorphology. For example, evidence for temporal horn enlargement (Zipursky *et al.*, 1994; Bogerts *et al.*, 1990) may not be surprising given dilatation is present in other portions of the ventricular system. However, it is also possible that enlargement of the lateral and third ventricles and the temporal horn is the result of volume abnormalities in more proximal regions, such as the basal ganglia, thalamus, and hippocampal formation, respectively.

The ventriculomegaly of schizophrenia is a group phenomenon and is mild compared to other disorders, such as alcoholism (Pfefferbaum *et al.*, 1992; Pfefferbaum, Sullivan, *et al.*, 1993) and Alzheimer's disease (Pfefferbaum *et al.*, 1990; Sullivan, Shear, *et al.*, 1993). Nonetheless, brain morphometry within the range of control measures does not preclude the possibility of anatomical pathology in a given patient. Evidence for a continuum in the degree of structural abnormality, in the context of normal anatomical variability, is suggested by results from a meta-analysis of the distribution pattern for ventricular-to-brain ratio (VBR) values in CT investigations of schizophrenia and controls (Daniel, Goldberg, Gibbons, & Weinberger, 1991). In each group, anatomical variation in VBR followed a normal distribution with no evidence for subgroups. However, when analyzed together, two groups emerged, and the schizophrenic patients were in the group with larger VBRs. Even when VBR was within the control range for a portion of schizophrenics, it tended to be in the range of upper normal measures. Most CT and MRI results are consistent with this characterization of volume distributions (Pfefferbaum *et al.*, 1988; Zipursky, Lim, Sullivan, Brown, & Pfefferbaum, 1992; Marsh & Weinberger, 1993; Zipursky *et al.*, 1994) as illustrated in Figure 2.

Widening of cortical sulci in schizophrenia has been observed less consistently than ventricular enlargement. Although both types of CSF-filled spaces may be abnormally large in schizophrenia, ventricular size is more readily and reliably measured than sulcal size, which is particularly difficult to achieve with

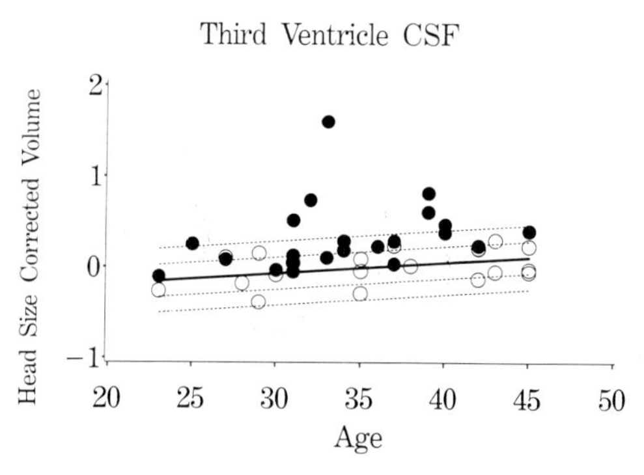

Figure 2. Plot of third ventricle volume, after adjusting for head size effects, versus age in 22 male schizophrenic patients (solid circles) and 20 healthy control males (open squares) (Zipursky *et al.*, 1993). Mean third ventricle volume in the patients was significantly increased relative to the controls ($p < .002$). While third ventricle volumes of controls and patients overlap to some extent and the majority of patients have values within the control range, the distribution of values suggests an overall upward shift in third ventricular volumes in schizophrenic patients compared to controls. This observation emphasizes the subtle nature of structural brain volume abnormalities in schizophrenia.

CT (Pfefferbaum, Lim, Rosenbloom, & Zipursky, 1990; Pfefferbaum, Sullivan, *et al.*, 1993; Raz *et al.*, 1987). MRI investigations examining the relationship between parenchymal abnormalities and ventricular or sulcal enlargement in schizophrenia are inconclusive. Although the cumulative data are insufficient to ascribe either a primary focal or diffuse etiopathogenesis or a common etiology for ventricular and sulcal enlargement, their combined presence in schizophrenia suggests a widespread pathological process affecting a variety of brain regions.

Widespread Tissue Volume Deficits. Reports from MRI studies of schizophrenia showing widespread cortical gray matter volume deficits without apparent involvement of white matter volumes (Harvey *et al.*, 1993; Zipursky *et al.*, 1992, 1994) complement neuroimaging evidence for widespread ventricular and sulcal enlargement (see Figure 3). These results are also consistent with some postmortem investigations which show overall reductions in subcortical and cortical gray matter, but not white matter (Pakkenberg, 1987), or abnormal prefrontal neuronal density (Selemon, Rajkowska, & Goldman-Rakic, 1995) in schizophrenics compared to age- and sex-matched controls. Furthermore, as reviewed below, imaging studies examining discrete brain regions have failed to identify one particular brain region which is specific to the disorder, consistent with the theory that clinical characteristics of schizophrenia indicate cerebral dysfunction in many brain regions. While most studies examining focal brain regions and tissue types have shown volume deficits in gray matter, and not white matter, in either the temporal or frontal cortex (Suddath *et al.*, 1989; Suddath, Christison, Fuller Torrey, Casanova, & Weinberger, 1990; Zipursky *et al.*, 1994), bilateral deficits in prefrontal white matter, and not gray matter, volumes are also described using MRI (Breier *et al.*, 1992). Widespread white matter pathology is potentially implicated by reports of corpus callosum abnormalities in MRI studies (reviewed by Pearlson & Marsh, 1993), although the findings are inconsistent. Nonetheless, postmortem studies also report increased numbers of vertical axons in the cingulate cortex (Benes, Sorensen, Vincent, Bird, & Sathi, 1992), indicating that white matter abnormalities may be present in schizophrenia. Thus, while there is considerable evidence for widespread or diverse cerebral dysmorphology, the underlying etiopathogenesis, in terms of the extent to which white matter or gray matter is affected, remains unsettled.

Regional Abnormalities. Many *in vivo* imaging studies have focused on abnormalities in specific anatomical regions or structures. Tissue volume deficits and metabolic differences in schizophrenic patients compared to healthy controls are most consistently reported in the temporal, frontal, and subcortical regions, even when other brain regions are also examined. However, subtle pathology has been detected in other functional regions or structures, including the cerebellum, corpus callosum, and brain stem (reviewed by Pearlson & Marsh, 1993), suggesting that the underlying pathophysiology in schizophrenia is not restricted to one area. This idea is further supported by the tendency for tissue volume deficits to occur in the context of ventricular and sulcal enlargement. In schizophrenia, therefore, different combinations of functional and structural abnormalities, as evident in neuroimaging investigations, may account for the heterogeneity of clinical presentations (Liddle, 1987).

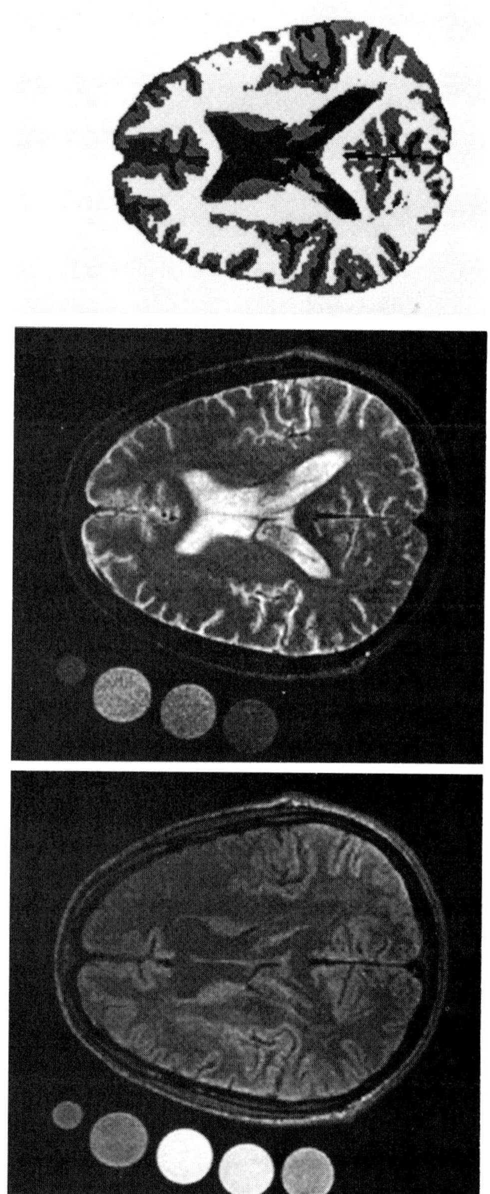

Figure 3. Axial spin echo MRI images (early echo [20 ms], left, late echo [80 ms], middle) in a 25-year-old schizophrenic man and the image segmentation of the raw images into CSF, gray matter, and white matter compartments (right).

Frontal Lobe Abnormalities. Frontal lobe dysfunction is particularly implicated in schizophrenia on the basis of features of the deficit state and impairments on neuropsychological testing (Goldberg *et al.*, 1991). While the neuropsychological deficits of schizophrenia are not limited to those subserved by the prefrontal cortex and its interconnections (Chelune, Heaton, Lehman, & Robinson, 1979; Kolb & Whishaw, 1983; Sullivan, Shear, Zipursky, Sagar, & Pfefferbaum, 1993), impairments in problem solving and abstract reasoning, for example, as measured by performance on the Wisconsin Card Sorting Test (WCST), are particularly disabling clinical features of schizophrenia and are thought to reflect prefrontal dysfunction (Milner, 1963; Sullivan, Mathalon, *et al.*, 1993). A number of functional imaging studies have thus sought to correlate physiological function of different brain regions with performance on cognitive tasks that specifically involve prefrontal cortex as well as tasks that involve other areas.

Assuming that neuroanatomical abnormalities and clinical symptoms reflect cerebral dysfunction in many and diverse regions, results from functional studies might be expected to parallel evidence from structural imaging and neuropsychological studies for widespread dysfunction. However, the majority of functional imaging studies suggest there is selective yet subtle frontal dysfunction in schizophrenia. While the initial functional imaging studies on schizophrenia using the nitrous oxide technique (Kety *et al.*, 1948) found normal CBF values in chronic schizophrenics, subsequent studies using ^{133}Xe inhalation techniques demonstrated diminished frontal blood flow in older schizophrenic patients compared to controls (Ingvar & Franzen, 1974), findings which have been replicated by others (reviewed by Berman & Weinberger, 1991). Following the ^{133}Xe inhalation studies, PET studies measuring rCBF in chronic, severely ill schizophrenics with ^{18}FDG and ^{11}C-glucose PET similarly demonstrated relative decreases in prefrontal blood metabolism, that is, "hypofrontality" (Buchsbaum *et al.*, 1982, 1992; Brodie *et al.*, 1984; DeLisi, Buchsbaum, *et al.*, 1985; Wolkin *et al.*, 1985; Cohen *et al.*, 1987; Gur, Resnick, Alavi, *et al.*, 1987; Kishimoto, Kuwahara, & Ohno, 1987; Volkow *et al.*, 1987; Clark *et al.*, 1989), although a few PET studies have failed to show hypofrontality (Sheppard *et al.*, 1983; Jernigan, Sargent, Pfefferbaum, Kusubov, & Stahl, 1985; Kling, Metter, Riege, & Kuhl, 1986; Volkow *et al.*, 1986) or reported hyperfrontality (Szechtman, Nahmias, Garnett, & Firnau, 1988; Cleghorn *et al.*, 1989).

Cognitive activation studies indicate that physiological function of the prefrontal region is not uniformly compromised in schizophrenia, but only when called upon during cognitive tasks which require specific function of this area. Furthermore, relative hypofrontality in schizophrenia appears inherently related to the schizophrenic illness itself and not necessarily to individual attributes, genetic predisposition, or a subgroup of schizophrenic patients. Using ^{133}Xe inhalation rCBF and ^{15}O PET techniques, prefrontal blood flow in healthy controls significantly increased during performance of the WCST, relative to that during sensorimotor control tasks (Weinberger, Berman, & Illowsky, 1988; Berman *et al.*, 1991). However, schizophrenics showed relative hypoperfusion of the prefrontal cortex during performance of the WCST in both neuroleptic-medicated (Berman, Zec, & Weinberger, 1986) and non-neuroleptic-treated (Berman *et al.*, 1986; Weinberger *et al.*, 1988) conditions. Such "hypofrontality" was not demonstrated consistently in schizophrenic patients during cognitive tasks which activated either nonspecific brain regions or specific regions other than the prefrontal cortex (Berman, Illowsky, & Weinberger, 1988). Similar results are seen in comparisons of monozygotic twins discordant and concordant for schizophrenia using the ^{133}Xe inhalation technique (Berman, Torrey, Daniel, & Weinberger, 1992). Even when the affected twin's WCST performance was better than its well sibling, the degree of prefrontal activation during the WCST was lower in each affected twin relative to its healthy co-twin. Furthermore, comparisons of monozygotic twins concordant for schizophrenia showed no significant relationships between lifetime neuroleptic dose and rCBF during resting and cognitive activation tasks.

Evidence for reduced frontal activation in schizophrenia has been further replicated by other groups using SPECT. Lewis (1991) demonstrated a significant relationship between relatively low perfusion in the left dorsolateral prefrontal cortex using ^{99m}Tc-HMPAO SPECT during a word fluency test in 25 schizophrenic men compared with 25 healthy age-matched male controls. Similarly, frontal hypoperfusion was observed with ^{133}Xe SPECT in response to the Tower of London puzzle, a problem-solving task, in both medication-free and neuroleptic-naive schizophrenic patients compared to controls (Andreasen *et al.*, 1992). However, hypoperfusion was reported in the left mesial frontal cortex, suggesting there may be areas of frontal dysfunction in addition to the prefrontal cortex.

Anatomical investigations of frontal regions do not correspond directly to the relative hypofrontality demonstrated in functional imaging studies on schizophrenia. Tissue deficits in frontal regions are suggested by reports of qualitative prefrontal sulcal prominence on CT (Shelton *et al.*, 1988), whereas there were no differences in occipital-parietal sulcal markings between controls and schizophrenics. MRI investigations have shown prefrontal and frontal gray matter deficits along with similar deficits throughout the neocortex (Zipursky *et al.*, 1992), prefrontal white matter deficits along with temporal lobe gray matter deficits (Breier *et al.*, 1992), and reduced frontal (Andreasen *et al.*, 1986; De Myer *et al.*, 1988) and prefrontal area (Raine *et al.*, 1992) in schizophrenics compared to healthy controls. Despite evidence for tissue volume deficits in other brain regions, many controlled investigations have not found significant differences in schizophrenia for frontal gray or white matter volumes (Kelsoe, Cadet, Pickar, & Weinberger, 1988; Suddath *et al.*, 1989, 1990; DeLisi, Hoff, *et al.*, 1991), frontal area (Andreasen, Ehrhardt, *et al.*, 1990; Nasrallah, Schwarzkopf, Olson, & Coffman, 1990; Rossi *et al.*, 1990; Uematsu & Kaiya, 1989), linear frontal lobe measures (Smith *et al.*, 1984, 1987), prefrontal lobular shape (Casanova *et al.*, 1990), or frontal white matter T1 values (Besson, Corrigan, Cherryman, & Smith, 1987). However, as suggested by activation studies, the frontal abnormality may be more apparent when called upon to function, thus implicating a role for dysfunctional networks between the frontal lobes and other regions. Morphometric methods may also lack the sensitivity for detecting subtle frontal tissue volume differences; some studies include large portions of the frontal lobe region in their measure, have not corrected for artifacts which affect tissue segmentation, or may have used MRI sequences which limited tissue resolution.

Temporal Neocortical Abnormalities. While many studies on schizophrenia have focused on the lack of physiological activation in frontal regions, similar deficits are revealed in the temporal lobes when the activation task or sampling technique is designed to probe for dysfunction in the temporal lobes. For example,

Gur, Resnick, Alavi, *et al.* (1987) reported greater left temporal lobe glucose metabolism in schizophrenics, a finding also reported by DeLisi *et al.* (1989). Cleghorn, Szechtman, *et al.* (1991) reported relative hypometabolism of ^{18}FDG in the posterior superior temporal gyrus area in a group of neuroleptic naive subjects experiencing auditory hallucinations during the scan. In contrast, Tamminga *et al.* (1992) demonstrated decreased ^{18}FDG metabolism in the hippocampus and anterior cingulate. Using high-resolution ^{18}FDG-PET, Nordahl *et al.* (1993) reported decreases in rCMRGlc in the posterior middle temporal lobe gyrus bilaterally in 18 patients with subtypes of paranoid and undifferentiated schizophrenia compared to 11 healthy controls. The 8 paranoid subtype patients, however, showed significant increases in left anterior temporal cortical metabolism compared to both the healthy controls and the 10 patients with undifferentiated schizophrenia and significant decreases in anterior cingulate rCMRGlc compared to the patients with undifferentiated schizophrenia (see Figure 4). These data suggest that clinical phenomenology may vary according to the extent of dysfunction in different brain regions.

Many MRI reports on the temporal-limbic abnormalities correspond to morphometric or cytoarchitectural pathology demonstrated in postmortem studies (reviewed by Hyde, Casanova, Kleinman, & Weinberger, 1991). Like the evidence for frontal lobe pathology, many of the temporal lobe abnormalities in schizophrenia occur in the context of pathology elsewhere, namely ventricular or sulcal enlargement (see Figure 5 for depiction of temporal-limbic structures). Anatomical pathology detected includes deficits in temporal lobe tissue bilaterally (Dauphinais *et al.*, 1990), on the right (Suddath *et al.*, 1989; Barta, Pearlson, Powers, Richards, & Tune, 1990), on the left (Rossi *et al.*, 1990; Suddath *et al.*, 1990), or bilateral temporal lobe gray matter (Suddath *et al.*, 1989; Marsh *et*

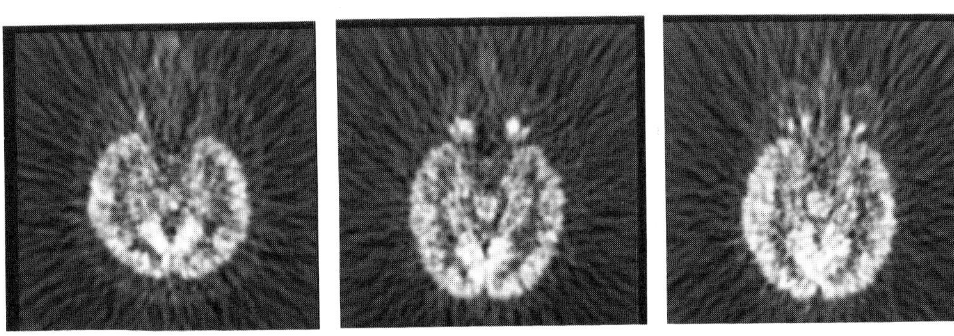

Figure 4. These axial images, acquired with ^{18}FDG-PET, are from two men with schizophrenia, one paranoid subtype (left) and one undifferentiated subtype (right), and from one 60-year-old healthy male control (center) and are representative of findings from a larger study (Nordahl *et al.*, in press). Using a high-resolution (2.6 mm full width half maximum) PET scanner, acquired images are parallel to the long axis of the temporal lobe and sample the middle temporal gyrus predominantly. Subjects were imaged in a dimly lit room while staring at the same visual stimulus (a nondescript blue screen) and in the presence of background fan and other machine noise. Note that the left hemisphere appears on the right and each subject shows a different pattern of glucose metabolism. Primary visual cortex activity is increased symmetrically in all subjects. Both schizophrenic patients show bilateral decreases in posterior temporal lobe rCMRGlc relative to the control. The paranoid subtype shows increased left anterior temporal rCMRGlc relative to the patient with undifferentiated schizophrenia and the control. (Images provided by Thomas E. Nordahl, M.D.)

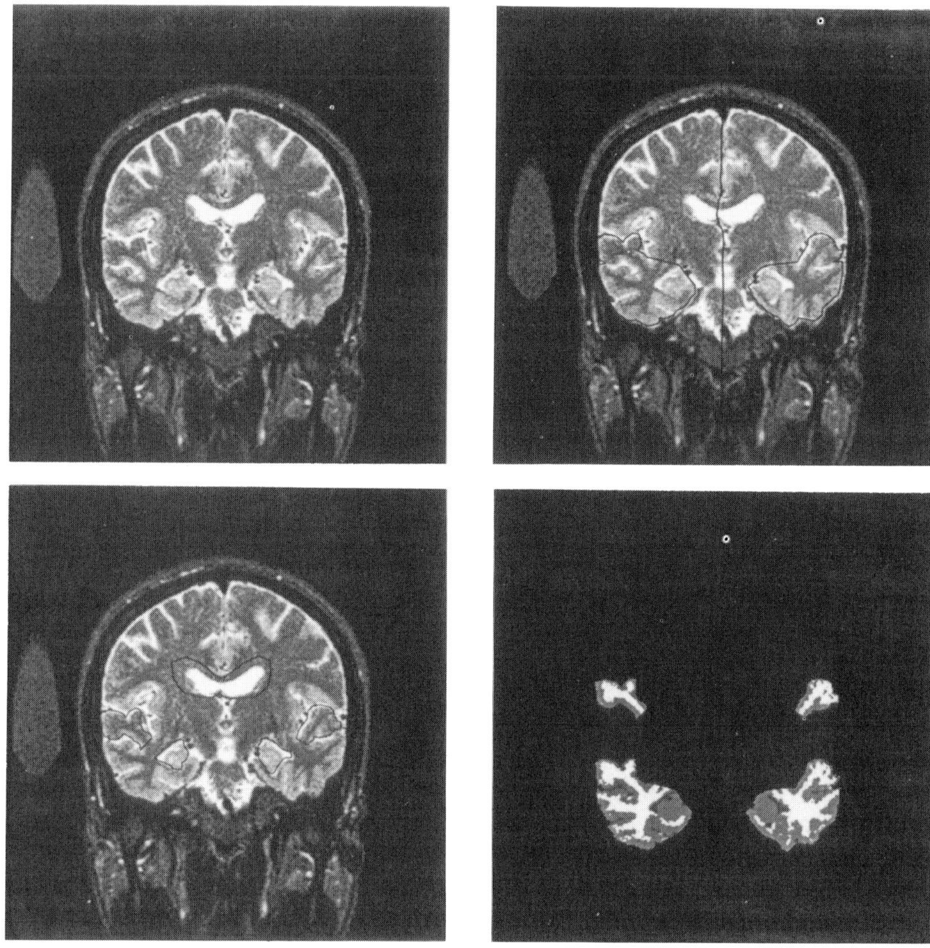

Figure 5. Coronal spin echo MRI image (from the schizophrenia patient shown in Figure 3) demonstrating temporal-limbic structures. The late echo image is in the upper left. Regions of interest (temporal lobe, superior temporal gyrus, lateral ventricles, hippocampus, and temporal horn) are outlined manually using computerized image analysis programs in the upper right and lower left. The results of image segmentation into CSF, gray matter, and white matter compartments are in the lower right.

al., 1996; Zipursky *et al.,* 1994). Specific neocortical volume reductions are also reported for superior temporal gyrus tissue or gray matter bilaterally (Barta *et al.,* 1990; Marsh *et al.,* 1996; Zipursky *et al.,* 1994) and for left, but not right, superior temporal gyrus gray matter (Shenton *et al.,* 1992). Others have failed to detect significant differences in temporal lobe size (Kelsoe *et al.,* 1988; Johnstone *et al.,* 1989; Young *et al.,* 1991; DeLisi, Hoff, Kushner, Calev, & Stritzke, 1992; DeLisi, Stritzke, *et al.,* 1992; Colombo *et al.,* 1993), although differences in morphometric methods may partially account for the failure to replicate results. In several of the negative studies (Kelsoe *et al.,* 1988; Johnstone *et al.,* 1989; Young *et al.,* 1991), measurements were derived from a single slice or thick (1 cm) slices were used.

Investigations of the superior temporal plane in schizophrenia have been prompted by theories implicating temporal lobe involvement, especially of the left hemisphere and superior temporal gyrus, as important to the pathogenesis of the illness (Flor-Henry, 1969, 1989; Crow, 1993). One approach for investigating superior temporal gyrus morphometry is examination of the lateral sulcus and planum temporale, an area comprising the posterior portion of the superior temporal gyrus, which is important, in part, for language processing. The planum temporale is also known for its hemispheric asymmetry, with the left planum temporale normally being larger than the right. There is evidence for such in MRI studies (Steinmetz *et al.*, 1989). Theories invoking predominant left hemisphere pathology in schizophrenia suggest that deviations in the normal pattern of asymmetry would be especially evident in the superior temporal plane if the pathogenetic process is selective to the left hemisphere. It remains unclear, however, how widespread cortical gray matter abnormalities relate to theories invoking selective brain regions in schizophrenia. Indeed, many of the results from neuroimaging studies reviewed above indicate that CNS abnormalities in schizophrenia are often bilateral or may even involve the right, but not left, hemisphere (reviewed by Cutting, 1992).

MRI studies show some evidence for abnormal neuroanatomical asymmetry in schizophrenia and there is similar evidence from postmortem studies, especially among male patients (Falkai *et al.*, 1992). Using coronal MRI images, Rossi *et al.* (1992) found a normal pattern of asymmetry in 12 healthy controls but not in 20 schizophrenic patients. The planum temporale area was smaller on the left and larger on the right in the patients relative to the controls, although neither difference was significant. Gender, aging, and head size effects were not examined. A separate group (A.L. Hoff, Riordan, *et al.*, 1992), after covarying structural measures for age and education, found that women with schizophreniform illnesses (n = 13) lacked the normal pattern of asymmetry for lateral sulcus measures from sagittal MRI images, whereas men with schizophreniform illness (n = 17) and men (n = 21) and women (n = 20) controls showed the normal pattern. While the evidence for left temporal lobe involvement in schizophrenia is compelling and certain clinical phenomena would seem to invoke heteromodal association areas (e.g., temporal neocortex), these results must still be considered in the context of evidence for abnormalities elsewhere, including the right hemisphere. In other words, even though dysmorphology of the planum temporale may have certain clinical implications, it is not clear that the planum temporale is preferentially affected over other brain regions. The MRI data on planum temporale measures is also difficult to interpret because of the differing morphometric methods as well as the considerable morphological variation between the right and left hemispheres among individuals. While healthy men tend to show the left greater than right pattern of planum temporale asymmetry, healthy women show this pattern less consistently (Witelson & Kigar, 1992). Further, with *in vivo* studies, there is no way to confirm whether the underlying cytoarchitecture is representative of the planum temporale or an adjacent region. Finally, it is difficult to draw firm conclusions regarding brain asymmetry in pathological states when relatively small numbers of controls and patients have been studied.

Medial Temporal Lobe Structures. Involvement of the limbic system in processing sensory information and affective regulation suggests that the medial temporal region and associated limbic structures may have a different functional role in schizophrenia compared to neocortical association areas. Prior to the advent of MRI, several postmortem studies demonstrated limbic dysmorphology such as reductions in the size of the hippocampus, amygdala, and parahippocampal gyrus; decreases in neuronal number or cell size in the hippocampus and entorhinal cortex; and cytoarchitectural abnormalities in the hippocampus, entorhinal cortex, and cingulate gyrus (reviewed by Bogerts, 1993). However, whether medial temporal lobe volume abnormalities are localized to the left hemisphere, specific structures, or specific portions or regions of the medial temporal lobe and the extent to which they are evident in younger living patients with schizophrenia is controversial. Several MRI studies describe tissue volume deficits of both the right and left anterior hippocampus (Suddath *et al.*, 1990) and amygdala-hippocampus complex (DeLisi, Dauphinais, & Gershon, 1988; Breier *et al.*, 1992; Buchanan *et al.*, 1993; Marsh, Suddath, Higgins, & Weinberger, 1994), whereas others found significant reductions only in the left hemisphere for the amygdala (Barta *et al.*, 1990) and left parahippocampal gyrus (Shenton *et al.*, 1992; Kawasaki *et al.*, 1993). When the mammillary bodies have been used to define anterior and posterior portions of the amygdala-hippocampal complex, some have found amygdala-hippocampus volume deficits limited to anterior sections on the left only (Shenton *et al.*, 1992), whereas others found significant volume reductions in posterior sections bilaterally in both first-episode and chronic schizophrenic patients (Bogerts *et al.*, 1990; 1993). In monozygotic twins discordant for schizophrenia, bilateral tissue volume deficits of the hippocampus in affected twins were present only in single MRI sections of the hippocampal head (Suddath *et al.*, 1990). Several studies failed to detect volume differences in hippocampal volume (DeLisi, Hoff, *et al.*, 1991a; Swayze, Andreasen, Alliger, Yuh, & Ehrhardt, 1992; Colombo *et al.*, 1993; Zipursky *et al.*, 1994; Marsh *et al.*, 1996), amygdala area (Swayze *et al.*, 1992), or total amygdala-hippocampus volumes (Kelsoe *et al.*, 1988) between schizophrenics and controls. Again, methodological differences, such as use of a single MRI slice, different slice thicknesses, or different morphometric techniques and image analysis software may account for discrepant results. For example, much of the evidence for hippocampal volume deficits in schizophrenia from neuroimaging studies is based on measures which include portions of both the amygdala and the hippocampus, whereas measures of the hippocampus alone tend not to show significant differences between patients and controls. In two separate MRI studies in our laboratory, despite the occurrence of widespread reductions in neocortical gray matter, there were no differences in hippocampal volumes between healthy male controls and male patients with schizophrenia (Zipursky *et al.*, 1994) or a more severely ill cohort of male patients with childhood or young-adult onset schizophrenia residing in a state hospital (Marsh *et al.*, 1996).

Subcortical Abnormalities. Functional imaging studies of the basal ganglia are characterized mainly by dopamine receptor density. In animal studies, several investigators have shown that most neuroleptic medications interfere with dopamine neurotransmission by blocking D2 dopamine receptors (Creese, Burt, & Snyder, 1976; Seeman, Lee, Chau-Wong, & Wong, 1976). A central controversy is whether D2 neuroreceptor density is abnormal in schizophrenic patients. While there is evidence for increased D2 receptor density in basal ganglia in postmortem investigations on schizophrenia (Seeman, Ulpian, & Bergeron, 1984) *in vivo* neuroimaging studies provide conflicting results. A PET study comparing neuroleptic-naive schizophrenic patients and age-matched healthy controls (Wong *et al.*, 1986) revealed a two- to threefold increase in D2 receptor density in the caudate and putamen using ^{11}C-N-methylspiperone (^{11}C-N-MSP), whereas another investigation (Farde, Hall, Ehrin, & Sedvall, 1986; Farde *et al.*, 1987) showed no group differences using ^{11}C-raclopride or ^{11}C-N-MSP (Farde *et al.*, 1990). A number of methodological differences, however, may account for these discrepant results (Andreasen *et al.*, 1988; Sedvall, 1992). By contrast, PET studies examining neuroleptic binding to D2 receptors and treatment efficacy, time course of action, and side effects have shown greater agreement. In ^{11}C-raclopride PET studies, all neuroleptic medications have been shown to occupy D2 basal ganglia receptors, an effect which is both dose-dependent and related to the neuroleptic half-life (Farde *et al.*, 1988). Neuroleptic-related extrapyramidal symptoms are also found to correlate with the degree of D2 receptor occupancy, ranging from 70–89% in patients with greater extrapyramidal side effects. In contrast, patients receiving clozapine, an atypical neuroleptic not associated with extrapyramidal side effects, had D2 receptor occupancy of 38–63%. While antipsychotic medications have also been shown to block both D1 and D2, clozapine has equal D1 and D2 blockade (Farde, Wiesel, A.L., & Sedvall 1989) and blocks a unique subset, the D4 receptor or the clozapine receptor (Seeman, 1992). PET neuroreceptor studies are thus useful for the study of potential new antipsychotic compounds.

Basal ganglia dysfunction has also been reported in PET studies on cerebral metabolism and blood flow. Using ^{15}O PET to compare 12 neuroleptic-free schizophrenic patients and matched controls, there are reports of lower rCBF and $rCMRO_2$ values in the basal ganglia, without concomitant hypofrontality as observed by others (Sheppard *et al.*, 1983). Gur *et al.* (Gur, Resnick, Alavi, *et al.*, 1987; Gur, Resnick, Gur, *et al.*, 1987) found a subcortical-cortical-cortical gradient in rCMRGlc as well as hemispheric asymmetry in schizophrenics both on and off neuroleptics, whereas others reported increased rCBF in the left globus pallidus (Early, Reiman, Raichle, & Spitznagel, 1989) or reduced rCMRGlc in the basal ganglia (Buchsbaum *et al.*, 1992) in never-medicated patients with schizophrenia and increases in basal ganglia rCMRGlc after neuroleptic treatment (DeLisi, Holcomb, *et al.*, 1985).

MRI has improved the ability to evaluate subcortical structures, especially the basal ganglia. Like postmortem evidence (reviewed by Bogerts, 1993), MRI studies on the basal ganglia and other subcortical structures are somewhat inconsistent. Compared to healthy controls, schizophrenic patients have been reported to have increased volumes of the left caudate nucleus by one group (Breier *et al.*, 1992), whereas another group (Jernigan, Zisook, *et al.*, 1991) reported a subtle enlargement of the lenticular nucleus (putamen plus globus pallidus) but not the caudate. Swayze *et al.* (1992) reported significant caudate and putamen enlargement in male patients only. Others have not found significant differences in lenticular size (Kelsoe *et al.*, 1988; DeLisi, Hoff, *et al.*, 1991) in schizophrenic patients compared to controls. DeLisi, Hoff, *et al.* (1991) reported increased caudate nucleus volume in chronic schizophrenic patients compared to patients presenting with first-episode schizophrenia. Such size increases in basal ganglia nuclei are possibly a consequence of medication (Chakos *et al.*, 1993), although the volume increases might be regarded as complementary to reported increases in dopamine receptor sites, an abnormality which also is not clearly related to medication effects or the illness itself. It is also possible that abnormalities in the basal ganglia are more likely to be manifest as physiological dysfunction rather than anatomical pathology. The role of the basal ganglia in neuronal circuitry involving the frontal and temporal-limbic regions has been discussed as possibly important to the pathogenesis and treatment of schizophrenia (Alexander, DeLong, & Strick, 1986; Carlsson, 1988; Buchsbaum, 1990; Robbins, 1990; Breier *et al.*, 1992).

Other Brain Regions. Examinations of cerebellar and corpus callosum morphology in schizophrenia are limited in number, but the evidence for pathology in these regions is also more modest. It is possible that the most pervasive pathophysiological processes in schizophrenia affect temporal-frontal-basal ganglia circuits and gray matter more than white matter in terms of neuroanatomical pathology, but that some patients will show disturbances and pathology reflective of other brain regions. Cerebellar tissue deficits were suggested by evidence of enlargement of the vermal sulci and fourth ventricle in CT studies (reviewed by Shelton & Weinberger, 1986), although MRI investigations have generally not supported these results (reviewed by Pearlson & Marsh, 1993; Rossi, Stratta, Mancini, Decataldo, & Casacchia, 1993). Similarly, MRI studies on corpus callosum size have not consistently shown significant differences between schizophrenic patients and healthy controls (reviewed by Pearlson & Marsh, 1993), although some indicate focal reductions in corpus callosum area (Woodruff, Pearlson, Geer, Barta, & Chilcoat, 1993). Since there appears to be considerable disease- and gender-related variability in corpus callosum morphology, the current morphometric methods may not be sufficiently sensitive to detect subtle group differences.

A few functional imaging studies report abnormalities in the parietal lobe and cerebellum. These include reports of parietal lobe glucose hypometabolism both in neuroleptic-treated patients using ^{11}C-glucose (Kishimoto, Kuwahara, *et al.*, 1987) and in unmedicated schizophrenic patients using ^{18}FDG-PET (Cleghorn *et al.*, 1989). One ^{18}FDG-PET study reported lower rCMRGlc in the cerebellum of chronic schizophrenic patients compared to controls (Volkow *et al.*, 1992).

Relationship of Neuroimaging Findings to Clinical Variables

Relationship to Clinical Symptoms. Heterogeneity of the clinical features of schizophrenia, in the context of the lack of a pathognomic lesion and evidence of widespread pathology, limits our understanding of the implications of evident CNS pathology in schizophrenia. Unlike neurological disorders in which the clinical features tend to be related to a pathognomic lesion, relationships between subtle structural or functional neuroimaging abnormalities in schizophrenia are often not clearly apparent. In an effort to ascertain clinicopathological correlations, one approach has been to focus on relationships between localized dysmorphology and clinical symptoms or treatment responsiveness. In general, cerebral dysmorphology (namely ventricular enlargement) tends to be associated with enduring features of the illness, that is, neuropsychological impairment, negative symptoms, and a less favorable outcome (reviewed by Lewis, 1990). In structural imaging, these relationships are not consistently evident. For example, Friedman, Knutson, Shurell, & Meltzer (1991) reported a correlation between the degree of prefrontal sulcal prominence on CT and poor response to 6 weeks of clozapine treatment, suggesting that neuroleptic responsivity may require the presence of more brain tissue and thus receptor sites. Furthermore, this relationship was not present for the VBR, indicating that the finding may be selective to the prefrontal region. In contrast, patients with deficit and nondeficit schizophrenia showed a counterintuitive pattern on MRI, with smaller prefrontal cortical volumes in the nondeficit group (Buchanan *et al.*, 1993).

A few recent MRI investigations reported significant relationships between volume reductions in left anterior superior temporal gyrus and severity of auditory hallucinations (Barta *et al.*, 1990) and between left posterior superior temporal gyrus gray matter and severity of thought disorder, as measured by the Thought Disorder Index (Shenton *et al.*, 1992). Preliminary evidence (Flaum, Swayze, Gupta, O'Leary, & Andreasen, 1993) of correlations between different brain regions and complexes of negative, positive, and disorganized symptoms suggests that regions other than the superior temporal gyrus also correlate with positive symptoms. In a study of 165 schizophrenia spectrum patients, positive symptom severity predicted enlargement of the lateral ventricle and the temporal horn and decreased superior temporal gyrus volume. Although multiple correlations were performed, among the positive symptoms, only hallucinations predicted superior temporal gyrus volume. There was also a unique relationship between negative symptoms and reduced amygdala-hippocampal volume (Flaum *et al.*, 1993). In contrast, Bogerts *et al.*, (1993) found a significant negative correlation between both right and left amygdala-hippocampal volumes and the severity of positive symptoms, but no relationship with negative symptoms. We have not been able to detect significant clinicopathological relationships between temporal lobe or superior temporal gyrus measures and factor scores derived from the Brief Psychiatric Rating Scale (BPRS) (Overall & Gorham, 1962) for hallucinatory behavior or conceptual disorganization in a sample of Veterans Hospital patients, but we did replicate negative correlations between left superior temporal gyrus gray matter volumes and psychotic symptoms in a sample of state hospital patients (Marsh *et al.*, 1996; Zipursky *et al.*, 1994).

PET studies have also explored relationships between clinical symptoms and perfusion abnormalities or glucose metabolic rate. Using ^{18}FDG-PET, Wolkin *et al.*, (1992) found a correlation between negative symptoms and prefrontal hypometabolism in unmedicated schizophrenics without any relationship to age, cerebral atrophy, or positive symptom severity. Earlier PET studies failed to find a relationship between clinical symptoms on the BPRS, illness severity, and a reduced anterior-posterior gradient using ^{18}FDG (DeLisi, Buchsbaum, *et al.*, 1985). Others (Tamminga *et al.*, 1992), however, report reduced rCMRGlc in the hippocampus and anterior cingulate in both deficit and nondeficit subgroups of schizophrenic patients compared to healthy controls. Kishimoto, Kuwahara, *et al.* (1987) suggested that a subgroup of schizophrenic patients with decreased parietal rCMRGlc tended to have delusions and hallucinations, whereas the subgroup with hypofrontality using ^{11}C-glucose had a flat, blunted affect. Along similar lines, DeLisi *et al.*, (1989) found a positive correlation between left temporal rCMRGlc and increased symptom severity using ^{18}FDG-PET.

Liddle and associates (Liddle, Friston, Hirsch, & Frackowiak, 1990; Liddle *et al.*, 1992) have addressed how the symptom complexes in schizophrenia (syndromes of disorganization, reality distortion, and psychomotor poverty) (Liddle, 1987) relate to resting ^{15}O PET-derived rCBF patterns. Although perfusion abnormalities were seen in several regions, there emerged a pattern of correlations between symptoms and abnormal regional perfusion, suggesting that different functional brain systems are involved in the production of different kinds of symptoms, although there are some regions which appear to be involved regardless of the type of symptom complex. Symptoms of psychomotor poverty and disorganization correlated negatively with blood flow in the prefrontal and left parietal association cortices and positively in the caudate nuclei. Symptoms of reality distortion were related to increased rCBF in the left mesial temporal region, left ventral striatum, and left prefrontotemporal regions, whereas negative correlations with rCBF were found for the right posterior cingulate and caudate. In a separate analysis (Friston, Liddle, Frith, Hirsh, & Frackowiak, 1992), increases in rCBF in the left parahippocampal gyrus and left globus pallidus region were found to be a common abnormality associated with all three behavioral subsyndromes.

Relationships of Imaging Abnormalities to Neuropsychological Deficits. Despite extensive literature on structural brain imaging and cognitive status in schizophrenia, there are few reports relating patterns of cognitive sparing and impairment to circumscribed brain structures. The most consistently reported correlations involve nonspecific measures such as the area or volume of the lateral ventricles. In particular, larger lateral ventricles measured with CT were observed to be associated with scores from traditional test batteries, including the Halstead-Reitan Battery (Donnelly, Weinberger, Waldman, & Wyatt, 1980; Pandurangi *et al.*, 1986), Wechsler Adult Intelligence Scale (Donnelly *et al.*, 1980; Pandurangi *et al.*, 1986), Luria-Nebraska Neuropsychological Battery (Golden *et al.*, 1980; Pfefferbaum *et al.*, 1988), and the Mini-Mental Status Examination (Goldberg *et al.*, 1988). Studies restricting CT measures to frontal brain regions reported predictable relationships, firstly, between enlarged frontal ventricle-to-brain ratios and performance on tests of immediate verbal memory, psychomotor speed, conceptual thinking, as well as general intellectual level (Keilp *et al.*, 1988), and secondly, between right frontal sulcal enlargement and greater number of perseverations on the Wisconsin Card Sorting Test (Rosse, Schwartz, Mastropaolo, Goldberg, & Deutsch, 1991) although relationships between prefrontal neuropsychological and prefrontal tissue volume deficits were not evident using MRI (Raine *et al.*, 1992).

In general, recent neuroimaging studies have been marginally more successful in revealing structure/function relationships. Among the significant correlations reported using MRI are those between atypical asymmetry of the lateral sulcus and better cognitive function in first-episode schizophrenic patients (A.L. Hoff, Riordan, *et al.*, 1992); between tests of attention (Digit Span) and problem solving (WCST) and third but not lateral ventricular volume (Bornstein, Schwarzkopf, Olson, & Nasrallah, 1992), also in first-episode schizophrenic patients; and between amplitude of the P300 brain potential recorded over the left temporal lobe (at electrode site T3) and the volume of the posterior superior temporal gyrus, the brain region underlying that electrode site (McCarley *et al.*, 1993). Functional brain imaging studies suggest greater abnormality of the anterior as well as subcortical regions. In the xenon inhalation studies by Weinberger and colleagues (Berman *et al.*, 1988), schizophrenic patients had reduced activation of blood flow selectively to the dorsolateral prefrontal cortex while performing the WCST, and performance on the WCST was related to the degree of activation of this area. Moreover, some evidence for selective frontal hypometabolism comes from the observation that the same patients showed the normal pattern of posterior activation while performing the Raven's Progressive Matrices, despite impaired performance on the test. A recent SPECT study showed newly diagnosed schizophrenic patients (with no or little medication treatment history) to have impaired striatal suppression, seen in controls, while performing the WCST (Rubin *et al.*, 1991), which the authors interpreted as a lack of corticostriatal feedback during prefrontal activation.

The relative lack of specific brain structure/cognitive function correlations may be related to the coarseness of either the neuroimaging or neuropsychological measures used in most studies. In addition, the observation of widespread cortical volume deficits in gray matter (Zipursky *et al.*, 1992; Harvey *et al.*, 1993) together with evidence for widespread deficits in multiple cognitive domains (e.g., Kolb & Whishaw, 1983; Gruzelier, Seymour, Wilson, Jolley, & Hirsch, 1988; Goldberg *et al.*, 1990; McKenna *et al.*, 1990; Braff *et al.*, 1991; Gur, Saykin, & Gur, 1991; Saykin *et al.*, 1991; Liddle & Morris, 1991; Gold *et al.*, 1992; Hoff, O'Donnell, Morris, & DeLisi, 1992b; Sullivan, Shear, Zipursky, *et al.*, 1996) may reduce the possibility of finding selective brain/behavior relationships of the type commonly observed in neurological patients with focal brain lesions.

Relationship of Imaging Findings to Aging and Course of Illness. An understanding of the relationships among aging, duration of illness, and the nature of structural abnormalities in schizophrenia would help to elucidate underlying mechanisms of the disease-related dysmorphology. If the abnormalities are static following the clinical onset of symptoms, a neurodevelopmental pathogenesis might be invoked, in which brain dysmorphology is the result of aberrant brain development occurring early in life. Alternatively, if the abnormalities are found to be progressive in nature after the clinical onset of schizophrenia, the underlying pathophysiology would be more consistent with a neurodegenerative process. Whether the volume abnormalities in schizophrenia observed with *in vivo* imaging are progressive is controversial.

A number of imaging studies suggest that structural brain abnormalities in schizophrenia, such as ventricular enlargement, are present at the time of symptomatic onset in most patients as a group (Degreef *et al.*, 1992; DeLisi, Stritzke, *et al.*, 1991; Schulz, Koller, Kishore, & Hamer, 1983; Weinberger, DeLisi, Perman, Targum, & Wyatt, 1982) and do not progress over and above that expected with normal aging in follow-up studies (Degreef *et al.*, 1991; DeLisi, Stritzke, *et al.*, 1991; 1992; Hoffman, Ballard, Turner, & Casey, 1991; Illowsky, Juliano, Bigelow, & Weinberger, 1988; Spoonheim, Iacono, & Beiser, 1991; Vita, Sacchetti, Valvasson, & Cazzullo, 1988b). Similarly, in cross-sectional studies, significant relationships between structural abnormalities and illness duration or age of illness onset are not established (Dauphinais *et al.*, 1990; Suddath *et al.*, 1989, 1990; Pfefferbaum *et al.*, 1988; Zipursky, Lim, Rosenbloom & Pfefferbaum, 1988; Zipursky *et al.*, 1992, 1994; Marsh, Harris, *et al.*, 1996; Marsh, Suddath, *et al.*, 1994). However, contradictory evidence suggests that, at least in some patients, pathological changes occur after the clinical onset of the illness (Woods & Yurgelun-Todd, 1991). In some follow-up studies, ventricular size was shown to progress over time for schizophrenic patients, as a group, relative to controls (Kemali, Maj, Galderisi, Milici, & Salvati, 1989; Woods *et al.*, 1990) or some individual patients (Illowsky *et al.*, 1988; Nasrallah, Olson, McCalley-Whitters, Chapman, & Jacoby, 1986). In several neuroimaging studies, methodological issues relating to head position in the imaging scanner may have contributed to significant differences between structural measures obtained on two occasions (cf., Shear *et al.*, 1995). The studies failing to show progression, however, do not preclude the possibility that progressive brain dysmorphology may occur in the later years of life in schizophrenic patients. Whether such a change is related to symptom duration or to enhanced vulnerability to the effects of normal aging must also be explored.

RELATIONSHIP OF IMAGING ABNORMALITIES TO GENDER. Clinical differences between men and women with schizophrenia, as groups, are well recognized (Goldstein, 1988; Lewis, 1992): women tend to have better premorbid adjustment, a later age of onset, and a more favorable course of illness. However, only a few investigations have addressed the biological correlates of these clinical differences, including potential variations in brain structural pathology. Since clinical phenomena used to diagnose schizophrenia are not gender-based, it is not clear why the essential pathophysiological mechanisms underlying schizophrenia would be necessarily different for men and women. However, sex hormones are known to affect morphological development and differentiation of the brain (see Chapter 5). Thus, to the extent that the pathophysiology of schizophrenia disrupts sex hormonal activity, it may also disrupt the normal pattern of cerebral morphology that distinguishes the sexes, most notably, the corpus callosum and planum temporale (see Hines & Green, 1991).

Neuroimaging studies on gender differences in schizophrenia have produced inconsistent results (DeLisi, Dauphinais, & Hauser, 1989). Regarding studies on corpus callosum morphology (see review by Pearlson & Marsh, 1993), measurements made from a single midsagittal image (Nasrallah *et al.*, 1986) found schizophrenic women to have a thicker corpus callosum in the anterior and middle regions, as compared with women controls, but no such differences emerged between male patients and controls. Raine *et al.* (1990) described in schizophrenia a reversal of the normal sex differences, in which schizophrenic women had a thicker corpus callosum than schizophrenic men, and control men had thicker corpus callosi than control women. Andreasen *et al.* (1986) reported smaller cranial and cerebral size in men but not women with schizophrenia, as compared with sex-matched control groups. An attempt to replicate this finding (Nasrallah *et al.*, 1990) produced the opposite effect. The implications of illness severity on brain structural measures were raised in interpretations of these results. CT and MRI suggest a greater frequency of structural abnormalities, such as ventricular enlargement, in men compared to women with schizophrenia (Andreasen, Swayze, Flaum, Yates, *et al.*, 1990; Flaum, Arndt, & Andreasen, 1990). In part, some of the inconsistent data on gender differences may be attributed to various methodological issues: measurements of brain structures were not uniformly corrected for head size differences among individuals and between men and women; measures may have been taken from a single image; small numbers of patients were studied; and control groups may have been inadequate for establishing the normal variance between and within each gender group. A large number of controls are necessary to establish the expected variance between the sexes and to assess reliably gender-based differences in brain morphology in schizophrenia.

NEUROLOGICAL MODELS OF SCHIZOPHRENIA

The anatomical and functional abnormalities in schizophrenia involve many brain regions and may lead to disruptions in distinct functional neural systems which underlie the various symptoms of schizophrenia. The pattern of widespread involvement is further supported by observations of diverse neuropsychiatric conditions affecting a variety of brain regions and presenting with schizophrenic-like disorders. In their comprehensive review, Davison and Bagley (1969) showed that the occurrence of schizophrenia-like syndromes in neurological conditions, including complex partial epilepsy, cerebral tumors, and CNS trauma, could not be localized to pathology in one particular brain region, although lesions in the temporal lobe and diencephalon were perhaps involved more consistently. In this light, the significance of temporal lobe pathology in schizophrenia, in either idiopathic cases or neuropsychiatric conditions, may be related to an association between the production of psychotic symptoms and pathology within the temporal lobe or its functional connections. In particular, Trimble (1991) has emphasized that the relationships between symptoms and temporal lobe pathology should not be regarded as distinctive or causal for schizophrenia. Regardless of diagnosis, he argues, certain symptoms, such as Schneiderian first-rank symptoms (e.g., auditory hallucinations in the third person or arguing with one another), are suggestive of temporal lobe pathology; other symptoms implicate pathology elsewhere. For example, frontal pathology may related to symptoms of apathy or amotivation, and caudate pathology or states of relative dopamine excess appears to be important in degenerative basal ganglia disorders with accompanying delusions and hallucinations (Beckson & Cummings, 1992).

Neuropsychiatric conditions provide models for investigating widespread tissue abnormalities in idiopathic schizophrenia. A recent MRI investigation of patients with congenital rubella infections and schizophrenia-like syndromes revealed marked and widespread deficits in cortical gray matter but not white matter volume, ventriculomegaly, and markedly smaller brain size, relative to healthy age-matched controls (Lim *et al.,* 1995). These data provide a potential model for examining the effects of perinatally acquired viral infections on the pathogenesis of brain dysmorphology (e.g., ventriculomegaly and gray matter volume deficits), as implicated in viral hypotheses of schizophrenia (e.g., Sham *et al.,* 1992). Metachromatic leukodystrophy, a demyelinating disease of the central and peripheral nervous systems, which can present with features of schizophrenia early in its course, has also been proposed as a model for idiopathic schizophrenia (Hyde *et al.,* 1991). This model emphasizes the role of disturbances in neuronal connectivity from the frontal cortex.

Temporal Lobe Epilepsy as a Neurological Model for Schizophrenia. The occurrence of transient psychotic symptoms or prolonged schizophrenia-like syndromes in some patients with temporal lobe epileptic foci has led to hypotheses on the relationship between epilepsy and schizophrenia (Slater & Beard, 1963; McKenna, Kane, & Parrish, 1985; Stevens, 1991). Accordingly, a small number of neuroimaging studies have sought similarities between these two diseases. An early qualitative CT analysis (Toone, Dawson, & Driver, 1982) reported no difference in the incidence of structural lesions between psychotic epilepsy patients with and without Present State Examination (PSE)-derived diagnoses of nuclear schizophrenia (a PSE category of schizophrenia which is based primarily on the presence of the first-rank symptoms of Schneider) (Wing, Cooper, & Sartorius, 1974). Similarly, a comparison of patients with both epilepsy and psychosis to patients with idiopathic schizophrenia (Perez, Trimble, Murray, & Reider, 1985) revealed no group differences on a series of linear measures of ventricular regions and qualitative assessments of cerebral sulci. Using MRI, epileptic patients with auditory hallucinations and nuclear schizophrenia were shown to have increased T1 times in the left temporal lobe compared to epilepsy patients who had nuclear schizophrenia without hallucinations (Conlon, Trimble, & Rogers, 1990), although there were no differences in T1 values in frontal, occipital, or temporal regions between epilepsy patients based on the presence or absence of psychiatric disorder. The biological significance of T1 relaxation time differences remains unclear, however. On measures of corpus callosum length and thickness using MRI, no group differences were evident between healthy controls and epilepsy patient with nuclear schizophrenia (Conlon & Trimble, 1988), a finding which conflicts with some studies showing corpus callosum abnormalities in idiopathic schizophrenia (Pearlson & Marsh, 1993). More recently, Barr *et al.* (1993) compared lateral ventricular and hippocampal volumes from MRI scans of 32 first-episode schizophrenic patients to 39 patients with temporal lobe epilepsy (21 left, 18 right) and 42 healthy controls. Both patient groups showed nonlateralized ventricular enlargement relative to the control group; furthermore, the left hippocampal volumes in the schizophrenic group were comparable in size to those of the epilepsy group with a left temporal lobe epileptogenic focus.

Two functional imaging studies have reported on schizophrenia-like psychosis in epilepsy. An interictal ^{15}O inhalation PET investigation comparing epilepsy patients with and without psychosis to healthy controls supports evidence for physiological dysfunction in multiple anterior brain regions in schizophrenia-like conditions (Gallhofer, Trimble, Frackowiak, Gibbs, & Jones, 1985). While both epileptic groups, relative to controls, had decreased rCBF and oxygen metabolism, especially in the temporal regions, the psychotic epilepsy patients, compared to nonpsychotic epileptics, had a lower oxygen extraction ratio (defined as the relationship between blood flow and oxygen utilization) in the frontal, temporal, limbic, and basal ganglia regions, but not the occipital lobe. Preliminary results from a recent ^{99m}Tc-HMPAO SPECT study (Marshall, Syed, Fenwick, & Lishman, 1993) revealed rCBF decreases in the right thalamus and left medial temporal region in five epilepsy patients with schizophrenia-like psychoses compared to five nonpsychotic epileptic patients matched for age, type of epilepsy, age at onset of epilepsy, and seizure focus. The finding of left-sided medial temporal abnormalities is potentially interesting since three of the five subjects in each group had a right-sided seizure focus.

Affective Disorders

The clinical features of affective disorders center on a primary alteration in mood and accompanying neurovegetative, cognitive, and motor dysfunction. Abnormalities of limbic circuitry, including its connections to frontal lobe sites, have been hypothesized to underlie mood and its dysfunction for most of this century (Papez, 1937; McLean, 1952; Nauta, 1971). Support for these ideas stems from biological research findings of neurochemical and neuroendocrine dysregulation as well as evidence from the outcome of psychiatric surgical techniques to relieve intractable mood disorders (Ballantine, Levy, Dagi, & Giriunas, 1977; Ballantine, Bouckoms, Thomas, & Giriunas, 1987; Bridges & Bartlett, 1977; Corkin, Twitchell, & Sullivan, 1979). In addition, neurological disorders affecting the basal ganglia and limbic system, such as Huntington's disease (Folstein *et al.,* 1979), Parkinson's disease (Cummings, 1992), cerebrovascular accidents (Starkstein & Robinson, 1989), and temporal lobe epilepsy (Altshuler, Devinsky, Post, & Theodore, 1990), are associated with the occurrence of affective syndromes and help direct neuroimaging investigations of the underlying pathology of idiopathic affective syndromes. While the nature of mood regulation and its disturbance remain a mystery, structural and functional imaging investigations suggest that selective brain regions are implicated in affective disorders and include the limbic system, subcortical nuclei, and frontal cortex.

The structural and functional neuroimaging data on affective disorders are limited by the relative dearth of studies. There are also methodological limitations. For example, many studies include heterogeneous groups with both unipolar and bipolar disorder and/or psychotic and nonpsychotic patients. Furthermore, in a number of studies, healthy controls were not used for comparisons; rather, affectively ill patients were studied as control groups for studies on schizophrenia. Despite the increased incidence of depression with aging, no study to date has rigorously investigated whether aging interacts with the evident dysmorphology in affective disorders.

Neuroimaging Findings in Affective Disorders

Ventricular Dilatation. A number of structural imaging studies on affective disorder report evidence of lateral ventriculomegaly (Nasrallah, Coffman & Olson, 1989), although the results are not consistent. CT investigations using healthy or nonpsychiatric patient groups as comparisons include reports of increased lateral VBR in unipolar depression (Scott, Golden, Ruedrich, & Bishop, 1983; Targum, Rosen, DeLisi, Weinberger, & Citrin, 1983; Luchins, Lewine, & Meltzer, 1984; Shima *et al.,* 1984; Pearlson, Garbacz, Moberg, Ahn, & DePaulo, 1985), lateral and/or third ventriculomegaly in bipolar disorder (Andreasen, Swayze, Flaum, Alliger, & Cohen, 1990; Dewan, Haldipur, Lane, *et al.,* 1988; Nasrallah, McCalley-Whitters, & Jacoby, 1982; Pearlson, Garbacz, Tompkins, *et al.,* 1984; Pearlson & Veroff, 1981), and increased linear measures of ventricular size among mixed groups of unipolar and bipolar disorder patients (Dolan, Calloway, & Mann 1985; Schlegel & Kretzschmer, 1987). While MRI studies have not consistently considered aging effects, there are reports of qualitative increases in ventricular size in bipolar patients (Figiel *et al.,* 1991) compared to a significantly younger control group and in elderly depressed patients (Rabins, Pearlson, Aylward, Kumar, & Dowell, 1991) compared to age-matched controls. Other structural imaging studies failed to show significant lateral ventricular size differences between healthy controls and elderly depressed patients using CT (Jacoby & Levy, 1980) or MRI (Coffey, Wilkinson, Weiner, Parashos, *et al.,* 1993), first-episode psychotic unipolar or bipolar disorder patients (Iacono *et al.,* 1988) using CT, or bipolar disorder patients with longer illness duration using MRI (Johnstone *et al.,* 1989; Swayze, Andreasen, Alliger, Yuh, & Ehrhardt, 1990).

Nonspecific Tissue Volume Deficits. Tissue volume deficits have been detected by a number of methods.

Structural Imaging Studies. Ventricular enlargement implicates complementary tissue volume deficits in brain tissue. Controlled MRI studies, however, provide little compelling evidence for sulcal widening in affective disorders. In a qualitative CT study, Dolan, Calloway, Thacker, and Mann (1986) found an increase in sulcal width, localized to frontal and temporal areas, in a mixed group of 74 unipolar and 27 bipolar depressed patients compared to 52 age-matched healthy controls; this relationship persisted after covarying for age. Moreover, the patient groups did not differ from each other in sulcal width ratings. A comparison of bipolar manic patients and controls (Nasrallah, Jacoby, & McCalley-Whitters, 1981) revealed modest evidence from CT images for an increase in the size of CSF-filled spaces in the cerebellar region. An MRI study of elderly depressed patients and control subjects reported significant differences in qualitative ratings of sulcal enlargement compared to age-matched controls (Rabins *et al.,* 1991), although results were not adjusted for age. Other studies did not detect significant differences between affectively ill patients and controls on cerebral atrophy ratings from CT scans of first-episode unipolar or bipolar psychotic patients (Iacono *et al.,* 1988) or from MRI scans of elderly depressed patients (Coffey, Wilkinson, Weiner, Parashos, *et al.,* 1993), even after correcting for age effects.

Functional Glucose Metabolic and rCBF Neuroimaging Studies. Functional neuroimaging investigations provide some evidence for physiological abnormalities involving widespread brain regions in mood disorders, although reductions in metabolic rate and blood flow may be more marked frontally. Mood state, affective disorder subgroup, and treatment response also seem to affect functional imaging results, with bipolar patients showing the most pronounced abnormalities. Using ^{133}Xe inhalation techniques in medication-free patients with major depression, several investigators found bilateral global hemispheric CBF reductions during resting states (Mathew *et al.,* 1980; Warren *et al.,* 1984; Sackeim *et al.,* 1990), whereas others found no group differences (Gur, Skolnick, Gur, & Caroff, 1984). In contrast, increased global CBF was reported in depressed patients compared to controls at rest (Rosenberg, Vostrup, Andersen, & Bolwig, 1988) and in the left anterior and right posterior hemispheres during a cognitive activation task (Gur *et al.,* 1984). Using ^{11}C glucose PET, Kishimoto, Takazu, *et al.* (1987) demonstrated diffuse cortical abnormalities in a mixed group of affectively ill patients (nine unipolar depressed, three unipolar depressed in remission, three manic) compared to seven healthy controls. Unipolar depressed patients had a generalized reduction in cortical amino acid pools whereas bipolar patients had global increases in the same pools. However, others have not detected differences between healthy controls and patients with major depression in global cortical blood flow at rest using ^{133}Xe inhalation (Gustafson, Risberg, & Silfverskiold, 1981; Goldstein *et al.,* 1985; Chabrol, Barrere, Guell, Bes, & Moron, 1986; Silfverskiold & Risberg, 1989; Berman *et al.,* 1993) or ^{99m}Tc-HMPAO SPECT (Maes *et al.,* 1993).

Global metabolic or blood flow abnormalities in bipolar disorder are also not consistently demonstrated. Buchsbaum *et al.* (Buchsbaum, DeLisi, Holcomb, & Cappalletti, 1984; Buchsbaum *et al.,* 1986) using ^{18}FDG-PET compared cerebral glucose uptake in medication-free schizophrenics and depressed patients (primarily bipolar) to controls who had their eyes closed while a mildly painful shock was applied to the arm. Both patient groups showed relative decreases in frontal CMRGlc and the anterioposterior gradient of CMRGlc compared to controls. By contrast, using an eyes-open paradigm and ^{18}FDG-PET, bipolar and bipolar mixed patients had lower whole brain glucose metabolic rates than unipolar depressed patients and controls (Baxter *et al.*, 1985, 1987). Their results also showed increased whole brain CMRGlc in bipolar patients when they were hypomanic and decreased whole brain CMRGlc when they were depressed (Baxter *et al.*, 1985). However, global differences in CBF have not been observed in bipolar disorder patients compared to controls using ^{133}Xe inhalation (Sackeim *et al.*, 1987; Silfverskiold & Risberg, 1989).

REGIONAL ABNORMALITIES. Several studies have found regional abnormalities associated with affective disorder.

Structural Neuroimaging Findings. It is not clear whether evidence for reduced glucose uptake and cerebral blood flow in affective disorders is a primary function of decreased cerebral activity or secondary to a decrease in neuronal cell number in the identified region. Whether gray matter or white matter is affected preferentially had also not been investigated comprehensively. Studies examining discrete neuroanatomical morphometry have focused on a limited number of regions and the results are inconsistent. A recent MRI study showed an increased ratio of total cerebral gray matter to white matter volume in 17 first-episode bipolar patients age matched to 16 healthy controls, although volumetric measures of cingulate gyrus, frontal lobe, and cerebral volume were not significantly different between groups (Strakowski, Wilson, *et al.*, 1993). Some MRI studies provide evidence for temporal lobe volume deficits in bipolar disorder (Altshuler *et al.*, 1991; Hauser, Altschuler, *et al.*, 1989), frontal lobe volume in major depression (primarily nonpsychotic) (Coffey, Wilkinson, Weiner, Parashos, *et al.*, 1993), and frontal cerebral width in major depression (presence of psychosis not indicated) (Krishnan *et al.*, 1992). Others show a trend toward reduced hippocampal size in bipolar patients (Swayze *et al.*, 1992). Negative controlled MRI studies did not detect significant group differences in temporal lobe (Johnstone *et al.*, 1989; Swayze *et al.*, 1992) or corpus callosum measures (Hauser, Dauphinais, *et al.*, 1989) in patients with bipolar disorder or major depression (primarily nonpsychotic) (Coffey, Wilkinson, Weiner, Parashos, *et al.*, 1993), sagittal frontal lobe area in bipolar patients (Coffman, Bornstein, Olson, Schwarzkopf, & Nasrallah, 1990), or amygdala-hippocampal volume in patients with major depression (Axelson *et al.*, 1993; Coffey, Wilkinson, Weiner, Parashos, *et al.*, 1993).

The few available studies of subcortical structures implicate volume deficits in major depression, but not in bipolar disorder. Positive MRI studies reveal bilateral decreases in caudate (Krishnan *et al.*, 1992) and putamen volumes (Husain *et al.*, 1991) in patients with major depression relative to age-matched controls. In contrast, bipolar patients and controls were not distinguishable from caudate or thalamic volumes on MRIs of first-episode patients (Strakowski, Wilson, *et al.*, 1993) or from caudate or putamen volumes in patients with bipolar disorder of longer duration (Swayze *et al.*, 1992).

MRI Signal Hyperintensities. Several studies have examined the occurrence of white matter and subcortical signal hyperintensities on MRI in psychiatric populations, including schizophrenia and affective disorder (reviewed by Brown, Lewine, Hudgins, & Risch, 1992). Although MRI signal hyperintensities are a nonspecific abnormality of unclear significance, their presence in specific locations in affective disorders may reflect damage to pathways connecting limbic and frontal regions to subcortical nuclei. In other contexts, subcortical and white matter hyperintensities are associated with a number of conditions (Coffey & Figiel, 1991), such as aging, vascular disease, demyelination, edema, flow enhancement, dilated perivascular spaces, and ischemic injury (Englund *et al.*, 1988). In both bipolar disorder and elderly depressed patients, most (Coffey *et al.*, 1988; Dupont *et al.*, 1987, 1990; Figiel *et al.*, 1991; Krishnan *et al.*, 1988; McDonald, Krishnan, Doraiswamy, & Blazer, 1991; Swayze *et al.*, 1990) but not all (Rabins *et al.*, 1991; Guze & Szuba, 1992) reports indicate a particularly high incidence of signal hyperintensities in the periventricular white matter and subcortical regions compared to controls. Their presence in some patients with late-life psychotic syndromes (Miller *et al.*, 1989) and increased severity or frequency in elderly depressed patients (Brown *et al.*, 1992; Coffey, Wilkinson, Weiner, Parashos, *et al.*, 1993) implicate a role for vascular pathology in these conditions.

Although subcortical hyperintensities on MRI have been reported in younger patients with bipolar disorder (Dupont *et al.*, 1987, 1990; Figiel *et al.*, 1991; Swayze *et al.*, 1990), a significantly increased frequency of subcortical signal hyperintensities was not observed in first-episode bipolar patients (Strakowski, Woods, *et al.*, 1993), suggesting that this abnormality may develop after the clinical onset of symptoms. However, in a small follow-up series, none of the seven patients with bipolar disorder and subcortical hyperintensities demonstrated changes after one year (Dupont *et al.*, 1990).

Functional Glucose Metabolic and rCBF Neuroimaging Studies. A number of functional imaging studies on affective disorders have specifically examined frontal lobe function. While studies by Baxter *et al.* (1985, 1987, 1989) indicated global hypometabolism in bipolar depressed patients, there was also evidence for laterality differences with greater reductions in CMRGlc in the left frontal lobe, namely the dorsal anterolateral prefrontal cortex, compared to the right hemisphere, regardless of affective subgroup. For example, using ^{18}FDG-PET in an eyes-open paradigm, Baxter and associates reported on three patients with left frontal hypometabolism (Baxter *et al.*, 1985), which improved in two patients after remission of symptoms. These findings are not entirely borne out by others in terms of the localization of frontal abnormalities, but overall there appears to be evidence for frontal hypofunction. While measures of resting total hemispheric ^{18}FDG-PET rCMRGlc did not differ between a group of 26 mixed affectively ill patients compared to seven healthy controls (Phelps *et al.*, 1984), there was a reduction in left frontotemporal rCMRGlc in the subgroup of unipolar depressed patients, suggesting left-right asymmetry. The degree of asymmetry increased with a worsening of mood and decreased with improvement in mood, as induced by methylphenidate. More recently, in patients with major depression, ^{15}O PET-derived measures of rCBF revealed decreases in the left anterior cingulate gyrus and left dorsolateral prefrontal cortex (Dolan *et al.*, 1992; Bench *et al.*, 1992). Similarly, Cohen *et al.* (1989) showed reduced relative metabolic rates using ^{18}FDG-PET in the middle prefrontal cortex in both affective disorder (bipolar disorder and major depression) and schizophrenic patients compared to controls during an auditory discrimination task, and Chabrol *et al.* (1986) reported on relative hypofrontality with ^{133}Xe inhalation at rest in adolescents with major depression, which normalized with antidepressant treatment. Using ^{133}Xe inhalation, Uytdenhoef *et al.* (1983) described increased left frontal rCBF and decreased right posterior rCBF in patients with major depression (type not specified) compared to euthymic bipolar patients, patients with minor depression, and healthy controls.

Although frontal hypofunction is apparent in both schizophrenia and affective disorders, it is not clear whether this abnormality is based on a common pathological process. Recently, this question was addressed by comparing rCBF measures during three cognitive states: rest, performance of the WCST, and performance of a numbers-matching task (Berman, Doran, Pickar, & Weinberger, 1993). Patients with chronic, treatment-resistant schizophrenia were compared to patients with moderate to severe treatment-resistant depressive illness and healthy controls. With two exceptions, all patients were medication-free for 4 weeks. Although affectively ill patients did not differ from normals in either global or regional CBF values under any condition, seven of the ten depressed patients showed a reversal of normal (left > right) rCBF asymmetry in the parietal region during the WCST compared to the opposite pattern in 15 of the 20 controls. In contrast, the schizophrenic patients had a selective deficit in prefrontal cortical blood flow only during the WCST and no abnormal asymmetries, (e.g., Berman & Weinberger, 1990). These results suggest that cerebral blood flow activation during a particular cognitive task is different in healthy controls, schizophrenia, and major depression and suggests physiological abnormalities in the two patient groups as well as differences between the two psychiatric groups.

Subcortical and temporal lobe abnormalities are also evident in functional neuroimaging investigations of affective disorders, although these regions have been explored less and different PET methodologies leave the findings inconclusive. Compared to controls and bipolar patients who showed global decreases in CMRGlc using ^{18}FDG-PET, unipolar depressed patients showed lower caudate rCMRGlc which increased with symptomatic improvement (Baxter *et al.*, 1985; Schwartz, Baxter, Mazziotta, Gerner, & Phelps, 1987). However, Buchsbaum *et al.* (1986) found reduced caudate rCMRGlc in both unipolar and bipolar depressed patients, especially on the right in the bipolar patients. Studies by Wong *et al.* (1987) suggested increases in caudate D2 receptor binding in bipolar patients compared to healthy controls, findings similar to those seen in schizophrenic patients by this group, and suggested abnormalities in dopaminergic systems in both conditions. ^{18}FDG-PET temporal lobe rCMRGlc was not significantly different from controls in a mixed group of affectively ill patients, although there were relative decreases in rCMRGlc in the moderate to severely depressed subgroup compared to controls, especially for the right temporal lobe (Post *et al.*, 1987).

Other Neuroimaging Findings. Other nonvolumetric imaging parameters, such as T1 relaxation times, have been examined in affective disorders. These studies show no consistent pattern of T1 abnormalities in major depression. Specifically, some studies reported shortened T1 times for the hippocampus of depressed patients relative to healthy controls, but no differences in T1 time for the thalamus and frontal and temporal lobe white matter (Krishnan *et al.*, 1991). Others reported increased T1 time in the frontal lobes of unipolar depressed patients, but not in bipolar disorder or healthy controls subjects (Dolan, Poynton, Bridges, & Trimble, 1990). Another group found T1 to be state-related in bipolar disorder, with normalization of a generalized increase in T1 time after lithium treatment (Rangel-Guerra, Periz Payan, Minkoff, & Todd, 1983). Again, it must be kept in mind that the magnitude of these T1 changes is not sufficiently large to implicate pathology and their physiological significance has yet to be established.

Relationship of Neuroimaging Findings to Clinical Variables

Relationships of Imaging Abnormalities to Clinical Symptoms. Relationships between neuroimaging abnormalities and clinical features of affective disorders are not clearly established. Similarities of some neuroimaging findings in affective disorders and schizophrenia suggest that structural dysmorphology may be related to the pathophysiology of psychosis rather than the underlying mood disorder. Several CT studies indicate that measures of VBR or cortical atrophy are greater in patients with psychotic depression than in nonpsychotic major depression (Targum *et al.*, 1983; Luchins *et al.*, 1984; Rothschild *et al.*, 1989; Schlegel *et al.*, 1989a; Shiraishi *et al.*, 1992; Schlegel & Kretzschmer, 1987), although this has not been found uniformly (Pearlson, Garbacz, Tompkins, *et al.*, 1984; Rabins *et al.*, 1991; Dewan, Haldipur, Boucher, Ramachandran, & Major, 1988). CT measures of ventricular size were shown to correlate with clinical features of major depression, such as psychomotor retardation (Schlegel *et al.*, 1989b). Elderly depressed patients with ventriculomegaly observed with CT had lower anxiety scale scores and higher depression scale scores compared to patients with normal-sized ventricles, although aging effects were not considered (Jacoby & Levy, 1980). While relationships between clinical symptoms and ventricular size were not examined, an MRI study did show significant correlations between measures of caudate volumes and Montgomery-Asberg Depression Scale scores (Krishnan *et al.*, 1992). In bipolar disorder, Pearlson, Garbacz, Breakey, Ahn, and DePaulo (1984) found relationships between ventricular size on CT and negative symptoms, unemployment, and total number of hospitalizations, whereas others (Nasrallah *et al.*, 1982; Roy-Byrne, Post, Kellner, Joffe, & Uhde, 1988) demonstrated an inverse association between VBR and numbers of hospitalizations and illness chronicity in bipolar and mixed bipolar/unipolar samples. A negative relationship between VBR and lithium response in bipolar disorder was reported by one group (Sacchetti *et al.*, 1987), but others have not found relationships between ventricular size and treatment response or past treatment with electroconvulsive therapy (ECT) (Luchins *et al.*, 1984; Pearlson *et al.*, 1984; Dolan *et al.*, 1985; Dewan, Haldipur, Boucher, *et al.*, 1988).

Relationships of Imaging Findings to Neuropsychological Measures. Cognitive impairments in affective disorders do not show clear relationships to focal structural abnormalities, although there is evidence for specific associations with functional imaging findings. Using CT, high scores on the Halstead-Reitan Battery Impairment Index were related to increased ventricular size in euthymic bipolar patients (Dewan, Haldipur, Boucher, *et al.*, 1988), but there was no evidence for discrete neuropsychological deficits. Likewise, in psychotic major depression, relationships were reported between ventricular enlargement and widespread cognitive deficits (Rothschild *et al.*, 1989). On MRI, psychotic bipolar patients demonstrated a relationship between diffuse and nonlateralized cognitive impairment and smaller midsagittal cerebral and frontal areas (Coffman *et al.*, 1990) compared to healthy controls. However, elderly depressed patients failed to show significant statistical relationships between CT indices of ventricular and sulcal size and cognitive impairment (Jacoby & Levy, 1980). ^{15}O

PET-derived measures of rCBF indicate a decrease in left anterior medial prefrontal cortex and an increase in vermal rCBF in depressed patients with cognitive impairments relative to depressed patients without cognitive deficits, based on Mini-Metal Status Examination scores (Bench *et al.*, 1992; Dolan *et al.*, 1992), whereas depressed patients, as a group, had decreased rCBF in the left anterior cingulate and left dorsolateral prefrontal cortex. Compared to healthy controls, the cognitively impaired depressed patients also had blood flow deficits in both dorsolateral prefrontal cortices, the left anterior cingulate gyrus, and the right insula.

Relationship of Imaging Abnormalities to Neuroendocrine Dysfunction. Pathology of the hypothalamic-pituitary-adrenal (HPA) axis, as demonstrated in part by an abnormal response of increased cortisol production during the Dexamethasone Suppression Test, is commonly seen in major depression. Identification of corticosteroid receptors on the hippocampal formation and evidence that sustained high cortisol levels can cause irreversible hippocampal damage in animals implicates a role for limbic structures in HPA regulation and suggests that brain dysmorphology in affective disorders may be related to abnormal amounts of circulating glucocorticoids (reviewed by Sapolsky & Plotsky, 1990). HPA axis dysfunction is also implicated because of the role of glucocorticoids on dopamine activity and cognitive function and hypotheses that psychotic symptoms, at least in major depression, may be related to enhanced cortisol activity (Schatzberg, Rothschild, Langlais, Bird, & Cole, 1985; Rothschild *et al.*, 1989). Disruption of the HPA axis could be secondary to inherent neuroanatomical pathology, or alternatively, the abnormally functioning HPA system could result in a tonic state of hypercortisolism which is related to deficits in brain tissue volumes. Recent MRI investigations report correlations between postdexamethasone cortisol level and amygdala-hippocampal volume in 19 middle-age patients with major depression compared to 30 older controls, after adjusting for age and gender (Axelson *et al.*, 1993). In similar comparisons, pituitary volume calculated from a midsagittal MRI image correlated with postdexamethasone cortisol level in depressed patients, after adjusting for age and gender (Axelson *et al.*, 1992). There is no consistent relationship reported between ventricular size and postdexamethasone cortisol level (Targum *et al.*, 1983; Schlegel & Kretzschmer, 1987; Rothschild *et al.*, 1989; Van den Bossche *et al.*, 1991; Coffey, Wilkinson, Weiner, Ritchie, & Aque, 1993) or other indices of hypercortisolemia (Kellner, Rubinow, Gold, & Post, 1983; Risch *et al.*, 1992).

Relationship of Imaging Abnormalities to Gender. Structural imaging studies examining gender differences suggest that men may have a greater incidence of abnormalities than women. However, this issue has not been examined extensively. Ventriculomegaly on CT images was more frequent in men than women with either unipolar or bipolar depression (Dolan *et al.*, 1985; Schlegel & Kretzschmer, 1987). Other studies report larger VBR in male depressed (Andreasen, Swayze, Flaum, Alliger, *et al.*, 1990) and bipolar (Swayze *et al.*, 1990) patients compared to controls but not difference between female patients and controls, even after accounting for age.

Relationships of Imaging Abnormalities to Clinical Course. As in schizophrenia, it is not clear in affective disorders whether associated neuroanatomical abnormalities are progressive or static or if dysfunction or dysmorphology occurs as the result of the underlying etiopathogenesis, repeated episodes of acute illness, or medical treatments. Clinical research, supported by kindling and neuroendocrine theories of depression, indicates that recurrent episodes of affective illness may lead to neurotoxic effects and a progressively more recalcitrant form of the illness (discussed by Altshuler, 1993).

While several studies on schizophrenia indicate that structural differences relative to controls are evident at the first episode of illness or early in the clinical course, first-episode and longitudinal studies on affective disorders are few. Using CT, Iacono *et al.* (1988) reported an increase in third ventricular width in first-episode schizophrenic patients, but no differences in measures of third ventricular width or VBR in 18 first-episode bipolar patients, compared to healthy controls. Recently, using MRI, Strakowski, Wilson, *et al.* (1993) demonstrated an increase in third ventricular volume and a trend toward increased VBR in first-episode bipolar patients compared to age-matched controls. In a prospective longitudinal study, 21 patients with affective disorder (type unspecified) demonstrated a significant increase in the mean CT-derived measure of VBR over the average follow-up period of 3 years, suggesting that the pathological process may be progressive (Vita, Sacchetti, & Cazzullo, 1988); in contrast, measures for the schizophrenic and healthy control comparison groups did not progress. Another study, however, did not find significant changes in structural measures in nine bipolar patients over an interval of 1 to 4.5 years (Woods *et al.*, 1990).

While some have suggested that structural pathology in bipolar disorder may be neurodevelopmental in nature (Nasrallah, 1991; Beckmann & Jakob, 1991), there is some evidence from cross-sectional structural imaging studies that the neuroanatomical abnormalities in affective disorders progress over time. Illness duration has been shown to correlate significantly with increases in linear measures of lateral and third ventricular size in a mixed group of patients with major depression and bipolar disorder (Schlegel & Kretzschmer, 1987) or of right temporal lobe volume deficits in men with bipolar disorder (Altshuler *et al.*, 1991). Others have not found a significant correlation between VBR (Dolan *et al.*, 1985; Roy-Byrne *et al.*, 1988), temporal lobe area (Hauser, Altschuler, *et al.*, 1989), or caudate volume (Krishnan *et al.*, 1992) with illness duration, after correcting for age.

Neurological Models of Affective Disorder

A number of neurological diseases, especially those with subcortical nuclei as the principal site of pathology, have associated mood disorders and thus may provide clues for understanding the pathogenesis of these conditions. Functional imaging studies, in particular, implicate dysfunction in frontal-subcortical networks and complement studies of idiopathic affective disorders. The "subcortical triad" of dyskinesia, depression, and dementia in Huntington's disease provides a compelling model for the pathology of affective disorders (Peyser & Folstein, 1990). In PET and MRI studies, Huntington's patients, regardless of mood state, showed caudate, putamen, and cingulate hypometabolism and similar linear measures of ventricular size on PET, whereas frontal lobe and paralimbic hypometabolism distinguished the four depressed Huntington's patients from five nondepressed Huntington's patients and seven age-matched healthy controls (Mayberg *et al.*, 1992). Along the same lines, a role for frontostriatal function in mood regulation is implicated (Mayberg *et al.*, 1990) showing inferior orbital frontal and caudate hypometabolism in a ^{18}FDG-PET study of five patients with Parkinson's disease and depression compared to six age-matched healthy controls and five Parkinson's patients without depression. In contrast, in secondary mania, Cummings (1986) suggested that when toxic agents are not involved, these conditions tend to involve focal lesions and degenerative disorders affecting deep subcortical structures such as the basal ganglia, thalamus, hypothalamus, and midbrain. He hypothesized that the hypothalamic structures might be responsible for the modulation of certain driven behaviors, such as sleep, libido, and appetite, behaviors which are frequently disturbed during manic episodes and herald symptoms of relapse in a stable patient. Another disease model for affective disorder is provided by the association of lesion locations with poststroke affective syndromes (Robinson *et al.*, 1984; Starkstein & Robinson, 1989). Notably, left hemispheric frontal cortical and basal ganglia strokes were associated with a greater frequency of major depression, suggesting that the underlying structural lesion, with its attendant functional consequences, is etiologically important to the development of the mood disturbance. Furthermore, a family history of psychiatric disorders was seen in patients with right hemisphere strokes and major depressive syndromes, suggesting that genetic vulnerability may be an important factor in right hemisphere cases. These findings also have some relevance to the increased incidence of subcortical hyperintensities on MRI, often thought to be related to vascular disease, in elderly patients who develop depressive illnesses (Figiel *et al.*, 1991).

Obsessive-Compulsive Disorder

The cardinal clinical features of obsessive-compulsive disorder (OCD)—intrusive, involuntary, recurrent, and stereotyped obsessive thoughts and compulsive behaviors—suggest a fundamental disturbance in CNS mechanisms underlying the sequencing and control of certain kinds of motor and cognitive behaviors. Unlike in schizophrenia, in which the underlying functional and structural deficits involve many and diverse functional brain regions, brain regions implicated in OCD appear to be relatively circumscribed. In general, hypotheses invoke primary dynamic abnormalities of the striatum, orbitofrontal cortex, and the subcortical-cortical connections between the two regions, although input from other brain regions may mediate activity of this circuit (Rapoport & Wise, 1988; Baxter, 1990; Insel, 1992). These theories derive their support from the occurrence of OCD in certain neurological disorders and the results of palliative neurosurgical procedures for intractable OCD which disrupt frontal and cingulate projections to the striatum (Ballantine *et al.*, 1977; McLean, 1952; Bridges & Bartlett, 1977). While results from structural neuroimaging investigations of idiopathic OCD are relatively unrevealing, PET studies on idiopathic OCD demonstrate more compelling findings.

Neuroimaging Findings in Idiopathic OCD

Neuroimaging investigations indicate that frontal-subcortical connections involving the caudate nucleus, orbitofrontal cortex, and cingulate gyrus are disrupted in idiopathic OCD. CT studies showed a modest decrease in caudate nucleus volume (Luxenberg *et al.*, 1988) and increased VBR (Behar *et al.*, 1984) in OCD patients compared to controls, although a third CT study revealed no significant group differences in VBR, cerebral asymmetry, or sulcal prominence (Insel, Donnelly, Lalakea, Alterman, & Murphy, 1983). MRI studies have not supported earlier CT studies; one study reported an increase in right caudate volume in OCD (Scarone *et al.*, 1992) and another reported no group differences in caudate size (Kellner *et al.*, 1991). Prolonged T1 relaxation times were observed in right frontal white matter (Garber, Ananth, Chiu, Griswold, & Oldendorf, 1989); however, no T1 differences occurred in the caudate, dorsolateral frontal cortex, lenticular nuclei, or medial temporal lobe. Recently, MRI signal intensity values, regarded as an estimate of T1, were found to be greater for the left versus the right caudate nucleus in OCD patients, whereas controls showed no asymmetry (Calabrese, Colombo, Bonfanti, Scotti, & Scarone, 1993). Taken together, the structural imaging methods have not been able to detect a consistent or definitive neuromorphological abnormality in OCD. However, sample sizes have been relatively small, most studies have measured only a minimum of potentially involved brain regions, and other important factors such as aging, head size, and gender have not been routinely considered.

Findings from dynamic imaging studies on OCD have provided greater insights than structural imaging studies into the pathophysiology underlying OCD. ^{18}FDG-PET studies on unmedicated patients with OCD consistently report relative increases of glucose metabolic activity in frontal lobe regions, namely the orbitofrontal cortex bilaterally (Baxter *et al.*, 1987, 1988; Nordahl *et al.*, 1989; Swedo, Schapiro, *et al.*, 1989) as well as the prefrontal and cingulate regions (Swedo, Schapiro, *et al.*, 1989). There are also reports of increased absolute glucose metabolism in the caudate nucleus (Baxter *et al.*, 1987, 1988), whereas others found no group differences in rCMRGlc in the caudate nucleus (Swedo *et al.*, 1992; Nordahl *et al.*, 1989) or cingulate gyrus (Baxter *et al.*, 1987; Nordahl *et al.*, 1989). Following medication treatment, ^{18}FDG-PET studies have shown decreased rCMRGlc in the caudate nucleus, cingulate gyrus, and thalamus (Baxter *et al.*, 1992); in the orbitofrontal cortex bilaterally, but not the caudate nucleus (Swedo *et al.*, 1992); and in the left caudate nucleus and right orbitofrontal cortex (Benkelfat *et al.*, 1990). Similarly, after behavioral treatment for OCD without adjunctive medications, there was a decrease in right caudate rCMRGlc (Baxter *et al.*, 1992), suggesting that medications alone do not influence pathophysiology. Correlational analyses show that clinical response to treatment relates to both orbitofrontal (Swedo *et al.*, 1992) and caudate (Benkelfat *et al.*, 1990; Baxter *et al.*, 1992) activity.

A ^{99m}Tc-HMPAO SPECT study revealed medial frontal, but not orbitofrontal, cortical rCBF to be increased in OCD (Machlin *et al.*, 1991), with a reduction in the medial frontal rCBF after treatment with fluoxetine (Hoehn-Saric, Pearlson, Harris, Machlin, & Camargo, 1991). Using the same ligand in medication-free patients, others (Rubin, Villanueva-Meyer, Ananth, Trajmar, & Mena, 1992) describe increased uptake in the dorsal parietal cortex bilaterally, left posterofrontal cortex, and the orbital frontal cortex bilaterally, but decreased rCBF in the caudate nucleus. While the above findings are compelling in their relative consistency, interpretations are confounded by the small sample sizes, frequent use of multiple comparisons of brain activity in several regions, and limited imaging resolution which may preclude identification of localized physiological abnormalities.

In general, functional imaging studies tend to show hypermetabolism of the orbitofrontal cortex and caudate nuclei. While this contrasts with the frontostriatal hypofunction seen in neurological disorders with symptomatic OCD, as described below, the data invoke several theories involving frontal-subcortical structures in OCD. In terms of relationships between CNS activity and mental functioning, the behavioral, cognitive, and affective components of the disorder might result from dysfunction of the normal excitatory and inhibitory influences exerted by the thalamo-cortical-striatal pathways, such as the "orbitofrontal loop" (Alexander *et al.,* 1986; Rapoport & Wise, 1988). Others suggest that primary damage to different topographic regions of the striatum influences distinct limbic and cortical circuits and is responsible for the different kinds of motor and/or cognitive behaviors in OCD (Baxter, 1990). Alternatively, orbitofrontal hyperfunction may reflect the increased effort exerted when patients attempt to resist or control intrusive thoughts or behaviors (Insel, 1992). In other words, reduced or otherwise abnormal function elsewhere in the brain, perhaps undetected, may result in compensatory hyperfunction in another region, which may be the local substrate of the pathological behavior.

Neurological Models of OCD

Manifestations of OCD in known neurological disorders are almost always associated with basal ganglia disease and support neuroimaging data implicating the caudate nucleus in OCD. Perhaps the most dramatic examples of symptomatic OCD occurred in patients who developed postencephalitic parkinsonism following the encephalitis lethargica epidemic during World War I (Schilder, 1938). Huntington's disease, neuroacanthocytosis (a degenerative disorder similar to Huntington's disease), and Sydenham's chorea (a movement disorder which can occur following rheumatic heart disease) are all associated with striatal degeneration and an increased frequency of OCD (Swedo, Rapoport, *et al.,* 1989). Toxic, hypoxic, or ischemic damage affecting the basal ganglia, especially the globus pallidus, can cause parkinsonian motor abnormalities as well as OCD. The underlying neuropathology of Gilles de la Tourette's syndrome is not established, although it is thought to be a caused by a basal ganglia disorder. In over half of the patients with Tourette's syndrome, the primary tic disturbance is accompanied by OCD. Increased incidence of neurological soft signs, including tics as well as difficulties with postural control, gait, and alternating movements, are also seen in patients with OCD (Hollander *et al.,* 1990). While OCD can occur in frontal lobe degenerative syndromes, patients with frontal lobe disorders without subcortical involvement rarely demonstrate OCD symptoms (Cummings & Cunningham, 1992). The clinical literature would thus suggest that subcortical dysfunction, at least, and possibly disruption in frontal-subcortical connections are necessary conditions for OCD symptoms to occur.

Neuroimaging investigations examining the pathophysiology of OCD in neurological disorders are relatively rare. Laplane *et al.* (1989) describe their ^{18}FDG-PET findings from eight patients with bilateral basal ganglia damage, mild extrapyramidal motor symptoms, obsessive-compulsive symptoms, and a frontal lobe–like syndrome characterized by a lack of spontaneous drive with preserved intellect. All patients had endured toxic or anoxic damage to the lenticular nuclei, as confirmed by CT or MRI studies. ^{18}FDG-PET in seven of the patients revealed normal glucose utilization in the cerebral cortex and decreased rCMRGlc in the striatum and prefrontal cortex relative to other brain regions. Since relative hypometabolism in the basal ganglia might reflect primary tissue loss, the data were regarded as suggestive of prefrontal hypometabolism and a pathological process affecting frontal-subcortical circuity.

Conclusions

Structural and functional imaging methodologies have proven to be powerful tools for exploring the relationships between brain structure and function in psychiatric illnesses. Because brain regions have some degree of functional specificity and far-reaching yet selective connectivity, it might be expected that abnormalities in brain function would correspond to identified neuroanatomical abnormalities and that the links to associated psychopathology would be evident. However, the genesis of a given mental state, with all its behavioral, cognitive, affective, perceptual, and meaningful components, is exceedingly complex and involves the dynamic function of multiple brain regions, thus rendering restricted structure-function relationships elusive and perhaps even unlikely. The value of studying neuropsychiatric conditions with known CNS pathology, then, is to provide models for examining selective aspects of psychiatric syndromes in the context of specific pathology. The identified pathological patterns in neuropsychiatric syndromes together may thus provide clues as to the pathognomonic substrates of idiopathic psychiatric conditions.

For each of the disorders reviewed, it is clear that the brain is involved and that brain structure and function relationships differ quantitatively from healthy controls. It is more difficult to conclude with confidence that any of the detected abnormalities are pathognomonic for a given illness or even that there is disruption of a primary and localized CNS structure-function relationship. Even though frontal lobe dysfunction is a common thread to all of these conditions, its presence may serve more as a hallmark of brain dysfunction rather than of specific frontal lobe pathology. In schizophrenia, the neuroanatomic differences are widespread and largely associated with frontal hypofunction in dynamic studies. Investigations on affective disorder are less comprehensive than those from the schizophrenia literature. These, too, however, are associated with neuroanatomical abnormalities in a variety of brain regions and frontal hypofunction. The occurrence of mood disorders in patients with primary basal ganglia disorders, and a few imaging studies on these groups, implicate subcortical-frontal pathology in idiopathic affective syndromes. Idiopathic OCD, in contrast to schizophrenia and affective disorders, is associated with frontal lobe hyperfunction, with only modest evidence to date for detectable structural pathology. Evidence for frontal lobe dysfunction, in the context of structural abnormalities in other regions, possibly indicates that certain clinical features require frontal lobe activity or, at least, the recruitment of neurons which involve frontal lobe function. This coordinated neural effort could occur through any number of afferent and efferent frontal cortical pathways involving subcortical, cerebellar, and brain stem structures as well as other cortical regions. The possibilities in pathological circuitry almost certainly contribute to the heterogeneity of the clinical manifestations of each psychiatric disease and to the elusiveness of identifying structural and functional pathology characteristic of one disease.

References

Alexander, G., DeLong, M., & Strick, P. (1986). Parallel organization of functionally segregated circuits linking basal ganglia and cortex. *Annual Review of Neuroscience, 9*, 357–381.

Altshuler, L. L. (1993). Bipolar disorder: Are repeated episodes associated with neuroanatomic and cognitive changes? *Biological Psychiatry, 33*, 563–565.

Altshuler, L. L., Conrad, A., Hauser, P., Li, X., Guze, B. H., Denikoff, K., Tourtelotte, W., & Post, R. (1991). Reduction of temporal lobe volume in bipolar disorder: A preliminary report of magnetic resonance imaging. *Archives of General Psychiatry, 48*, 482–483.

Altshuler, L. L., Devinsky, O., Post, R. M., & Theodore, W. (1990). Depression, anxiety, and temporal lobe epilepsy: Laterality of focus and symptoms. *Archives of Neurology, 47*, 284–288.

Andreasen, N. C., Carson, R., Diksic, M., Evans, A., Farde, L., Gjedde, A., Hakim, A., Lal, S., Nair, N., Sedvall, G., Tune, L., & Wong, D. F. (1988). Workshop on schizophrenia, PET, and dopamine D2 receptors in the human neostriatum. *Schizophrenia Bulletin, 14*, 471–484.

Andreasen, N. C., Ehrhardt, J. C., Swayze, V. W., Alliger, R. J., Yuh, W. T. C., Cohen, G., & Ziebell, S. (1990). Magnetic resonance imaging of the brain in schizophrenia: The pathophysiologic significance of structural abnormalities. *Archives of General Psychiatry, 47*, 35–44.

Andreasen, N., Nasrallah, A., Dunn, V., Olson, S. C., Grove, W. M., Ehrhardt, J. C., Coffman, J. A., & Crossett, J. H. W. (1986). Structural abnormalities in the frontal system in schizophrenia. *Archives of General Psychiatry, 43*, 136–144.

Andreasen, N. C., Rezai, K., Alliger, R., Swayze, V. W., Flaum, M., Kirchner, P., Cohen, G., & O'Leary, D. S. (1992). Hypofrontality in neuroleptic-naive patients and in patients with chronic schizophrenia—Assessment with Xenon-133 single-photon emission computed tomography and the Tower of London. *Archives of General Psychiatry, 49*, 943–958.

Andreasen, N. C., Swayze, V., Flaum, M., Alliger, R., & Cohen, G. (1990). Ventricular abnormalities in affective disorder—clinical and demographic correlates. *American Journal of Psychiatry, 147*, 893–900.

Andreasen, N. C., Swayze, V. W., Flaum, M., Yates, W. R., Arndt, S., & McChesney, C. (1990). Ventricular enlargement in schizophrenia evaluated with computed tomographic scanning—effects of gender, age, and stage of illness. *Archives of General Psychiatry, 47*, 1008–1015.

Arndt, S., Cohen, G., Alliger, R. J., Swayze, V. W., & Andreasen, N. C. (1991). Problems with ratio and proportion measures of imaged cerebral structures. *Psychiatry Research: Neuroimaging, 40*, 79–90.

Axelson, D. A., Doraiswamy, P. M., Boyko, O. B., Escalona, P. R., McDonald, W. M., Ritchie, J. C., Patterson, L. J., Ellinwood, E. H. J., Nemeroff, C. B., & Krishan, K. R. R. (1992). In vivo assessment of pituitary volume with magnetic resonance imaging and systematic stereology: Relationship to dexamethasone suppression test results in patients. *Psychiatry Research, 44*, 63–70.

Axelson, D. A., Doraiswamy, P. M., Mcdonald, W. M., Boyko, O. B., Tupler, L. A., Patterson, L. J., Nemeroff, C. B., Ellinwood, E. H., & Krishnan, K. R. R. (1993). Hypercortisolemia and hippocampal changes in depression. *Psychiatry Research, 47*, 163–173.

Bachneff, S. A. (1991). Positron emission tomography and magnetic resonance imaging: A review and a local circuit neurons hypo(dys)function hypothesis of schizophrenia. *Biological Psychiatry, 30*, 857–886.

Ballantine, H. T., Bouckoms, A. J., Thomas, E. K., & Giriunas, I. E. (1987). Treatment of psychiatric illness by stereotactic cingulotomy. *Biological Psychiatry, 22*, 807–819.

Ballantine, H. T., Levy, B. S., Dagi, T. F., & Giriunas, I. E. (1977). Cingulotomy for psychiatric illness: Report of 13 years' experience. In W. H. Sweet, S. Obrador, and J. G. Martin-Rodriguez (Eds.), *Neurosurgical Treatment in Psychiatry, Pain, and Epilepsy* (pp. 333–353). Baltimore: University Park Press.

Barr, W. B., Ashtari, M., Degreef, G., Bogerts, B., Bilder, R. M., Schaul, N., & Lieberman, J. (1993). Brain morphometric comparison of first episode schizophrenia and temporal lobe epilepsy. *Schizophrenia Research, 9*, 192.

Barta, P. E., Pearlson, G. D., Powers, R. E., Richards, S. S., & Tune, L. E. (1990). Auditory hallucinations and smaller superior temporal gyral volume in schizophrenia. *American Journal of Psychiatry, 147*, 1457–1462.

Baxter, L., Schwartz, J., Mazziotta, J., Phelps, M. E., Pahl, J. J., Buzze, B. E.. & Fairbanks, L. (1988). Cerebral glucose metabolic rates in obsessive-compulsive disorder. *American Journal of Psychiatry, 145*, 1560–1563.

Baxter, L. E., Schwartz, J. M., Phelps, M. E., Mazziotta, J. C., Gore, B. H., Selin, C. E., Gerner, R. H., and Sumida, R. H. (1985). Reduction of prefrontal cortex glucose metabolism is common to three types of depression. *Archives of General Psychiatry, 46,* 243–250.

Baxter, L. R. (1990). Brain imaging as a tool in establishing a theory of brain pathology in obsessive compulsive disorder. *Journal of Clinical Psychiatry, 51,* 22–25.

Baxter, L. R., Phelps, M. E., Mazziotta, J. C., Guze, B. H., Schwartz, J. M., & Selin, C. E. (1987). Local cerebral glucose metabolic rates in obsessive-compulsive disorder: A comparison with rates in unipolar depression and in normal controls. *Archives of General Psychiatry, 44,* 211–218.

Baxter, L. R., Phelps, M. E., Mazziotta, J. C., Schwartz, J. M., Gerner, R. H., Selin, C. E., & Sumida, R. M. (1985). Cerebral metabolic rates for glucose in mood disorders. *Archives of General Psychiatry, 42,* 441–447.

Baxter, L. R., Schwartz, J. M., Bergman, K. S., Szuba, M. P., Guze, B. H., Mazziotta, J. C., Alazraki, A., Selin, C. E., Ferng, H. K., Munford, P., & Phelps, M. E. (1992). Caudate glucose metabolic rate changes with both drug and behavior therapy for obsessive-compulsive disorder. *Archives of General Psychiatry, 49,* 681–689.

Baxter, L. R., Schwartz, J. M., Phelps, M. E., Mazziotta, J. C., Guze, B. H., Selin, C. E., Gerner, R. H., & Sumida, R. M. (1989). Reduction of prefrontal cortex glucose metabolism common to three types of depression. *Archives of General Psychiatry, 46,* 243–250.

Beckmann, H., & Jakob, H. (1991). Prenatal disturbances of nerve cell migration in the entorhinal region: A common vulnerability factor in functional psychoses? *Journal of Neural Transmission, 84,* 155–164.

Beckson, M., & Cummings, J. L. (1992). Psychosis in basal ganglia disorders. *Neuropsychiatry, Neuropsychology, and Behavioral Neurology, 5,* 126–131.

Behar, D., Rapoport, J. L., Berg, C. J., Denckla, M. B., Mann, L., Cox, C., Fedio, P., Zahn, T., & Wolfman, M. G. (1984). Computerized tomography and neuropsychological test measures in adolescents with obsessive-compulsive disorder. *American Journal of Psychiatry, 141,* 363–369.

Bench, C. J., Dolan, R. J., Friston, K. J., & Frackowiak, R. S. J. (1990). Positron emission tomography in the study of brain metabolism in psychiatric and neuropsychiatric disorders. *British Journal of Psychiatry, 157,* 82–95.

Bench, C. J., Friston, K. J., Brown, R. G., Scott, L. C., Frackowiak, R. S. J., & Dolan, R. J. (1992). The anatomy of melancholia—focal abnormalities of cerebral blood flow in major depression. *Psychological Medicine, 22,* 607–615.

Benes, F., Sorensen, I., Vincent, S., Bird, E., & Sathi, M. (1992). Increased density of glutamate-immunoreactive vertical processes in superficial laminae in cingulate cortex of schizophrenic brain. *Cerebral Cortex, 2,* 503–512.

Benkelfat, C., Nordahl, T. E., Semple, W. E., King, C., Murphy, D. L., & Cohen, R. M. (1990). Local cerebral glucose metabolic rates in obsessive-compulsive disorder: Patients treated with clomipramine. *Archives of General Psychiatry, 47,* 840–848.

Berman, K. F., Doran, A. R., Pickar, D., & Weinberger, D. R. (1993). Is the mechanism of prefrontal hypofunction in depression the same as in schizophrenia? Regional cerebral blood flow during cognitive activation. *British Journal of Psychiatry, 162,* 183–192.

Berman, K. F., Illowsky, B. P., & Weinberger, D. R. (1988). Physiological dysfunction of dorsolateral prefrontal cortex in schizophrenia. IV. Further evidence for regional and behavioral specificity. *Archives of General Psychiatry, 45,* 616–622.

Berman, K. F., Randolph, C., Gold, J., Holt, D., Jones, D. W., Goldberg, T. E., Carson, R. C., Herscovitch, P., and Weinberger, D. R. (1991). Physiological activation of frontal lobe studied with positron emission tomography and oxygen-15 water during working memory tasks. *Journal of Cerebral Blood Flow and Metabolism, 11* (Suppl. 2), S851.

Berman, K. F., Torrey, E. F., Daniel, D. G., & Weinberger, D. R. (1992). Regional cerebral blood flow in monozygotic twins discordant and concordant for schizophrenia. *Archives of General Psychiatry, 49,* 927–934.

Berman, K. F., & Weinberger, D. R. (1990). Lateralisation of cortical function during cognitive tasks—regional cerebral blood flow studies of normal individuals and patients with schizophrenia. *Journal of Neurology, Neurosurgery, and Psychiatry, 53,* 150–160.

Berman, K. F., & Weinberger, D. R. (1991). Functional localization in the brain in schizophrenia. In A. Tasman & S. M. Goldfinger (Eds.), *Review of Psychiatry* (pp. 24–59). Washington, DC: American Psychiatric Press.

Berman, K. F., Zec, R. F., & Weinberger, D. R. (1986). Physiologic dysfunction of dorsolateral prefrontal cortex in schizophrenia. II. Role of neuroleptic treatment, attention, and mental effort. *Archives of General Psychiatry, 43,* 126–135.

Berman, S. M., Whipple, S. C., Fitch, R. J., & Noble, E. P. (1993). P3 in young boys as a predictor of adolescent substance abuse. *Alcohol, 10,* 69–76.

Besson, J. A. O., Corrigan, F. M., Cherryman, G. R., & Smith, F. W. (1987). Nuclear magnetic resonance brain imaging in chronic schizophrenia. *British Journal of Psychiatry, 150,* 161–163.

Bogerts, B. (1993). Recent advances in the neuropathology of schizophrenia. *Schizophrenia Bulletin, 19,* 431–445.

Bogerts, B., Ashtari, M., Degreef, G., Alvir, J. M., Bilder, R. M., & Lieberman, J. A. (1990). Reduced temporal limbic structure volumes on magnetic resonance images in first episode schizophrenia. *Psychiatry Research: Neuroimaging, 35,* 1–13.

Bogerts, B., Lieberman, J. A., Ashtari, M., Bilder, R. M., Degreef, G., Lerner, G., Johns, C., & Masiar, S. (1993). Hippocampus amygdala volumes and psychopathology in chronic schizophrenia. *Biological Psychiatry, 33,* 236–246.

Bornstein, R. A., Schwarzkopf, S. B., Olson, S. C., & Nasrallah, H. A. (1992). Third-ventricle enlargement and neuropsychological deficit in schizophrenia. *Biological Psychiatry, 31,* 954–961.

Braff, D. L., Heaton, R., Kuck, J., Cullum, M., Moranville, J. Grant, I., & Zisook, S. (1991). The generalized pattern of neuropsychological deficits in outpatients with chronic schizophrenia with heterogeneous Wisconsin Card Sorting Test results. *Archives of General Psychiatry, 48,* 891–898.

Breier, A., Buchanan, R. W., Elkashef, A., Munson, R. C., Kirkpatrick, B., & Gellad, F. (1992). Brain morphology and schizophrenia—a magnetic resonance imaging study of limbic, prefrontal cortex, and caudate structures. *Archives of General Psychiatry, 49,* 921–926.

Bridges, P. K., & Bartlett, J. R. (1977). Psychosurgery: Yesterday and today. *British Journal of Psychiatry, 131,* 249–260.

Brodie, J. D., Christman, D. R., Corona, J. F., Fowler, J. S., Gomez-Mont, F., Jaeger, J., Micheels, P. A., Rotrosen, J., Russell, J. A., Volkow, N., Wikler, A., Wolf, A., and Wolkin, A. (1984). Patterns of metabolic activity in the treatment of schizophrenia. *Annals of Neurology, 15,* S166–S169.

Brown, F. W., Lewine, R. J., Hudgins, P. A., and Risch, S. C. (1992). White matter hyperintensity signals in psychiatric and nonpsychiatric subjects. *American Journal of Psychiatry, 149,* 620–625.

Buchanan, R. W., Breier, A., Kirkpatrick, B., Elkashef, A., Munson, R. C., Gellad, F., & Carpenter, W. T. (1993). Structural abnormalities in deficit and nondeficit schizophrenia. *American Journal of Psychiatry, 150,* 59–65.

Buchsbaum, M. S. (1990). The frontal lobes, basal ganglia, and temporal lobes as sites for schizophrenia. *Schizophrenia Bulletin, 16,* 379–389.

Buchsbaum, M., DeLisi, L., Holcomb, H., & Cappalletti, J. (1984). Anteroposterior gradients in cerebral glucose use in schizophrenia and affective disorders. *Archives of General Psychiatry, 41,* 1159–1166.

Buchsbaum, M. S., Haier, R. J., Potkin, S. G., Nuechterlein, K., Bracha, H. S., Katz, M., Lohr, J., Wu, J., Lottenberg, S., Jerabek, P. A., Trenary, M., Tafalla, R., Reynolds, C., & Bunney, W. E. (1992). Frontostriatal disorder of cerebral metabolism in never-medicated schizophrenics. *Archives of General Psychiatry, 49,* 935–942.

Buchsbaum, M. S., Ingvar, D. H., Kessler, R. M., Waters, R. N., Cappelletti, J., Van Kammen, D. P., King, A. C., & Johnson, J. L. (1982). Cerebral glucography with positron tomography in normals and patients with schizophrenia. *Archives of General Psychiatry, 39,* 251–259.

Buchsbaum, M. S., Nuechterlein, K. H., Haier, R. J., Wu, J., Sicotte, N., Hazlett, E., Asarnow, R., Potkin, S., & Guich, S. (1990). Glucose metabolic rate in normals and schizophrenics during the continuous performance test assessed by positron emission tomography. *British Journal of Psychiatry, 156,* 216–227.

Buchsbaum, M. S., Wu, J., DeLisi, L. E. Holcomb, H., Kessler, R., Johnson, J., King A. C., Hazlett, K., Langston, K., & Post, R. M. (1986). Frontal cortex and basal ganglia metabolic rates assessed by positron emission tomography with [^{18}F]2-deoxy-glucose in affective illness. *Journal of Affective Disorders, 10,* 137–152.

Calabrese, G., Colombo, C., Bonfanti, A., Scotti, G., & Scarone, S. (1993). Caudate nucleus abnormalities in obsessive-compulsive disorder: Measurements of MRI signal intensity. *Psychiatry Research: Neuroimaging, 50,* 89–92.

Carlsson, A. (1988). The current status of the dopamine hypothesis of schizophrenia. *Neuropsychopharmacology, 1,* 179–203.

Casanova, M. F., Goldberg, T. E., Suddath, R. L., Daniel, D. G., Rawlings, R., Lloyd, D. G., Loats, H. L., Kleinman, J. E., & Weinberger, D. R. (1990). Quantitative shape analysis of the temporal and prefrontal lobes of schizophrenic patients: A magnetic resonance image study. *Journal of Neuropsychiatry and Clinical Neurosciences, 2,* 363–372.

Chabrol, H., Barrere, M., Guell, A., Bes, A., & Moron, P. (1986). Hyperfrontality of cerebral blood flow in depressed adolescents (letter). *American Journal of Psychiatry, 143,* 263–264.

Chakos, M. H., Lieberman, J. A., Bilder, R. M., Lerner, G., Bogerts, B., and Ashtari, M. (1993). Prospective MRI study of caudate pathomorphology in first episode schizophrenia. *Schizophrenia Research, 9,* 196.

Chelune, G. J., Heaton, R. K. Lehman, R. A., & Robinson, A. (1979). Level versus pattern of neuropsychological performance among schizophrenics and diffusely brain damaged patients. *Journal of Consulting and Clinical Psychology, 47,* 155–163.

Clark, C., Klonoff, H., Tyhurst, J. S., Li, D., Martin, W., & Pate, B. D. (1989). Regional cerebral glucose metabolism in three sets of identical twins with psychotic symptoms. *Canadian Journal of Psychiatry, 34,* 263–270.

Cleghorn, J., Garnett, E., Nahmias, C., Firnau, G., Brown, G., Kaplan, R., Szechtman, H., & Szechtman, B. (1989). Increased frontal and reduced parietal glucose metabolism in acute untreated schizophrenia. *Psychiatry Research, 28,* 119–133.

Cleghorn, J. M., Szechtman, H., Garnett, E. S., Brown, G. M., Nahmias, C., Szechtman, B., Kaplan, R., & Franco, S. (1991). Neuroleptic effects on regional brain metabolism in first episode schizophrenics. *Schizophrenia Research, 5,* 208–209.

Cleghorn, J. M., Zipursky, R. B., & List, S. J. (1991). Structural and functional brain imaging in schizophrenia. *Journal of Psychiatry and Neuroscience, 16,* 53–74.

Coffey, C. E., & Figiel, G. S. (1991). Neuropsychiatric significance of subcortical encephalomalacia. In B. J. Carroll, & J. E. Barrett (Eds.), *Psychopathology and the Brain* (pp. 243–246). New York: Raven Press.

Coffey, C. E., Figiel, G. S., Djang, W. T., Cress, M., Saunders, W. B., & Weiner, R. D. (1988). Leukoencephalopathy in elderly depressed patients referred for ECT. *Biological Psychiatry, 24,* 1188–1196.

Coffey, C. E., Wilkinson, W. E., Parashos, I. A., Soady, S. A. R., Sullivan, R. J., Patterson, L. J., Figiel, G. S., Webb, M. C., Spritzer, C. E., & Djang, W. T. (1992). Quantitative cerebral anatomy of the aging human brain—a cross-sectional study using magnetic resonance imaging. *Neurology, 42,* 527–536.

Coffey, C. E., Wilkinson, W. E., Weiner, R. D., Parashos, I.A., Djang, W. T., Webb, M. C., Figiel, G. S., & Spritzer, C. E. (1993). Quantitative cerebral anatomy in depression: A controlled magnetic resonance imaging study. *Archives of General Psychiatry, 50,* 7–16.

Coffey, C. E., Wilkinson, W. E., Weiner, R. D., Ritchie, J. C., & Aque, M. (1993). The dexamethasone suppression test and quantitative cerebral anatomy in depression. *Biological Psychiatry, 33,* 442–449.

Coffman, J. A., Bornstein, R. A., Olson, S. C., Schwarzkopf, S. B., & Nasrallah, H. A. (1990). Cognitive impairment and cerebral structure by MRI in bipolar disorder. *Biological Psychiatry, 27,* 1188–1196.

Cohen, R. M., & Nordahl, T. E. (1988). Brain imaging techniques. In F. G. Howells (Ed.), *Modern Perspectives in Clinical Psychiatry* (pp. 102–129). New York: Bruner/Mazel.

Cohen, R., Semple, W., Gross, M., Nordahl, T., DeLisi, L., Holcomb, H., King, A., Morihisa, J., & Pickar, D. (1987). Dysfunction in a prefrontal substrate of sustained attention in schizophrenia. *Life Sciences, 140,* 2031–2039.

Cohen, R. M., Semple, W. E., Gross, M., Nordahl, T. E., King, A. C., Pickar, D., & Post, R. M. (1989). Evidence for common alterations in cerebral glucose metabolism in major affective disorders and schizophrenia. *Neuropsychopharmacology, 2,* 241–254.

Colombo, C., Abbruzzese, M., Livian, S., Scotti, G., Locatelli, M., Bonfanti, A., & Scarone, S. (1993). Memory functions and temporal-limbic morphology in schizophrenia. *Psychiatry Research: Neuroimaging, 50,* 45–56.

Conlon, P., & Trimble, M. R. (1988). A study of the corpus callosum in epileptic patients using magnetic resonance imaging. *Biological Psychiatry, 24,* 852–857.

Conlon, P., Trimble, M. R., & Rogers, D. (1990). A study of epileptic psychosis using magnetic resonance imaging. *British Journal of Psychiatry, 156,* 231–235.

Corkin, S., Twitchell, T. E., & Sullivan, E. V. (1979). Safety and efficacy of cingulotomy for pain and psychiatric disorder. In E. R. Hitchcock, H. T. Ballantine, & B. A. Meyerson (Eds.), *Modern Concepts in Psychiatric Surgery* (pp. 253–272). Elsevier/North-Holland Biomedical Press.

Creese, I., Burt, D. R., & Snyder, S. H. (1976). Dopamine receptor binding clinical and pharmacological potencies of anti-schizophrenic drugs. *Science, 192,* 481–483.

Crow, T. J. (1993). Origins of psychosis and the evolution of human language and communication. In N. Brunello, J. Mendelewicz, & G. Racagni (Eds.), *New Generation of Antipsychotic Drugs: Novel Mechanisms of Action* (pp. 39–61). Basel: Karger.

Cummings, J. L. (1986). Organic psychoses: Delusional disorders and secondary mania. *Psychiatric Clinics of North America, 9,* 293–311.

Cummings, J. L. (1992). Depression and Parkinson's disease: A review. *American Journal of Psychiatry, 149,* 443–454.

Cummings, J. L., & Cunningham, K. (1992). Obsessive-compulsive disorder in Huntington's disease. *Biological Psychiatry, 31,* 263–270.

Cutting, J. (1992). The role of right hemisphere dysfunction in psychiatric disorders. *British Journal of Psychiatry, 160,* 583–588.

Daniel, D. G., Goldberg, T. E., Gibbons, R. D., & Weinberger, D. R. (1991). Lack of a bimodal distribution of ventricular size in schizophrenia—a Gaussian mixture analysis of 1056 cases and controls. *Biological Psychiatry, 30,* 887–903.

Dauphinais, I. D., DeLisi, L. E., Crow, T. J., Alexandropoulos, K., Colter, N., Tuma, I., & Gershon, E. S. (1990). Reduction in temporal lobe size in siblings with schizophrenia: A magnetic resonance imaging study. *Psychiatry Research: Neuroimaging, 35,* 137–147.

Davis, P. J. M., & Wright, E. A. (1977). A new method for measuring cranial cavity volume and its application to the assessment of cerebral atrophy at autopsy. *Neuropathology and Applied Neurobiology, 3,* 341–358.

Davison, K., & Bagley, C. R. (1969). Schizophrenia-like psychoses associated with organic disorders of the central nervous system: A review of the literature. *British Journal of Psychiatry Special Publication, 4,* 113–187.

Degreef, G., Ashtari, M., Bogerts, B., Bilder, R. M., Jody, D. N., Alvir, J. M. J., & Lieberman, J. A. (1992). Volumes of ventricular system subdivisions measured from magnetic resonance images in first-episode schizophrenic patients. *Archives of General Psychiatry, 49,* 531–537.

Degreef, G., Ashtari, M., Wu, H. W., Borenstein, M., Geisler, S., & Lieberman, J. (1991). Follow-up MRI study in first episode schizophrenia. *Schizophrenia Research, 5,* 204–206.

Dekaban, A., & Sadowsky, D. (1978). Changes in brain weights during the span of human life: Relation of brain weights to body heights and body weights. *Annals of Neurology, 4,* 345–356.

DeLisi, L. S., Buchsbaum, M. S., Holcomb, H. H., Dowling-Zimmerman, S., Pickar, D., Boronow, J., Morihisa, J. M., van Kammen, D. P., Carpenter, W., Kessler, R., & Cohen, R. M. (1985). Clinical correlates of decreased anteroposterior metabolic gradients in positron emission tomography (PET) of schizophrenic patients. *American Journal of Psychiatry, 142,* 78–81.

DeLisi, L. E., Buchsbaum, M. S., Holcomb, H. H., Langston, K. C., King, A. C., Kessler, R., Pickar, D., Carpenter, W. T., Morihisa, J. M., Margolin, R., & Weinberger, D. R. (1989). Increased temporal lobe glucose use in chronic schizophrenic patients. *Biological Psychiatry, 25,* 835–851.

DeLisi, L. E., Dauphinais, D., & Gershon, E. S. (1988). Perinatal complications and reduced size of brain limbic structures in familial schizophrenia. *Schizophrenia Bulletin, 14,* 185–191.

DeLisi, L. E., Dauphinais, I. D., & Hauser, P. (1989). Gender differences in the brain: Are they relevant to the pathogenesis of schizophrenia? *Comprehensive Psychiatry, 30,* 197–208.

DeLisi, L. E., Hoff, A. L., Kushner, M., Calev, A., & Stritzke, P. (1992). Left ventricular enlargement associated with diagnostic outcome of schizophreniform disorder. *Biological Psychiatry, 32,* 199–201.

DeLisi, L. E., Hoff, A. L., Schwartz, J. E., Shields, G. W., Halthore, S. N., Gupta, S. M., Henn, F. A., & Anand, A. K. (1991). Brain morphology in first-episode schizophrenic-like psychotic patients: A quantitative magnetic resonance imaging study. *Biological Psychiatry, 29,* 159–175.

DeLisi, L. S., Holcomb, H. H., Cohen, R. M., Pickar, D., Carpenter, W., Morihisa, J. M., King, A. C., Kessler, R., & Buchsbaum, M. S. (1985). Positron emission tomography (PET) in schizophrenic patients with and without neuroleptic medication. *Journal of Cerebral Blood Flow and Metabolism, 5,* 201–206.

DeLisi, L. E., Stritzke, P., Riordan, H., Holan, V., Boccio, A., Kushner, M., McClelland, J., Van Eyl, O., & Anand, A. (1992). The timing of brain morphological changes in schizophrenia and their relationship to clinical outcome. *Biological Psychiatry, 31,* 241–254.

DeLisi, L. E., Stritzke, P. H., Holan, V., Anand, A., Boccio, A., Kuschner, M., Riordan, H., McClelland, J., & Van Eyle, O. (1991). Brain morphological changes in first episode cases of schizophrenia: Are they progressive? *Schizophrenia Research, 5,* 206–208.

De Myer, M. K., Gilmor, R. L., Hendrie, H. C., DeMyer, W. E., Augustyn, G. T., & Jackson, R. K. (1988). Magnetic resonance brain images in schizophrenic and normal subjects: Influence of diagnosis and education. *Schizophrenia Bulletin, 14,* 22–37.

Dewan, M. J., Haldipur, C. V., Boucher, M. F., Ramachandran, T., & Major, L. F. (1988). Bipolar affective disorder II. EEG, neuropsychological, and clinical correlates of CT abnormality. *Acta Psychiatrica Scandinavica, 77,* 677–682.

Dewan, M. J., Haldipur, C. V., Lane, E. E., Ispahani, A., Boucher, M. F., & Major, L. F. (1988). Bipolar affective disorder I. Comprehensive quantitative computed tomography. *Acta Psychiatrica Scandinavica, 77,* 670–676.

Dolan, R. J., Bench, C. J., Brown, R. G., Scott, L. C., Friston, K. J., & Frackowiak, R. S. J. (1992). Regional cerebral blood flow abnormalities in depressed patients with cognitive impairment. *Journal of Neurology, Neurosurgery, and Psychiatry, 55,* 768–773.

Dolan, R. J., Calloway, S. P., & Mann, A. H. (1985). Cerebral ventricular size in depressed subjects. *Psychological Medicine, 15,* 873–878.

Dolan, R. J., Calloway, S. P., Thacker, P. F., & Mann, A. H. (1986). The cerebral cortical appearance in depressed subjects. *Psychological Medicine, 16,* 775–779.

Dolan, R. J., Poynton, A. M., Bridges, P. K., & Trimble, M. R. (1990). Altered magnetic resonance white-matter T1 values in patients with affective disorder. *British Journal of Psychiatry, 157,* 107–110.

Donnelly, E. F., Weinberger, D. R., Waldman, I. N., & Wyatt, R. J. (1980). Cognitive impairment associated with morphological brain abnormalities on computed tomography in chronic schizophrenic patients. *Journal of Nervous and Mental Disease, 168,* 305–308.

Dorst, J. P. (1971). Changes of the skull during childhood. In T. H. Newton & D. G. Potts (Eds.), *Radiology of the skull and brain: The skull.* (pp. 118–131) St. Louis, MO: Mosby.

Duara, R., Gross-Glenn, K., Barker, W. W., Chang, J. Y., Apicella, A., Loewenstein, D., & Boothe, T. (1987). Behavioral activation and the variability of cerebral glucose metabolic measurements. *Journal of Cerebral Blood Flow and Metabolism, 7,* 266–271.

Dupont, R. M., Jernigan, T. L., Butters, N., Delis, D., Hesselink, J. R., Heindel, W., & Gillin, J. C. (1990). Subcortical abnormalities detected in bipolar affective disorder using magnetic resonance imaging: Clinical and neuropsychological significance. *Archives of General Psychiatry, 47,* 55–59.

Dupont, R. M., Jernigan, T. L., Gillin, J. C., Butters, N., Delis, D. C., & Hesselink, J. R. (1987). Subcortical signal hyperintensities in bipolar patients detected by MRI (letter). *Psychiatry Research, 21,* 357–358.

Early, T. S., Reiman, E. M., Raichle, M. E., & Spitznagel, E. L. (1989). Left globus pallidus abnormality in never-medicated patients with schizophrenia. *Proceedings of the National Academy of Sciences, of USA, 84,* 561–563.

Englund, E., Brun, A., & Alling, C. (1988). White matter changes in dementia of Alzheimer's type. *Brain, 111,* 1425–1439.

Falkai, P., Bogerts, B., Greve, B., Pfeiffer, U., Machus, B., Folsch-Reetz, B., Majtenyi, C., & Ovary, I. (1992). Loss of sylvian fissure asymmetry in schizophrenia—A quantitative post mortem study. *Schizophrenia Research, 7,* 23–32.

Farde, L., Hall, H., Ehrin, E., & Sedvall, G. (1986). Quantitative analysis of D2-dopamine receptor binding in the living human brain by PET. *Science, 231,* 258–261.

Farde, L., Wiesel, F.-A., Halldin, C., & Sedvall, G. (1989). Central D_1- and D_2- dopamine receptor occupancy in schizophrenic patients treated with antipsychotic drugs. *Archives of General Psychiatry, 45,* 71–76.

Farde, L., Wiesel, F.-A., Hall, H., Halldin, C., Stone-Elander, S., & Sedvall, G. (1987). No D2 receptor increase in PET study of schizophrenia. *Archives of General Psychiatry, 44,* 671–672.

Farde, L., Wiesel, F. A., Halldin, C., & Sedvall, G. (1988). Central D2-dopamine receptor occupancy in schizophrenic patients treated with antipsychotic drugs. *Archives of General Psychiatry, 45,* 71–76.

Farde, L., Wiesel, F. A., Stone-Elander, S., Halldin, C., Nordstrom, A. L., Hall, H., & Sedvall, G. (1990). D2 dopamine receptors in neuroleptic-naive schizophrenic patients: A positron emission tomography study with ^{11}C-raclopride. *Archives of General Psychiatry, 47,* 213–219.

Figiel, G. S., Krishnan, K. R. R., Rao, V. P., Doraiswamy, M., Ellinwood, E. H., Nemeroff, C. B.,

Evans, D., & Boyko, O. (1991). Subcortical hyperintensities on brain magnetic resonance imaging: A comparison of normal and bipolar subjects. *Journal of Neuropsychiatry and Clinical Neurosciences, 3,* 18–22.

Flaum, M., Arndt, S., & Andreasen, N. C. (1990). The role of gender in studies of ventricle enlargement in schizophrenia: A predominantly male effect. *American Journal of Psychiatry, 147,* 1327–1332.

Flaum, M. A., Swayze, V. W., Gupta, S., O'Leary, D. S., & Andreasen, N. C. (1993). Symptom domains and neural substrates of schizophrenia: A magnetic resonance imaging study. *Schizophrenia Research, 9,* 198–199.

Flor-Henry, P. (1969). Psychosis and temporal lobe epilepsy: A controlled investigation. *Epilepsia, 10,* 363–395.

Flor-Henry, P. (1989). Psychopathology and hemispheric specialization: Left hemisphere dysfunction in schizophrenia, psychopathy, hysteria and obsessional syndrome. In F. Boller & J. Graffman (Eds.), *Handbook of neuropsychology* (pp. 477–494). Amsterdam: Elsevier.

Folstein, S. E., Folstein, M. F., & McHugh, P. R. (1979). Psychiatric syndromes in Huntington's disease. *Advances in Neurology, 23,* 281–289.

Friedman, L., Knutson, L., Shurell, M., & Meltzer, H. Y. (1991). Prefrontal sulcal prominence is inversely related to response to clozapine in schizophrenia. *Biological Psychiatry, 29,* 865–877.

Friston, K. J., Liddle, P. F., Frith, C. D., Hirsch, S. R., & Frackowiak, R. S. J. (1992). The left medial temporal region and schizophrenia—a PET study. *Brain, 115,* 367–382.

Gallhofer, B., Trimble, M. R., Frackowiak, R., Gibbs, J., & Jones, T. (1985). A study of cerebral blood flow and metabolism in epileptic psychosis using positron emission tomography and oxygen. *Journal of Neurology, Neurosurgery, and Psychiatry, 48,* 201–206.

Garber, J., Ananth, J. V., Chiu, L. C., Griswold, V. J., & Oldendorf, W. H. (1989). Nuclear magnetic resonance study of obsessive-compulsive disorder. *American Journal of Psychiatry, 146,* 1001–1005.

Glejser, H. (1969). A new test for heteroscedacity. *Journal of the American Statistical Association, 64,* 316–323.

Gold, J., Goldberg, T., & Weinberger, D. (1992). Prefrontal function and schizophrenic symptoms. *Neuropsychiatry, Neuropsychology, and Behavioral Neurology, 5,* 253–261.

Goldberg, T. E., Gold, J. M., & Braff, D. L. (1991). Neuropsychological functioning and time-linked information processing in schizophrenia. In A. Tasman & S.M. Goldfinger (Eds.), *Review of psychiatry* (pp. 60–78). Washington, DC: American Psychiatric Press.

Goldberg, T. E., Kleinman, J. E., Daniel, D. G., Myslobodsky, M. S., Ragland, J. D., & Weinberger, D. R. (1988). Dementia praecox revisited age disorientation, mental status, and venticular enlargement. *British Journal of Psychiatry, 153,* 187–190.

Goldberg, T. E., Ragland, J. D., Torrey, E. F., Gold, J. M., Bigelow, L. B., and Weinberger, D. R. (1990). Neuropsychological assessment of monozygotic twins discordant for schizophrenia. *Archives of General Psychiatry, 47,* 1066–1072.

Golden, C. J., Moses, J. A., Zelazowski, R., Graber, B., Zatz, L. M., Horvath, T. B., & Berger, P. A. (1980). Cerebral ventricular size and neuropsychological impairment in young chronic schizophrenics. *Archives of General Psychiatry, 37,* 619–623.

Goldstein, J. M. (1988). Gender differences in the course of schizophrenia. *American Journal of Psychiatry, 145,* 684–689.

Goldstein, P. C., Brown, G. G., Welch, K. M. A., Marcus, A., Ewing, J. R., & Rosenbaum, G. (1985). Age-related decline of rCBF in schizophrenia and major affective disorder. *Journal of Cerebral Blood Flow and Metabolism,* 5(Suppl.):S203–S204.

Gruzelier, J., Seymour, K., Wilson, L., Jolley, A., & Hirsch, S. (1988). Impairments on neuropsychologic tests of temporohippocampal and frontohippocampal functions and word fluency in remitting schizophrenia and affective disorders. *Archives of General Psychiatry, 45,* 623–629.

Gur, R. C., Saykin, A. J., & Gur, R. E. (1991). Neuropsychological study of schizophrenia. *Schizophrenia Research, 1,* 153–162.

Gur, R. E., Resnick, S. M., Alavi, A., Gur, R. C., Caroff, S., Dann, R., Silver, F. L., Saykin, A. J., Chawluk, J. B., Kushner, M., & Reivich, M. (1987). Regional brain function in schizophrenia: I. A positron emission tomography study. *Archives of General Psychiatry, 44,* 119–125.

Gur, R. E., Resnick, S. M., Gur, R. C., Alavi, A., Caroff, S., Kushner, M., & Reivich, M. (1987). Regional brain function in schizophrenia: II. Repeated evaluation with positron emission tomography. *Archives of General Psychiatry, 44,* 126–129.

Gur, R. E., Skolnick, B. E., Gur, R. C., & Caroff, S. (1984). Brain function in psychiatric disorders. II.

Regional cerebral blood flow in medicated unipolar depressives. *Archives of General Psychiatry, 41,* 695–699.

Gustafson, L., Risberg, J., & Silfverskiold, P. (1981). Cerebral blood flow in dementia and depression. *Lancet, 1,* 275.

Guze, B. H., & Szuba, M. P. (1992). Leukoencephalopathy and major depression: A preliminary report. *Psychiatry Research, 45,* 169–175.

Harvey, I., Ron, M., du Bouley, G., Wicks, D., Lewis, S., & Murray, R. M. (1993). Reduction of cortical volume in schizophrenia on magnetic resonance imaging. *Psychological Medicine, 23,* 591–604.

Hauser, P., Altshuler, L. L., Berrettini, W., Dauphenais, I. D., Gelernter, J., & Post, R. M. (1989). Temporal lobe measurement in primary affective disorder by magnetic resonance imaging. *Journal of Neuropsychiatry and Clinical Neurosciences, 1,* 128–134.

Hauser, P., Dauphinais, D., Berrettini, W., DeLisi, L. E., Gelernter, J., & Post, R. M. (1989). Corpus callosum dimensions measured by magnetic resonance imaging in bipolar affective disorder and schizophrenia. *Biological Psychiatry, 26,* 659–668.

Hines, M., & Green, R. (1991). Human hormonal and neural correlates of sex-typed behaviors. In A. Tasman & S. M. Goldfinger (Eds.), *American Psychiatric Press Review of Psychiatry* (pp. 536–555). Washington, DC: American Psychiatric Press.

Hoehn-Saric, R., Pearlson, G. D., Harris, G. J., Machlin, S. R., & Camargo, E. E. (1991). Effects of fluoxetine on regional cerebral blood flow in obsessive-compulsive patients. *American Journal of Psychiatry, 148,* 1243–1245.

Hoff, A. L., Riordan, H., O'Donnell, D., Stritzke, P., Neale, C., Boccio, A., Anand, A. K., & DeLisi, L. E. (1992). Anomalous lateral sulcus asymmetry and cognitive function in first-episode schizophrenia. *Schizophrenia Bulletin, 18,* 257–268.

Hoff, A. R., O'Donnell, D. W., Morris, L., & DeLisi, L. E. (1992). Neuropsychological functioning in first-episode schizophreniform patients. *American Journal of Psychiatry, 149,* 898–903.

Hoffman, W. F., Ballard, L., Turner, E. H., & Casey, D. E. (1991). Three-year follow-up of older schizophrenics: Extrapyramidal syndromes, psychiatric symptoms, and ventricular brain ratio. *Biological Psychiatry, 30,* 913–926.

Hollander, E., Schiffman, E., Cohen, B., Riverastein, M. A., Rosen, W., Gorman, J. M., Fyer, A. J., Papp, L., & Liebowitz, M. R. (1990). Signs of central nervous system dysfunction in obsessive-compulsive disorder. *Archives of General Psychiatry, 47,* 27–32.

Husain, M. M., McDonald, W. M., Doraiswamy, P. M., Figiel, G. S., Na, C., Escalona, P. R., Boyko, O. B., Nemeroff, C. B., & Krishnan, K. R. K. (1991). A magnetic resonance imaging study of putamen nuclei in major depression. *Psychiatry Research, 40,* 95–99.

Hyde, T. M., Casanova, M. F., Kleinman, J. E., & Weinberger, D. R. (1991). Neuroanatomical and neurochemical pathology in schizophrenia. In A. Tasman & S. M. Goldfinger (Eds.), *American Psychiatric Press Review of Psychiatry* (pp. 7–23). Washington, DC: American Psychiatric Press.

Iacono, W. G., Smith, G. N., Moreau, M., Beiser, M., Fleming, J. A. E., Lin, T. L., & Flak, B. (1988). Ventricular and sulcal size at the onset of psychosis. *American Journal of Psychiatry, 145,* 820–824.

Illowsky, B., Juliano, D. M., Bigelow, L. B., & Weinberger, D. R. (1988). Stability of CT scan findings in schizophrenia: Results of an 8-year follow-up study. *Journal of Neurology, Neurosurgery, and Psychiatry, 51,* 209–213.

Ingvar, D. H., & Franzen, G. (1974). Abnormalities of cerebral blood flow distribution in patients with chronic schizophrenia. *Acta Psychiatrica Scandinavica, 50,* 425–462.

Insel, T. R. (1992). Toward a neuroanatomy of obsessive-compulsive disorder. *Archives of General Psychiatry, 49,* 739–744.

Insel, T. R., Donnelly, E. F., Lalakea, M. L., Alterman, I. S., & Murphy, D. S., (1983). Neurological and neuropsychological studies of patients with obsessive-compulsive disorder. *Biological Psychiatry, 18,* 741–751.

Jacoby, R. J., & Levy, R. (1980). Computed tomography in the elderly. 3. Affective disorder. *British Journal of Psychiatry, 136,* 270–275.

Jernigan, T. L., Archibald, S. L., Berhow, M. T., Sowell, E. R., Foster, D. S., & Hesselink, J. R. (1991). Cerebral structure on MRI. 1. Localization of age-related changes. *Biological Psychiatry, 29,* 55–67.

Jernigan, T. L., Sargent, T., Pfefferbaum, A., Kusubov, N., & Stahl, S. (1985). 18 Fluorodeoxyglucose PET in schizophrenia. *Psychiatry Research, 16,* 317–329.

Jernigan, T. L., Zatz, L. M., Moses, J. A., & Berger, P. A. (1982). Computed tomography in schizophrenics and normal volunteers: I. Fluid volume. *Archives of General Psychiatry, 39,* 765–770.

Jernigan, T. L., Zisook, S., Heaton, R. K., Moranville, J. T., Hesselink, J. R., & Braff, D. L. (1991).

Magnetic resonance imaging abnormalities in lenticular nuclei and cerebral cortex in schizophrenia. *Archives of General Psychiatry, 48,* 881–890.
Johnstone, E. C., Crow, T. J., Frith, D. C., Husband, J., & Kreel, L. (1976). Cerebral ventricular size and cognitive impairment in schizophrenia. *Lancet, 2,* 924–926.
Johnstone, E. C., Owens, D. G., Crow, T. J., Alexandropolis, K., Bydder, G., & Colter, N. (1989). Temporal lobe structure as determined by nuclear magnetic resonance in schizophrenia and bipolar affective disorder. *Journal of Neurology, Neurosurgery, and Psychiatry, 52,* 736–741.
Kawasaki, Y., Maeda, Y., Urata, K., Higashima, M., Yamguchi, N., Suzuki, M., Takashima, T., & Ide, Y. (1993). A quantitative magnetic resonance imaging study of patients with schizophrenia. *European Archives of Psychiatry and Clinical Neuroscience, 242,* 268–272.
Keilp, J. G., Sweeney, J. A., Jacobsen, P., Solomon, C., St. Louis, L., Deck, M., Frances, A., & Mann, J. J. (1988). Cognitive impairment in schizophrenia: Specific relations to ventricular size and negative symptomatology. *Biological Psychiatry, 24,* 47–55.
Kellner, C. H., Jolley, R. R., Holgate, R. C., Austin, L., Lydiard, R. B., Laraia, M., & Ballenger, J. C. (1991). Brain MRI in obsessive-compulsive disorder. *Psychiatry Research, 36,* 45–49.
Kellner, C. H., Rubinow, D. R., Gold, P. W., & Post, R. M. (1983). Relationship of cortisol hypersecretion to brain CT scan alterations in depressed patients. *Psychiatry Research, 8,* 191–197.
Kelsoe, J. R., Cadet, J. L., Pickar, D., & Weinberger, D. R. (1988). Quantitative neuroanatomy in schizophrenia. *Archives of General Psychiatry, 45,* 533–541.
Kemali, D., Maj, M., Galderisi, S., Milici, N., & Salvati, A. (1989). Ventricle-to-brain ratio in schizophrenia: A controlled follow-up study. *Biological Psychiatry, 26,* 753–756.
Kessler, J., Herholz, K., Grond, M., & Heiss, W. D. (1991). Impaired metabolic activation in Alzheimer's disease—PET study during continuous visual recognition. *Neuropsychologia, 29,* 229–243.
Kety, S. S., Woodford, R. B., Harmel, M. H., Freyhan, F. A., Appel, K. E., & Schmidt, C. F. (1948). Cerebral blood flow and metabolism in schizophrenia—the effects of barbituate semi-narcosis, insulin coma, and electroshock. *American Journal of Psychiatry, 104,* 765–770.
Kishimoto, H., Kuwahara, H., & Ohno, S. (1987). Three subtypes of chronic schizophrenia identified using 11-C-glucose positron emission tomography. *Psychiatry Research, 21,* 285–292.
Kishimoto, H., Takazu, O., Ohno, S., Yamaguchi, T., Fujita, H., Kuwahara, H., Ishii, T., Matsushita, M., Yokoi, S., & Iio, M. (1987). ^{11}C-glucose metabolism in manic and depressed patients. *Psychiatry Research, 22,* 81–88.
Kling, A. S., Metter, E. J., Riege, W. H., & Kuhl, D. E. (1986). Comparison of PET measurement of local brain glucose metabolism and CAT measurement of brain atrophy in chronic schizophrenia and depression. *American Journal of Psychiatry, 143,* 175–180.
Kolb, B., & Whishaw, I. Q. (1983). Performance of schizophrenic patients on tests sensitive to left or right frontal, temporal, or parietal function in neurological patients. *Journal of Nervous and Mental Disease, 171,* 435–443.
Krishnan, K. R. K., Doraiswamy, P. M., Figiel, G. S., Husain, M. M., Shah, S. A., Na, C., Boyko, O. G., McDonald, W. M., Nemeroff, C. B., & Ellinswood, E. H. (1991). Hippocampal abnormalities in depression. *Journal of Neuropsychiatry and Clinical Neurosciences, 3,* 387–391.
Krishnan, K. R. R., Goli, V., Ellinwood, E. H., France, R. D., Blazer, D. G., & Nemeroff, C. B. (1988). Leukoencephalopathy in patients diagnosed as major depressive. *Biological Psychiatry, 23,* 519–522.
Krishnan, K. R. R., McDonald, W. M., Escalona, P. R., Doraiswamy, P. M., Na, C., Husain, M. M., Figiel, G. S., Boyko, O. B., Ellinwood, E. H., & Nemeroff, C. B. (1992). Magnetic resonance imaging of the caudate nuclei in depression: Preliminary observations. *Archives of General Psychiatry, 49,* 553–557.
Laplane, D., Levasseur, M., Pillon, B., Dubois, B., Baulac, M., Mazoyer, B., Tran Dinh, S., Sette, G., Danze, F., & Baron, J. C. (1989). Obsessive-compulsive and other behavioral changes with bilateral basal ganglia lesions. *Brain, 112,* 699–725.
Lewis, S. (1992). Sex and schizophrenia: Vive la difference. *British Journal of Psychiatry, 161,* 445–450.
Lewis, S. W. (1990). Computerized tomography in schizophrenia. 15 years on. *British Journal of Psychiatry,* 157 (Suppl.), 16–24.
Lewis, S. W., Ford, R. A., Syed, G. M. S., Reveley, A. M., & Toone, B. K. (1991). A controlled study of ^{99}m technetium single photon emission tomography in chronic schizophrenia. *Psychological Medicine, 22,* 27–35.
Liddle, P. F. (1987). The syndromes of chronic schizophrenia: A re-examination of the positive-negative dichotomy. *British Journal of Psychiatry, 151,* 145–151.

Liddle, P. F., Friston, K. J., Frith, C. D., Hirsch, S. R., Jones, T., & Frackowiak, R. S. J. (1992). Patterns of cerebral blood flow in schizophrenia. *British Journal of Psychiatry, 160,* 179–186.

Liddle, P. F., Friston, K. J., Hirsch, S. R., & Frackowiak, R. S. J. (1990). Regional cerebral metabolic activity in chronic schizophrenia. *Schizophrenia Research, 3,* 23–24.

Liddle, P. F., & Morris, D. L. (1991). Schizophrenic syndromes and frontal lobe performance. *British Journal of Psychiatry, 158,* 340–345.

Lim, K. O., Beal, M., Harvey, Jr., R. L., Myers, T., Lane, B., Sullivan, E. V., Faustman, W. O., & Pfefferbaum, A. (1995). Brain dysmorphology in adults with congenital rubella plus schizophrenialike symptoms. *Biological Psychiatry, 37,* 764–776.

Lim, K. O., Zipursky, R. B., Watts, M. C., & Pfefferbaum, A. (1992). Decreased grey matter in normal aging: An in vivo MR study. *Journal of Gerontology, 47,* B26–B30.

Luchins, D. J., Lewine, R. R. J., & Meltzer, H. Y. (1984). Lateral ventricular size, psychopathology, and medication response in the psychoses. *Biological Psychiatry, 19,* 29–44.

Luxenberg, J. S., Swedo, S. E., Flament, M. F., Friedland, R. P., Rapoport, J., & Rapoport, S. I. (1988). Neuroanatomical abnormalities in obsessive-compulsive disorder detected with quantitative x-ray computed tomography. *American Journal of Psychiatry, 145,* 1089–1093.

Machlin, S. R., Harris, G. J., Pearlson, G. D., Hoehn-Saric, R., Jeffery, P., & Camargo, E. E. (1991). Elevated medial-frontal cerebral blood flow in obsessive-compulsive patients—a SPECT study. *American Journal of Psychiatry, 148,* 1240–1242.

Maes, M., Dierckx, R., Meltzer, H. Y., Ingels, M., Schotte, C., Vandewoude, M., Calabrese, J., Cosyns, P. (1993). Regional cerebral blood flow in unipolar depression measured with Tc-99m-HMPAO single photon emission computed tomography: Negative findings. *Psychiatry Research: Neuroimaging, 50,* 77–88.

Manschreck, T. C. (1986). Motor abnormalities in schizophrenic disorders. In H. A. Nasrallah & D. R. Weinberger (Eds.), *Handbook of schizophrenia: Vol. 1. The neurology of schizophrenia* (pp. 65–96). New York: Elsevier.

Marsh, L., Harris, D., Lim, K. O., Beal, M., Hoff, A. L., Minn, K., Csernansky, J. G., DeMent, S., Faustman, W. O., Sullivan, E. V., & Pfefferbaum, A. (submitted for publication). Structural MRI abnormalities in adults with severe chronic schizophrenia and an early age onset of clinical onset.

Marsh, L., Suddath, R. L., Higgins, N., & Weinberger, D. R. (1994). Medial temporal lobe structures in schizophrenia: Relationship of size to duration of illness. *Schizophrenia Research, 11*(3), 225–238.

Marsh, L., & Weinberger, D. R. (1993). Structural brain changes in schizophrenia: The issue of subgroups. In K. Maurer (Ed.), *Imaging of the brain in psychiatry and related fields* (pp. 19–26). New York: Springer-Verlag.

Marshall, E. J., Syed, G. M. S., Fenwick, P. B. C., & Lishman, W. A. (1993). A pilot study of schizophrenia-like psychosis in epilepsy using single-photon emission computerised tomography. *British Journal of Psychiatry, 163,* 32–36.

Mathalon, D., Sullivan, E. V., Rawles, J. M., & Pfefferbaum, A. (1993). Correction for head size in brain imaging measurements. *Psychiatry Research: Neuroimaging, 50,* 121–139.

Mathew, R. J., Meyer, J. S., Francis, D. J., Semchuk, K. M., Mortel, K., & Claghorn, J. L. (1980). Cerebral blood flow in depression. *American Journal of Psychiatry, 137,* 1449–1450.

Mayberg, H. S., Starkstein, S. E., Peyser, C. E., Brandt, J., Dannals, R. F., & Folstein, S. E. (1992). Paralimbic frontal lobe hypometabolism in depression associated with Huntington's disease. *Neurology, 42,* 1791–1797.

Mayberg, H. S., Starkstein, S. E., Sadzot, B., Preziosi, T., Andrezejewski, P. L., Dannals, R. F., Wagner, H. N., & Robinson, R. G. (1990). Selective hypometabolism in the inferior frontal lobe in depressed patients with Parkinson's disease. *Annals of Neurology, 28,* 57–64.

McCarley, R. W., Shenton, M. E., O'Donnell, B. F., Faux, S. F., Kikinis, R., Nestor, P. G., & Jolesz, F. A. (1993). Auditory P300 abnormalities and left posterior superior temporal gyrus volume reduction in schizophrenia. *Archives of General Psychiatry, 50,* 190–197.

McDonald, W. M., Krishnan, K. R. R., Doraiswamy, P. M., & Blazer, D. G. (1991). Occurrence of subcortical hyperintensities in elderly subjects with mania. *Psychiatry Research: Neuroimaging, 40,* 211–220.

McKenna, P. J., Kane, J. M., & Parrish, K. (1985). Psychotic syndromes in epilepsy. *American Journal of Psychiatry, 142,* 895–904.

McKenna, P. J., Tamlyn, D., Lund, C. E., Mortimer, A. M., Hammond, S., & Baddeley, A. D. (1990). Amnesic syndrome in schizophrenia. *Psychological Medicine, 20,* 967–972.

McLean, P. D. (1952). Some psychiatric implications of physiological studies on frontotemporal

portion of limbic system (visceral brain). *Electroencephalography and Clinical Neurophysiology, 4,* 407–418.
Mesulam, M. M., & Geschwind, N. (1983). On the possible role of neocortex and its limbic connections in the process of attention and schizophrenia. *Psychiatry Research, 14,* 249–261.
Miller, B., Lesser, I., Boone, K., Goldberg, M., Hill, E., Miller, M., Benson, D., & Mehringer, M. (1989). Brain white-matter lesions and psychosis. *British Journal of Psychiatry, 155,* 73–78.
Milner, B. (1963). Effects of different brain lesions on card sorting. *Archives of Neurology, 9,* 90–100.
Murphy, D. G. M., Decarli, C., Schapiro, M. B., Rapoport, S. I., & Horwitz, B. (1992). Age-related differences in volumes of subcortical nuclei, brain matter, and cerebrospinal fluid in healthy men as measured with magnetic resonance imaging. *Archives of Neurology, 49,* 839–845.
Nasrallah, H. A. (1991). Neurodevelopmental aspects of bipolar affective disorder. *Biological Psychiatry, 29,* 1–2.
Nasrallah, H., Andreasen, S., Coffman, J., Olson, S., Dunn, V., Erhardt, J., & Chapman, S. (1986). A controlled magnetic resonance imaging study of corpus callosum thickness in schizophrenia. *Biological Psychiatry, 21,* 274–282.
Nasrallah, H. A., Coffman, J. A., & Olson, S. C. (1989). Structural brain-imaging findings in affective disorders: An overview. *Journal of Neuropsychiatry and Clinical Neurosciences, 1,* 21–26.
Nasrallah, H. A., Jacoby, C. G., & McCalley-Whitters, M. (1981). Cerebellar atrophy in schizophrenia and mania. *Lancet, 1,* 1102.
Nasrallah, H. A., McCalley-Whitters, M., & Jacoby, C. G. (1982). Cerebral ventricular enlargement in young manic males. *Journal of Affective Disorders, 4,* 15–19.
Nasrallah, H. A., Olson, S. C., McCalley-Whitters, M., Chapman, S., & Jacoby, C. G. (1986). Cerebral ventricular enlargement in schizophrenia: A preliminary follow-up study. *Archives of General Psychiatry, 43,* 157–159.
Nasrallah, H. A., Schwarzkopf, S. B., Olson, S. C., & Coffman, J. A. (1990). Gender differences in schizophrenia on MRI brain scans. *Schizophrenia Bulletin, 16,* 205–209.
Nauta, W. J. H. (1971). The problem of the frontal lobe: A reinterpretation. *Journal of Psychiatric Research, 8,* 167–187.
Nordahl, T. E., Benkelfat, C., Semple, W., Gross, M., King, A. C., & Cohen, R. M. (1989). Cerebral glucose metabolic rates in obsessive-compulsive disorder. *Neuropsychopharmacology, 2,* 23–28.
Nordahl, T. E., Kusubov, N., Carter, C., Salamat, M. S., Cummings, A. M., O'Shora-Celaya, L., Eberling, J., Robertson, L., Huesman, R., Jagust, W., & Budinger, T. F. (in press). Temporal lobe metabolic differences in medication-free out-patients with schizophrenia via the PET-600. *Neuropsychopharmacology.*
Overall, J. E., & Gorham, D. R. (1962). The Brief Psychiatric Rating Scale. *Psychological Reports, 10,* 799–812.
Pakkenberg, B. (1987). Post-mortem study of chronic schizophrenic brains. *British Journal of Psychiatry, 151,* 744–752.
Pandurangi, A. K., Dewan, M. J., Boucher, M., Levy, B., Ramachandran, T., Bartell, K., Bick, P. A., Phelps, B. H., and Major, L. (1986). A comprehensive study of chronic schizophrenic patients II. Biological, neuropsychological, and clinical correlates of CT abnormality. *Acta Psychiatrica Scandinavica, 73,* 161–171.
Papez, J. W. (1937). A proposed mechanism of emotion. *Archives of Neurology, 38,* 725–743.
Pearlson, G. D., Garbacz, D. J., Breakey, W. R., Ahn, H. S., & DePaulo, J. R. (1984). Lateral ventricular enlargement associated with persistent unemployment and negative symptoms in both schizophrenia and bipolar disorder. *Psychiatry Research, 12,* 1–9.
Pearlson, G. D., Garbacz, D. J., Moberg, P. J., Ahn, H. S., & DePaulo, J. R. (1985). Symptomatic, familial, perinatal, and social correlates of computerized axial tomography (CAT) changes in schizophrenics and bipolars. *Journal of Nervous and Mental Disease, 173,* 42–50.
Pearlson, G. D., Garbacz, D. J., Tompkins, R. H., Ahn, H. S., Gutterman, D. F., Veroff, A. E., & DePaulo, J. R. (1984). Clinical correlates of lateral ventricular enlargement in bipolar affective disorder. *American Journal of Psychiatry, 141,* 253–256.
Pearlson, G. D., & Marsh, L. (1993). Magnetic resonance imaging in psychiatry. In J. U. Oldham, M. B. Riba, & A. Tasmas (Eds.), *Annual Review of Psychiatry, Vol 12* (pp. 347–381). Washington, DC: American Psychiatric Association Press.
Pearlson, G. D., & Veroff, A. E. (1981). Computerized tomographic scan changes in manic-depressive illness (letter). *Lancet, 2,* 470.
Perez, M. M., Trimble, M. R., Murray, N. M. F., & Reider, I. (1985). Epileptic psychosis: An evaluation of present state examination profiles. *British Journal of Psychiatry, 146,* 155–163.

Peyser, C. E., & Folstein, S. E. (1990). Huntington's disease as a model for mood disorders. Clues from neurochemistry and neuropathology. *Molecular and Chemical Neuropathology, 12,* 99–119.

Pfefferbaum, A., Lim, K. O., Rosenbloom, M., & Zipursky, R. B. (1990). Brain magnetic resonance imaging: Approaches for investigating schizophrenia. *Schizophrenia Bulletin, 16,* 453–476.

Pfefferbaum, A., Lim, K. O., Zipursky, R. B., Mathalon, D. H., Lane, B., Ha, C. N., Rosenbloom, M. J., & Sullivan, E. V. (1992). Brain gray and white matter volume loss accelerates with aging in chronic alcoholics: A quantitative MRI study. *Alcoholism: Clinical and Experimental Research, 16,* 1078–1089.

Pfefferbaum, A., Mathalon, D. H., Sullivan, E. V., Rawles, J. M., Zipursky, R. B., & Lim, K. O. (1994). A quantitative magnetic resonance imaging study of changes in brain morphology from infancy to late adulthood. *Archives of Neurology, 51,* 874–887.

Pfefferbaum, A., Sullivan, E. V., Jernigan, T. L., Zipursky, R. B., Rosenbloom, M. J., Yesavage, J. A., & Tinklenberg, J. R. (1990). A quantitative analysis of CT and cognitive measures in normal aging and Alzheimer's disease. *Psychiatry Research: Neuroimaging, 35,* 115–136.

Pfefferbaum, A., Sullivan, E. V., Rosenbloom, M. J., Shear, P. K., Mathalon, D. H., & Lim, K. O. (1993). Increase in brain cerebrospinal fluid volume is greater in older than younger alcoholic patients: A replication study and CT/MRI comparison. *Psychiatry Research: Neuroimaging, 50,* 257–274.

Pfefferbaum, A., Zatz, L., & Jernigan, T. L. (1986). Computer-interactive method for quantifying cerebrospinal fluid and tissue in brain CT scans: Effects of aging. *Journal of Computer Assisted Tomography, 10,* 571–578.

Pfefferbaum, A., & Zipursky, R. B. (1991). Neuroimaging studies of schizophrenia. *Schizophrenia Research, 4,* 193–208.

Pfefferbaum, A., Zipursky, R. B., Lim, K. O., Zatz, L. M., Stahl, S. M., & Jernigan, T. L. (1988). Computed tomographic evidence for generalized sulcal and ventricular enlargement in schizophrenia. *Archives of General Psychiatry, 45,* 633–640.

Phelps, M. E., Kuhl, D. E., & Mazziotta, J. C. (1981). Metabolic mapping of the brain's response to visual stimulation: Studies in humans. *Science, 211,* 1445–1448.

Phelps, M. E., Mazziotta, J. C., Baxter, L., & Gerner, R. (1984). Positron emission tomographic study of affective disorders: Problems and strategies. *Annals of Neurology, 15* (Suppl. 1):S149–S156.

Posner, M. I., Early, T. S., Reiman, E., Pardo, P. J., & Dhawan, M. (1988). Asymmetries in hemispheric control of attention in schizophrenia. *Archives of General Psychiatry, 45,* 814–821.

Post, R. M., DeLisi, L. E., Holcomb, H. H. Uhde, T. W., Cohen, R., & Buchsbaum, M. S. (1987). Glucose utilization in the temporal cortex of affectively ill patients: Positron emission tomography. *Biological Psychiatry, 22,* 545–553.

Rabins, P. V., Pearlson, G. D., Aylward, E., Kumar, A. J., & Dowell, K. (1991). Cortical magnetic resonance imaging changes in elderly inpatients with major depression. *American Journal of Psychiatry, 148,* 617–620.

Raine, A., Harrison, G. N., Reynolds, G. P., Sheard, C., Cooper, J. E., & Medley, I. (1990). Structural and functional characteristics of the corpus callosum in schizophrenics, psychiatric controls, and normal controls—a magnetic resonance imaging and neuropsychological evaluation. *Archives of General Psychiatry, 47,* 1060–1064.

Raine, A., Lencz, T., Reynolds, G., Harrison, G., Sheard, C., Medley, I., Reynolds, L., & Cooper, J. (1992). An evaluation of structural and functional prefrontal deficits in schizophrenia: MRI and neuropsychological measures. *Psychiatry Research: Neuroimaging, 45,* 123–137.

Rangel-Guerra, R. A., Periz Payan, H., Minkoff, L., & Todd, L. E. (1983). Nuclear magnetic resonance in bipolar affective disorders. *American Journal of Neuroradiology, 4,* 229–231.

Rapoport, J. L., & Wise, S. P. (1988). Obsessive-compulsive disorder: Evidence for basal ganglia dysfunction. *Psychopharmacology Bulletin, 24,* 380–384.

Raz, S., Raz, N., & Bigler, E. D. (1988). Ventriculomegaly in schizophrenia: Is the choice of controls important? *Psychiatry Research, 24,* 71–77.

Raz, S., Raz, N., Weinberger, D. R., Boronow, J., Pickar, D., Bigler, E., & Turkheimer, E. (1987). Morphological brain abnormalities in schizophrenia determined by computed tomography: A problem of measurement? *Psychiatry Research, 22,* 91–98.

Risch, S. C., Lewine, R. J., Kalin, N. H., Jewart, R. D., Risby, E. D., Caudle, J. M., Stipetic, M., Turner, J., Eccard, M. B., & Pollard, W. E. (1992). Limbic-hypothalamic-pituitary-adrenal axis activity and ventricular-to-brain ratio studies in affective illness and schizophrenia. *Neuropsychopharmacology, 6,* 95–100.

Robbins, T. W. (1990). The case for frontostriatal dysfunction in schizophrenia. *Schizophrenia Bulletin, 16,* 391–402.

Robinson, R. G., Kubos, K. L., Starr, L. B., Rao, K., & Price, T. R. (1984). Mood disorders in stroke patients: Importance of location of lesion. *Brain, 107,* 81–93.

Rosenberg, R., Vostrup, S., Andersen, A., & Bolwig, T. G. (1988). Effect of ECT on cerebral blood flow in melancholia assessed with SPECT. *Convulsive Therapy, 4,* 62–73.

Rosse, R. B., Schwartz, B. L., Mastropaolo, J., Goldberg, R. L., & Deutsch, S. I. (1991). Subtype diagnosis in schizophrenia and its relation to neuropsychological and computerized tomography measures. *Biological Psychiatry, 30,* 63–72.

Rossi, A., Stratta, P., Dalbenzio, L., Tartaro, A., Schiazza, G., Dimichele, V., Bolino, F., & Casacchia, M. (1990). Reduced temporal lobe areas in schizophrenia—preliminary evidences from a controlled multiplanar magnetic resonance imaging study. *Biological Psychiatry, 27,* 61–68.

Rossi, A., Stratta, P., Mancini, F., Decataldo, S., & Casacchia, M. (1993). Cerebellar vermal size in schizophrenia—a male effect. *Biological Psychiatry, 33,* 354–357.

Rossi, A., Stratta, P., Mattei, P., Cupillari, M., Bozzao, A., Gallucci, M., & Casacchia, M. (1992). Planum temporale in schizophrenia—A magnetic resonance study. *Schizophrenia Research, 7,* 19–22.

Rothschild, A. J., Benes, F., Hebben, N., Woods, B., Luciana, M., Bakanas, E., Samson, J., & Schatzberg, A. (1989). Relationships between brain CT scan findings and cortisol in psychotic and nonpsychotic depressed patients. *Biological Psychiatry, 26,* 565–575.

Roy-Byrne, P. P., Post, R. M., Kellner, C. H., Joffe, R. T., & Uhde, T. W. (1988). Ventricular-brain ratio and life course of illness in patients with affective disorder. *Psychiatry Research, 23,* 277–284.

Rubin, P., Holm, S., Friberg, L., Videbech, P., Andersen, H. S., Bendsen, B. B., Stromso, N., Larsen, J. K., Lassen, N. A., & Hemmingsen, R. (1991). Altered modulation of prefrontal and subcortical brain activity in newly diagnosed schizophrenia and schizophreniform disorder—a regional cerebral blood flow study. *Archives of General Psychiatry, 48,* 987–995.

Rubin, R. T., Villanueva-Meyer, J., Ananth, J., Trajmar, P. G., & Mena, I. (1992). Regional xenon 133 cerebral blood flow and cerebral technetium 99m HMPAO uptake in unmedicated patients with obsessive-compulsive disorder and matched normal control subjects. *Archives of General Psychiatry, 49,* 695–702.

Ryvlin, P., Philippon, B., Cinotti, L., Froment, J. C., Le Bars, D., & Mauguiere, F. (1992). Functional neuroimaging strategy in temporal lobe epilepsy—A comparative study of 18FDG-PET and Tc-99m-HMPAO-SPECT. *Annals of Neurology, 31,* 650–656.

Sacchetti, E., Vita, A., Conte, G., Pennati, A., Alciati, A., Calzeroni, A., Invernizzi, G., & Cazzullo, C. L. (1987). Cerebral ventricular size and clinical response to lithium prophylaxis in major effective disorder. *International Journal of Neuroscience, 32,* 51.

Sackeim, H. A., Prohovnik, I., Apter, S., Lucas, L., Decina, P., Mukherjee, S., Prudic, J., & Malitz, S. (1987). Regional cerebral blood flow in affective disorders: Relations to phenomenology and effects of treatment. In R. Takahashi, P. Flor-Henry, J. Gruzelier, & S. Niwa (Eds.), *Cerebral dynamics, laterality, and psychopathology.* New York: Elsevier.

Sackeim, H. A., Prohovnik, I., Moeller, J. R., Brown, R. P., Apter, S., Prudic, J., Devanand, D. P., & Mukherjee, S. (1990). Regional cerebral blood flow in mood disorders I. Comparison of major depressives and normal controls at rest. *Arhives of General Psychiatry, 47,* 60–70.

Sapolsky, R. M., & Plotsky, P. M. (1990). Hypercortisolism and its possible neural basis. *Biological Psychiatry, 27,* 937–952.

Saykin, A. J., Gur, R. C., Gur, R. E., Mozley, P. D., Mozley, L. H., Resnick, S. M., Kester, D. B., & Stafiniak, P. (1991). Neuropsychological function in schizophrenia—selective impairment in memory and learning. *Archives of General Psychiatry, 48,* 618–624.

Scarone, S., Colombo, C., Livian, S., Abbruzzese, M., Ronchi, P., Locatelli, M., Scotti, G., & Smeraldi, E. (1992). Increased right caudate nucleus size in obsessive-compulsive disorder: Detection with magnetic resonance imaging. *Psychiatry Research: Neuroimaging, 45,* 115–121.

Schatzberg, A. F., Rothschild, A. J., Langlais, P. J., Bird, E. D., & Cole, J. O. (1985). A corticosteroid/dopamine hypothesis for psychotic depression and related states. *Journal of Psychiatric Research, 19,* 57–64.

Schilder, P. (1938). The organic background of obsessions and compulsions. *American Journal of Psychiatry, 94,* 1397–1416.

Schlegel, S., & Kretzschmer, K. (1987). Computed tomography in affective disorders. Part I: Ventricular and sulcal measurements. *Biological Psychiatry, 22,* 4–14.

Schlegel, S., Maier, W., Philipp, M., Aldenhoff, J. B., Heuser, I., Kretzschmar, K., & Benkert, O. (1989a). Computed tomography in depression: Association between ventricular size and psychopathology. *Psychiatry Research, 29,* 221–230.

Schlegel, S., Maier, W., Philipp, M., Aldenhoff, J. B., Heuser, I., Kretzschmer, K., & Benkert, O. (1989b). Computed tomography in depression: Association between ventricular size and psychopathology. *Psychiatry Research, 29,* 221–230.

Schulz, S. C., Koller, M. M., Kishore, P. R., & Hamer, R. M. (1983). Ventricular enlargement in teenage patients with schizophrenia spectrum disorder. *American Journal of Psychiatry, 140,* 1592–1595.

Schwartz, J. M., Baxter, L. M., Mazziotta, J. C., Gerner, R. H., & Phelps, M. E. (1987). The differential diagnosis of depression. Relevance of positron emission tomography studies of cerebral glucose metabolism to the bipolar-unipolar dichotomy. *Journal of the American Medical Association, 258,* 1368–1374.

Scott, M. L., Golden, C. J., Ruedrich, S. L., & Bishop, R. J. (1983). Ventricular enlargement in major depression. *Psychiatry Research, 8,* 91–93.

Sedvall, G. (1992). The current status of PET scanning with respect to schizophrenia. *Neuropsychopharmacology, 7,* 41–54.

Sedvall, G., Farde, L., Persson, A., & Wiesel, F.-A. (1986). Imaging of neurotransmitter receptors in the living human brain. *Archives of General Psychiatry, 43,* 995–1005.

Seeman, P. (1992). Dopamine receptor sequences: Therapeutic levels of neuroleptics occupy D2-receptors, clozapine occupies D4. *Neuropsychopharmacology, 7,* 261–284.

Seeman, P., Lee, T., Chau-Wong, M., & Wong, K. (1976). Antipsychotic drug doses and neuroleptic/dopamine receptors. *Nature, 261,* 717–719.

Seeman, P., Ulpian, C., & Bergeron, C. (1984). Bimodal distribution of dopamine receptor densities in brain of schizophrenics. *Science, 225,* 728–730.

Selemon, L. D., Rajkowska, G., & Goldman-Rakic, P. S. (1995). Abnormally high neuronal density in schizophrenic cortex: A morphometric analysis of prefrontal area 9 and occipital area 17. *Archives of General Psychiatry, 52,* 805–818.

Sham, P. K., O'Callaghan, E., Takei, N., Murray, G. K., Hare, E. H., & Murray, R. M. (1992). Schizophrenia following pre-natal exposure to influenza epidemics between 1939–1960. *British Journal of Psychiatry, 160,* 461–466.

Shear, P. K., Sullivan, E. V., Mathalon, D. H., Lim, K. O., Davis, L. F., Yesavage, J. A., Tinklenberg, J. R., & Pfefferbaum, A. (1995). Longitudinal volumetric computed tomographic analysis of regional brain changes in normal aging and Alzheimer's disease. *Archives of Neurology, 52,* 392–402.

Shelton, R. C., Karson, C. N., Doran, A. R., Pickar, D., Bigelow, L. B., & Weinberger, D. R. (1988). Cerebral structural pathology in schizophrenia: Evidence for a selective prefrontal cortical defect. *American Journal of Psychiatry, 145,* 154–163.

Shelton, R. C., & Weinberger, D. R. (1986). X-ray computerized tomography studies in schizophrenia: A review and synthesis. In H. A. Nasrallah & D. Weinberger (Eds.), *The neurology of schizophrenia* (pp. 207–250). Amsterdam: Elsevier.

Shenton, M. E., Kikinis, R., Jolesz, F. A., Pollak, S. D., LeMay, M., Wible, C. G., Hokama, H., Martin, J., Metcalf, D., Coleman, M., & McCarley, R. W. (1992). Abnormalities of the left temporal lobe and thought disorder in schizophrenia—a quantitative magnetic resonance imaging study. *New England Journal of Medicine, 327,* 604–612.

Sheppard, G., Gruzelier, J., Manchanda, R., Hirsch, S. R., Wise, R., Frackowiak, R., & Jones, T. (1983). ^{15}O positron emission tomographic scanning in predominantly never-treated acute schizophrenic patients. *Lancet, 1,* 1448–1452.

Shima, S., Shikano, T., Kitamura, T., Masuda, Y., Tsukomo, T., Kanba, S., & Asai, M. (1984). Depression and ventricular enlargement. *Acta Psychiatrica Scandinavica, 70,* 275–277.

Shiraishi, H., Koizumi, J., Hori, M., Terashima, Y., Suzuki, T., Saito, K., Mizukami, K., Tanaka, Y., & Yamaguchi, N. (1992). A computerized tomographic study in patients with delusional and non-delusional depression. *Japanese Journal of Psychiatry and Neurology, 46,* 99–105.

Silfverskiold, P., & Risberg, J. (1989). Regional cerebral blood flow in depression and mania. *Archives of General Psychiatry, 46,* 253–259.

Slater, E., & Beard, A. W. (1963). The schizophrenia-like psychoses of epilepsy. i. Psychiatric aspects. *British Journal of Psychiatry, 109,* 95–150.

Smith, G. N., & Iacono, W. G. (1986). Lateral ventricular size in schizophrenia and choice of control group. *Lancet, 1,* 1450.

Smith, R. C., Baumgartner, R., & Calderon, M. (1987). Magnetic resonance imaging studies of the brains of schizophrenic patients. *Psychiatry Research, 20,* 33–46.

Smith, R. C., Calderon, M., Ravichandran, G. K., Largen, J., Vroulis, G., Shvartsburd, Gordon, J., & Schoolar, J. C. (1984). Nuclear magnetic resonance in schizophrenia: A preliminary study. *Psychiatry Research, 12,* 137–147.

Spoonheim, S. R., Iacono, W. G., & Beiser, M. (1991). Stability of ventricular size after the onset of psychosis in schizophrenia. *Psychiatry Research: Neuroimaging, 40,* 21–30.

Starkstein, S. E., & Robinson, R. G. (1989). Affective disorders and cerebral vascular disease. *British Journal of Psychiatry, 154,* 170–182.

Steinmetz, H., Rademacher, J., Huang, Y. X., Hefter, H., Zilles, K., Thron, A., & Freund, H. J. (1989). Cerebral asymmetry: MR planimetry of the human planum temporale. *Journal of Computer Assisted Tomography, 13,* 996–1005.

Stevens, J. R. (1991). Psychosis and the temporal lobe. *Advances in Neurology, 55,* 79–96.

Strakowski, S. M., Wilson, D. R., Tohen, M., Woods, B. T., Douglass, A. W., and Stoll, A. L. (1993). Structural brain abnormalities in first-episode mania. *Biological Psychiatry, 33,* 602–609.

Strakowski, S. M., Woods, B. T., Tohen, M., Wilson, D. R., Douglass, A. W., & Stoll, A. L. (1993). MRI subcortical signal hyperintensities in mania at first hospitalization. *Biological Psychiatry, 33,* 204–206.

Suddath, R., Casanova, M. F., Goldberg, T. E., Daniel, D. G., Kelsoe, J. R., & Weinberger, D. R. (1989). Temporal lobe pathology in schizophrenia. *American Journal of Psychiatry, 146,* 464–472.

Suddath, R. L., Christison, G. W., Fuller Torrey, E., Casanova, M. F., & Weinberger, D. R. (1990). Anatomical abnormalities in the brains of monozygotic twins discordant for schizophrenia. *New England Journal of Medicine, 322,* 789–794.

Sullivan, E. V., Mathalon, D. H., Zipursky, R. B., Kersteen-Zucker, Z., Knight, R. T., & Pfefferbaum, A. (1993). Factors of the Wisconsin Card Sorting Test as a test of dorsolateral prefrontal cortical function in schizophrenia and alcoholism. *Psychiatry Research, 46,* 175–200.

Sullivan, E. V., Shear, P. K., Mathalon, D. H., Lim, K. O., Yesavage, J. A., Tinklenberg, J. R., & Pfefferbaum, A. (1993). Greater abnormalities of brain cerebrospinal fluid volumes in younger than older patients with Alzheimer's disease. *Archives of Neurology, 50,* 359–373.

Sullivan, E. V., Shear, P. K., Lim, K. O., Zipursky, R. B., & Pfefferbaum, A. (1996). Cognitive and motor impairments are related to gray matter volume deficits in schizophrenia. *Biological Psychiatry, 39,* 234–240.

Swayze, V. W., Andreasen, N. C., Alliger, R. J., Ehrhardt, J. C., & Yuh, W. T. C. (1990). Structural brain abnormalities in bipolar affective disorder—ventricular enlargement and focal signal hyperintensities. *Archives of General Psychiatry, 47,* 1054–1059.

Swayze, V. W., Andreasen, N. C., Alliger, R. J., Yuh, W. T. C., & Ehrhardt, J. C. (1992). Subcortical and temporal structures in affective disorder and schizophrenia—A magnetic resonance imaging study. *Biological Psychiatry, 31,* 221–240.

Swedo, S. E., Pietrini, P., Leonard, H. L., Schapiro, M. B., Rettew, D. C., Goldberger, E. L., Rapoport, S. I., Rapoport, J. L., & Grady, C. L. (1992). Cerebral glucose metabolism in childhood-onset obsessive-compulsive disorder. *Archives of General Psychiatry, 49,* 690–694.

Swedo, S. E., Rapoport, J. L., Cheslow, D. L., Leonard, H. L., Ayoub, E. M., Hosier, D. M., & Wald, E. R. (1989). High prevalence of obsessive-compulsive symptoms in patients with Sydenham's chorea. *American Journal of Psychiatry, 146,* 246–249.

Swedo, S. E., Schapiro, M. B., Grady, C. L., Cheslow, D. L., Leonard, H. L., Kumar, A., Friedland, R. P., Rapoport, S. I., & Rapoport, J. L. (1989). Cerebral glucose metabolism in childhood-onset obsessive-compulsive disorder. *Archives of General Psychiatry, 46,* 518–523.

Szechtman, H., Nahmias, C., Garnett, E. S., & Firnau, G. (1988). Effect of neuroleptics on altered cerebral glucose metabolism in schizophrenia. *Archives of General Psychiatry, 45,* 523–541.

Tamminga, C. A., Thaker, G. K., Buchanan, R., Kirkpatrick, B., Alphs, L. D., Chase, T. N., & Carpenter, W. T. (1992). Limbic system abnormalities identified in schizophrenia using positron emission tomography with fluorodeoxyglucose and neocortical alterations with deficit syndrome. *Archives of General Psychiatry, 49,* 522–530.

Targum, S., Rosen, L., DeLisi, L., Weinberger, D., & Citrin, C. (1983). Cerebral ventricular size in major depressive disorder: Association with delusional symptoms. *Biological Psychiatry, 18,* 329–336.

Taylor, M. A. (1991). The role of the cerebellum in the pathogenesis of schizophrenia. *Neuropsychiatry, Neuropsychology, and Behavioral Neurology, 4,* 251–280.

Toone, B. K., Dawson, J., & Driver, M. V. (1982). Psychoses of epilepsy: A radiological evaluation. *British Journal of Psychiatry, 140,* 244–248.

Torrey, E. F., & Peterson, M. R. (1974). Schizophrenia and the limbic system. *Lancet, 2,* 942–946.

Trimble, M. R. (1991). Can schizophrenia be localized? *Journal of Neuropsychiatry and Clinical Neurosciences, 3,* 89–94.

Uematsu, M., & Kaiya, H. (1989). Midsagittal cortical pathomorphology of schizophrenia: A magnetic resonance imaging study. *Psychiatry Research, 30,* 11–20.

Uytdenhoef, P., Portelange, P., Jacquy, J., Charles, G., Linkowski, P., & Mendlewicz, J. (1983). Regional cerebral blood flow and lateralized hemispheric dysfunction in depression. *British Journal of Psychiatry, 143,* 128–132.

Van den Bossche, B., Maes, M., Brussaard, C., Schotte, C., Cosyns, P., De Moor, J., & De Schepper, A. (1991). Computed tomography of the brain in unipolar depression. *Journal of Affective Disorders, 21,* 67–74.

Vita, A., Sacchetti, E., & Cazzullo, C. L. (1988). A CT scan follow-up study of cerebral ventricular size in schizophrenia and major affective disorders. *Schizophrenia Research, 1,* 165–166.

Vita, A., Sacchetti, E., Valvassori, G., & Cazzullo, C. L. (1988). Brain morphology in schizophrenia: A 2- to 5-year CT scan follow-up study. *Acta Psychiatrica Scandinavica, 78,* 618–621.

Volkow, N. D., Brodie, J. D., Wolf, A. P., Gomez-Mont, F., Canaro, R., Van Gelder, P., Russel, J. A., & Overall, J. (1986). Brain organization in schizophrenia. *Journal of Cerebral Blood Flow and Metabolism, 6,* 441–446.

Volkow, N. D., Levy, A., Brodie, J. D., Wolf, A. P., Canaro, R., Van Gelder, P., & Henn, F. (1992). Low cerebellar metabolism in medicated patients with chronic schizophrenia. *American Journal of Psychiatry, 149,* 686–688.

Volkow, N. D., Wolf, A. P., Van Gelder, P., Brodie, J. A., Overall, J. E., Canaro, R., & Gomez-Mont, F. (1987). Phenomenological correlates of metabolic activity in 18 patients with chronic schizophrenia. *American Journal of Psychiatry, 144,* 151–158.

Warren, L. R., Butler, R. B., Katholi, C. R., McFarland, C. E., Crews, E. L., & Halsey, J. H. (1984). Focal changes in cerebral blood flow produced by monetary incentive during a mental mathematics task in normal and depressed subjects. *Brain and Cognition, 3,* 71–85.

Weinberger, D. R. (1987). Implications of normal brain development for the pathogenesis of schizophrenia. *Archives of General Psychiatry, 44,* 660–669.

Weinberger, D. R., Berman, K. F., & Illowsky, B. P. (1988). Physiological dysfunction of dorsolateral prefrontal cortex in schizophrenia. III. A new cohort and evidence for a monoaminergic mechanism. *Archives of General Psychiatry, 45,* 609–615.

Weinberger, D. R., DeLisi, L. E., Perman, G. P., Targum, S., & Wyatt, R. J. (1982). Computed tomography in schizophreniform disorder and other acute psychiatric disorders. *Archives of General Psychiatry, 39,* 778–783.

Wing, J. K., Cooper, J. E., & Sartorius, V. (1974). *The description and classification of psychiatric symptoms.* London: Cambridge University Press.

Witelson, S. F., & Kigar, D. L. (1992). Sylvian fissure morphology and asymmetry in men and women—bilateral differences in relation to handedness in men. *Journal of Comparative Neurology, 323,* 326–340.

Wolkin, A., Jaeger, J., Brodie, J. D., Wolf, A. P., Fowler, J., Rotrosen, J., Gomez-Mont, F., & Canaro, R. (1985). Persistence of cerebral metabolic abnormalities in chronic schizophrenia as determined by positron emission tomography. *American Journal of Psychiatry, 142,* 564–571.

Wolkin, A., Sanfilipo, M., Wolf, A. P., Angrist, B., Brodie, J. D., & Rotrosen, J. (1992). Negative symptoms and hypofrontality in chronic schizophrenia. *Archives of General Psychiatry, 49,* 959–965.

Wong, D. F., Pearlson, G. D., Tune, L., Villemagne, V., Ross, C., Dannals, R. F., Links, J., Ravert, H., Wilson, A., Wagner, H. N., & Gjedde, A. (1987). *In vivo* measurement of D2 receptor abnormalities in drug naive and treated manic-depressive patients. *Journal of Nuclear Medicine, 28,* 611.

Wong, D. F., Wagner, H. N. J., Tune, L. E., & Dannals, R. F. (1986). Positron emission tomography reveals elevated D2 dopamine receptors in drug-naive schizophrenics. *Science, 234,* 1558–1563.

Woodruff, P. W. R., Pearlson, G. D., Geer, M. J., Barta, P. E., & Chilcoat, H. D. (1993). A computerized magnetic resonance imaging study of corpus callosum morphology in schizophrenia. *Psychological Medicine, 23,* 45–56.

Woods, B. T., & Yurgelun-Todd, D. (1991). Brain volume loss in schizophrenia: When does it occur and is it progressive? *Schizophrenia Research, 5,* 202–204.

Woods, B. T., Yurgelun-Todd, D., Benes, F. M., Frankenburg, F. R., Pope, H. G., & McSparren, J.

(1990). Progressive ventricular enlargement in schizophrenia—comparison to bipolar affective disorder and correlation with clinical course. *Biological Psychiatry, 27,* 341–352.

Young, A. H., Blackwood, D. H. R., Roxborough, H., McQueen, J. K., Martin, M. J., & Kean, D. (1991). A magnetic resonance imaging study of schizophrenia—brain structure and clinical symptoms. *British Journal of Psychiatry, 158,* 158–164.

Zatz, L., Jernigan, T. L., & Ahumada, A. J. (1982). Changes on computed cranial tomography with aging: Intracranial fluid volume. *American Journal of Neuroradiology, 3,* 1–11.

Zipursky, R. B., Lim, K. O., Rosenbloom, M., & Pfefferbaum, A. (1988). Differing effects of duration of illness on CT findings in alcoholism and schizophrenia. *Psychopharmacology Bulletin, 24,* 495–500.

Zipursky, R. B., Lim, K. O., Sullivan, E. V., Brown, B. W., & Pfefferbaum, A. (1992). Widespread cerebral gray matter volume deficits in schizophrenia. *Archives of General Psychiatry, 49,* 195–205.

Zipursky, R. B., Marsh, L., Lim, K. O., DeMent, S., Shear, P. K., Sullivan, E. V., Murphy, G. M., Csernansky, J. G., & Pfefferbaum, A. (1994). Volumetric MRI assessment of temporal lobe structures in schizophrenia. *Biological Psychiatry, 35,* 501–516.

5

Gender Differences in Brain Morphometry and Function

ELANA FARACE AND ERIC TURKHEIMER

INTRODUCTION

It should come as no surprise that the brains of men and women differ in both size and shape. Many aspects of human anatomy are sexually dimorphic. Men are, on average, taller and heavier than women, not to mention shaped differently. Although differences in size and shape that are unrelated to function might be of interest to an anatomist or neurosurgeon, gender differences in brain structure are of interest to behavioral scientists primarily to the extent that they provide clues to differences in the function of the brain. The considerable interest generated by gender differences in brain anatomy stems from the fact that anatomical gender differences appear to be related to some of the more important dimensions of human brain function: lateralization, verbal versus spatial ability, and recovery from brain injury.

This chapter will review morphometric and other anatomical gender differences in the brain as they relate to what is known about gender differences in mental ability, lateralization, and recovery of function. First, we summarize evidence concerning gender differences in cognitive ability in normals. Second, we review gender differences in lateralization of function in both normal and brain-damaged populations. Third, we review gender differences in the morphometry of anatomical structures of the brain. Finally, we explore possible relationships

---

ELANA FARACE AND ERIC TURKHEIMER Department of Psychology, University of Virginia, Charlottesville, Virginia 22903-2477.

Neuroimaging II: Clinical Applications, edited by Erin D. Bigler. Plenum Press, New York, 1996.

between gender differences in cognitive function and brain anatomy. Throughout, we emphasize findings based on neuroimaging methodologies; other results, such as the voluminous experimental literature involving lateralized stimulus presentation in normal subjects, will be summarized, and the reader referred to relevant reviews.

Gender Differences in Ability, Lateralization, and Anatomy

Ability Differences in Normal Men and Women

Small, but apparently reliable, gender differences in verbal and spatial cognitive ability have been observed in normal subjects. Women show better performance than men on verbal tasks such as generating synonyms, word fluency, computation, and anagrams, whereas men show superior performance on many visual-spatial tasks, such as mental rotation, spatial perception, and spatiotemporal tasks (Halpern, 1992; Maccoby & Jacklin, 1974). Meta-analyses have suggested that the gender difference in many verbal and spatial abilities is stable over the course of development (Hyde & Linn, 1988; Linn & Peterson, 1985) and recent historical time (Masters & Sanders, 1993). The stability of the gender difference in verbal and spatial abilities suggests that congenital gender differences in brain structure may contribute to differences in performance. Although this chapter will focus on gender differences in child-rearing practices contribute to observed gender differences in ability (e.g., Sherman, 1967, and the "Bent Twig Hypothesis").

Gender Differences in Lateralized Function

It is well established that most humans have verbal abilities represented primarily in the left hemisphere and spatial abilities primarily in the right hemisphere (Levy-Agresti & Sperry, 1968; Springer & Deutsch, 1989). Men and women may differ in the degree to which verbal and spatial functions are lateralized. This difference was reported as early as 1880 by Chrichton-Browne, who noted "the tendency to symmetry in the two halves of the cerebrum is stronger in women than in men," (p. 65) and is fairly well documented today, although the magnitude of the gender difference remains controversial (Kingsburg, LaBarba, & Bowers, 1987; Levy & Reid, 1978; McGlone, 1977, 1980; Turkheimer & Farace, 1992).

Lateralization of function has recently been called into question on both theoretical and empirical grounds. Efron (1990), in an influential book, has advanced the argument that "lateralization of function" has grown from a specific explanation of particular research findings to a broad explanatory principle that can explain almost anything and in so doing has lost most of its scientific value. Efron argues that functions like "language" or "spatial ability" are too molar to be localized anywhere in the brain. Efron would prefer that researchers limit themselves to observing correlations between lesions to particular regions of the brain and certain deficits in performance and that they refrain from reaching conclusions about the localization of complex functions on the basis of these correlations:

> The existence of this correlation [between left fronto-temporal lesions and language disturbance] undoubtedly indicates that this brain region contains neural circuits that

> support language functions, but it does not necessarily mean that the function called "language" is performed by these circuits. It is more likely that these circuits perform a variety of functions, some of which are *required* for normal language function. (p. 9, italics in original)

Efron is endorsing a research strategy with a goal of identifying the modular computational architecture of complex abilities. These computational modules may be related to the "functions" of specific brain structures, but we should not expect immediately observable aspects of cognitive function to map onto gross brain anatomy in a direct way.

Although Efron (1990) is correct that it would be naive to expect to find functions like "language" represented by particular brain structures, it does not necessarily follow that lateralization of function has no value as an explanatory construct. Individual differences in handedness is a fact, as is the correlation between left hemisphere lesions and aphasia. It is possible that the neural circuits that support language functions are not distribuιed the same way in different individuals. A bottom-up explanation of lateralized function, proceeding from physiology to anatomy to low-level cognitive structures, would do a great deal to elucidate the complexity of observations that have been reported in higher cognitive functions. In the absence of such explanations, however, it is natural to propose models that attempt to explain known findings in terms of incompletely observed hypothetical constructs.

If lateralization of function is a hypothetical construct, the appropriate scientific task is to build a coherent empirical theory that ties together related observations in disparate contexts, that is, to establish construct validity (Cronbach & Meehl, 1955). It remains to be seen whether such a process produces "a closed conceptual system resembling the Ptolemaic theory of planetary motions, where any new observation was accounted for by postulating yet another planetary cycle" (Efron, 1990, pp. 26–27) or a useful theory leading to decisive new experiments.

Experimental Studies of Lateralization in Normals

A large body of experimental research has investigated gender differences in lateralization. As these studies do not employ neuroimaging methodologies, they can only be summarized here. This topic has been extensively reviewed, most recently by Halpern (1992).

Lateralized Stimulus Presentation. The most commonly employed experimental methodology for inferring lateralization of function is to present separate stimuli simultaneously to the left and right ears or to the left and right visual fields. Because of the lateralized organization of the auditory and visual perceptual systems (which, unfortunately, is by no means simple, especially in the auditory system; see Efron, 1990), responses to lateralized stimuli may be related to lateralized organization of the brain.

This large literature has been summarized by Bryden (1982), in several chapters of Beaumont (1982), and by Hellige (1990). Bryden (1982) suggests that studies of lateralized auditory presentation of verbal and spatial stimuli are about equally divided between those showing no gender differences and those showing greater lateralization in men. A smaller number of studies using verbal stimuli in visual modality show a weaker tendency for greater lateralization in men. A much more skeptical review of studies employing visual stimuli can be found in Fairweather (1982).

INTERFERENCE TASKS. Interference tasks in which normal subjects verbalize while they manipulate objects have also been used to demonstrate greater bilateral representation of bilateral verbal abilities in women. Wolff and Cohen (1980) performed a study in which right-handed subjects performed simultaneous reading and finger-tapping tasks. Verbalizations disrupted right-hand performance in men but disrupted left-hand performance in women. Elliott, Weeks, Lindley, and Jones (1986) performed a similar study using a unimanual force production task as interference for simultaneous verbalizations in right-handed men and women. Elliott *et al.* (1986) also found that the speech task disrupted right-hand performance in men and left-hand performance in women. Kingsburg *et al.* (1987) reported a study in which men and women solved a block design problem with one hand while either simultaneously finger tapping with the other hand or vocalizing words. Finger tapping had no differential effect on performance in either gender. However, vocalization disrupted women's performance using either hand, suggesting that verbal abilities are more bilateral in women.

IMAGING STUDIES OF LATERALIZATION OF FUNCTION

Gender differences in lateralization of function have also been studied using functional imaging techniques, ranging from electroencephalography (EEG) to positron emission tomography (PET). These methods have the advantage of providing relatively direct information about brain activity. They are also less invasive than surgical procedures, although several of them do require administration of radioactive tracers.

ELECTROENCEPHALOGRAPHY. EEG was used as a measure of cortical activation before the term "functional imaging" was coined, and it is a forerunner of contemporary neuroimaging methodologies. Although EEG has limited spatial resolution, it is a useful means of testing hypotheses about the activation of large structures such as the cerebral hemispheres. In EEG studies of gender differences in lateralization of function, measurements are taken at sites in the left and right hemispheres while subjects complete tasks designed to selectively activate one hemisphere. Comparisons of ratios of left to right activation during the two types of tasks are used as an indication of lateralization of function.

Like most studies that will be reviewed in this chapter, EEG studies of lateralization are beset by small samples and methodological variability, and as a consequence consistent results are difficult to come by. Nonetheless, most studies have shown that men have greater differential activation of the hemispheres between verbal and nonverbal conditions (Glass, Butler, & Carter, 1984; Ray, Morell, Frediani, & Tucker, 1976; Trotman & Hammond, 1979). The most widely cited failure to demonstrate this gender difference (Davidson, Schwartz, Pugash, & Bromfield, 1976) was based on a very small sample (five men and four women, although Ray *et al.*, 1976, was not much larger) in which male and female subjects did not differ in their degree of lateralization; the lateralization effect was significant for female subjects but not for male subjects.

The studies cited above were based on the power function of EEG spectra, which is related to the mean level of desynchronization during a task. The EEG data can also be analyzed for "coherence," a measure of the correlation in desynchronization across frequencies at inter- or intrahemispheric sites. Female subjects have been shown to have greater degrees of interhemispheric coherence on both verbal and nonverbal tasks (Beaumont, Mayes, & Rugg, 1978; Flor-Henry & Koles, 1982), consistent with a greater degree of interhemispheric cooperation in women.

Regional Cerebral Blood Flow. Regional cerebral blood flow (rCBF) is a method of topographical brain mapping. Subjects receive a radioligand, either by inhalation or by injection, which clears the blood-brain barrier. Activity, as measured by stationary scintillation counters, is a measure of cerebral blood flow (Gur & Gur, 1990). As summarized by Gur and Gur (1990), the largest gender difference detected by rCBF is that women have higher overall levels of blood flow than men, for reasons that remain unclear. Few studies have examined gender differences in laterality using rCBF. Gur *et al.* (1982) performed rCBF with ^{133}Xe inhalation in 62 normal subjects (15 right-handed men, 15 right-handed women, 15 left-handed men, 15 left-handed women). During the testing, verbal (Miller Analogies Test) and spatial (Benton's line orientation) tasks were projected on an overhead screen. Cognitive activity resulted in an increased blood flow to the hemispheres. The two-way interaction between task and hemisphere was significant, with blood flow greater to the left hemisphere during the verbal task and to the right hemisphere during the spatial task. Interpretation of this finding was complicated by higher-order interactions. The laterality effect was most pronounced in right-handed women and left-handed men. Left-handed women showed no difference between flow to the left and right hemispheres in the verbal task, and right-handed men showed greater flow to the left hemisphere during verbal and spatial tasks.

Wood, Flowers, and Naylor (1991) reported a study of rCBF in 60 normal subjects (30 men, 30 women) using a statistical method analogous to the analysis of coherence in EEG spectra. Instead of examining hemispheric differences during verbal and spatial processing, they analyzed correlations among the activation of eight left hemisphere and eight right hemisphere regions of interest during a spelling task. There was a positive relationship between Broca's (language reception) area and Wernicke's (language production) area in men but not in women. In women, they found bitemporal coupling, in which left temporal regions were activated in parallel with right temporal regions. The linkage of intrahemispheric regions in men and interhemispheric regions in women is suggestive of greater bilateral processing in women.

Positron Emission Tomography. PET quantifies brain metabolism in three dimensions. Initial studies of gender differences in glucose metabolic rates suggested that women had higher overall rates than men (Baxter *et al.*, 1987), but later studies showed that the difference appeared to result from head size differences (Yoshii *et al.*, 1988), and the most recent studies, using more accurate quantitation methods, have found no gender differences at all (Miura *et al.*, 1990).

Using a correlational method similar to that employed by Wood *et al.* (1991), Azari and co-workers (1992) performed a ^{18}F-flourodeoxyglucose PET study of normal subjects (15 women, 18 men, less than 40 years old). Subjects were scanned at resting state (eyes covered, ears plugged). Correlations were computed among 198 regions of interest, 99 per hemisphere. There were no gender differences in overall activation, or in the difference in activation between the hemispheres. When the correlation matrices among the regions of interest were compared, the associations that were stronger in women than in men involved interhemispheric regions, while the stronger associations in men were intrahemispheric.

Gender Differences in Patient Populations

The functional organization of the brain can also be explored by studying the consequences of localized brain dysfunction, which may be the result of disease (stroke or trauma), neurosurgery for tumor or epilepsy, or clinical procedures which involve temporary brain impairment or manipulation.

Clinically Induced Hemispheric Dysfunction. The independent functions of the individual cerebral hemispheres can be studied in commissurotomized patients who have had their corpus callosa severed as a means of controlling intractable epilepsy (Gazzaniga, 1985). Bogen, DeZure, Tenhouten, and Marsh (1972) tested ten commissurotomized patients (six men, four women) with the WAIS Similarities subset and the Street test, a spatial test involving incomplete figures and requiring a verbal answer. Although there were no gender differences for performance on the Similarities test, there was a trend for men to be more impaired than women on the Street test. (The difference was significant if one male subject who "possesses some form of interhemispheric communication not present in the other patients" is omitted.) Assuming that the Similarities subtest can be solved using verbal skills alone, whereas the Street test requires both spatial and verbal skills, the findings provide some support for the hypothesis that women are more likely than men to have both abilities represented in each hemisphere, permitting them to solve the items without interhemispheric transfer.

In the Wada procedure, one hemisphere is anesthetized with sodium amytal as a means of determining lateral dominance prior to neurosurgical procedures (Wada & Rasmussen, 1962). McGlone and Fox (1982) used the Wada procedure in a study of gender differences in the representation of verbal functions in right-handed adult epileptics with left-hemisphere dominance for language. Baseline oral fluency performance was compared to performance during sodium amytal injections in the left and right hemispheres. Fluency performance was more impaired during left hemisphere injections in both genders, but the difference between left and right hemisphere performance was significantly greater in men than in women. Men demonstrated an improvement over baseline during right hemisphere injections; women showed a slight decrement in performance. Once again, these results are consistent with the hypothesis that women are more likely to have verbal abilities represented in both hemispheres.

Stimulation Mapping. Another method of investigating cerebral organization as a means of preparing for neurosurgical procedures involves localized stimulation with a small electrical probe. The resultant behaviors, sensations, or disruptions of function are indicative of the function of the stimulated location. Ojemann (1983) reviewed the use of stimulation mapping as a localization technique. Ojemann, Ojemann, Lettich, and Berger (1989) studied stimulation mapping in the left, dominant hemisphere in 117 patients (56 men, 61 women). Language was mapped by an analysis of deficits in object naming during stimulation. Although an earlier study including a subset of the same subjects had suggested that men had greater distribution of language sites in the left hemisphere than women (Mateer, Polen, & Ojemann, 1982), in the larger sample the result was limited to men with low verbal IQ (VIQ), who were more likely than women to have language sites in the left parietal lobes. Women were slightly more likely to have frontal language sites. Ojemann's findings suggest that within the dominant hemisphere, women may be more focally organized for language than men.

Lesion Studies. Much of our knowledge of the functioning of the brains of lower animals follows from experiments in which brains are surgically lesioned under strict experimental control. Lesion studies in humans are based on naturally occurring lesions, such as those produced by cerebrovascular accident, cerebral trauma, or neurosurgery for tumor or epilepsy. Inference of normal function from lesion sequelae is not a simple matter, even in animals, and presents formidable methodological difficulties in humans. Naturally occurring lesions are large and unevenly distributed in the brain. Quantitation of lesion location presents thorny statistical issues (Turkheimer, Yeo, Jones, & Bigler, 1990). Moreover, as Sundet (1988) has pointed out, there are particular methodological problems with studying gender differences in a stroke population. Men and women are at unequal risk for stroke, have different recovery profiles, and there are gender differences in the effects of stroke on aging. Also, women may remain in the hospital longer than men, probably because men are more likely to have a caregiver spouse present when they get home (Sundet, 1988).

Incidence and Severity of Aphasia. The question of whether there are gender differences in the incidence, type, severity, or recovery rates of aphasia has a checkered history, marked by provocative findings that have not been replicated well. The original observation was provided by McGlone (1977), who examined the incidence of aphasia in a sample of 92 patients with unilateral lesions. Of the 45 left hemisphere–lesioned patients, 16 were found to be aphasic, with men having a much higher rate of aphasia than women (12 of 25 men and 4 of 31 women). No right hemisphere–lesioned patients were aphasic. Several studies failed to replicate this finding, as detailed by Yeo (1989). Pizzamiglio, Mammucari, and Razzano (1985) did find evidence that female global aphasics' language comprehension improved significantly more than that of male global aphasics after three months of language therapy.

A second possibility was explored by Kimura (1983), who tested 143 men and 73 women with left hemisphere lesions, including a subgroup of 81 patients with hemisphere damage located in the anterior or posterior regions, as well as 169 patients with right hemisphere lesions. There was a greater incidence of aphasia following left hemisphere lesions in men than in women. Kimura (1983) also found that women with left hemisphere lesions in the anterior region had a greater incidence of aphasia than women with posterior left hemisphere damage. No anterior-posterior differences were found in men. Kimura (1983) hypothesized that gender differences in the incidence of aphasia may result from intrahemispheric differences in language organization in women. Unfortunately, this finding has not been well replicated either (e.g., Kertesz & Benke, 1989), although Sundet (1988) reports some limited support in the form of a finding that men have more severe symptoms of fluent aphasias resulting from left posterior lesions.

Bilateral representation of spatial abilities would be suggested if women have more trouble with spatial tasks following a left hemisphere lesion. Accordingly, Friedland and Kershner (1986) tested four male and four female right-handed adult Broca's aphasia (left hemisphere stroke) patients with a constructional stick design task and found that men performed relatively better than women on the task. The authors suggested that men may rely more on their right hemisphere for spatial tasks and that women may have some degree of left hemisphere involvement in spatial tasks.

Verbal and Performance IQ. A great deal of data has been collected on gender differences in the effects of unilateral lesions on verbal and performance IQ (VIQ and PIQ, respectively). While VIQ and PIQ are far from perfect measures of verbal and especially spatial abilities (Kaufman, 1990), measured intelligence has intrinsic clinical importance, and the pattern of results in lesion studies of VIQ and PIQ are generally analogous to those found using more specific verbal and spatial measures.

Gender differences in the effects of unilateral lesions on VIQ and PIQ have been investigated since Lansdell's (1962) finding of a gender difference in epileptic patients' spatial abilities following temporal lobectomy. Lansdell (1962) found that VIQ-PIQ difference scores were larger in men than in women following both right and left hemisphere lesions. His findings were subsequently replicated in a variety of brain-injured populations, including patients suffering from stroke, head injury, and tumor (Warrington, James, & Maciejewski, 1986; Whelan & Walker, 1988).

A meta-analysis of 12 studies reporting separate effects of right and left hemisphere lesions on VIQ and PIQ (Turkheimer & Farace, 1992) revealed the usual gender difference in VIQ-PIQ differences. Further analysis suggested that the effect was composed of independent effects of right and left hemisphere lesions on VIQ and PIQ: greater VIQ deficits in women than in men following right hemisphere lesions and greater PIQ deficits in women than in men following left hemisphere lesions.

Although gender differences in VIQ and PIQ following unilateral brain damage are now fairly well established (Kaufman, 1990), the etiology of the effect remains unclear. Three major hypotheses have been forwarded. The first, proposed by McGlone (1977), is that women have a more bilateral representation of verbal processes than men. The second, proposed by Kimura (1983) in the context of her research on aphasia, is that the gender difference is the result of differences in the intrahemispheric functional organization of the left hemisphere, with verbal areas located more anteriorly in women than in men. The third hypothesis is a proposal by Inglis and Lawson (1982) that gender differences are caused by differences in problem-solving strategy: women suffer greater PIQ deficits following left hemisphere lesions as a result of greater use of verbal strategies in solving PIQ items.

Our laboratory has conducted an investigation of the relationships among IQ, gender, and focal lesion laterality (Turkheimer, Farace, Yeo, & Bigler, 1993) using a quantitative method for the analysis of covariation between lesion location and IQ scores (Turkheimer *et al.*, 1990). Results of these analyses were used to test the three explanatory hypotheses, using statistical models fit with LISREL-VII (Jöreskog & Sörbom, 1988). The data did not fit the McGlone (1977) or the Kimura (1983) hypotheses, but they did fit the Inglis and Lawson (1982) hypothesis, in that women suffered greater PIQ deficits than men following left hemisphere lesions.

Inglis and Lawson (1982) suggest that the PIQ deficits of women with left hemisphere lesions might be caused by a strategy difference in solving PIQ items, in that women use more verbal strategies for solving PIQ items. However, other explanations of the phenomenon are possible. Greater PIQ deficits in women with left hemisphere lesions could result from gender differences in functional brain organization in the absence of strategy differences, if women have more nonverbal abilities relevant to PIQ represented in the left hemisphere.

The strategy versus functional organization question has been investigated more directly (Farace, Turkheimer, & Lee, submitted) in a study of gender differences in the effects of verbal and spatial interference on a PIQ subtest, Picture Arrangement, in 149 (81 men, 68 women) normal subjects. Subjects solved picture arrangement items while simultaneously attempting to memorize verbal or spatial distractors. There was no gender difference found in the effects of type of distraction on the picture arrangement task, suggesting that men and women did not differ in their problem-solving strategy for picture arrangement.

Gender Differences in Brain Anatomy

Although sexual dimorphisms have been observed in the brains of lower animals and humans for some time (see Goy & McEwen, 1980, for review of early work), recent advances in neuroimaging technology have opened many new avenues for investigation, especially in humans. The results of these investigations parallel those of the functional studies reviewed above, for better and for worse: many provocative results have been reported, which in general appear to point in the direction of gender differences along the right-left axis of the brain; but the most provocative results have proved difficult to replicate, and it has not been possible to formulate a theory of gender differences in brain anatomy that explains the various anatomical findings, much less relate them to the equally complex body of findings regarding lateralization of function. Our review of gender differences in brain anatomy with focus on neuroimaging findings in humans; animal and postmortem work will be summarized as necessary.

Differences in Nuclei and Cell Architecture

Exploration of gender differences in the brain will begin at the cellular level. Several studies have examined gender differences in brain nuclei, that is, clusters of neuronal cell bodies that group together (Diamond, Scheibel, & Elson, 1985).

Hypothalamus. Most studies of gender differences at the cellular level have centered on the hypothalamus, a structure thought to govern sexual behavior (Hines & Green, 1991). The hypothalamus contains at least four different nuclei which can be mapped (Allen, Hines, Shryne, & Gorski, 1989). Several gender differences have been shown in hypothalamic nuclei.

Swaab and Fliers (1985) performed a study of hypothalamic tissue collected postmortem from 13 men and 18 women of unknown handedness. They found a sexually dimorphic cell group in the preoptic hypothalamus. The volume of the nucleus was 2.5 ± 0.6 times greater in men than in women and had 2.2 ± 0.5 times as many cells. This nucleus was called the sexually dimorphic nucleus of the preoptic area (SDN-POA) of the hypothalamus. The function of the nucleus is unknown, although from studies of other mammals it is thought to be associated with gonadotropin release, maternal behavior, and sexual behavior (Allen *et al.*, 1989).

Hofman and Swaab (1989) studied hypothalamic tissue from 30 subjects (13 men and 17 women) in the SDN-POA of the hypothalamus. They characterized this nucleus as an oval-shaped section of cells that are more densely packed than the surrounding region. This nucleus is located in the middle of the preoptic area. The authors found that the nucleus had a larger cross-section and a larger volume (2.2 times larger) in men, and the number of cells was 2.1 times higher. However, if brain weight was taken into account statistically, there was no remaining gender difference.

Allen *et al.* (1989) performed a tissue study of the preoptic-anterior hypothalamic area (PO-AHA) in 22 age-matched brains (11 men, 11 women). They found four cell groups in the PO-AHA, for which they proposed the term interstitial nuclei of the anterior hypothalamus (INAH 1–4). They reported that INAH-1 is probably equivalent to the Swaab and Fliers (1985) and Hofman and Swaab (1989) SDN-POA of the hypothalamus. Allen *et al.* (1989) found that INAH-1, INAH-4, and the supraoptic nucleus (a control nucleus chosen for its lack of gender differences in other species) were not sexually dimorphic, but INAH-2 and INAH-3 did show gender differences. INAH-2 was 2 times larger and INAH-3 was 2.8 times larger in men than in women, regardless of age. They also noted that women's values varied cyclically with their menstrual cycles.

Swaab and Fliers (1985), Hofman and Swaab (1989), and Allen *et al.* (1989) disagree on exactly which cell groups on the PO-AHA are sexually dimorphic, although they do agree that male cell groups in this area tend to be larger. Part of their difficulty may be a lack of standardization of the histochemical procedure for dividing the area into nuclei. Differences between the studies may also be confounded by age; these nuclei are known to shrink with age (Allen *et al.*, 1989), and the women in the Swaab and Fliers (1985) study were 10 years older on average than the men. In the Allen *et al.* (1989) study, male and female subjects were matched for age.

It is interesting to note that Simon Levay's (1991) finding of sexual orientation differences in the nuclei of the hypothalamus also occurred in INAH-3. Like Allen *et al.* (1989), Levay (1991) found that INAH-3 was more than twice as large in the heterosexual men as in the women. In addition, Levay (1991) found that INAH-3 was also more than twice as large in the heterosexual men as in the homosexual men. This suggests that these nuclei are correlated with sexual orientation, because both the women (presumed heterosexual) and the homosexual men were oriented toward men. This study has some methodological problems, such as the number of subjects with AIDS and uncertainty in the classification of sexual orientation of its groups, but it shows support for the involvement of the nuclei of the hypothalamus in sexual behavior.

Cells in Wernicke's Area. Jacobs, Schall, and Scheibel (1993) reported a study of gender differences in cell architecture in Wernicke's area, an area of the left hemisphere where language functions are thought to be represented (Springer & Deutsch, 1989). The authors performed a study of basilar dendrites and supragranular pyramidal cells using Golgi staining in tissue gained postmortem from 20 normal subjects (ten men and ten women, all right-handed). They found slightly higher overall dendritic density in women.

Differences in Larger Anatomical Structures

Historically, studies of gender differences in the brain were conducted using brains available postmortem. Even after the introduction of advanced brain imaging techniques such as computed tomography (CT), magnetic resonance imaging (MRI), and PET, postmortem studies are still often performed, and they continue to offer some advantages over imaging studies: the spatial resolution is decidedly superior, and the possibility of direct study of tissue, such as cell body staining, can make human work more directly comparable to animal work. It is also easier to measure brain weight and tissue composition. However, in postmortem studies, researchers rarely have control of the type of behavioral data collected before death, making studies of structure/function relationships difficult.

Limitations of spatial resolution make brain imaging techniques most useful for the study of larger brain structures, and as a result, imaging studies usually focus on the cerebral cortex and interhemispheric connections. Brain imaging techniques have the advantage of allowing researchers to study live human subjects noninvasively. Concurrent behavioral and background data can be collected from subjects, which allows brain images to be correlated with behavioral measures. Also, PET has the additional benefit of producing images which reflect brain function during a task.

Gender Differences in Brain Size

Men, on average, have larger and heavier brains than women. This unsurprising observation leads to the question of whether men's brains are larger relative to the corresponding gender difference in body size. This question is more difficult to answer than it appears at first. Of the many available statistical methods for "controlling" the effects of body size when considering group differences in brain size (Reist, 1985), two have been explored in gender difference studies: analysis of covariance (ANCOVA) and brain/body ratios. In ANCOVA, brain weights are adjusted by subtracting the portion of brain weight that can be predicted from a linear regression on body size. The remainder is the portion of brain size that is unrelated to body size, and gender comparisons on this variable provide a statistical estimate of the gender difference in brain size that is independent of body size. Ratios of brain size to body size simply rescale brain measurements in terms of body measurements.

Analysis of covariance is only justified when the regression of brain size on body size is equivalent in the two genders and only makes a difference if brain size and body size are correlated with each other. Peters (1991) argues that neither condition holds in the study of human brain anatomy: correlations between brain size and body size are quite low, and the correlation tends to be higher in men than in women. Peters (1991) suggests that the question of relative brain size in the genders is not meaningful in terms of anatomy or function and does not endorse adjusting brain sizes at all.

Other investigators have reported that use of brain/body ratios does reduce or eliminate gender differences in brain size. Ho, Roessmann, Straumfjord, and Monroe (1980a, 1980b) found gender differences in raw brain weight in both black and white subjects, but not in ratios of brain weight to body height, weight, or surface area. Correlations between brain weight and the body size measures were between .1 and .3. Ankney (1992) criticized the use of ratios by Ho *et al.* (1980a, 1980b) on statistical grounds and also because the resulting ratios were negatively correlated with body size. For any given body size, men in the Ho *et al.* study had higher ratios than women, but because men were larger overall, the mean ratios were roughly equal in men and women. Ankney (1992) used available data to approximate an ANCOVA of the data in Ho *et al.* (1980a, 1980b) and found that gender differences remained, which is not surprising given the small correlations between body and brain size in the sample.

As we will see again in studies of the size of the corpus callosum relative to the size of the entire brain, the fundamental methodological question of how smaller brain structures should be scaled in terms of the larger structures they compose has not been resolved. Although a considerable literature on this problem exists in systematic biology (Reist, 1985), animal physiology (Packard & Boardman, 1988), and sociology (Schuessler, 1973), it appears to have been explored only superficially in human neuroanatomy. The problem is especially difficult in this area because there appears to be no systematic allometric relationship between overall body size and brain size or between brain size and the size of smaller brain structures; understanding of such fundamental growth functions are a necessary first step to understanding group differences in size.

Several conclusions do appear to be established: a) Men have larger brains than women; b) this difference is of roughly the same magnitude as the gender difference in body height or weight; and c) for men and women of a given height or weight, men have larger brains. Conclusions (b) and (c) appear contradictory, but they are not (Ankney, 1992).

A related methodological problem involves the meaning of the size of brain structures on a cellular level. Fiber tracts could be the same absolute size but be very different structurally. Differences in the number of axons, axon density, axon size, the myelination of axons, and interfiber distance could all affect the size of a fiber tract. Differences in axonal structure, as opposed to the size of the fiber tract, may govern differences in function, as some arrangements of axons are particularly advantageous for transfer of certain kinds of information (Innocenti, 1984).

Studies of sex differences in the corpus callosum of the rat show differences in the composition of the corpus callosum that are not evident from gross size measures. Juraska and Kopcik (1988) found that although there were no sex differences in the size of the corpus callosum, female rats had more unmyelinated axons overall, and females from complex environments had more myelinated axons than comparably housed males. Also, plasticity was found to be a result of increases in axon number in females and of larger axons in males.

What is the relationship between the size of anatomical structures and axonal composition in humans? Witelson (1991a) has suggested that variation in the size of the midsagittal area of the corpus callosum reflects a gender difference in the number of axons connecting the right and left hemispheres, especially the temporoparietal regions. Aboitiz, Scheibel, Fisher, and Zaidel (1992a) studied fiber densities in 20 corpus callosa (ten men and ten women) gathered postmortem. There were no gender differences in fiber density and no correlations between fiber densities and callosal area, suggesting that differences in callosal size are related to differences in number of axons.

Gender Differences in Anatomical Asymmetry

The observation of gender differences in functional lateralization, combined with reports of asymmetrical morphometry of the cerebral hemispheres (Geschwind & Levitsky, 1968) and the more recent availability of neuroimaging data, have led to considerable interest in the question of whether there are gender differences in the degree of anatomical asymmetry. Geschwind and Levitsky (1968) found that the planum temporale was larger in the left hemisphere of 65% of 100 brains collected postmortem. Later studies have focused on the planum temporale and other frontotemporal regions, especially the sylvian fissure.

As we will see again in studies of the corpus callosum, the results of anatomical asymmetry studies are equivocal and seriously confounded by handedness (Kertesz, Polk, Black, & Howell, 1990; Witelson & Kigar, 1992). Positive results were reported by Bear, Schiff, Saver, Greenberg, and Freeman (1986) in a study of 30 male and 36 female brains examined with CT. At the most anterior extension of the frontal lobes, both genders were wider in the right hemisphere; at the occipital pole, both genders were larger in the left hemisphere. Men showed significantly greater differences between the left and right hemispheres on both measures.

Koff, Naeser, Pieniadz, Foundas, and Levine (1986), using very similar data, did not find sex differences. Koff *et al.* studied CT images from 87 male and 85 females subjects. Unlike Bear *et al.* (1986), who used a continuous measure of asymmetry, Koff *et al.* used the more traditional measure developed by LeMay (1977). Subjects were trichotomized as either symmetrical or left- or right-greater asymmetrical, and chi square was used as a test statistic for gender differences. Among right-handers, more men than women showed right anterior asymmetry and left posterior asymmetry, but the differences did not reach statistical significance. Continuous measures of asymmetry might have provided greater statistical power (Turkheimer, 1989).

Corpus Callosum. The corpus callosum is the myelinated axonal structure which connects the left and right hemispheres. It is believed to be the main structure through which information is transferred between the hemispheres, connecting both homologous and heterologous areas (Innocenti, 1984). Corpus callosa are collected for study either by dissection following autopsy or by measurement in sagittal section using an MRI image.

DeLacoste-Utamsing and Holloway (1982) published the first study regarding gender differences in the size or shape of the corpus callosum. They studied nine male and five female brains collected postmortem. Analyses of the maximal width of the splenium (the most posterior portion of the corpus callosum) revealed a gender difference of almost 10 standard errors. The gender difference in the bulbousness of the posterior portion of the corpus callosum was obvious to the naked eye.

Since the original report by DeLacoste-Utamsing and Holloway (1982) there have been many attempts at replication, with very mixed results. Patterns of results are difficult to discern because of small sample sizes and because of the great interstudy variability in the anatomical divisions and statistical methods that are employed. Several generalizations do appear to be possible, however.

DeLacoste-Utamsing and Holloway's (1982) finding that women's callosa were larger than men's in raw units does not appear to have been borne out, except in the same laboratory (Holloway & DeLacoste, 1986). In fact, men's callosa are almost always larger. The only exception is found in Byne, Bleier, and Houston (1988), who found that in subjects under 40, men had larger absolute callosal areas, but in subjects older than 40, women's measurements were larger by more than 3 standard errors.

This interaction of age with gender effects in the corpus callosum appears to be quite general. Mean ages of subjects were not provided in the original DeLacoste-Utamsing and Holloway (1982) report, but in the Holloway and DeLacoste (1986) replication, women and men had mean ages of 73 and 63.7 years, respectively. Witelson (1989, 1991a) has also reported that callosal size decreases with chronological age in men but not in women. Witelson (1991b) has theorized that there is a gender difference in neuronal loss in the corpus callosum during aging. Cowell, Allen, Zalatimo, and Denenberg (1992) have shown that women's corpus callosum areas continue to increase until ages 41–50, whereas those for men peak by age 20 and decline thereafter. No evidence of gender differences in the corpus callosum has been found in children, infants, or fetuses (Bell & Variend, 1985; Clarke, Kraftsik, Van Der Loos, & Innocenti, 1989). There is also evidence of gender differences in rate of other types of brain atrophy, with men showing greater declines with age (Gur *et al.*, 1991; Dekaban & Sadowsky, 1978).

Another variable that appears to have an effect on the outcome of corpus callosum investigations is handedness. Interactions among gender, lateralization, and handedness are extremely complex. Witelson (1985, 1989) found that in men but not in women, there was a difference between consistent right-handers (CRH) and inconsistent right-handers (nCRH) in the area of the posterior body of the corpus callosum. In CRH subjects, female brains were larger; in nCRH subjects, male brains were larger. Similar findings have been reported by Denenberg, Kertesz, and Cowell (1991) and by Habib *et al.* (1991). The original report by DeLacoste-Utamsing and Holloway (1982) did not report the handedness of the subjects.

Measurement issues also play a significant role in complicating the results of corpus callosum studies. As was discussed above in the section on sex differences in overall brain size when analyzing differences in the corpus callosum. The available choices include unadjusted raw measurements, ANCOVA, and ratios of callosal measurement to brain weight. The correlation between brain weight and callosum size is usually small.

DeLacoste-Utamsing and Holloway (1982) reported that women's splenia were significantly larger in unadjusted units. Total unadjusted callosal area was equal in the genders, but women's callosa were larger "relative to brain weight" (p. 1432). Subsequent studies have been virtually unanimous in demonstrating that women's callosa are not larger, in unadjusted units or when adjusted using ANCOVA; in fact, male brains are usually larger (Aboitiz, Scheibel, & Zeidel,

1992; Bell & Variend, 1985; Demeter, Ringo, & Doty, 1988; O'Kusky *et al.*, 1988; Weber & Weis, 1986; Weis, Weber, Wenger, & Kimbacher, 1989). When callosal sizes are expressed as proportions of brain size, women's callosa are often found to be larger (Allen, Richey, Chai, & Gorski, 1991; Kertesz, Polk, Howell, & Black, 1987; Reinarz, Coffman, Smoker, & Godersky, 1988). Other studies have failed to find gender differences in proportional callosal measurements, however (Emory, Williams, Cole, Amparo, & Meyer, 1991; Habib *et al.*, 1991; Oppenheim, Lee, Nass, & Gazzaniga, 1987).

The original DeLacoste and Holloway (1982) study focused on the splenium, the most posterior portion of the corpus callosum. Unfortunately, there is no obvious anatomical basis for dividing the corpus callosum, and there is little agreement among researchers on terms for subdivisions of the callosum. Witelson (1989) devised a system comprising seven segments of the callosum proceeding from anterior to posterior: rostrum, genu, rostral body, anterior midbody, posterior midbody, isthmus, and splenium (see Figure 1). Denenberg *et al.* (1991) performed a principal component analysis of 99 callosal widths and arrived at a similar system of seven noncontiguous regions extending from posterior to anterior.

Attempts to replicate the original DeLacoste-Utamsing and Holloway (1982) finding in the posterior midbody or splenium have not been successful in terms of the absolute size of the structures: in unadjusted units, women's structures do not appear to be larger (Elster, DiPersio, & Moody, 1990). Once again, issues of morphometric standardization complicate matters. When the sizes of subregions of the corpus callosum are expressed as a proportion of the total area of the corpus callosum, those of women are often found to be larger (Clarke *et al.*, 1989; Reinarz *et al.*, 1988; Steinmetz *et al.*, 1992; Witelson, 1989).

Proportional analyses of subregions of the callosa are essentially analyses of shape and are partial replications of DeLacoste and Holloway's (1982) original observation of greater "bulbosity" in the splenia. Some studies have undertaken to quantify bulbosity, with mostly negative results (Allen *et al.*, 1991, found that women's splenia were more bulbous; negative results were reported by Bell &

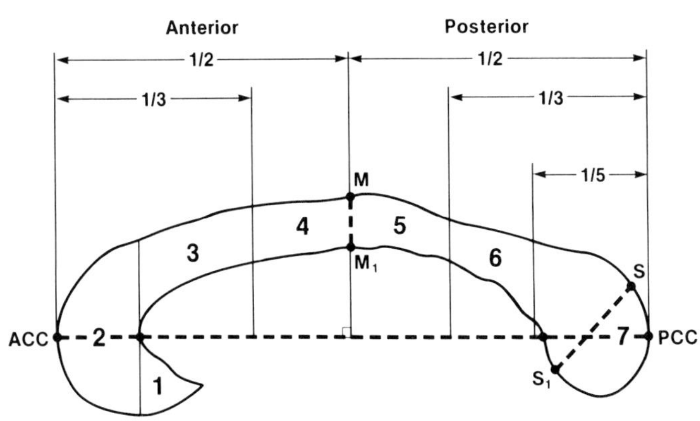

Figure 1. Drawing of the midsagittal view of the corpus callosum, divided into anterior and posterior halves; anterior, middle, and posterior thirds; and the posterior one-fifth region roughly equivalent to the splenium (region 7). Region 1 is roughly congruent to the rostrum, region 2 the genu, regions 3 and 4 together the anterior body, and regions 5 and 6 together the posterior body. Region 6 is also referred to as the isthmus. (Adapted from Witelson & Goldsmith, 1991.)

Variend, 1985; Byne *et al.*, 1988; Clarke *et al.*, 1989; Weber & Weis, 1986; Weis *et al.*, 1989).

Interpretation of the results of studies of the corpus callosum are complicated by the many unresolved methodological issues that have been discussed. It appears clear that gender differences, if they exist, interact with both age and handedness, which therefore should be routinely included in future studies. The problem of how to standardize the size of subunits of structures for the size of the larger structures of which they are a part deserves more careful consideration than it has received to date. The question has turned out to be surprisingly complex in other areas of research (Packard & Boardman, 1988; Reist, 1985; Schuessler, 1973, 1974). It also appears that the standard morphometric technique that has evolved, the assessment of a single midsagittal section of the corpus callosum, is an inadequate means of quantifying the size and shape of such a complex three-dimensional structure.

Several aspects of this body of research—methodological variance among studies, confounds from external variables, considerable interindividual variability in the characteristic of interest, and small sample sizes—suggest that meta-analysis may be a particularly useful method of extracting a signal from the noisy data that have been reported.

Anterior Commissure and Massa Intermedia. In addition to the corpus callosum, two other major fiber tracts connect the hemispheres: the anterior commissure, which connects the parietal lobes (Diamond *et al.*, 1985), and the massa intermedia, which crosses the third ventricle between the two thalami (Hines & Green, 1991).

Allen and Gorski (1986, 1991) performed a postmortem quantitative analysis of the anterior commissure in 100 age-matched men and women. The anterior commissure was 11.8% larger in women, although male brains were significantly heavier. Demeter *et al.* (1988) found no gender differences in a smaller (21 men, 11 women) sample of postmortem brains. Male brains were substantially larger, however, and no correction was made for brain size because the correlation between brain size and anterior commissure size failed to reach significance ($r = .37$).

The massa intermedia is not present in all humans. Morel (1948) and Allen and Gorski (1991) report that it is more often present in women than in men. Morel (1948) reported an agenesis rate of 14.3% in women and 27.9% in men in an epidemiological study of 823 patients (355 men, 468 women). Allen and Gorski (1991) also found that the massa intermedia was more often present in women than in men. In addition, they found that among subjects who had a massa intermedia, women had significantly larger ones.

Structure–Function Relationships

We have reviewed three domains of gender differences in cognition and neuroanatomy: differences in ability, differences in lateralization of function, and differences in the morphometry of the brain. All three domains are controversial and suggestive of the possibility of gender differences in a complex hypothetical construct called "lateralization." As we have noted, there is some well-founded dissatisfaction with the current level of empirical specification of this construct (Efron, 1990). The task facing researchers in this area, therefore, is the establishment of the construct validity of lateralization (Cronbach & Meehl, 1955). To establish construct validity, researchers will have to formulate empirical theories of the nature of lateralization, especially theories relating observations in different domains.

The structure of this review suggests that we should look for three types of relationships: between ability and lateralization of function, between ability and neuroanatomy, and between lateralization of function and neuroanatomy. Only the first of these has been addressed in detail. There is a considerable literature addressing the relationship between ability, especially spatial ability in male subjects, and asymmetrical EEG activation (Ray, Newcombe, Semon, & Cole, 1981; Tucker, 1976), lateralized stimulus presentation (Voyer & Bryden, 1990), handedness (Harshman, Hampson, & Berenbaum, 1983), or physical measures of asymmetry such as finger ridge count (Kimura & Carson, 1993). As these studies generally do not include neuroimaging data, they will not be reviewed here. By and large, they have suggested that individual differences in lateralization are related to ability along the same dimensions on which gender differences are observed. That is, individuals with relatively stronger indications of lateralized function show patterns of ability more similar to those found in men. The most important question—whether gender differences in patterns of ability can be *explained* by gender differences in lateralization of function—remains to be resolved.

Ability and Morphometry

Researchers are just beginning to address the question of whether functional and anatomical gender differences are related. If it could be shown that within-gender individual differences in ability and lateralization were related to within-gender anatomical variation in hemispheric asymmetry or callosal morphometry, the empirical status of the construct of lateralization would be significantly enhanced.

The only study of ability and the anatomy of interhemispheric tracts was reported by Lansdell and Davie (1972). Information on the presence or absence of the massa intermedia was available in 74 male and female subjects who had been tested with the Wechsler-Bellevue intelligence test. The massa intermedia was present in 40% of the men and 33% of women (difference is nonsignificant, *ns*). In men but not in women, there was a small but significant deficit in nonverbal IQ among subjects in whom the massa intermedia was present.

There is a long history of studies demonstrating modest relationships between head size and general intelligence, although recent work has suggested that stronger relationships may emerge now that brain size can be measured directly instead of being inferred from skull diameter (Willerman, Schultz, Rutledge, & Bigler, 1991). More relevant to the concerns of this chapter are findings that relate anatomical asymmetries from CT or MRI to differences between VIQ and PIQ scores. These correlations have been reported to be as high as .57 (Yeo, Turkheimer, Raz, & Bigler, 1987). Willerman, Schultz, Rutledge, and Bigler (1992) have shown similar relationships in men but not in women.

In the most comprehensive study of relationships among gender, ability, lateralization of function, and cortical anatomy that has been undertaken to date, Hines, Chiu, McAdams, Bentler, and Lipcamon (1992) obtained measures of verbal fluency, visuospatial ability, dichotic listening asymmetry, and regional callosal anatomy from 28 female subjects. A structural equation model was employed to study relationships among ability, lateralization, and callosal anatomy. The size of the splenium was positively related to verbal fluency and negatively related to language lateralization. Other results, relating language lateralization to callosal morphometry, will be described below.

Although the sample size was very small given the complexity of the statistical analysis, these results are remarkably consistent with a model postulating interhemispheric transfer as the basis for bilateral processing of verbal fluency tasks in women. Women with larger splenia were less functionally lateralized and performed better on the verbal fluency tasks. It will be very interesting to see whether these findings replicate in larger samples and whether they will prove to be useful explanations of gender differences in verbal fluency and spatial ability.

Lateralization of Anatomy and Function

Gender differences in the size of the cerebral hemispheres and in the size and shape of interhemispheric commissures are of little neuropsychological interest if they are not related to functional measures of lateralization.

One area of research has compared male and female subjects on behavioral indicators of speed of interhemispheric transfer. Potter and Graves (1988) examined male and female subjects with tasks requiring comparison of stimuli presented to opposite sides of the body midline. Women had lower error rates when stimuli were presented to the left and right visual fields. Burton, Pepperrell, and Stredwick (1991) did not replicate this result and suggested that Potter and Graves' (1988) findings may have resulted from a gender difference in speed-accuracy trade-off on the task.

Four studies have examined relationships among callosal anatomy and dichotic listening. Kertesz *et al.* (1987) reported no correlation between callosal area and a laterality index from a dichotic listening task. O'Kusky *et al.* (1988), however, reported a correlation of −.31 between midsagittal callosum area and a laterality index similar to the one employed by Kertesz *et al.* (1987). Hines *et al.* (1992) employed structural equation modeling to study the relationship between splenial area and a lateralization index from a dichotic listening task and reported a standardized path coefficient of −.41.

An important study of the relationship between callosal morphometry and lateralization of function has recently been published by Clarke, Lufkin, and Zaidel (1993). Clarke *et al.* obtained callosal measurements from 60 normal graduate students (15 each of men and women, right- and left-handed) who were administered a dichotic listening task in which difference syllables were presented simultaneously to the left and right ears. Four commissurotomized patients were also given the dichotic listening task. Analysis of the commissurotomy patients demonstrated that left ear performance is dependent on interhemispheric transfer, that is, the patients demonstrated a very strong right ear advantage. All but one right-handed subject had a similar right ear advantage.

Correlations between callosal area measurements and left and right ear accuracy showed a surprising result: among right-handed subjects and male subjects, there were no correlations between callosal measurements and either left ear performance or the difference between right and left ear performance, but there were substantial *negative* correlations between callosal size and right ear performance.

Clarke *et al.* (1993) interpret the results of their experiment in predominantly negative terms: "However, we see that variations in morphometric measures of corpus callosum size in normal subjects do not predict individual differences in the effectiveness of transfer of auditory information between the cerebral hemispheres. Neither does the corpus callosum turn out to predict behavioral measures of hemispheric specialization" (p. 555). They also note, however, that the results are very consistent with a hypothesis involving functional inhibition or interference between the hemispheres.

It appears to us that the most viable conclusion is not that callosum size does not predict individual differences in the efficiency of interhemispheric transfer, but rather that efficient hemispheric transfer does not facilitate performance on all tasks. Faced with the difficult task of processing conflicting information in two channels, the optimal strategy for maximizing "superior ear" performance is to ignore one channel as completely as possible, as demonstrated by the performance of the commissurotomy patients. Efficient interhemispheric transfer makes such a strategy more difficult. We also note that, in men, correlations between callosal size and right-minus-left score were consistently negative (median = $-.27$), and the limited power of a statistical test of a small correlation in 30 subjects may not justify strong negative conclusions about the relationship between callosum size and functional laterality measures.

Perusal of these studies also reveals a subtle methodological issue in dichotic listening procedures that may be relevant. In Kertesz *et al.* (1987), the one study that found no relationship whatsoever between callosum measurements and laterality measures, trial subjects were instructed on each dichotic listening to report only the right or left ear stimulus. In the other three studies, subjects were instructed to attempt to repeat the stimuli presented to *both* ears. One might expect the interference effects described by Clarke *et al.* (1993) to be stronger when subjects are attempting to listen to both ears.

Conclusion

The research literature concerning gender differences in cognition, and their possible basis in brain differences, is vast and often contradictory. Upon concluding this review, it appears that one reason that the results of research in this area are so difficult to interpret is that the problem has been somewhat ill specified. The search for gender differences in functional asymmetry has proceeded in parallel with efforts to establish the validity of the hypothetical construct of lateralization of function. When contradictory results are observed, it is not clear whether the genders do not differ on a valid dimension of lateralized function or whether the construct of "laterality" is itself flawed.

Although it is not often explicitly stated (Bryden, 1982), the hypothetical construct of lateralization of function represents an empirical hypothesis that there exists an individual differences dimension that is manifest in anatomy, physiology, and cognition. At one extreme of this dimension, individuals are hypothesized to rely relatively more strongly on computational strategies requiring focused application of powerful processors located primarily in a single hemisphere; at the other extreme, individuals rely relatively more strongly on the cooperative action of multiple processors in both hemispheres.

We are careful to say that individuals rely "relatively more strongly," rather than "primarily," on one strategy or the other to emphasize that the degree of variation on the hypothesized trait is unknown and probably not especially large. It is probably true that all normal humans have two cerebral hemispheres that are to some degree specialized, but which nevertheless must cooperate to complete any task. Individual or group differences, if they exist at all, must be seen in the context of a more or less universal human functional architecture. Talk of "male" and "female" brains, "lateralized" and "unilateral" computational strategies, or "left" and "right" hemisphere tasks, therefore, is unlikely to lead to scientific progress.

One conclusion that can be drawn from the above is that observable gender differences in lateralization of function are likely to be quite small. Men and women may differ, to an unknown degree, on a complex dimension of the unimaginably complex universe of human cognitive ability. Observable indicators of this dimension are likely to be quite imperfect, as we have seen again and again in this review. For example, the size of the corpus callosum is of interest not because of any intrinsic importance of the size of brain structures, but because it may be an *indicator* of the functional importance of the callosum. The same could be said of EEG measures, the relative volume of the hemispheres, ear superiority in dichotic listening tasks, or hemisphere superiority in dichaptic visual tasks.

The first step toward better understanding of gender differences in cognition is the establishment of the validity of the construct on which the genders may differ. To accomplish this, studies will need to employ multiple indicators of lateralization in relatively large samples. The Hines *et al.* (1992) study is a good example of the kind of work that needs to be done. Hines *et al.* focused their attention on a single gender, leaving the question of gender differences to wait until a valid model of lateralization is developed. In a similar vein, it would probably be a good idea to focus on right-handed subjects while the construct validity of lateralization is being established. Future studies could endeavor to address some of the shortcomings of Hines *et al.* by increasing the size of the sample and including multiple behavioral indicators of lateralization.

New functional imaging methods are certain to play an important role in this process. Functional imaging provides the most valid indicators of lateralized function. Some of the most dramatic indications of gender differences in lateralized cognition have emerged from EEG, rCBF, and PET studies. Functional MRI promises to be especially important because of its superior temporal and spatial resolution and because its eventual lower expense may permit the study of much larger samples than have been available with PET or other radioactive techniques.

References

Aboitiz, F., Scheibel, A. B., Fisher, R. S., & Zaidel, E. (1992a). Fiber composition of human corpus callosum. *Brain Research, 598,* 143–153.

Aboitiz, F., Scheibel, A. B., & Zaidel, E. (1992). Morphometry of the sylvian fissure and the corpus callosum, with emphasis on sex differences. *Brain, 115,* 1521–1541.

Allen, L. S., & Gorski, R. A. (1986). Sexual dimorphism of the human anterior commissure. *Anatomical Record, 214,* 3A.
Allen, L. S., & Gorski, R. A. (1991). Sexual dimorphism of the anterior commissure and massa intermedia of the human brain. *Journal of Comparative Neurology, 312,* 97–104.
Allen, L. S., Hines, M., Shryne, J. E., & Gorski, R. A. (1989). Two sexually dimorphic cell groups in the human brain. *Journal of Neuroscience, 9,* 497–506.
Allen, L. S., Richey, M. F., Chai, Y. M., & Gorski, R. A. (1991). Sex differences in the corpus callosum of the living human being. *Journal of Neuroscience, 11,* 933–942.
Ankney, C. D. (1992). Sex differences in relative brain size: The mismeasure of woman, too? *Intelligence, 16,* 329–336.
Azari, N. P., Rapoport, S. I., Grady, C. L., DeCarli, C., Haxby, J. V., Schapiro, M. B., & Horwitz, B. (1992). Gender differences in correlations of cerebral glucose metabolic rates in young normal adults. *Brain Research, 574,* 198–208.
Baxter, L. R., Mazziotta, J. C., Phelps, M. E., Selin, C. E., Guze, B. H., & Fairbanks, L. (1987). Cerebral glucose metabolic rates in normal human females versus normal males. *Psychiatry Research, 21,* 237–245.
Bear, D., Schiff, D., Saver, J., Greenberg, M., & Freeman, R. (1986). Quantitative analysis of cerebral asymmetries. *Archives of Neurology, 43,* 598–603.
Beaumont, J. G. (Ed.) (1982). *Divided visual field studies of cerebral organization.* London: Academic Press.
Beaumont, J. G., Mayes, A. R., & Rugg, M. D. (1978). Asymmetry in EEG alpha coherence and power: Effects of task and sex. *Electroencephalography and Clinical Neurophysiology, 45,* 393–401.
Bell, A. D., & Variend, S. (1985). Failure to demonstrate sexual dimorphism of the corpus callosum in childhood. *Journal of Anatomy, 143,* 143–147.
Bogen, J. E., DeZure, R., Tenhouten, W. D., & Marsh, J. F. (1972). The other side of the brain IV. The A/P ratio. Bulletin of the Los Angeles Neurological Societies, 37, 49–61.
Bryden, M. P. (1982). *Laterality: Functional asymmetry in the intact brain.* New York: Academic Press.
Burton, A., Pepperrell, S., & Stredwick, J. (1991). Interhemispheric transfer in males and females. *Cortex, 27,* 425–429.
Byne, W., Bleier, R., & Houston, L. (1988). Variations in human corpus callosum do not predict gender: A study using magnetic resonance imaging. *Behavioral Neuroscience, 102,* 222–227.
Chrichton-Browne, J. (1880). On the weight of the brain and its component parts in the insane. *Brain, 2,* 42–67.
Clarke, S., Kraftsik, R., Van Der Loos, H., & Innocenti, G. M. (1989). Forms and measures in adult and developing human corpus callosum: Is there sexual dimorphism? *Journal of Comparative Neurology, 280,* 213–230.
Clarke, J. M., Lufkin, R. B., & Zaidel, E. (1993). Corpus callosum morphometry and dichotic listening performance: Individual differences in functional interhemispheric inhibition? *Neuropsychologia, 31,* 547–557.
Cowell, P. E., Allen, L. S., Zalatimo, N. S., & Denenberg, V. H. (1992). A developmental study of sex and age interactions in the human corpus callosum. *Developmental Brain Research, 66,* 187–192.
Cronbach, L. J., & Meehl, P. E (1955). Construct validity in psychological tests. *Psychological Bulletin, 52,* 281–302.
Davidson, R. J., Schwartz, G. E., Pugash, E., & Bromfield, E. (1976). Sex differences in patterns of EEG asymmetry. *Biological Psychology, 4,* 119–138.
Dekaban, A. S., & Sadowsky, D. (1978). Changes in brain weights during the span of human life: Relation of brain weights to body heights and body weights. *Annals of Neurology, 4,* 345–356.
DeLacoste-Utamsing, C., & Holloway, R. L. (1982). Sexual dimorphism in the human corpus callosum. *Science, 216,* 1431–1432.
Demeter, S., Ringo, J. L., & Doty, R. W. (1988). Morphometric analysis of the human corpus callosum and anterior commissure. *Human Neurobiology, 6,* 219–226.
Denenberg, V. H., Kertesz, A., & Cowell, P. E. (1991). A factor analysis of the human's corpus callosum. *Brain Research, 548,* 126–132.
Diamond, M. C., Scheibel, A. B., & Elson, L. M. (1985). *The human brain coloring book.* New York: Barnes & Noble Books.
Efron, R. (1990). *The decline and fall of hemispheric specialization.* Hillsdale, NJ: Lawrence Erlbaum Associates.
Elliott, D., Weeks, D. J., Lindley, S., & Jones, R. (1986). Sex differences in dual-task interference between speaking and a manual force-production task. *Perceptual and Motor Skills, 62,* 3–8.
Elster, A. D., DiPersio, D. A., & Moody, D. M. (1990). Sexual dimorphism of the human corpus

callosum studied by magnetic reasonance imaging: Fact, fallacy and statistical confidence. *Brain Development, 12,* 321–325.

Emory, L. E., Williams, D. H., Cole, C. M., Amparo, E. G., & Meyer, W. J. (1991). Anatomic variation of the corpus callosum in persons with gender dysphoria. *Archives of Sexual Behavior, 20,* 409–417.

Fairweather, H. (1982). Sex differences. In J. G. Beaumont (Ed.), *Divided visual field studies of cerebral organization* (pp. 147–194). London: Academic Press.

Farace, E., Turkheimer, E., & Lee, G. (submitted). Are there gender differences in problem solving strategy for performance IQ?

Flor-Henry, P., & Koles, Z. J. (1982). EEG characteristics of normal subjects: A comparison of men and women and of dextrals and sinistrals. *Research Communications in Psychology, Psychiatry and Behavior, 7,* 21–38.

Friedland, J., & Kershner, J. (1986). Sex-linked left lateralized central processor for hierarchically-structured material? Evidence from Broca's aphasia. *Neuropsychologia, 24,* 411–415.

Gazzaniga, M. S. (1985). *The social brain.* New York: Basic Books, Inc.

Geschwind, N., & Levitsky, W. (1968). Human brain: Left-right asymmetries in temporal speech region. *Science, 161,* 186–187.

Glass, A., Butler, S. R., & Carter, J. C. (1984). Hemispheric asymmetry of EEG alpha activation: Effects of gender and familial handedness. *Biological Psychology, 19,* 169–187.

Goy, R. W., & McEwen, B. S. (1980). *Sexual differentiation of the brain.* Cambridge, MA: MIT Press.

Gur, R. C., Gur, R. E., Obrist, W. D., Hungerbuhler, J. P., Younkin, D., Rosen, A. D., Skolnick, B. E., & Reivich, M. (1982). Sex and handedness differences in cerebral blood flow during rest and cognitive activity. *Science, 217,* 659–661.

Gur, R. C., Mozley, P. D., Resnick, S. M., Gottlieb, G. L., Kohn, M., Zimmerman, R., Herman, G., Atlas, S., Grossman, R., Berretta, D., Erwin, R., & Gur, R. E. (1991). Gender differences in age effect on brain atrophy measured by magnetic resonance imaging. *Proceedings of the National Academy of Sciences, USA, 88,* 2845–2849.

Gur, R. E., & Gur, R. C. (1990). Gender differences in regional cerebral blood flow. *Schizophrenia Bulletin, 16,* 247–254.

Habib, M., Gayraud, D., Oliva, A., Regis, J., Salmon, G., & Khalil, R. (1991). Effects of handedness and sex on the morphology of the corpus callosum: A study with brain magnetic resonance imaging. *Brain and Cognition, 16,* 41–61.

Halpern, D. F. (1992). *Sex differences in cognitive abilities* (2nd ed.). Hillsdale, NJ: Lawrence Erlbaum Associates.

Harshman, R. A., Hampson, E., & Berenbaum, S. A. (1983). Individual differences in cognitive abilities and brain organization, Part I: Sex and handedness differences in ability. *Canadian Journal of Psychology, 37,* 144–192.

Hellige, J. B. (1990). Hemispheric asymmetry. *Annual Review of Psychology, 41,* 55–80.

Hines, M., Chiu, L., McAdams, L. A., Bentler, P. M., & Lipcamon, J. (1992). Cognition and the corpus callosum: Verbal fluency, visuospatial ability, and language lateralization related to mid-sagittal surface areas of the callosal subregions. *Behavioral Neuroscience, 106,* 3–14.

Hines, M., & Green, R. (1991). Human hormonal and neural correlates of sex-typed behaviors. In A. Tasman & S. M. Goldfinger (Eds.), *Review of psychiatry, 10* (pp. 536–555). Washington, DC: American Psychiatric Press.

Ho, K.-C., Roessmann, U., Straumfjord, J. V., & Monroe, G. (1980a). Analysis of brain weight I. Adult brain weight in relation to sex, race, and age. *Archives of Pathology and Laboratory Medicine, 104,* 635–639.

Ho, K.-C., Roessmann, U., Straumfjord, J. V., & Monroe, G. (1980b). Analysis of brain weight II. Adult brain weight in relation to body height, weight, and surface area. *Archives of Pathology and Laboratory Medicine, 104,* 640–645.

Hofman, A., & Swaab, D. F. (1989). The sexually dimorphic nucleus of the preoptic area in the human brain: A comparative morphometric study. *Journal of Anatomy, 164,* 55–72.

Holloway, R. L., & DeLacoste, M. C. (1986). Sexual dimorphism in the human corpus callosum: An extension and replication study. *Human Neurobiology, 5,* 87–91.

Hyde, J. S., & Linn, M. C. (1988). Gender differences in verbal ability: A meta-analysis. *Psychological Bulletin, 104,* 53–69.

Inglis, J., & Lawson, J. S. (1982). A meta-analysis of sex differences in the effects of unilateral brain damage on intelligence test results. *Canadian Journal of Psychology, 36,* 670–683.

Innocenti, G. M. (1984). General organization of the callosal connection in the cerebral cortex. In E. G. Jones & A. Peters (Eds.), *Cerebral cortex* (pp. 291–303). New York: Plenum Press.

Jacobs, B., Schall, M., & Scheibel, A. B. (1993). A quantitative dendritic analysis of Wernicke's area in humans. II: Gender, hemispheric, and environmental factors. *Journal of Comparative Neurology, 327,* 97–111.

Jöreskog, K. G., & Sörbom, D. (1988). *LISREL-VII: A guide to the program and applications.* Chicago: SPSS, Inc.

Juraska, J. M., & Kopcik, J. R. (1988). Sex and environmental influences in the size and ultrastructure of the rat corpus callosum. *Brain Research, 450,* 1–8.

Kaufman, A. S. (1990). V-P IQ discrepancies in brain-damaged adults: Interactions with gender, race, and other patient variables. In *Assessing adolescent and adult intelligence* (pp. 301–343). Boston: Allyn and Bacon.

Kertesz, A., & Benke, T. (1989). Sex equality in intrahemispheric language organization. *Brain and Language, 37,* 401–408.

Kertesz, A., Polk, M. Black, S. E., & Howell, J. (1990). Sex, handedness, and the morphometry of cerebral asymmetries on magnetic resonance imaging. *Brain Research, 530,* 40–48.

Kertesz, A., Polk, M., Howell, J., & Black, S. E. (1987). Cerebral dominance, sex, and callosal size in MRI. *Neurology, 37,* 1385–1388.

Kimura, D. (1983). Sex differences in cerebral organization for speech and praxic functions. *Canadian Journal of Psychology, 37,* 19–35.

Kimura, D., & Carson, M. W. (1993). *Cognitive pattern and directional asymmetry of finger-ridge count* (Research Bulletin No. 716). London, Canada: University of Western Ontario, Department of Psychology.

Kingsburg, S. A., LaBarba, R. C., & Bowers, C. A. (1987). Sex differences in lateralization for spatial abilities. *Bulletin of the Psychonomics Society, 25,* 247–250.

Koff, E., Naeser, M. A., Pieniadz, J. M., Foundas, A. L., & Levine, H. L. (1986). Computed tomographic scan hemispheric asymmetries in right- and left-handed male and female subjects. *Archives of Neurology, 43,* 487–491.

Lansdell, H. (1962). A sex difference in the effects of temporal lobe neurosurgery on design preferences. *Nature, 194,* 852.

Lansdell, H., & Davie, J. C. (1972). Mass intermedia: Possible relation to intelligence. *Neuropsychologia, 10,* 207–210.

LeMay, M. (1977). Asymmetries of the skull and handedness. *Journal of the Neurological Sciences, 32,* 243–253.

Levay, S. (1991). A difference in hypothalamic structure between heterosexual and homosexual men. *Science, 253,* 1034–1037.

Levy, J., & Reid, M. (1978). Variations in cerebral organization as a function of handedness, hand posture in writing, and sex. *Journal of Experimental Psychology: General, 107,* 119–144.

Levy-Agresti, J., & Sperry, R. W. (1968). Differential perceptual capacities in major and minor hemispheres. *Proceedings of the National Academy of Sciences, USA, 61,* 1151.

Linn, M. C., & Peterson, A. C. (1985). Emergence and characterizations of sex differences in spatial ability: A meta-analysis. *Child Development, 56,* 1479–1498.

Maccoby, E. E., & Jacklin, C. N. (1974). *The psychology of sex differences.* Stanford, CA: Stanford University Press.

Masters, M. S., & Sanders, B. (1993). Is the gender difference in mental rotation disappearing? *Behavior Genetics, 23,* 337–341.

Mateer, C., Polen, S., & Ojemann, G. (1982). Sexual variation in cortical localization of naming as determined by stimulation mapping. *Behavioral and Brain Sciences, 5,* 310–311.

McGlone, J. (1977). Sex differences in the cerebral organization of verbal functions in patients with unilateral brain lesions. *Brain, 100,* 775–793.

McGlone, J. (1980). Sex differences in human brain asymmetry: A critical survey. *The Behavioral and Brain Sciences, 3,* 215–263.

McGlone, J., & Fox, A. J. (1982). Evidence from sodium amytal studies of greater asymmetry of verbal representation in men compared to women. In H. Akimoto, H. Kazamatsuri, M. Seino, & A. Ward (Eds.), *Advances in epileptology: XIIIth Epilepsy International Symposium* (pp. 389–391). New York: Raven Press.

Miura, S. A., Schapiro, M. B., Grady, C. L., Kumar, A., Salerno, J. A., Kozachuk, W. E., Wagner, E., Rapoport, S. I., & Horwitz, B. (1990). *Journal of Neuroscience Research, 27,* 500–504.

Morel, F. (1948). La massa intermedia ou commissure grise. *Acta Anatomica, 4,* 203–207.

Ojemann, G. A. (1983). Brain organization for language from the perspective of electrical stimulation mapping. *The Behavioral and Brain Sciences, 2,* 189–230.

Ojemann, G. O., Ojemann, J., Lettich, E., & Berger, M. (1989). Cortical language localization in left, dominant hemisphere: An electrical stimulation mapping investigation in 117 patients. *Journal of Neurosurgery, 71,* 316–326.

O'Kusky, J., Strauss, E., Kosaka, B., Wada, J., Li, D., Druhan, M., & Petrie, J. (1988). The corpus callosum is larger in right-hemisphere cerebral speech dominance. *Annals of Neurology, 24,* 379–383.

Oppenheim, J. S., Lee, B. C. P., Nass, R., & Gazzaniga, M. S. (1987). No sex-related differences in human corpus callosum based on magnetic resonance imagery. *Annals of Neurology, 21,* 604–606.

Packard, G. C., & Boardman, T. J. (1988). The misuse of ratios, indices, and percentages in ecophysiological research. *Physiological Zoology, 61,* 1–9.

Peters, M. (1991). Sex differences in human brain size and the general meaning of differences in brain size. *Canadian Journal of Psychology, 45,* 507–522.

Pizzamiglio, L., Mammucari, A., & Razzano, C. (1985). Evidence for sex differences in brain organization in recovery in aphasia. *Brain and Language, 25,* 213–223.

Potter, S. M., & Graves, R. E. (1988). Is interhemispheric transfer related to handedness and gender? *Neuropsychologia, 26,* 319–325.

Ray, W. J., Morell, M., Frediani, A. W., & Tucker, D. (1976). Sex differences and lateral specialization of hemispheric functioning. *Neuropsychologia, 14,* 391–394.

Ray, W. J., Newcombe, N., Semon, J., & Cole, P. (1981). Spatial abilities, sex differences and EEG functioning. *Neuropsychologia, 19,* 719–722.

Reinarz, S. J., Coffman, C. E., Smoker, W. R. K., & Godersky, J. C. (1988). MR imaging of the corpus callosum: Normal and pathologic findings and correlation with CT. *American Journal of Roentgenology, 151,* 791–798.

Reist, J. D. (1985). An empirical evaluation of several univariate methods that adjust for size variation in morphometric data. *Canadian Journal of Zoology, 63,* 1429–1439.

Schuessler, K. (1973). Ratio variables in path analysis. In A. S. Goldberger & O. D. Duncan (Eds.), *Structural equation models in the social sciences* (pp. 201–228). New York: Seminar Press.

Schuessler, K. (1974). Analysis of ratio variables: Opportunities and pitfalls. *American Journal of Sociology, 80,* 379–396.

Sherman, J. A. (1967). Problems of sex differences in space perception and aspects of intellectual development. *Psychological Review, 74,* 290–299.

Springer, S. P., & Deutsch, G. (1989). *Left brain, right brain.* New York: W. H. Freeman & Co.

Steinmetz, H., Jäncke, L., Kleinschmidt, A., Schlaug, G., Volkmann, J., & Huang, Y. (1992). Sex but no hand difference in the isthmus of the corpus callosum. *Neurology, 42,* 749–752.

Sundet, K. (1988). Sex differences in the severity and type of aphasia. *Scandinavian Journal of Psychology, 29,* 168–179.

Swaab, D. F., & Fliers, E. (1985). A sexually dimorphic nucleus in the human brain. *Science, 228,* 1112–1115.

Trotman, S. A., & Hammond, G. R. (1979). Sex differences in task-dependent EEG asymmetries. *Psychophysiology, 16,* 429–431.

Tucker, D. M. (1976). Sex differences in hemispheric specialization for synthetic visuospatial functions. *Neuropsychologia, 14,* 447–454.

Turkheimer, E. (1989). Techniques of quantitative measurement of morphological structures of the central nervous system. In E. D. Bigler, R. A., Yeo, & E. Turkheimer (Eds.), *Neuropsychological function and brain imaging* (pp. 47–64). New York: Plenum Press.

Turkheimer, E., & Farace, E. (1992). A reanalysis of gender differences in IQ scores following unilateral brain lesions. *Psychological Assessment, 4,* 498–501.

Turkheimer, E., Farace, E., Yeo, R. A., & Bigler, E. D. (1993). Quantitative analysis of gender differences in the effects of lateralized lesions on verbal and performance IQ. *Intelligence, 17,* 461–474.

Turkheimer, E., Yeo, R. A., Jones, C. L., & Bigler, E. D. (1990). Quantitative assessment of covariation between neuropsychological function and location of naturally occurring lesions in humans. *Journal of Clinical and Experimental Neuropsychology, 12,* 549–565.

Voyer, D., & Bryden, M. P. (1990). Gender, level of spatial ability, and lateralization of mental rotation. *Brain and Cognition, 13,* 18–29.

Wada, J. M., & Rasmussen, T. (1962). Intracarotid injection of sodium amytal for the lateralization of cerebral dominance: Experimental and clinical observations. *Journal of Neurosurgery, 17,* 266–282.

Warrington, E., James, M., & Maciejewski, C. (1986). The WAIS as a lateralizing and localizing instrument: A study of 656 patients with unilateral cerebral lesions. *Neuropsychologia, 24,* 223–239.

Weber, G., & Weis, S. (1986). Morphometric analysis of the human corpus callosum fails to reveal sex-related differences. *Journal für Hirnforschung, 27,* 237–240.

Weis, S., Weber, G., Wenger, E., & Kimbacher, M. (1989). The controversy about a sexual dimorphism of the human corpus callosum. *International Journal of Neuroscience, 47,* 169–173.

Whelan, T., & Walker, M. (1988). Effects of sex and lesion locus on measures of intelligence. *Journal of Consulting and Clinical Psychology, 56,* 633–635.

Willerman, L., Schultz, R., Rutledge, J. N., & Bigler, E. D. (1991). In vivo brain size and intelligence. *Intelligence, 15,* 223–228.

Willerman, L., Schultz, R., Rutledge, J. N., & Bigler, E. D. (1992). Hemisphere size asymmetry predicts relative verbal and nonverbal intelligence differently in the sexes: An MRI study of structure-function relations. *Intelligence, 16,* 315–328.

Witelson, S. F. (1985). The brain connection: The corpus callosum is larger in left-handers. *Science, 229,* 665–668.

Witelson, S. F. (1989). Hand and sex differences in the isthmus and genu of the human corpus callosum. *Brain, 112,* 799–835.

Witelson, S. F. (1991a). Neural sexual mosaicism: Sexual differentiation of the human temporo-parietal region for functional asymmetry. *Psychoneuroendocrinology, 16,* 131–153.

Witelson, S. F. (1991b). Sex differences in neuroanatomical changes with aging. *New England Journal of Medicine, 325,* 211–212.

Witelson, S. F., & Goldsmith, C. H. (1991). The relationship of hand preference to anatomy of the corpus callosum in men. *Brain Research, 545,* 175–182.

Witelson, S. F., & Kigar, D. L. (1992). Sylvian fissure morphology and asymmetry in men and women: Bilateral differences in relation to handedness in men. *Journal of Comparative Neurology, 323,* 326–340.

Wolff, P. H., & Cohen, C. (1980). Dual task performance during bimanual coordination. *Cortex, 16,* 119–133.

Wood, F. B., Flowers, D. L., & Naylor, C. E. (1991). Cerebral laterality in functional neuroimaging. In F. L. Kitterle (Ed.), *Cerebral laterality: Theory and research* (pp. 103–115). Hillsdale, NJ: Lawrence Erlbaum Associates.

Yeo, R. A. (1989). Individual differences. In E. D. Bigler, R. A. Yeo, & E. Turkheimer (Eds.), *Neuropsychological function and brain imaging* (pp. 295–316). New York: Plenum Press.

Yeo, R. A., Turkheimer, E., Raz, N., & Bigler, E. D. (1987). Volumetric asymmetries of the human brain: Intellectual correlates. *Brain and Cognition, 6,* 15–23.

Yoshii, F., Barker, W. W., Chang, J. Y., Loewenstein, D., Apicella, A., Smith, D., Boothe, T., Ginsberg, M. D., Pascal, S., & Duara, R. (1988). Sensitivity of cerebral glucose metabolism to age, gender, brain volume, brain atrophy and cerebral risk factors. *Journal of Cerebral Blood Flow and Metabolism, 8,* 654–661.

6

Neuroanatomy of the Aging Brain Observed *in Vivo*

A Review of Structural MRI Findings

NAFTALI RAZ

INTRODUCTION

There is little doubt that aging is accompanied by significant changes in brain structure. The question is whether the brain ages as a uniform whole or as a system of differentially vulnerable components. As neuropathologists investigating the aged human brain are "often struck by the focal nature of . . . cortical atrophy" (Katzman & Terry, 1983, p. 35), it seems that brain aging is, at least in its early stages, localized and selective. Cumulative evidence from autopsy studies suggests that in addition to diffuse changes in cerebral structure associated with old age, intensified regional cerebral atrophy is observed in humans and other mammals (see Coleman & Flood, 1987; Flood & Coleman, 1988; and Kemper, 1994, for comprehensive reviews). Postmortem findings indicate that certain brain regions such as the prefrontal, inferior parietal, and entorhinal cortices, as well as the hippocampal formation, are especially susceptible to aging (Arriagada & Hyman, 1990; Haug, 1985; Mani, Lohr, & Jeste, 1986; Terry, De Teresa, & Hansen, 1987). The results of neuroanatomical studies in nonhuman primates suggest that the visual and motor cortices are virtually unaffected by aging (Vincent, Peters, & Tigges, 1989; Tigges, Herndon, & Peters, 1990), although some human postmortem data reveal age-related atrophy in the macular

NAFTALI RAZ Department of Psychology, The University of Memphis, Memphis, Tennessee 38152.

Neuroimaging II: Clinical Applications, edited by Erin D. Bigler. Plenum Press, New York, 1996.

projection area of the striate cortex (Devaney & Johnson, 1980). In rodents, frontal lobes, amygdala, caudate-putamen complex, entorhinal cortex, and cerebellar hemispheres are also know as atrophy-prone areas (Shimada, Hosokawa, Ohta, Akiguchi, & Takeda, 1994). Selective vulnerability of brain tissue to aging is consistent with the notion of *pathoclisis* proposed more than 40 years ago by Vogt and Vogt (1951), who argued that some gray matter structures (grisea) exhibited increased susceptibility to pathogens.

As long as the findings of selective and localized cerebral aging were limited to autopsy material, their relevance to age-related cognitive and behavioral changes was very difficult to establish, hence the great interest generated by the introduction of noninvasive neuroimaging. With the introduction of computerized tomography (CT), age-related cerebral changes could be explored *in vivo* and related to behavioral measures. The strength of CT is the clear discrimination of cerebrospinal fluid (CSF) and brain parenchyma, whereas its weakness lies in relatively poor gray matter–white matter resolution. Not surprisingly, the most robust findings yielded by CT brain morphometry were confined to CSF-filled cavities. The CT-based brain morphometry helped establish age-related lateral and third ventriculomegaly and enlargement of the cerebral sulci (see Freedman, Knoefel, Naeser, & Levine, 1984, for a review). Thus, the strengths and limitations of a specific imaging technology may dictate both the questions and the answers regarding brain aging. It is apparent that CT was just not sensitive enough to test in healthy people the regional aging hypotheses that were generated by postmortem studies.

The advent of MRI, with its safety, speed, and neuroanatomical precision, ushered in a new era of *in vivo* neuroanatomy. For the first time, not only the whole brain and CSF-filled spaces, but also distinct gyri and nuclei, could be imaged, identified, demarcated, and measured in intact, behaving humans. In addition to clear visualization of neuroanatomy, the imaging technology based on nuclear magnetic resonance (NMR) yielded a plethora of findings on the biophysical, biochemical, and metabolic properties of brain tissue. Most of the recent exciting developments in applications of NMR to functional neuroscience are beyond the scope of this chapter and are reviewed elsewhere in this volume (see Chapter 12, this volume, and a comprehensive volume by Thatcher, Hallet, Zeffiro, John, & Huerta, 1994). In this chapter, I limit the account to neuroanatomical (structural) findings generated by application of MRI to the study of normal aging.

Methods of MRI-Based *in Vivo* Neuroanatomy

Magnetic resonance imaging yields a vast amount of data. To make their interpretation possible, these data should be processed and transformed. There are three major ways in which the information from *in vivo* MRI can be used. First, neuroanatomy and neuropathology can be assessed *qualitatively*, in the tradition of descriptive and clinical neuroradiology. Second, computer assisted morphometry yields areas and volumes of brain regions for *quantitative* evaluation of specific cerebral structures. Third, direct indices of macromolecular behavior of water in brain tissue may be obtained via computation of the relaxation time constants (T1 and T2) of circumscribed cerebral regions of interest (ROIs).

In contrast to the qualitative analysis of MRI scans, which is based on expert judgments regarding brain attributes observed on MR images, computer assisted morphometry can be used for quantification of specific brain structures. To date, several approaches to identification and measurement of cerebral ROIs from MR images have been developed, and some of them have been applied to the study of the aging brain. These techniques differ in the degree of automation, precision, and flexibility. Some employ software that follow a few neuroanatomic landmarks such as the major fissures, which automatically divide the brain into somewhat arbitrary ROIs (Jernigan, Press, & Hesselink, 1990; Jernigan *et al.*, 1991; Pfefferbaum *et al.*, 1992). Others rely on neuroanatomical definition and computerized tracing of the ROIs performed solely by trained human operators (N. Raz, Torres, Spencer, White, & Acker, 1992; N. Raz, Torres, & Acker, 1992; N. Raz, Torres, Spencer, & Acker, 1993; N. Raz *et al.*, 1993). There is a trade-off between speed and objectivity of highly automated techniques and the flexibility and neuroanatomical validity of the manual computer assisted approach. Substantial advantages of automated techniques such as automatic standardization of image coordinates become less attractive with the introduction of 3-D volume acquisition techniques. The fast volume acquisition sequences provide sufficient data to allow reslicing the image in any geometric plane, repositioning it in any standard system of coordinates, and, if necessary, correcting the subject's head tilt, pitch, and rotation. An example of such a 3-D technique is the spoiled gradient recall (SPGR) sequence implemented on the Signa scanners (General Electric Company, Milwaukee, WI). A set of typical MR images acquired with T1-weighted SPGR is presented in Figure 1. Initially, the images were acquired in the axial plane. They have been reformatted off-line and realigned according to the standard neurosurgical coordinates (Talairach & Tournoux, 1988). The midsagittal section was aligned along the anterior commissure–posterior commissure (AC–PC) line. The brain image was also aligned by the orbits and ear canals in the coronal plane and by the orbits in the axial plane. In the reformatted MR images shown in Figure 1, specific ROIs are outlined as in quantitative morphometry.

From the areas of the outlined ROIs and the interslice interval (1.5 mm in this set of examples), the volume of a given structure is computed. Higher interrater reliabilities attained with this technique are reported.

The third way of extracting information from MRI is examining the regional tissue properties indexed by the values of relaxation times. This method, unlike qualitative inspection and quantitative morphometry, involves no consideration of the shape of appearance of cerebral structures. It is based on the fact that in hydrogen NMR, the intensity of the signal received from a given brain locale, in addition to its dependence on the proton density, is determined by two processes: spin-lattice and spin-spin relaxation of water protons. The mathematical expression of this relationship is a biexponential function with two time constants, T1 and T2, which represent spin-lattice and spin-spin relaxation, respectively. Because the image contrast is defined by T1 and T2, it is possible to classify brain tissue with reference to the relaxation times. The magnitude of the time constants depends on a number of factors that reflect various facets of macromolecular behavior of brain water. Nevertheless, for a given temperature, frequency, and magnetic field strength, the primary determinant of the relaxation times is the interaction between three fractions of water: bulk, rotational bound, and irrotational bound (Fullerton, 1992; Mathur-DeVré, 1984; Unger,

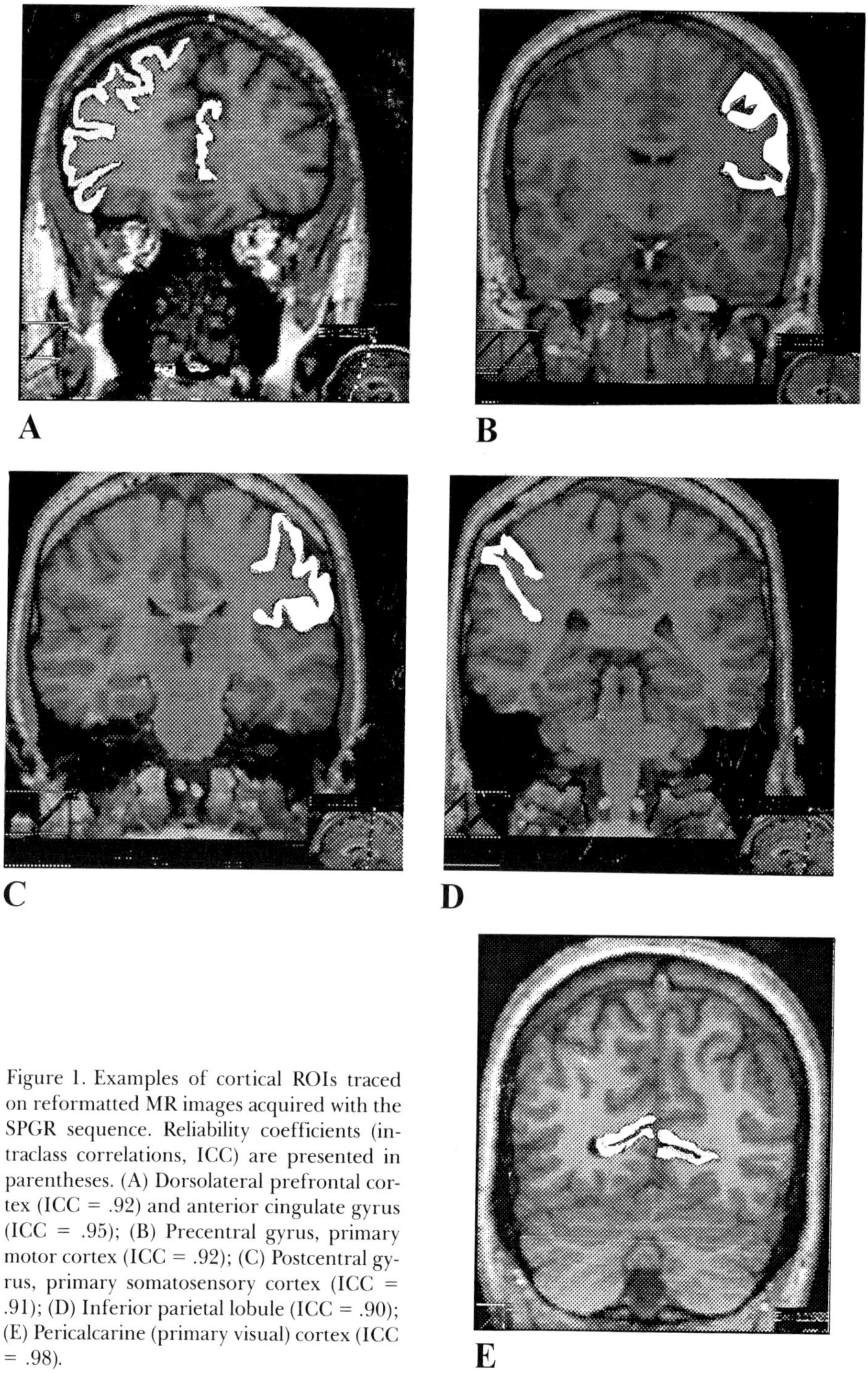

Figure 1. Examples of cortical ROIs traced on reformatted MR images acquired with the SPGR sequence. Reliability coefficients (intraclass correlations, ICC) are presented in parentheses. (A) Dorsolateral prefrontal cortex (ICC = .92) and anterior cingulate gyrus (ICC = .95); (B) Precentral gyrus, primary motor cortex (ICC = .92); (C) Postcentral gyrus, primary somatosensory cortex (ICC = .91); (D) Inferior parietal lobule (ICC = .90); (E) Pericalcarine (primary visual) cortex (ICC = .98).

Littlefield, & Gado, 1988). Because of this dependence, both T1 and T2 are sensitive to changes in brain cellular structure (Bottomley, Foster, Argersinger, & Pfeifer, 1984). Hence, one may expect that regional measures of relaxation times can yield valuable information about the aging brain. An example of a coronal MRI image with local values of T1 measured on-line is presented in Figure 2.

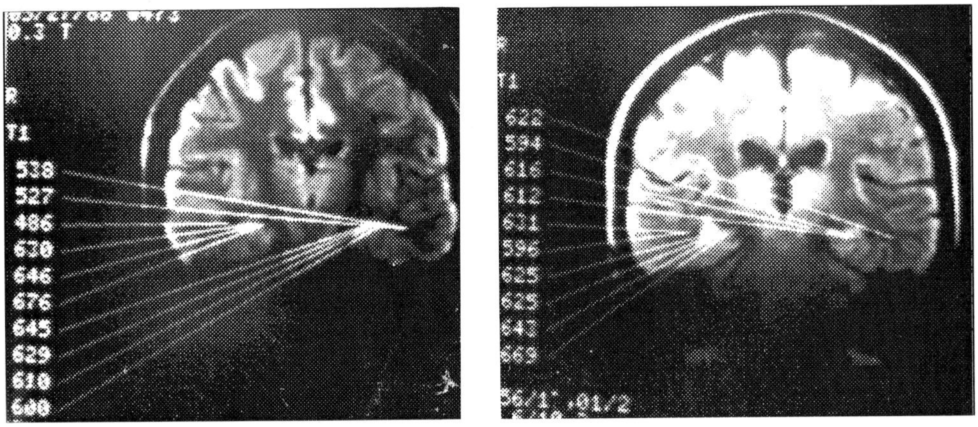

Figure 2. An example of on-line computation of spin-lattice relaxation time (T1) in temporal lobe ROIs of a 19-year-old female (left) and a 71-year-old female (right). The measures were performed on a 0.3 T scanner (Fonar Corporation, Melville, NY).

Review of the Findings

All three methods discussed in the preceding section have been applied to the study of brain aging *in vivo*. The findings are discussed below in the context of differential brain aging hypotheses.

White Matter

Myelin is crucial for speedy and error-free transmission of information between the central nervous system (CNS) and the periphery as well as within the brain. Demyelinating diseases such as multiple sclerosis are frequently accompanied by significant cognitive impairment (Rao, 1986), and myelination of CNS pathways is a prerequisite for normal cognitive development (Gibson, 1991). A feature of the aging brain most frequently noted upon visual inspection of the T2-weighted MR images is the *white matter hyperintensities* (WMH) or *leukoaraiosis* (Hachinsky, Potter, & Merskey, 1987). The WMH appear in a variety of shapes: bright irregular patches in subcortical white matter, periventricular rims and caps, and distributed punctate spots. An MR image illustrating leukoaraiosis is presented in Figure 3.

Since the introduction of MRI into the practice of clinical radiology, white matter lesions were noticed in asymptomatic elderly and linked to cerebrovascular risk factors (Bradley, Waluch, Brant-Zawadzki, Yadley, & Wykoff, 1984). It has been established that WMH lesions observed on the MR images represent actual neuropathological changes in the brain, and T2-weighted MRI is a highly sensitive tool for their detection (Brafman *et al.*, 1988a, 1988b). Nevertheless, the originally proposed link between this phenomenon and cerebrovascular disease is somewhat tenuous.

As the clinical and pathological data accumulate, it becomes more apparent that although cerebrovascular risk factors may play a role in producing WMH in the aging brain, age *per se* emerges as the best predictor of white matter lesions (Agartz, Marions, Sääf, Wahlund, & Wetterberg, 1992; Awad, Spetzler, Hodak, Awad, & Carey, 1986; Breteler *et al.*, 1994; Fazekas, 1989; Horikoshi, Yagi, & Fukamachi, 1993; Hunt *et al.*, 1989; Kertesz *et al.*, 1988; Kozachuk *et al.*, 1990;

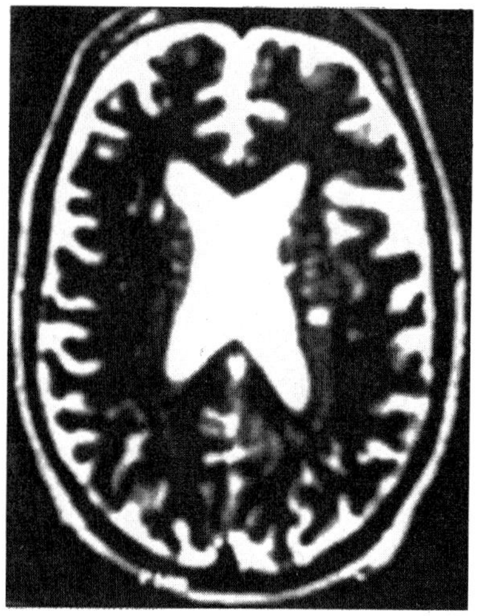

Figure 3. Leukoaraiosis in an asymptomatic 69-year-old female. Acquisition parameters: TR/TE = 3000/108 msec, NEX = 1, matrix = 256 × 256, FOV = 20 cm.

Meguro *et al.*, 1993; Sarpel, Chaudry, & Hindo, 1987; Shimada, Kawamoto, Matsubayashi, & Ozawa, 1990; Schmidt *et al.*, 1993). According to several studies, long-standing hypertension is linked to a small but significant increase in WMH incidence (Gerard & Weisberg, 1986; Schmidt *et al.*, 1993; van Swieten *et al.*, 1991; Yamashita *et al.*, 1991), yet other studies reveal no associated clinical significance (Agartz *et al.*, 1992; Fein *et al.*, 1990; Hunt *et al.*, 1989; Kobari, Meyer, Ichijo, & Oravez, 1990). Some investigators have suggested that ambulatory measures of blood pressure are better predictors of WMH prevalence in asymptomatic elderly than "casual" single-sample office assessment (Shimada *et al.* 1990) and that regional (e.g., frontal) WMH may be linked to an increase in arterial blood pressure (Yamashita *et al.*, 1991).The WMH are observed with equal frequency in normal elderly and in stroke patients matched for age, although the severity of the lesions is greater in the patients (Kozachuk *et al.*, 1990). The WMH are ubiquitous in elderly patients with idiopathic dizziness, but not more frequent than in matched controls (Day *et al.*, 1990). According to some accounts, WMH are equally prevalent in healthy elderly and in patients with probable Alzheimer's disease (AD) (Leys *et al.*, 1990), whereas other authors report high frequency of WMH in AD (Filley *et al.*, 1989; Lechner, Niederkorn, & Schmidt, 1991).

The role of diabetes mellitus, a condition associated with generalized brain atrophy in the elderly (Araki *et al.*, 1994; McCall, 1992; Soininen, Puranen, Helkala, Laakso, & Riekkinen, 1992), as an age-independent predictor of WMH is unclear. In earlier studies, diabetes emerged as the major risk factor for WMH, independent of other cerebrovascular risk considerations (Hendrie, Farlow, Austrom, Edwards, & Williams, 1989; Schmidt *et al.*, 1992). In more recent, better controlled, prospective investigations, however, the role of diabetes as a risk factor for WMH in the elderly has been questioned (Araki *et al.*, 1994; Breteler *et al.*, 1994; Schmidt *et al.*, 1993).

No specific regional distribution of WMH has been noted within the hemispheric white matter (Agartz *et al.*, 1992; Boone *et al.*, 1992; Hunt *et al.*, 1989). Yet neuropathological studies suggest that although the appearance of WMH from different cerebral locales is similar, the underlying histopathological basis is not. In the brains of the elderly who died of "non-neurological causes," distinct neuropathology was associated with each type of WMH lesion. Patches of signal hyperintensity on the MR image reflect myelin pallor and expanded perivascular spaces. Periventricular rims were observed in conjunction with loss of the ependymal lining. Caps correspond to the loci of myelin pallor, gliosis, and arteriosclerosis (Grafton *et al.*, 1991; Chimovitz, Estes, Furlan, & Awad, 1992). It has been hypothesized that the pathology underlying most of the WMH emerges in a chain of events which starts with age-related loss of neurons that results in axonal degeneration and subsequent invasion of the CSF into perivascular (Virchow-Robin) spaces (Ball, 1989).

In general, leukoaraiosis appears to be related to the water content of the white matter, which increases dramatically with advancing age (Samorajski & Rolsten, 1973; Wiggins, *et al.*, 1988). Several factors contribute to that increase: focal edema due to microinfarctions (Tomonaga, Yamanuchi, Tohgi, & Kameyama, 1982), perivascular demyelination (Ansari & Loch, 1975), hypoxic-ischemic damage (Ginsberg, Hedley-Whyte, & Richardson, 1975), and loss of glycolipids and proteins from the myelin sheath (Kirkpatrick & Hayman, 1987; Wiggins *et al.*, 1988). The dominant influence on the white matter T1 and T2 values is exercised by cholesterol, which represents about 30% of lipid content of the myelin (Koenig, 1991). Therefore, loss of cholesterol from myelinated fibers may play an important role in age-related prolongation of the white matter relaxation times.

Age-related increase in the spin-lattice and spin-spin relaxation constants of the cerebral white matter has been replicated in a number of studies. The results of these investigations can be compared using product-moment correlation between age and time constants. The comparison shows that across a variety of ROIs, the findings are quite consistent for both T1 and T2: $r = .26$ (Breger *et al.*, 1991), $r = .63$ (temporal lobe T1, N. Raz, Millman, & Sarpel, 1990), $r = .64$ (frontal lobe T1), and $r = .56$ (frontal lobe T2) (Wahlund *et al.*, 1990). Agartz and colleagues (Agartz, Sääf, Wahlund, & Wetterburg, 1991) also reported significant age-related prolongation of T1 but not T2 times in the cerebral white matter. The slope of the age-related trend of T1 was steeper for the frontal than for the occipital lobe. Interestingly, in two samples (N. Raz *et al.*, 1990; Agartz *et al.*, 1991), the relationship between age and the T1 relaxation times was nonlinear, with the slope being steeper for advanced age. A similar nonlinear trend for accumulation of WMH was observed by Jernigan *et al.* (1990). Significant quadratic trends observed in these samples suggest that for healthy elderly there may be a threshold of white matter deterioration with accumulation of white matter lesions in excess of this level leading to a clinical expression of pathology.

Cortical Gray Matter

Although qualitative judgments about the extent of cortical atrophy are useful for clinical decisions, they are seldom used in research. On visual inspection, cerebral gray matter of healthy individuals reveals no striking age-related changes such as WMH. In contrast to the white matter lesions, presence of hypointense spots on the MR images of the cerebral cortex is so rare that it can be treated as a pathognomonic sign of brain disease (Matsubayashoni, Shimada, Kawamoto, & Ozawa, 1992; Shimada *et al.*, 1990).

Direct assessment of the relaxation time constants of the cortical gray matter also has not been as revealing as the measurements in white matter. A set of processes in the gray matter promotes a trend toward shortening of the relaxation time constants. These processes, which eventually result in introduction of dense structures into the gray matter, include emergence of Hirano bodies, formation of neurofibrillary tangles and neuritic plaques, gliosis, and granulovacuolar degeneration (Terry *et al.*, 1987; Wisniewski & Terry, 1973). The relationship between aging and relaxation constants of the gray matter is not as clear as for the white matter, and there is virtually no systematic data on this topic. The interactions between age-related changes in gray and white matter are complex. Nevertheless, the net result of the described changes in the cellular structure and macromolecular composition is reduction in gray-white differences in the aging brain as observed on CT and MR images (George, de Leon, Ferris, & Kricheff, 1981; N. Raz *et al.*, 1990).

Quantitative MRI-based morphometry relies on anatomical or geometric definitions of brain structures and ROIs. The main question in reviewing the findings generated by this approach is whether they demonstrate regional *pathoclitic* differences in the aging of human brain and, particularly, the cerebral cortex. Although there are two few studies to merit a full-scale meta-analysis, some conclusions can be drawn on the basis of quantitative evaluation and comparison of the results. To integrate the results of the studies on MRI-derived gross anatomy of the aging brain within a common metric, the reported results were converted to correlation coefficients (r) and are summarized in Table 1.

In some studies, brain measures were adjusted for cranial size either by expressing them as a percentage of cranial volume or by statistically partialing out the cranial measures. The results from 15 samples indicate that there is age-related loss of brain parenchyma (median $r = -.39$ for raw brain volume and $r = -.57$ for adjusted). Because none of the reviewed studies was longitudinal, the age differences reported here may be influenced by a secular trend in brain and cranium size, although this influence is not particularly strong (Miller & Corsellis, 1977). Interestingly, about the same magnitude of reduction in brain size with age was observed some 90 years ago in a series of postmortem studies (Pearl, 1905). Age-relatd expansion of CSF-filled spaces observed in postmortem and CT studies has been clearly confirmed by MRI investigations. The effect of age on cortical sulci, lateral ventricles, and the third ventricle show little between-sample variation around the median of $r = .60$ for raw and adjusted measures alike.

Two studies in which global gray matter volume was measured are in agreement that there is a generalized age-related loss of cortical gray matter (Jernigan *et al.*, 1990; Pfefferbaum *et al.*, 1992). Unfortunately, regional measures of the cortical gray matter volume were reported only by Pfefferbaum *et al.* (1992) and N. Raz and co-workers (N. Raz, Torres, Spencer, & Acker, 1993; N. Raz *et al.*, 1994). Coffey *et al.*, 1992) measured the volume of anatomically defined regions that included both gray and white matter, thus averaging and attenuating possible regional differences. A similar approach of combining gray and white matter was taken by N. Raz, Torres, Spencer, and Acker (1993) in dealing with the occipital lobe. Jernigan *et al.* (1991) have combined a number of heterogenous cortical areas into automatically defined ROIs and consequently averaged cortical areas that might respond to aging differentially. Indeed, when two regions of the parietal lobe are measured separately, different correlations between age and the regional volume are observed: the magnitude of correlation between age and the size of the postcentral gyrus never exceed 0.19, whereas for

TABLE 1. QUANTITATIVE MRI STUDIES OF THE AGING BRAIN: WHOLE BRAIN, CORTICAL GRAY AND WHITE MATTER, AND CSF-FILLED CAVITIES

		Effect size (correlation with age)										
		Brain							CSF			
Reference	Adjusted/raw	Whole	Gray	White	Frontal	Temporal	Parietal	Occipital	Sulcal	Lateral	Third	Combined
Condon *et al.*, 1988	Adjusted	−.50[a]										
Yoshii *et al.*, 1988	Raw	.11										
Jernigan *et al.*, 1990	Adjusted	−.62	−.80						.58			.65
Gur *et al.*, 1991	Raw	−.32							.74	.63		
	Adjusted	−.56							.76			
Jernigan *et al*, 1991	Adjusted			.04								
Malko *et al*, 1991	Raw								.55			
Tanna *et al.*, 1991	Adjusted	−.57										.68
Coffey *et al.*, 1992	Adjusted	−.48			−.47[b]	−.37[b]				.71	.75	.73
Murphy *et al.*, 1992	Adjusted	−.57							.58	.39	.37	.38
Pfefferbaum *et al.*, 1992	Raw	−.18	−.51	.21	−.51[b]	−.57[b]	−.47[b]	−.38[b]	.51	.50	.53	.51
Shah *et al.*, 1992	Raw	−.44										
Raz, Torres, Spencer, &	Adjusted	−.22			−.29		−.32[c]	−.02[b]		.63		
Acker, (Study 1) 1993	Raw	−.40			−.43		−.31[c]	−.28[b]		.60		
Raz, Torres, Spencer, &	Adjusted				−.61	−.14[d]	−.02[e]	−.15[b]				
Acker (Study 2) 1994	Raw				−.58	−.14[d]	−.14[e]	−.14[b]				
DeCarli *et al.*, 1994	Adjusted	−.71			−.58	−.10						
	Raw	−.52			−.43	−.22						
N. Raz *et al.*, 1994	Raw	−.39			−.61		−.03	.07		.60	.69	.64
Median effect size	Adjusted	−.57	−.80	.04	−.53	−.14	−.17	−.08	.58	.63	.56	.66
	Raw	−.39	−.51	.21	−.51	−.22	−.17	−.21	.51	.60	.61	.58

[a]Estimated from p values.
[b]Combined gray and white matter.
[c]Combined IPL and postcentral gyrus.
[d]Combined superior temporal and parahippocampal gyri.
[e]Postcentral gyrus only.

the inferior parietal lobule, all correlations were above 0.34 (N. Raz *et al.*, 1993, 1994). Combining these areas into one ROI obscures the evidence of their differential aging. When the temporal lobe areas were measured separately (N. Raz, Torres, Spencer, & Acker, 1993), the correlation between age and the volume of the parahippocampal gyrus was essentially zero ($r = -.04$), whereas the superior temporal gyrus showed a weak trend for age-related shrinkage ($r = -.23$). It must be noted that the superior temporal gyrus itself is a very heterogenous structure that includes primary sensory (auditory) cortex as well as high-order polymodal cortices. Primary motor cortex (precentral gyrus) has been measured in one sample and exhibited no significant age-related shrinkage ($r = -.14$; N. Raz *et al.*, 1994).

Across five samples (Coffey *et al.*, 1992; Pfefferbaum *et al.*, 1992; Raz, Torres, Spencer, & Acker, 1993; DeCarli *et al.*, 1994; N. Raz *et al.*, 1994), there was a discernible trend for age-related pathoclisis of the cerebral cortex. It appears that association (especially prefrontal) cortices and the lobes containing them tend to shrink with age, whereas the primary sensory areas or the ROIs dominated by them show little differences across the age span. The robustness of this trend, based on a handful of studies, needs to be determined. However, the observed pattern of cerebral aging is in agreement with the postmortem findings. In addition, a recent study of brain glucose metabolism provided convergent evidence of selective age-related vulnerability of frontal and parietal association cortices (Azari *et al.*, 1992).

In contrast to a clearly established age-dependent alteration of the macromolecular properties of cerebral white matter, there is no global change in the volume of the white matter in the aging brain. Nevertheless, when regional measures of white matter volume are reported (N. Raz, Torres, Spencer, & Acker, 1993; N. Raz *et al.*, 1994), there is a trend for age-related shrinkage. Logically, with attrition of the gray matter, axonal atrophy should follow. This trend, however, may be detectable only in the regions of the greatest neuronal loss, which happen to be the areas such as dorsolateral prefrontal cortex (N. Raz, Torres, Spencer, & Acker, 1993; N. Raz *et al.*, 1993, 1994). Aging shows little effect on the corpus callosum, a heavily myelinated collection of fibers that connect cortical areas in two hemispheres. In 19 studies, the median correlation between the cross-sectional midsagittal area of the corpus callosum and age is rather small: $r = -.19$ (Driesen & Raz, 1995).

Limbic-Diencephalic Structures

To gerontologists, the limbic system is an area of special interest because of the role its components play in declarative memory, one of the cognitive functions most prominently affected by age (Craik & Jennings, 1992; Verhaeghen, Marcoens, & Goosense, 1993). This cerebral system consists of the hippocampal formation, anterior thalamus, and mammillary bodies, as well as the entorhinal and cingulate cortices, the amygdala, the medial dorsal nucleus of the thalamus, and the fornix (Zola-Morgan & Squire, 1991; Squire, 1992). Sample MR images in Figure 4 illustrate some of the limbic and diencephalic structures; the results of the MRI-based morphometry of limbic and diencephalic structures, the brain stem, and the cerebellum are presented in Table 2.

Given the importance of these brain areas for mnemonic functions (Squire, Knowlton, & Musen, 1993), it is somewhat surprising that the age-related

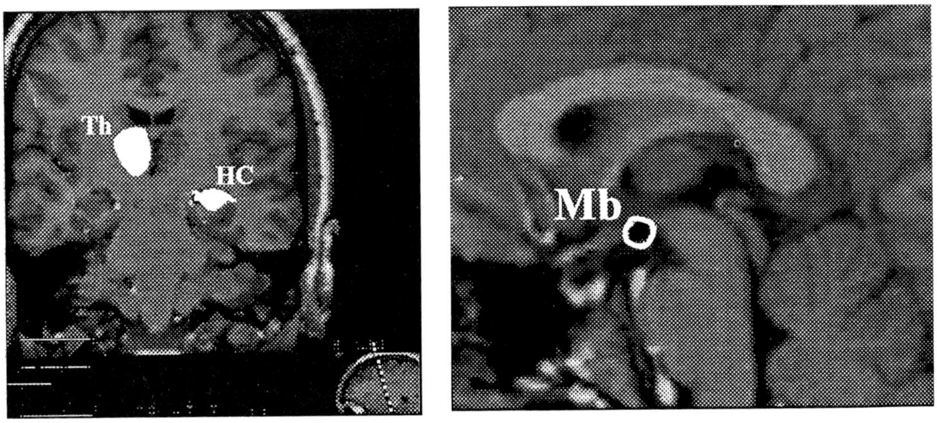

A B

Figure 4. MR images of limbic and diencephalic structures. (A) Hippocampus (ICC = .98) and thalamus; (B) Mammillary body (ICC = .92).

shrinkage in limbic-diencephalic ROIs is minimal. In the hippocampal formation, only a weak trend toward age-related shortening of the gray matter T1 was observed: r = −.25 (N. Raz *et al.*, 1990). In addition, there was no correlation between age and the volume of two other limbic regions, the parahippocampal gyrus (a part of the entorhinal cortex) and the anterior cingulate gyrus. Before we can reach any conclusions on the relative stability of the limbic-diencephalic areas with age, several factors need to be taken into account.

First, the precision of volume estimates of some structures such as the

TABLE 2. QUANTITATIVE MRI STUDIES ON THE AGING BRAIN: LIMBIC AND DIENCEPHALIC STRUCTURES

		Effect size (correlation with age)			
Reference	Raw/adjusted	Hippocampus	Thalamus	Mammillary body	Anterior cingulate
Charness & DeLaPaz, 1987	Raw			−.08	
Jernigan *et al.*, 1991	Adjusted	−.38	−.12[b]		−.13[c]
Lim *et al.*, 1991	Adjusted	−.22			
Coffey *et al.*, 1992	Adjusted	−.27			
Jack *et al.*, 1992	Raw	−.58[a]			
Raz, Torres, & Acker,	Raw			−.51	
1992	Raw	−.03			−.18
Raz, Torres, Spencer, & Acker, 1993	Raw	−.23			−.15
Raz, unpublished	Raw		−.37		
N. Raz *et al.*, 1994	Raw	−.28			−.30
Raz & Acker, 1994	Raw			−.39	
Median effect size	Adjusted	−.27	−.12	—	−.13
	Raw	−.26	−.37	−.39	−.18

[a]From plotted data.
[b]Anterior and posterior averaged.
[c]Includes anterior insula.
[d]From tabulated data.

hippocampus is affected by use of only a small number of slices. The two to three slices used by Raz, Torres, Spencer, and Acker (1993) in the computation of the hippocampal volume may not be sufficient for obtaining a stable estimate; eight systematically sampled slices per structure is the recommended minimum (Mayhew & Olsen, 1991). Indeed, in other samples, some of the more stable estimates of hippocampal volume based on greater number of slices reveal moderate age-related atrophy (Jack, Petersen, O'Brien, & Tangalos, 1992), although others did not (N. Raz *et al.,* 1994). It is possible that despite high reliability of the hippocampal measures, the validity of the volume estimates was limited. Second, except for the Jack *et al.* (1992) and N. Raz *et al.* (1994) studies, the cross-section of the hippocampus was measured in a somewhat tilted coronal plane; this could affect the validity of the measures. These studies, however, do not particularly agree on the magnitude of estimated age effects on the hippocampus.

A possible explanation for the lack of signs of age-related atrophy on MRI of these limbic structures is that brain changes may be restricted to specific subareas within the measured structures. For example, in normal aging, as in Alzheimer's disease, only specific cortical laminae of the parahippocampal gyrus may be affected (Damasio, Van Hoesen, & Hyman, 1990). Such specific laminar atrophy would be difficult to detect with MRI morphometry. Contrary to postmortem findings (Arriagada & Hyman, 1990; Kemper, 1994), no age-related changes were observed in the volume of the parahippocampal gyrus.

In contrast to the equivocal findings in the other limbic-diencephalic areas, a relatively consistent age-related shrinkage of the mammillary bodies was observed in healthy volunteers as well as in neurological patients with negative MRI scans. The median correlation with age in three studies was $r = -.48$. It is unclear why the Charness and DeLaPaz (1987) study yielded no age effect, the inclusion of patients with space-occupying lesions (though not "impinging on the mammillary bodies") may be one of the reasons.

Basal Ganglia

Motor activity changes in the course of normal aging (Ingram, 1988; Mortimer, 1988). Because of the role played by the basal ganglia in planning, execution, and control of movement (Alheid, Switzer, & Heimer, 1990), it is possible that age-related changes in striatal morphology are linked to the alterations in motor function. Age-dependent loss of neurons in the putamen and the caudate nucleus, as well as reduction of the volume of the caudate and the lentiform nuclei (all outlined in Figure 5), have been documented in postmortem investigations (see N. Raz, Torres, & Acker, 1995).

A meta-analysis of neuroimaging and postmortem studies (N. Raz *et al.,* 1995) augmented by additional measures conducted in our laboratory reveals a modest but reliable influence of age on the caudate nucleus volume: the median correlation between age and the size of the caudate nucleus is $r = -.48$. Age-related declines in the volume of the putamen *per se* and in the lentiform nucleus (putamen with globus pallidus) are smaller than age-related shrinkage of the caudate nucleus: the median correlation with age is $r = -.35$. It is possible that age-related deterioration of the basal ganglia is more prominent in males than in females, and the Age × Sex interaction ought to be examined in every study on cerebral aging. In the N. Raz *et al.* (1995) sample, the correlation with age was $r = -.60$ in males and only $r = .02$ in females. The data on the aging of the globus

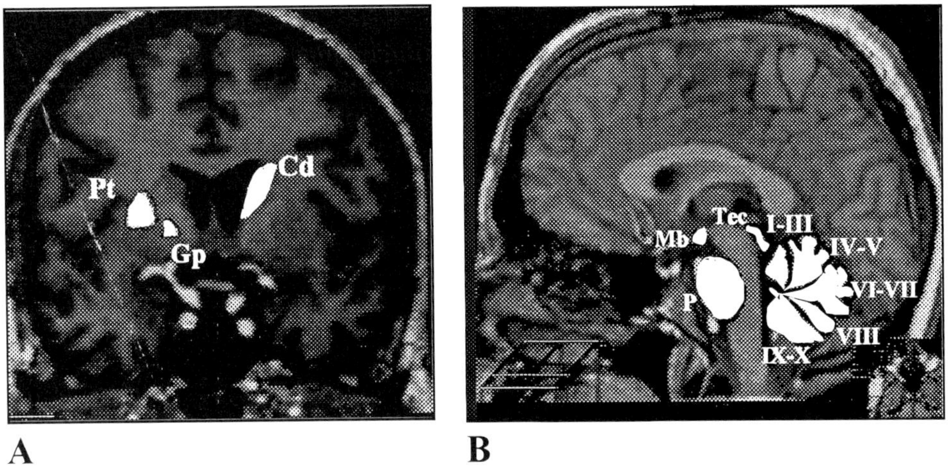

Figure 5. Examples of striatal and mesencephalic ROIs. (A) Caudate nucleus (ICC = .97); putamen (ICC = .92); globus pallidus (ICC = .89); (B) Ventral pons (ICC = .96); tectum (ICC = .85); cerebellar vermis: lobules I–III (ICC = .96), lobules IV–V (ICC = .98), lobules VI—VII (ICC = .98), lobule VIII (ICC = .94) and lobules IX–X (ICC = .89).

pallidus are rather scarce. The results of MRI-based morphometry (N. Raz, unpublished) suggest no age-related reduction in the volume of the globus pallidus.

Even though the paleostriatum may not shrink with age, the influence of aging on this structure may be expressed in other ways. Investigation of macromolecular properties of the aging basal ganglia showed that in the globus pallidus, T2 values are inversely proportional to the tissue iron content, which changes dramatically with age. As a result of age-related iron accumulation, hypodense spots are frequently observed on the T2-weighted MR images of asymptomatic middle-aged and elderly individuals (Pujol *et al.*, 1992; Schenker, Meier, Wichman, Boesiger, & Valvanis, 1993). The pallidal iron deposits which are absent at birth reach their asymptotic concentration in the third decade of life, hence there is very little change in T2 past that age (Schenker *et al.*, 1993). Both iron concentration in the globus pallidus and local T2 are exponentially related to age. Age-related iron accumulation in the globus pallidus creates areas of hypointensity that may distort volume estimates. Indeed, the extent of hypointense area of this basal ganglion is associated with age ($r = .39$, computed from published raw data; Milton, Atlas, Lexa, Mozley, & Gur, 1991). The reduction in signal intensity within the globus pallidus was also significantly correlated with age in a sample of healthy volunteers ($r = -.69$, Pujol *et al.*, 1992).

Cerebellum

Age-related degeneration of the cerebellum has been consistently reported in postmortem studies. The autopsy evidence suggests that aging is associated with a general reduction in cerebellar weight (Ellis, 1920) and attrition of Purkinje cells in the cerebellar hemispheres and the vermis (Ellis, 1920; Hall, Miller, & Corsellis, 1975; Torvik, Torp, & Lindboe, 1986). A midsagittal MR image showing the cerebellar vermis with highlighted lobules I-III, IV-V, VI-VII, VIII, and IX-X is presented in Figure 5b. The findings of quantitative MRI studies of cerebellum, pons, and midbrain are summarized in Table 3.

To date, four quantitative MRI studies of the regional neuroanatomy of the cerebellar vermis in healthy and neurologically asymptomatic elderly have been conducted. In addition, two reports provide data on age-related differences in the whole vermis. Comparison is hampered by the fact that in each study, the vermis is divided somewhat differently. Schaefer and colleagues (1991) have measured lobules I-V, VI-VII, and VIII-X in neuroradiologically normal individuals. They found that the midsagittal area of the anterior vermis (lobules I-V) normalized by the cranial size remained stable across the age span, whereas the posterior vermis shrunk with age. N. Raz *et al.* (N. Raz, Torres, Spencer, White, & Acker, 1992) have observed a ventral-dorsal gradient of age-related shrinkage of the cerebellar vermis in a combined sample of healthy volunteers and neuroradiologically normal patients. The age-associated area differences were the smallest in the most ventral (extreme anterior and posterior) lobules I-III and IX-X. The atrophy was more apparent in the intermediate segments (lobules IV, V, and VIII) and reached its highest magnitude in the most dorsal segment of the vermis—declive-folium-tuber or lobules VI-VII. A similar pattern of age-related correlations has been observed in the vermis of healthy volunteers by Raz and Acker (1994), although the magnitude of the correlations was considerably attenuated (see Table 3). Shah and colleagues (1991) found no age-related differences in the posterior vermis (lobules VI-IX) of normal controls and observed a negative effect of aging on the anterior vermis (lobules I-V). Weis and colleagues (Weis, Kimbacher, Wenger, & Neuhold, 1993) noted a modest age-related shrinkage of the whole vermis.

Age-related differences in the total cerebellar volume were examined in two studies (Escalona *et al.*, 1991; N. Raz & Acker, 1994) that revealed negative correlations between age and the total volume of the cerebellum, including hemispheric gray and white matter as well as the vermis. The only *in vivo* study of the dentate nucleus revealed that the size of signal hypointensity areas on the MR images that correspond to iron deposits was unrelated to age ($r = -.09$, computed from the raw data published by Milton *et al.*, 1991).

Tectum, Midbrain, and Pons

Autopsy findings indicate that in contrast to age-related shrinkage of the cerebellum, the neighboring brain stem structures (pons and medulla) maintain their size throughout the life span (Blinkov & Glezer, 1968). The results of MRI morphometry suggest that there are no age-related changes in the cross-sectional area of the tectum (see Figure 5b) (N. Raz, Torres, & Acker, 1992; N. Raz & Acker, 1994). However, in two studies (Doraiswami *et al.*, 1992; Weis, Kimbacher, Wenger, & Neuhold, 1993) age-related shrinkage of the midbrain was found. The interpretation of these findings is hindered by the fact that the measured ROI included substantia nigra and a number of other nuclei from which monoaminergic fibers originate. Considerable age-related atrophy exhibited by these nuclei (McGeer *et al.*, 1990) could affect the volume estimates of the midbrain as a whole. The inference based on MRI planimetry is complicated by the possibility that there may be no association between changes in the volume of a nucleus and alterations of its cytoarchitecture and neurochemical properties. For instance, the diameter of the red nucleus, a midbrain structure involved in

TABLE 3. QUANTITATIVE MRI STUDIES ON THE AGING BRAIN: MESENCEPHALON, PONS, AND CEREBELLUM

		Effect size (correlation with age)							
					Cerebellar vermis				Cerebellar
Reference	Raw/adjusted	Pons	Tectum	Midbrain	I–V	VI–VII	VIII–X	Total	hemispheres
Hayakawa *et al.*, 1989	Raw	.15						−.12[a]	
Escalona *et al.*, 1991	Adjusted								−.27
Schaefer *et al.*, 1991	Adjusted				−.23[b]	−.40[b]	−.60[b]		
Shah *et al.*, 1991	Adjusted				−.33	---------- −.07 ----------			
Doraiswami *et al.*, 1992	Adjusted			−.39					
Raz, Torres, Spencer, Winter, & Acker, 1992	Raw	.14			−.29	−.50	−.33	−.54	
Raz, Torres, & Acker, 1992	Raw		−.08						
Weis *et al.*, 1993	Raw	−.06		−.45				−.28	
Raz & Acker, 1994	Raw	.07	−.08		.04	−.31	−.13	−.20	−.54
Median effect size	Adjusted	—	—	−.39	−.28	−.40	−.60	—	−.27
	Raw	.10	−.08	−.45	−.12	−.41	−.23	−.24	−.54

[a]From plotted data.
[b]From tabulated data.

motor circuits, did not correlate with age (r = .01), whereas the area of hypointensity formed by iron deposits in this nucleus showed a weak age-related decline (r = −.27; computed from published data, Milton *et al.*, 1991).

The ventral pons (highlighted in Figure 5B) is volumetrically stable across the age span. The median correlation with age in four samples is small but positive (r = .14). Weis *et al.* (1993) also reported a lack of age differences in the midsagittal area of the medulla (r = .09). In addition, Shah *et al.* (1992) found no association between age and the ventral pons area or between age and the area of medulla. Age-related stability of the pons has been recently replicated in yet another sample of healthy elderly (Carper, Kaye, & Janowsky, 1993). No correlations, however, were reported in either study.

From Images to Brains: Interpretations and Speculations

The emerging picture of differential brain aging observed *in vivo* can be graphically summarized in a map as presented in Figure 6. In this map, the areas are marked in accordance with their vulnerability to aging as expressed in correlations between the size of a given ROI and age.

These findings contradict the notion of global age-related degeneration fostered by widely replicated nonspecific findings such as age-related ventriculomegaly (Freedman *et al.*, 1984). Brain regions apparently age at an uneven pace, although the causes and mechanisms of such selective aging are unknown. The observed shrinkage of the cortical ribbon most likely reflects loss of neuronal bodies, and not of the glial cells. While the postmortem evidence of age-related neuronal loss, neuronal shrinkage, and dendritic atrophy is abundant (Coleman & Flood, 1987; Flood & Coleman, 1988), there are no age-related declines in glial population (Devaney & Johnson, 1980; Peters, Josephson, & Vincent, 1991). A number of neurochemical, structural, developmental, and phylogenetic factors have been proposed to explain the topography of age-related changes in the cerebral cortex, but to date no unified model has emerged.

The loss of cortical and diencephalic neurons may be mediated by age-related neurochemical changes which may emerge with atrophy of the subcortical sources of catecholamines (McGeer *et al.*, 1990). Indeed, the prefrontal cortex, one of the major recipients of subcortical dopaminergic and noradrenergic projections (Fallon & Loughlin, 1987), shows significant age-related declines, whereas the primary visual cortex—a region with sparse catecholaminergic innervation (Levitt, Rakic, & Goldman-Rakic, 1984)—is spared by aging. There is also a significant age-related drop in the population of some dopaminergic (D_1) and noradrenergic (α_1 and β_1, but not β_2) receptors in the prefrontal cortex (De Keyser, De Backer, Vauquelin, & Ebinger, 1990; Kalaria *et al.*, 1989; Kalaria & Andorn, 1991).

The caudate and putamen receive dense dopaminergic afferents from the substantia nigra, pars compacta (Alheid *et al.*, 1990). Age-related attrition of the source neurons where these afferents originate (McGeer, 1978) may result in degeneration of their striatal targets. In humans, direct projections from the substantia nigra reach the prefrontal cortex (Sarnat & Netsky, 1981), a brain region which is also known for its increased susceptibility to aging (Kemper, 1994). To date, however, no reliable MRI-based measures of the substantia nigra

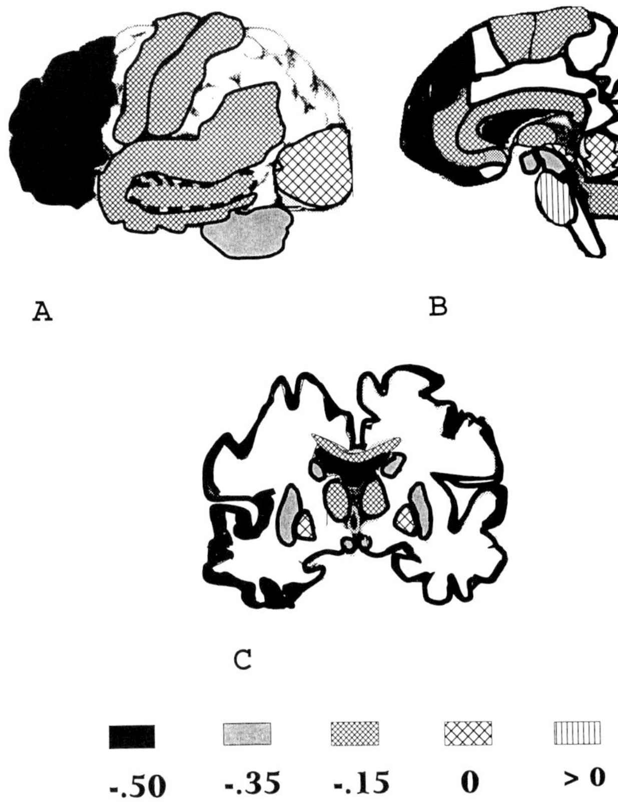

Figure 6. A series of maps summarizing the topography of age-related differences in the size of brain structures. Shading represents the magnitude of the median negative correlation with age as indicated in the bottom of the figure. (A) Lateral view, (B) midsagittal view, and (C) coronal view through the medial thalamus and the mammillary bodies.

have been reported, and the question of age-related shrinkage of the substantia nigra in healthy elderly remains open. The volume of nigral hypointense regions generated presumably by accumulation of iron yielded no age-related trends ($r = -.10$, computed from published data, Milton *et al.*, 1991). It is unclear whether the dopaminergic function of the neostriatum declines with age at all. Some postmortem evidence affirmative (McGeer, McGeer, & Suzuki, 1977), whereas two *in vivo* positron emission tomography investigations (Martin, Palmer, Patlak, & Calne, 1989; Sawle *et al.*, 1990) yielded utterly discordant results.

A model of brain aging based on the central cholinergic dysfunction (Bartus, Dean, Pontecorvo, & Flickeret, 1985) predicts selective age-related deterioration of structures that receive dense cholinergic projections from the nucleus basalis of Meynert—the hippocampus, parahippocampal and cingulate gyri, and insula (Mesulam, Rosen, & Mufson, 1984). According to the MRI evidence, most of these structures manifest no significant age-related shrinkage. Thus, the MRI findings do not support the predictions of the cholinergic model of brain aging. One major caveat to this argument is, of course, the lack of MRI measures of the nucleus basalis. When this structure can be measured *in vivo,* a stronger test of the cholinergic hypothesis of brain aging will be possible.

According to the excitotoxic model, brain aging is a result of cumulative damage inflicted upon areas that are rich in excitatory amino acid (EAA) receptors. Neurons in such areas, for example, pyramidal neurons in the hippocampus, are highly susceptible to the cumulative effects of hypoxia and ischemia which may occur even in asymptomatic elderly (Cotman & Monaghan, 1987). However, because there are virtually no data on age-related and regional differences in cortical EAA activity (Morgan & May, 1990), the results of the studies reviewed here cannot be related to the excitotoxic model of cortical aging.

The discussed neurochemical models of brain aging are not mutually exclusive, and several neurotransmitter systems can interact in creating localized areas of pathoclisis. Several patterns of cortical damage may be observed in the general population of aging brains. Yet, in each individual, only one of these patterns may predominate. With all these caveats in mind, the suggested configuration of cortical aging is more consistent with the catecholaminergic deafferentation model than with other neurochemical models.

Differential declines of specific cerebral areas may be precipitated by atrophy in the structures to which the areas are connected. A good example is the mammillary body, which lies at the crossroads of the corticopontocerebellar pathway. It receives input from the hippocampus, ventral tegmentum, and medial prefrontal cortex and sends projections to the neocerebellar vermis (lobules VI and VII) (Nieuwenhuys, Voogd, & Van Huijsen, 1988; Allen & Hopkins, 1989, 1990). All these structures show measurable age-related declines (Kemper, 1994; N. Raz, Torres, & Acker, 1992). The mammillary body receives virtually no monoaminergic innervation and is a target of hippocampal glutaminergic efferents running from the subiculum through the fornix (Nieuwenhuys, 1985). This connection makes possible degeneration of the mammillary body due to excitotoxic activity via the fornix. Such transneuronal degeneration has been reported in some cases of ischemic damage to the hippocampus (Lindboe, Erichsen, & Strøm, 1989). Thus, excitotoxic rather than catecholaminergic change may underlie the age-related trophy of the mammillary bodies.

The pattern of connectivity of the tectum may explain to some extent the contradictory findings regarding its response to aging. The tectum consists of two pairs of nuclei, superior and inferior colliculi, which serve as gateways for the passage of sensory information to the higher cortical processing centers (Stein, 1984). The colliculi belong to separate neural systems. The inferior colliculus receives projections from the lower tiers of the auditory brain stem and sends projections to the medial geniculate body in the thalamus. The connections of the superior colliculus are more diverse. They include the ventromedial and the pulvinar nuclei of the thalamus; primary visual, motor, and auditory cortices; and the frontal eye fields (Nieuwenhuys *et al.*, 1988; Stein, 1984). These areas are not particularly vulnerable to the effects of aging. Nevertheless, the superior colliculus receives direct noradrenergic efferents from the locus ceruleus and serotonergic efferents from the nucleus raphe dorsalis (Nieuwenhuys, 1985). Both nuclei show significant age-related cell loss which could result in transaxonal degeneration in the superior colliculus (McGeer *et al.*, 1990).

The age-sensitive and age-resistant areas of the vermis are also dissimilar in their connectivity. The lingula and the nodulus are the most specialized structures committed to processing vestibular information, and they have extensive peripheral innervation. The declive, the folium, and the tuber receive projections from the tectum as well as corticopontocerebellar fibers with virtually no connections to lower segments of the CNS (Baloh, 1984; Nieuwenhuys *et al.*, 1988).

The selectivity of brain aging may have evolutionary underpinnings. Age-sensitive association areas of the neocortex, such as dorsolateral prefrontal and inferior parietal areas, are, phylogenetically, the latest additions to the brain, whereas the primary sensory areas have changed little in the course of mammalian evolution (Armstrong, 1990; Sarnat & Netsky, 1981). In evolutionary terms, the sensory cortices are "the most conservative in the human brain" (Armstrong, 1990, p. 7). They lag behind the prefrontal and posterior association areas in the degree of corticalization; that is, they are smaller than would be expected for a primate cortex of a given size. The neocerebellar vermis (lobules VI-VII), which is especially vulnerable to the effects of aging, is also one of the most recent evolutionary developments in the cerebellum. In contrast, the lingula and the nodulus, probably the oldest parts of the vermis (Larsell & Jansen, 1972), are somewhat less sensitive to aging. In the basal ganglia, the neostriatal components are vulnerable to aging, whereas the phylogenetically older paleostriatum is not. These contrasting local responses of the brain to the process of aging may follow a pattern: the last to come (in evolution) is the first to go (in senium).

In the evolutionary context, a number of characteristics set primary and tertiary cortical areas apart. For example, in primates, between-species differences in longevity are associated with increase in the relative size of the cerebellum and the neocortex except for primary visual areas (Allman, McLaughlin, & Hakeem, 1993). Brain areas exhibiting the highest incidence of Alzheimer-type pathology also tend to be phylogenetically more recent than the areas spared by the disease (Arnold, Hyman, Flory, Damasio, & Van Hoesen, 1991; Rapoport, 1990). According to Rapoport (1990) the reason for the differential vulnerability of selected brain systems is that these systems or "ensembles of functionally and anatomically connected brain regions" underwent "integrated phylogeny" during recent evolution. During this process, changes were introduced in the primate genome, especially in chromosome 21. Rapoport speculates that these changes precipitated the emergence of Alzheimer's disease (AD) and the neuropathologically similar Down syndrome (DS). Although recent *in vivo* findings suggest that adult patients with DS exhibit some regional brain abnormalities similar to those observed in considerably older normal individuals, a number of neuroanatomic features such as grossly reduced pontine size sets this group of patients apart from normal elderly and AD patients (N. Raz *et al.*, 1995).

The pattern of selective brain aging and brain topography of Alzheimer's pathology resembles the distribution of growth-associated protein (GAP-43) in the brain. The highest concentrations of GAP-43 have been observed in the association cortices, whereas only a minimal amount was detected in the primary sensory cortices (Neve, Finch, Bird, & Benowitz, 1988). The findings of Neve and colleagues may be interpreted as evidence for increased plasticity of the listed cortical regions. Such malleability could serve the needs of information storage and processing, which increased in the course of primate evolution. The down side of the enhanced plasticity may be increased sensitivity and thus increased vulnerability to putative age-associated pathogens.

Brain regions that are especially vulnerable to aging are also developmentally less precocious than the areas of relative sparing. For instance, while the primary visual cortex reaches its mature size at a very early age, prefrontal association areas grow at a much slower pace (Huttenlocher, 1990). The thickness of the cortical ribbon in the primary somatosensory cortex reaches its maximum by the end of the first year of life, whereas in the dorsolateral prefrontal cortex, the growth continues into late adolescence (Semenova, Vasilieva, Tsekhmistrenko, & Shumeiko, 1989). Relatively slow intrauterine development of the cerebellum is followed by a spurt in the first two years of life, whereas the midbrain structures reach their adult size *in utero* or promptly after birth (Koop, Rilling, Herrmann, & Kretschmann, 1986). The neocerebellar (lobules VI-VII), but not the anterior (lobules I-V) vermis tends to grow even beyond this spurt period into early adulthood (Jernigan & Tallal, 1990). If the brain regions that are slow to develop maintain neural plasticity throughout adulthood, they may be more vulnerable to putative pathogens than more stable regions. There is some support for the notion that the neural plasticity of the association cortices is prolonged relative to the sensory areas (Neve *et al.*, 1988). It is possible that increased vulnerability of the discussed cortical areas is the price for their increased malleability, but the mechanisms that could mediate such a trade-off are unknown.

A note on sex differences in brain aging is in order. Although gender-specific patterns of age-related changes in cerebral structures were reported in only a few studies, the direction of the difference is consistent. Almost invariably, males fare worse than females. They show steeper age-related declines in total gray matter volume (Gur *et al.*, 1991) and greater age-dependent increase in relative cerebral CSF content (Condon, Grant, Hadley, & Lawrence, 1988). Recently, steeper age-related declines in temporal lobe volume were reported for males (Cowell, Turetsky, Gur, & Gur, 1993). In some samples, age-related shrinkage of the basal ganglia was restricted to males (Raz *et al.*, 1995). These gender-related differences, albeit small and not always statistically significant, are consistent. In general, males are more vulnerable than females to a wide spectrum of CNS insults (Gualtieri & Hicks, 1985). Such selective vulnerability is observed at the early perinatal period and is sometimes explained in terms of differential maturation of the sexes (S. Raz *et al.*, 1994; S. Raz *et al.*, 1995). On the other hand, a reverse sex difference was reported in some recent studies. For example, Breteler and colleagues (1994) observed significantly higher prevalence of WMH in their female subjects. Raz and Acker (1994) found a significantly steeper linear decline in the area of posterior vermis in their female subjects than in their male counterparts. These findings pose a challenge, but regardless of the mechanism, gender differences established so early may dictate divergent trajectories of aging later in life.

Functional Implications of Regional Brain Aging

Age-related impairment and age-related disease are diagnosed by their overt expressions—symptoms and patterns of performance. Therefore, it is important to establish whether a specific alteration in brain structure or function, no matter how spectacular, is associated with specific behavioral change. To date, there are only a handful of studies documenting and charting age-related brain changes *in vivo*. Hence, it is not surprising that there are even fewer attempts to relate changes in brain structure to age-related differences in cognitive performance.

Several groups of investigators have examined neuropsychological performance in the elderly in relation to the observed burden of age-related changes in the cerebral white matter. Although in some studies (Filley *et al.*, 1989; Hunt *et al.*, 1989; Rao, Mittenberg, Bernardin, Haughton, & Leo, 1989; Yamashita *et al.*, 1991) no significant association was observed between cognitive variables and WMH indices, in other samples leukoaraiosis predicted cognitive decline (Breteler *et al.*, 1994; Junqué *et al.*, 1990). Boone and colleagues (1992) suggested that the association between cognition and the state of cerebral white matter is a threshold phenomenon. Their findings indicate that asymptomatic elderly experience a drop in cognitive performance if the white matter lesions accumulate and reach a certain threshold level. The implication is that at least when white matter is concerned, a logistic function describes the link between cognitive and cerebral aging better than the more frequently used linear models. According to a recent report, slowing of information processing, regarded as a hallmark of cognitive aging (Meyerson, Hale, Wagstaff, Poon, & Smith, 1990), was especially severe in subjects with substantial leukoaraiosis (Schmidt *et al.*, 1993; Ylikoski *et al.*, 1993). Thus, it is possible that more specific speeded tasks are more sensitive to age-related brain phenomena, such as alteration of the white matter, than global tests of higher cognitive functions.

Other age-related brain changes were shown to be associated with cognitive aging. Performance on age-sensitive cognitive tests correlated with prolongation of the white matter T1 value and reduction in the gray-white ratio of T1 (N. Raz *et al.*, 1990) as well as with the loss of cerebral parenchyma (N. Raz *et al.*, 1993). In this sample, however, most of the variance contributed to cognitive performance by the measures of brain integrity was not unique but, rather, shared with age. Wahlund *et al.* (1990) found no significant correlations between relaxation times and cognitive performance in a sample of healthy elderly. Correlating more specific indices of performance with T2 of the globus pallidus revealed several significant associations. Correlations between T2 and finger-tapping speed was $r = .45$. Shortened T2 in the globus pallidus was also associated with difficulties in response inhibition (as shown by Stroop task performance) and in incidental learning. Because the authors did not provide age-corrected data, it is unclear what proportion of the unique variance is attributable to changes in the globus pallidus *per se* and how much is shared with age. Other cognitive functions, such as procedural learning, that are known to be impaired by lesions in the neostriatum (Heindel, Salmon, Shults, Walicke, & Butters, 1989) have not yet been examined in conjunction with age-related changes in the basal ganglia.

Regarding the aging of the cerebellar vermis, it is unclear whether regional vermian atrophy has functional implications. Future studies should concentrate on elucidating the relationship between cognitive and behavioral functions and differential age-dependent tissue loss in vermian regions. To date, a rather broad array of behaviors has been associated with the cerebellum. The cerebellum and its nuclei may play an important role in classical conditioning (Thompson, 1986), selective attention (Courchesne, 1987), habituation (Leaton & Supple, 1986), and procedural learning (Glickstein & Yeo, 1990; Seitz, Roland, Bohm, Greite, & Stone-Elander, 1990). It has been intimated that even more complex problem-solving activities customarily allocated to the cerebral cortex may rely on cerebellar activity (Leiner, Leiner, & Dow, 1986, 1989; but see Daum *et al.*, 1993).

Exploration of differential brain aging has some aspects that may serve more than just gerontological purposes. It can provide a useful model for study of cerebral localization of functions. A classic way of testing hypotheses about localization of abilities to specific cerebral structures is a lesion experiment. A cognitive neuroscientist concerned with human performance is prevented from applying direct structural manipulation to the subjects. The luxury of controlled lesions, which is taken for granted by psychobiologists who work with animal models, is unknown in human neuropsychology. Only sporadic and uncontrolled events such as cerebral insults, neoplastic space-occupying lesions, and neurosurgical interventions result in structural alterations that can be correlated with cognitive and behavioral measures. There are several disadvantages to such an approach. Population selection is restricted by age (strokes, perinatal brain hemorrhages), highly unusual premorbid brain history (epileptic surgery patients), and uncontrolled influences of mechanical forces on the distant brain regions (space-occupying lesions).

Many hypotheses based on the findings from human lesion studies have never been tested in the normal population. For such a test, one needs a human model in which a natural manipulation on the brain involves neither grossly damaging premorbid conditions nor the contaminating sequelae of brain disease and the side effects of its treatment. A substantial degree of variability in brain structure and cognitive function is necessary for a meaningful application of quasi-experimental methodology used in studies on individual differences. Normal uncomplicated aging seems to be well suited for this role. It is a unique structural manipulation on the human brain that helps to avoid some of the pitfalls inherent in studying the cases of focal lesions and degenerative diseases. If the brain indeed ages selectively, and if in healthy elderly who are free of behavioral and cognitive pathology, a pattern of differential brain shrinkage can be demonstrated *in vivo*, studying brain-behavior relationships throughout the lifespan can provide a better understanding of individual differences in the neural underpinnings of cognition.

Acknowledgments

This work was supported in part by the National Institute of Aging grant AG-11230 and by the Center of Excellence grant from the State of Tennessee to the Department of Psychology, The University of Memphis.

References

Agartz, I., Marions, O., Sääf, J., Wahlund, L.-O., & Wetterberg, L. (1992). Visual ratings of magnetic resonance images of human cerebrospinal fluid spaces and white brain matter: Relation to sex and age in healthy volunteers. *Magnetic Resonance Imaging, 10,* 135–142.

Agartz, I., Sääf, J., Wahlund, L.-O., & Wetterberg, L. (1991). T1 and T2 relaxation time estimates in the normal human brain. *Radiology, 181,* 537–543.

Alheid, G. E., Switzer III, R. C., & Heimer, L. (1990). Basal ganglia. In G. Paxinos (Ed.), *The human nervous system* (pp. 483–582). San Diego: Academic Press.

Allen, G. V., & Hopkins, D. A. (1989). Mamillary body in the rat: Topography and synaptology of projections from the subicular complex, prefrontal cortex, and midbrain tegmentum. *Journal of Comparative Neurology, 286,* 214–231.

Allen, G. V., & Hopkins, D. A. (1990). Topography and synaptology of mammillary body projections to the mesencephalon and pons in the rat. *Journal of Comparative Neurology, 301,* 311–336.

Allman, J. M., McLaughlin, T., & Hakeem, A. (1993). Brain structures and life span in primate species. *Proceedings of the National Academy of Science, USA, 90,* 118–122.

Ansari, K. A., & Loch, J. (1975). Decreased myelin basic protein content of the aged human brain. *Neurology, 25,* 1045–1050.
Araki, Y., Nomura, M., Tanaka, H., Yamamoto, H., Tsukaguchi, I., & Nakamura, H. (1994). MRI of the brain in diabetes mellitus. *Neuroradiology, 36,* 101–103.
Armstrong, E. (1990). Evolution of the brain. In G. Paxinos (Ed.), *The human nervous system* (pp. 1–16). San Diego:Academic Press.
Arnold, S., Hyman, B., Flory, J., Damasio, A., & Van Hoesen, G. (1991). The topographical and neuroanatomical distribution of neurofibrillary tangles and neuritic plaques in the cerebral cortex of patients with Alzheimer's disease. *Cerebral Cortex, 1,* 103–116.
Arriagada, P. V., & Hyman, B. T. (1990). Entorhinal cortex is an early site of neurofibrillary tangle pathology in non-demented elderly. *Society for Neuroscience Abstracts, 16,* 1265.
Awad, A., Spetzler, R. F., Hodak, J. A., Awad, C. A., & Carey, R. (1986). Incidental subcortical lesions identified on magnetic resonance imaging in the elderly, I: Correlation with age and cerebrovascular risk factors. *Stroke, 17,* 1084–1089.
Azari, N. P., Rapoport, S. I., Salerno, J. A., Grady, C. L., Gonzales-Aviles, A., Schapiro, M. B., & Horwitz, B. (1992). Interregional correlations of resting cerebral glucose metabolism in old and young women. *Brain Research, 552,* 279–290.
Ball, M. J. (1989). "Leukoaraiosis" explained. *Lancet, 2,* 612–613.
Baloh, R. W. (1984). Neurotology of aging: Vestibular system. In M. L. Albert (Ed.), *Clinical neurology of aging* (pp. 345–361). New York: Oxford University Press.
Bartus, R. T., Dean, R. L., Pontecorvo, M. J., & Flicker, C. (1985). The cholinergic hypothesis: A historical overview, current perspective, and future directions. *Annals of the New York Academy of Sciences, 444,* 332–358.
Blinkov, S. M., & Glezer, I. I. (1968). *The human brain in figures and tables.* New York: Basic Books.
Boone, K. B., Miller, B. L., Lesser, I. M., Mehringer, M., Hill-Gutierrez, E., Goldberg, M. A., & Berman, N. G. (1992). Neuropsychological correlates of white-matter lesions in healthy elderly subjects: A threshold effect. *Archives of Neurology, 49,* 549–554.
Bottomley, P. A., Foster, T. H., Argersinger, R. E., & Pfeifer, L. M. (1984). Review of normal tissue hydrogen NMR relaxation times and relaxation mechanisms from 1-100 MHz: Dependence on tissue type, NMR frequency, temperature, species, excision, and age. *Medical Physics, 11,* 425–448.
Bradley, W. G., Waluch, V., Brant-Zawadzki, M., Yadley, R. A., & Wykoff, R. R. (1984). Patchy, periventricular white matter lesions in the elderly: A common observation during NMR imaging. *Noninvasive Medical Imaging, 1,* 35–41.
Brafman, B. H., Zimmerman, R. A., Trojanowski, J. Q., Gonatas, N. K., Hickey, W. F., Schlaefer, W. W. (1988a). Brain MR: Pathologic correlations with gross and histopathology. 1. Lacunar infarction and Virchow-Robin spaces. *American Journal of Roentgenology, 151,* 551–558.
Brafman, B. H., Zimmerman, R. A., Trojanowski, J. Q., Gonatas, N. K., Hickey, W. F., Schlaefer, W. W. (1988b). Brain MR: Pathologic correlations with gross and histopathology. 2. Hyperintense white-matter foci in the elderly. *American Journal of Roentgenology, 151,* 559–566.
Breger, R. K., Yetkin, F. Z., Fischer, M. E., Papke, R. A., Haughton, V. M., & Rimm, A. A. (1991). T1 and T2 in the cerebrum: Correlation with age, gender, and demographic factors. *Radiology, 181,* 545–547.
Breteler, M. M. B., van Swieten, J. C., Bots, M. L., Grobbee, D. E., Claus, J. J., van den Hout, J. H. W., van Harskamp, F., Tanghe, H. L. J., de Jong, P. T. V. M., van Gijn, J., & Hofman, A. (1994). Cerebral white matter lesions, vascular risk factors, and cognitive functions in a population-based study: The Rotterdam study. *Neurology, 44,* 1246–1252.
Brody, H. (1973). Aging in the vertebrate brain. In M. Rockstein (Ed.), *Development and aging in the nervous system* (pp. 121–134). New York: Academic Press.
Carper, R. A., Kaye, J. A., & Janowsky, J. S. (1993). Quantitative MRI analysis of normal and abnormal brain aging. *Society for Neuroscience Abstracts, 19,* 179.
Charness, M. E., & DeLaPaz, R. L. (1987). Mammillary body atrophy in Wernicke's encephalopathy: Antemortem identification using magnetic resonance imaging. *Annals of Neurology, 22,* 595–600.
Chimovitz, M., Estes, M., Furlan, A. J., & Awad, I. A. (1992). Further observations on the pathology of subcortical lesions identified on magnetic resonance imaging. *Archives of Neurology, 49,* 747–752.
Coffey, C. E., Wilkinson, W. E., Parashos, I. A., Soady, S. A. R., Sullivan, R. J., Patterson, L. J., Figiel, G. S., Webb, M. C., Spritzer, C. E., & Djang, W. T. (1992). Quantitative cerebral anatomy of the aging human brain: A cross-sectional study using magnetic resonance imaging. *Neurology, 42,* 527–536.

Coleman, P. D., & Flood, D. G. (1987). Neuron numbers and dendritic extent in normal aging and Alzheimer's disease. *Neurobiology of Aging, 8,* 521–545.
Condon, B., Grant, R., Hadley, D., & Lawrence, A. (1988). Brain and intracranial cavity volumes: In vivo determination by MRI. *Acta Neurologica Scandinavica, 78,* 387–393.
Cotman, C. W., & Monaghan, D. T. (1987). Chemistry and anatomy of excitatory amino acid systems. In H. Y. Meltzer (Ed.), *Psychopharmacology: The third generation of progress* (pp. 197–210). New York: Raven Press.
Courchesne, E. (1987). A neurophysiological view of autism. In E. Schopler & G. Mesibov (Eds.), *Neurobiological issues in autism* (pp. 285–384). New York: Plenum Press.
Cowell, P., Turetsky, B. I., Gur, R. C., & Gur, R. E. (1993). Sex differences in aging of the human frontal and temporal lobes. *Society for Neuroscience Abstracts, 19,* 179.
Craik, F. I. M., & Jennings, J. (1992). Human memory. In F. I. M. Craik & T. A. Salthouse (Eds.), *Handbook on aging and cognition* (pp. 51–110). Hillsdale, NJ: Erlbaum.
Damasio, A. R., Van Hoesen, G. W., & Hyman, B. T. (1990). Reflections on the selectivity of neuropathological changes in Alzheimer's disease. In M. F. Schwartz (Ed.), *Modular deficits in Alzheimer-type dementia* (pp. 83–100). Cambridge, MA: MIT Press.
Daum, I., Ackermann, H., Schugens, M. M., Reimond, C., Dichgans, J., & Birnbaumer, N. (1993). The cerebellum and cognitive functions in humans. *Behavioral Neuroscience, 107,* 411–419.
Day, J. J., Freer, C. E., Dixon, A. K., Coni, N., Hall, L. D., Sims, C., & Gehlhaar, E. W. (1990). Magnetic resonance imaging of the brain and brain-stem in elderly patients with dizziness. *Age and Ageing, 19,* 144–150.
DeCarli, C., Murphy, D. G. M., Gillette, J. A., Haxby, J. V., Teichberg, D., Schapiro, M. B., & Horwitz, B. (1994). Lack of age-related differences in temporal lobe volume of very healthy adults. *American Journal of Neuroradiology, 15,* 689–696.
De Keyser, J., De Backer, J.-P., Vauquelin, G., & Ebinger, G. (1990). The effect of aging on the D_1 dopamine receptors in human frontal cortex. *Brain Research, 528,* 308–310.
Devaney, K. O., & Johnson, H. A. (1980). Neuronal loss in the aging visual cortex of man. *Journal of Gerontology, 35,* 836–841.
Doraiswami, P. M., Na, C., Husain, M., Figiel, G. S., McDonald, W. M., Ellinwood, E. H., Boyko, O. B., & Krishnan, K. R. (1992). Morphometric changes in the human midbrain with normal aging: MR and stereologic findings. *American Journal of Neuroradiology, 13,* 383–386.
Driesen, N. R., & Raz, N. (1995). The influence of sex, age, and handedness on corpus callosum morphology: A meta-analysis. *Psychobiology, 23,* 240–247.
Ellis, R. S. (1920). Norms for some structural changes in human cerebellum from birth to old age. *Journal of Comparative Neurology, 32,* 1–33.
Escalona, P. R., McDonald, W. M., Doraiswamy, P. M., Boyko, O. B., Husain, M. M., Figiel, G. S., Laskowitz, D., Ellinwood, E. H., & Krishnan, K. R. R. (1991). In vivo stereological assessment of human cerebellar volume: Effects of gender and age. *American Journal of Neuroradiology, 12,* 927–929.
Fallon, J. H., & Loughlin, S. E. (1987). Monoamine innervation of cerebral cortex and a theory of the role of monoamines in cerebral cortex and basal ganglia. In E. Jones & A. Peters (Eds.), *Cerebral cortex: Vol. 6. Further aspects of cortical functions, including hippocampus* (pp. 41–128). New York: Plenum Press.
Fazekas, F. (1989). Magnetic resonance signal abnormalities in asymptomatic individuals: Their incidence and functional correlates. *European Neurology, 29,* 164–168.
Fein, G., Van Dyke, C., Davenport, L., Turetsky, B., Brant-Zawadzki, M., Zatz, L., Dillon, W., & Valk, P. (1990). Preservation of normal cognitive functioning in elderly subjects with extensive white-matter lesions of long duration. *Archives of Neurology, 47,* 220–223.
Filley, C. M., Davis, K. A., Schmitz, S. P., Stears, J. C., Heaton, R. K., Kelly, J., Culig, K. M., & Scherrzinger, A. L. (1989). Neuropsychological performance and magnetic resonance imaging in Alzheimer's disease and normal aging. *Neuropsychiatry, Neuropsychology, and Behavioral Neurology, 2,* 81–91.
Flood, D. G., & Coleman, P. D. (1988). Neuron numbers and size in aging brain: Comparison of human, monkey, and rodent data. *Neurobiology of Aging, 9,* 453–463.
Freedman, M., Knoefel, J., Naeser, M., & Levine, H. (1984). Computerized axial tomography in aging. In M. L. Albert (Ed.), *Clinicial neurology of aging* (pp. 139–148). New York: Oxford University Press.
Fullerton, G. (1992). Physiologic basis of magnetic relaxation. In D. D. Stark & W. G. Bradley (Eds.), *Magnetic resonance imaging:* Vol. 1 (pp. 88–167). St. Louis, MO: Mosby Year Book.
George, A., de Leon, M., Ferris, S., & Kricheff, I. I. (1981). Parenchymal CT correlates of senile

dementia (Alzheimer disease): Loss of gray-white matter discriminability. *American Journal of Neuroradiology, 2,* 205–213.

Gerard, G., & Weisberg, L. A. (1986). MRI periventricular lesions in adults. *Neurology, 36,* 998–1001.

Gibson, K. (1991). Myelination and behavioral development: A comparative perspective on questions of neoteny, altriciality, and intelligence. In K. R. Gibson & A. C. Petersen (Eds.), *Brain maturation and cognitive development: Comparative and cross-cultural perspectives.* New York: Aldine De Gruyter.

Ginsberg, M. D., Hedley-Whyte, T., & Richardson, E. P. (1975). Hypoxic-ischemic leukoencephalopathy in man. *Archives of Neurology, 33,* 5–14.

Glickstein, M., & Yeo, C. (1990). The cerebellum and motor learning. *Journal of Cognitive Neuroscience, 2,* 69–80.

Grafton, S. T., Sumi, S. M., Stimac, G. K., Alvord, E. C., Shaw, C.-M., & Nochlin, D. (1991). Comparison of postmortem magnetic resonance imaging and neuropathologic findings in the cerebral white matter. *Archives of Neurology, 48,* 293–298.

Gualtieri, T., & Hicks, R. (1985). An immunoreactive theory of selective male affliction. *The Behavioral and Brain Sciences, 8,* 427–441.

Gur, R. C., Mozley, P. D., Resnick, S. M., Gotlieb, G. L., Kohn, M., Zimmerman, R., Herman, G., Atlas, S., Grossman, R., Beretta, D., Erwin, R., & Gur, R. E. (1991). Gender differences in age effect on brain atrophy measured by magnetic resonance imaging. *Proceedings of the National Academy of Sciences of the United States of America, 88,* 2845–2849.

Hachinsky, V. C., Potter, P., & Merskey, H. (1987). Leuko-araiosis. *Archives of Neurology, 44,* 21–23.

Hall, T. C., Miller, K. H., & Corsellis, J. A. N. (1975). Variations in human Purkinje cell population according to age and sex. *Neuropathology and Applied Neurobiology, 1,* 267–292.

Haug, H. (1985). Are neurons of the human cerebral cortex really lost during aging? A morphometric examination. In J. Tarber & W. H. Gispen (Eds.), *Senile dementia of Alzheimer type* (pp. 150–163). Berlin: Springer-Verlag.

Hayakawa, K., Konishi, Y., Matsuda, T., Kuriyama, M., Konishi, K., Yamashita, K., Okumura, R., & Hamanaka, D. (1989). Development and aging of the brain midline structures: Assessment with MRI imaging. *Radiology, 172,* 171–177.

Heindel, W. C., Salmon, D. P., Shults, C. W., Walicke, P. A., & Butters, N. (1989). Neuropsychological evidence for multiple implicit memory systems: A comparison of Alzheimer's, Huntington's, and Parkinson's disease patients. *Journal of Neuroscience, 9,* 582–587.

Hendrie, H. C., Farlow, M. R., Austrom, M. G., Edwards, M. K., & Williams, M. A. (1989). Foci of increased T2 signal intensity on brain MR scans of healthy elderly subjects. *American Journal of Neuroradiology, 10,* 703–707.

Horikoshi, T., Yagi, S., & Fukamachi, A. (1993). Incidental high-intensity foci in white matter on T2-weighted magnetic resonance imaging. *Neuroradiology, 35,* 151–155.

Hunt, A. L., Orrison, W. W., Yeo, R. A., Haaland, K. Y., Rhyne, R. L., Garry, P. J., & Rosenberg, G. A. (1989). Clinical significance of MRI white matter lesions in the elderly. *Neurology, 39,* 1470–1474.

Huttenlocher, P. R. (1990). Morphometric study of human cerebral cortex development. *Neuropsychologia, 28,* 517–527.

Ingram, D. K. (1988). Motor performance variability during aging in rodents: Assessment of reliability and validity of individual differences. *Annals of the New York Academy of Sciences, 515,* 70–96.

Jack, C. R., Petersen, R. C., O'Brien, P. C., & Tangalos, E. G. (1992). MR-based hippocampal volumetry in the diagnosis of Alzheimer's disease. *Neurology, 42,* 183–188.

Jernigan, T. L., Archibald, S. L., Berhow, M. T., Sowell, E. R., Foster, D. S., & Hesselink, J. R. (1991). Cerebral structure on MRI, part I: Localization of age-related changes. *Biological Psychiatry, 29,* 55–67.

Jernigan, T. L., Press, G. A., & Hesselink, J. R. (1990). Methods for measuring brain morphologic features on magnetic resonance images: Validation and normal aging. *Archives of Neurology, 45,* 404–408.

Jernigan, T. L., & Tallal, P. (1990). Late childhood changes in brain morphology observed with MRI. *Developmental Medicine and Child Neurology, 32,* 379–385.

Junqué, C., Pujol, J., Vendrell, P., Bruna, O., Jódar, M., Ribas, J. C. Viñas, J., Capdevila, A., & Martí-Vilalta, J. L. (1990). Leuko-araiosis on magnetic resonance imaging and speed of mental processing. *Archives of Neurology, 47,* 150–156.

Kalaria, R. N., & Andorn, A. C. (1991). Adrenergic receptors in aging and Alzheimer's disease: Decreased α_2-receptors demonstrated by [^{3}H]p-aminoclonidine binding in prefrontal cortex. *Neurobiology of Aging, 12,* 131–136.

Kalaria, R. N., Andorn, A. C., Tabaton, M., Whitehouse, P. J., Harik, S. I., & Unnerstall, J. R. (1989). Adrenergic receptors in aging and Alzheimer's disease: Increased β_2-receptors in prefrontal cortex and hippocampus. *Journal of Neurochemistry, 53,* 1772–1781.
Katzman, R., & Terry, R. (1983). Normal aging of the nervous system. In R. Katzman & R. Terry (Eds.), *The neurology of aging* (pp. 15–50). Philadelphia: F. A. Davis.
Kemper, T. L. (1994). Neuroanatomical and neuropathological changes during aging and in dementia. In: M. L. Albert & E. J. E. Knoepfel (Eds.), *Clinical neurology of aging* (2nd ed., pp. 3–67). New York: Oxford University Press.
Kertsez, A., Black, S. E., Tokar, G., Benke, T., Carr, T., & Nicholson, R. (1988). Periventricular and subcortical hyperintensities on magnetic resonance imaging: Rims, caps and unidentified bright objects. *Archives of Neurology, 45,* 404–408.
Kirkpatrick, J. B., & Hayman, L. A. (1987). White-matter lesions in MR imaging of clinically healthy brains of elderly subjects: Possible pathological basis. *Radiology, 162,* 509–511.
Kobari, M., Meyer, J. S., Ichijo, M., & Oravez, W. T. (1990). Leukoaraiosis: Correlations of MR and CT findings with blood flow, atrophy, and cognition. *American Journal of Neuroradiology, 11,* 273–281.
Koenig, S. H. (1991). Cholesterol of myelin is the determinant of gray-white matter contrast in MRI of brain. *Magnetic Resonance in Medicine, 20,* 285–291.
Koop, M., Rilling, G., Herrmann, A., & Kretschmann, H.-J. (1986). Volumetric development of the fetal telencephalon, cerebral cortex, diencephalon, and rhombencephalon including the cerebellum in man. *Bibliotheca Anatomica, 28,* 53–78.
Kozachuk, W. E., DeCarli, C., Schapiro, M. B., Wagner, E. E., Rapoport, S. I., & Horwitz, B. (1990). White matter hyperintensities in dementia of Alzheimer's type and in healthy subjects without cerebrovascular risk factors. *Archives of Neurology, 47,* 1306–1310.
Larsell, O., & Jansen, J. (1972). *The comparative anatomy and histology of the cerebellum. Vol. 3. The human cerebellum, cerebellar connections and cerebellar cortex.* Minneapolis: University of Minnesota Press.
Leaton, R. N., & Supple, W. F. (1986). Cerebellar vermis: Essential for long-term habituation of the acoustic startle response. *Science, 232,* 513–515.
Lechner, H., Niederkorn, K., & Schmidt, R. (1991). Does cerebrovascular insufficiency contribute to Alzheimer's disease? *Annals of the New York Academy of Sciences, 640,* 74–79.
Leiner, H. C., Leiner, A. L., & Dow, R. S. (1986). Does the cerebellum contribute to mental skills? *Behavioral Neuroscience, 100,* 443–454.
Leiner, H. C., Leiner, A. L., & Dow, R. S. (1989). Reappraising the cerebellum: What does the hindbrain contribute to the forebrain? *Behavioral Neuroscience, 103,* 998–1008.
Levitt, P., Rakic, P., & Goldman-Rakic, P. S. (1984). Region-specific distribution of catecholamine afferents in primate cerebral cortex: A fluorescence histochemical analysis. *Journal of Comparative Neurology, 225,* 1–14.
Leys, D., Soetaert, G., Petit, H., Fauquette, A., Pruvo, J.-P., & Steinling, M. (1990). Periventricular and white matter magnetic resonance imaging hyperintensities do not differ between Alzheimer's disease and normal aging. *Archives of Neurology, 47,* 524–527.
Lim, K., Zipurski, R. B., Murphy, G. M., & Pfefferbaum, A. (1990). In vivo quantification of the limbic system using MRI: Effects of normal aging. *Psychiatry Research: Neuroimaging, 35,* 15–26.
Lindboe, C. F., Erichsen, A. A., & Strøm, E. H. (1989). Atrophy and sponginess of the mammillary bodies with neuronal sparing: Not only inactive Wernicke's encephalopathy. *Acta Pathologica, Microbiologica et Immunologica Scandinavica, 97,* 667–670.
Malko, J. A., Hoffman, J. C., & Green, R. C. (1991). MR measurement of intracranial CSF volume in 41 elderly normal volunteers. *American Journal of Neuroradiology, 12,* 371–374.
Mani, R. B., Lohr, J. B., & Jeste, D. V. (1986). Hippocampal pyramidal cells in the human: A quantitative study of neuronal loss in sectors CA1 through CA4. *Experimental Neurology, 94,* 29–40.
Martin, W. R. W., Palmer, M. R., Patlak, C. S., & Calne, D. B. (1989). Nigrostriatal function in humans studied with positron emission tomography. *Annals of Neurology, 26,* 535–542.
Mathur-DeVré, R. (1984). Biomedical implications of the relaxation behavior of water related to NMR imaging. *British Journal of Radiology, 58,* 817–823.
Matsubayashi, K., Shimada, K., Kawamoto, A., & Ozawa, T. (1992). Incidental brain lesions on magnetic resonance imaging and neurobehavioral function in the apparently healthy elderly. *Stroke, 23,* 175–180.
Mayhew, T. M., & Olsen, D. R. (1991). Magnetic resonance imaging (MRI) and model-free estimates of brain volume determined using the Cavalieri principle. *Journal of Anatomy, 178,* 133–144.
McCall, A. (1992). The impact of diabetes on the CNS. *Diabetes, 41,* 557–570.

McGeer, E. G. (1978). Aging and neurotransmitter metabolism in the human brain. In R. Katzman, R. Terry, & K. L. Bick (Eds.), *Alzheimer's disease: Senile dementia and related disorders, aging series: Vol. 7* (pp. 427–445). New York: Raven Press.
McGeer, P. L., & McGeer, E. G. (1989). Amino acid neurotransmitters. In G. J. Siegel, B. W. Agranoff, R. W. Albers, & P. W. Molinoff (Eds.), *Basic neurochemistry: Molecular, cellular, and medical aspects* (pp. 311–332). New York: Raven Press.
McGeer, P. L., McGeer, E. G., Akiyama, H., Itagaki, S., Harrop, R., & Peppard, R. (1990). Neuronal degeneration and memory loss in Alzheimer's disease and aging. In J. Eccles & O. Creutzfeldt (Eds.), *The principles of design and operation of the brain* (pp. 410–431). Berlin: Springer.
McGeer, P. L., McGeer, E. G., & Suzuki, J. S. (1977). Aging and extrapyramidal function. *Archives of Neurology, 34,* 33–35.
Meguro, K., Yamaguchi, T., Hishinuma, T., Miyazawa, H., Ono, S., Yamada, K., & Matsuzawa, T. (1993). Periventricular hyperintensity on magnetic resonance imaging correlated with brain ageing and atrophy. *Neuroradiology, 35,* 125–129.
Mesulam, M.-M., Rosen, A. D., & Mufson, E. J. (1984). Regional variations in cortical cholinergic innervation chemoarchitectonics of acetylcholinesterase containing fibers in the macaque brain. *Brain Research, 311,* 254–258.
Meyerson, J., Hale, S., Wagstaff, D., Poon, L. W., & Smith, G. A. (1990). The information-loss model: A mathematical theory of age-related cognitive slowing. *Psychological Review, 97,* 475–487.
Miller, A. K. H., & Corsellis, J. A. N. (1977). Evidence for a secular increase in human brain weight during the past century. *Annals of Human Biology, 4,* 253–257.
Milton, W. J., Atlas, S. W., Lexa, F. J., Mozley, P. D., & Gur, R. E. (1991). Deep gray matter hypointensity patterns with aging in healthy adults: MR imaging at 1.5 T. *Radiology, 181,* 715–719.
Morgan, D. G., & May, P. C. (1990). Age-related changes in synaptic neurochemistry. In E. L. Schneider & J. W. Rowe (Eds.), *Handbook of the biology of aging* (3rd ed., pp. 219–254). San Diego: Academic Press.
Mortimer, J. (1988). Human motor behavior and aging. *Annals of the New York Academy of Sciences, 515,* 54–65.
Murphy, D., DeCarli, C. S., Williams, W., Rapoport, S. I., Schapiro, M. B., & Horwitz, B. (1992). Age-related differences in volumes of subcortical nuclei, brain matter, and cerebro-spinal fluid in healthy men as measured with magnetic resonance imaging (MRI). *Archives of Neurology, 49,* 839–845.
Neve, R. L., Finch, E. A., Bird, E. D., & Benowitz, L. I. (1988). Growth-associated protein GAP-43 is expressed selectively in associative regions of the human brain. *Proceedings of the National Academy of Sciences of the United States of America, 85,* 3638–3642.
Nieuwenhuys, R. (1985). *Chemical neuroanatomy of the human central nervous system.* Berlin: Springer.
Nieuwenhuys, R., Voogd, J., & Van Huijsen, C. (1988). *Central nervous system: A synopsis and atlas.* Berlin: Springer.
Pearl, T. (1905). Biometric studies on man. Variation and correlation in brain weight. *Biometrica, 4,* 13–104.
Peters, A., Josephson, K., & Vincent, S. L. (1991). Effects of aging on the neuroglial cells and pericytes within area 17 of the rhesus monkey cerebral cortex. *Anatomical Record, 229,* 384–398.
Pfefferbaum, A., Lim, K. O., Zipursky, R. B., Mathalon, D. H., Rosenbloom, M. J., Lane, B., Ha, C. N., & Sullivan, E. V. (1992). Brain gray and white matter volume loss accelerated with aging in chronic alcoholics: A quantitative MRI study. *Alcoholism: Clinical and Experimental Research, 16,* 1078–1089.
Pujol, J., Junqué, C., Vendrell, P., Grau, J., Martí-Vilalta, J., Olivé, C., & Gili, J. (1992). Biological significance of iron-related magnetic resonance imaging changes in the brain. *Archives of Neurology, 49,* 711–717.
Rao, S. M. (1986). Neuropsychology of multiple sclerosis: A critical review. *Journal of Clinical and Experimental Neuropsychology, 8,* 503–542.
Rao, S. M., Mittenberg, W., Bernardin, L., Haughton, V., & Leo, G. J. (1989). Neuropsychological test findings in subjects with leukoaraiosis. *Archives of Neurology, 46,* 40–44.
Rapoport, S. I. (1990). Topography of Alzheimer's disease: Involvement of association neocortices and connected regions; Pathological, metabolic and cognitive correlations; relation to evolution. In S. I. Rapoport, H. Pettit, D. Leys, & Y. Christen (Eds.), *Imaging, cerebral topography, and Alzheimer's disease.* Berlin: Springer-Verlag.
Raz, N., & Acker, J. D. (1994, September). Differential aging of subcortical structures observed *in vivo:* A prospective study. Paper presented at the Annual Meeting of the European Neuroscience Association, Vienna, Austria.

Raz, N., Briggs, S. D., Gunning, F. M., Head, D. P., McQuain, J. D., Loken, W. J., Thornton, A. E., & Acker, J. D. (1994, November). Differential aging of cerebral cortex observed *in vivo:* Selective vulnerability of the prefrontal gray matter. Paper presented at the 24th Annual Meeting of the Society for Neuroscience, Miami Beach, FL.

Raz, N., Millman, D., & Sarpel, G. (1990). Cerebral correlates of cognitive aging: Grey-white matter differentiation in the medial temporal lobes, and fluid vs. crystallized abilities. *Psychobiology, 18,* 475–481.

Raz, N., Torres, I. J., & Acker, J. D. (1992). Age-related shrinkage of the mammillary bodies: Evidence from *in vivo* MRI. *NeuroReport, 3,* 713–716.

Raz, N., Torres, I., & Acker, J. D. (1995). Age, gender, and hemispheric differences in human striatum: A quantitative review and new data from *in vivo* MRI morphometry. *Neurobiology of Learning and Memory, 63,* 133–142.

Raz, N., Torres, I. J., Briggs, S. D., Spencer, W. D., Thornton, A. E., Loken, W., Gunning, F. M., McQuain, J. D., Driesen, N. R., & Acker, J. D. (1995). Selective neuroanatomical abnormalities in Down syndrome and their cognitive correlates: Evidence from MRI morphometry. *Neurology, 45,* 356–366.

Raz, N., Torres, I. J., Spencer, W. D., & Acker, J. D. (1993). Pathoclysis in aging human cerebral cortex: Evidence from *in vivo* MRI morphometry. *Psychobiology, 21,* 151–160.

Raz, N., Torres, I. J., Spencer, W. D., Baertschie, J. C., Millman, D., & Sarpel, G. (1993). Neuroanatomical correlates of age-sensitive and age-invariant cognitive abilities: An *in vivo* MRI investigation. *Intelligence, 17,* 407–422.

Raz, N., Torres, I. J., Spencer, W. D., White, K., & Acker, J. D. (1992). Age-related regional differences in cerebellar vermis observed *in vivo. Archives of Neurology, 49,* 611–617.

Raz, S., Lauterbach, M. D., Hopkins, T., Porter, C. L., Riggs, W. W., Magill, L. H., & Sander, C. J. (1995). A female advantage in cognitive recovery from early cerebral insult. *Developmental Psychology, 31,* 958–966.

Raz, S., Goldstein, R., Shah, F., Hopkins, T., Riggs, W. W., Magill, L. H., & Sander, C. J. (1994). Sex differences in early vulnerability to cerebral injury and their neurodevelopmental implications. *Psychobiology, 22,* 244–253.

Samorajski, T., & Rolsten, C. (1973). Age and regional differences in the chemical composition of brains of mice, monkeys and humans. In D. H. Ford (Ed.), *Neurobiological aspects of maturation and aging. Progress in Brain Research, 40,* 253–265.

Sarnat, H. B., & Netsky, M. G. (1981). *Evolution of the nervous system* (2nd ed.). New York: Oxford University Press.

Sarpel, G., Chaudry, F., & Hindo, W. (1987). Magnetic resonance imaging periventricular hyperintensity in a veterans administration hospital population. *Archives of Neurology, 44,* 725–728.

Sawle, G. V., Colebatch, J. G., Shah, A., Brooks, D. J., Marsden, C. D., & Frackowiak, R. S. (1990). Striatal function in normal aging: Implications for Parkinson's disease. *Annals of Neurology, 28,* 799–804.

Schaefer, G. B., Thompson, J. N., Bodensteiner, J. B., Gingold, M., Wilson, M., & Wilson, D. (1991). Age-related changes in the relative growth of the posterior fossa. *Journal of Child Neurology, 6,* 15–19.

Schenker, C., Meier, D., Wichman, W., Boesiger, P., & Valvanis, A. (1993). Age distribution and iron dependency of the T2 relaxation time in the globus pallidus and putamen. *Neuroradiology, 35,* 119–124.

Schmidt, R., Fazekas, F., Kleinert, K., Offenbacher, H., Gindl, K., Payer, F., Freidl, W., Niederkorn, K., & Lechner, H. (1992). Magnetic resonance imaging signal hyperintensities in the deep and subcortical white matter. *Archives of Neurology, 49,* 825–827.

Schmidt, R., Fazekas, F., Offenbacher, H., Dusek, E., Zach, E., Reinhart, B., Greishofer, P., Freidl, W., Eber, B., Schumacher, M., Koch, M., & Lechner, H. (1993). Neuropsychological correlates of MRI white matter hyperintensities: A study of 150 normal volunteers. *Archives of Neurology, 49,* 825–827.

Seitz, R. J., Roland, P. E., Bohm, C., Greitz, T., & Stone-Elander, S. (1990). Motor learning in man: A positron emission tomographic study. *NeuroReport, 1,* 57–66.

Semenova, L. K., Vasilieva, V. A., Tsekhmistrenko, T. A., & Shumeiko, N. S. (1989). Osobennosti ansamblevoy organizatsii kory bolshovo mozga tcheloveka ot rozhdeniya do 20 let [Pecularities of ensemble organization of the human cerebral cortex from birth up to 20 years of age]. *Arkhiv Anatomii, Ghistologhighii i Embriologhii, 97,* 15–24.

Shah, S. A., Doraiswami, P. M., Husain, M. M., Escalona, P. R., Na, C., Figiel, G. S., Patterson, L. J., Ellinwood, E. H., McDonald, W. M., Boyko, O. B., Nemeroff, C. B., & Krishnan, K. R. R. (1992).

Posterior fossa abnormalities in major depression: A controlled magnetic resonance imaging study. *Acta Psychiatrica Scandinavica, 85,* 474–479.
Shah, S. A., Doraiswami, P. M., Husain, M. M., Figiel, G. S., Boyko, O. B., McDonald, W. M., Ellinwood, E. H., & Krishnan, K. R. R. (1991). Assessment of posterior fossa structures with midsagittal MRI: The effects of age. *Neurobiology of Aging, 12,* 371–374.
Shimada, A., Hosokawa, M., Ohta, A., Akiguchi, I., & Takeda, T. (1994). Localization of atrophy-prone areas in the aging mouse brain: Comparison between the brain atrophy model SAM-P/10 and the normal control SAM-R/1. *Neuroscience, 59,* 859–869.
Shimada, K., Kawamoto, A., Matsubayashi, K., & Ozawa, T. (1990). Silent cerebrovascular disease in the elderly: Correlation with ambulatory pressure: *Hypertension, 16,* 692–699.
Soininen, H., Puranen, M., Helkala, E.-L., Laakso, M., & Riekkinen, P. J. (1992). Diabetes mellitus and brain atrophy: A computed tomography study in an elderly population. *Neurobiology of Aging, 13,* 717–721.
Squire, L. R. (1992). Memory and the hippocampus: A synthesis from findings with rates, monkeys, and humans. *Psychological Review, 99,* 195–231.
Squire, L. R., Knowlton, B., & Musen, G. (1993). The structure and organization of memory. *Annual Review of Psychology, 44,* 453–495.
Stein, B. (1984). Development of the superior colliculus. *Annual Review of Neuroscience, 7,* 95–125.
Talairach, J., & Tournoux, P. (1988). *Co-planar stereotaxic atlas of the human brain.* Stuttgart: G. Thieme.
Tanna, N. K., Kohn, M. I., Horwich, D. N., Jolles, P. R., Zimmerman, R. A., Alves, W. M., & Alavi, A. (1991). Analysis of brain and cerebrospinal fluid volumes with MR imaging: Impact on PET data correction for atrophy. *Radiology, 178,* 123–130.
Terry, R. D., DeTeresa, R., & Hansen, L. A. (1987). Neocortical cell counts in normal human adult aging. *Annals of Neurology, 21,* 530–539.
Thatcher, R. W., Hallett, M., Zeffiro, T., John, E. R., & Huerta, M. (1994). *Functional neuroimaging: Technical foundations.* San Diego: Academic Press.
Thompson, R. F. (1986). The neurobiology of learning and memory. *Science, 233,* 941–947.
Tigges, J., Herndon, J. G., & Peters, A. (1990). Neuronal population of area 4 during the life span of the rhesus monkey. *Neurobiology of Aging, 11,* 201–208.
Tomonaga, M., Yamanuchi, H., Tohgi, H., & Kameyama, M. (1982). Clinicopathological of progressive subcortical vascular encephalopathy (Binswanger type) in the elderly. *Journal of American Geriatric Society, 30,* 524–529.
Torvik, A., Torp, S., & Lindboe, C. F. (1986). Atrophy of the cerebellar vermis in ageing: A morphometric and histological study. *Journal of Neurological Sciences, 76,* 283–294.
Unger, E., Littlefield, A. J., & Gado, M. (1988). Water content and water structure in CT and MR signals: Possible influence in detection of early stroke. *American Journal of Neuroradiology, 9,* 687–691.
van Swieten, J. C., Geyskes, G. G., Derix, M. M. A., Peeck, B. M., Ramos, L. M. P., van Latum, J. C., & van Gijn, J. (1991). Hypertension in the elderly is associated with white matter lesions and cognitive decline. *Annals of Neurology, 30,* 825–830.
Verhaeghen, P., Marcoens, A., & Goosense, L. (1993). Facts and fiction about memory aging: A quantitative integration of research findings. *Journal of Gerontology: Psychological Sciences, 48,* P157–P171.
Vincent, S. L., Peters, A., & Tigges, J. (1989). Effects of aging on the neurons within the area 17 of the rhesus monkey cerebral cortex. *Anatomical Record, 223,* 329–341.
Vogt, C., & Vogt, O. (1951). Importance of neuroanatomy in the field of neuropathology. *Neurology, 1,* 205–218.
Wahlund, L.-O., Agartz, I., Almqvist, O., Basun, H., Forssell, L., Saaf, J., & Wetterberg, L. (1990). The brain in healthy aged individuals: MR imaging. *Radiology, 174,* 675–679.
Weis, S., Kimbacher, M., Wenger, E., & Neuhold, A. (1993). Morphometric analysis of the corpus callosum using MR: Correlation of measurements with aging in healthy individuals. *American Journal of Neuroradiology, 14,* 637–645.
Wiggins, R. C., Corman, A., Rolsten, C., Samorajski, T., Ballinger, W. E., & Freund, G. (1988). Effects of aging and alcohol on the biochemical composition of histologically normal human brain. *Metabolic Brain Diseases, 3,* 67–80.
Wisniewski, H. M., & Terry, R. D. (1973). Morphology of the aging brain, human and animal. In D. H. Ford (Ed.), *Neurobiological aspects of maturation and aging.* Amsterdam: Elsevier.
Yamashita, K., Kobayashi, S., Fukuda, H., Koide, H., Okada, K., & Tsunematsu, T. (1991). [The relationship between cerebral white matter changes, mental function and blood pressure in normal elderly]. *Japanese Journal of Geriatrics, 28,* 546–550.

Ylikoski, R., Ylikoski, A., Erkinjuntti, T., Sulkava, R., Raininki, R., & Tilvis, R. (1993). White matter changes in healthy elderly persons correlate with attention and speed of mental processing. *Archives of Neurology, 50,* 818–824.

Yoshii, F., Barker, W. W., Chang, J. Y., Lowenstein, D., Apicella, A., Smith, D., Boothe, T., Ginsburg, M. D., Pascal, S., & Duara, R. (1988). Sensitivity of cerebral glucose metabolism to age, gender, brain volume, brain atrophy, and cerebrovascular risk factors. *Journal of Cerebral Blood Flow and Metabolism, 8,* 654–661.

Zola-Morgan, S., & Squire, L. R. (1991). The medial temporal lobe memory system. *Science, 253,* 1380–1386.

II

Clinical Syndromes

7

Positron Emission Tomography Neuroimaging in Dementia

KATHLEEN A. WELSH-BOHMER AND JOHN M. HOFFMAN

INTRODUCTION

Positron emission tomography (PET) neuroimaging is an advanced nuclear medicine technology that allows visualization of functional brain activity *in vivo*. The current advances in the technology have allowed a high degree of temporal and spatial resolution, permitting the measurement of transient brain events (seconds) over relatively small brain regions. As such, the method has been a valuable tool for neuropsychological investigations of human cognition. In addition, PET imaging has been used clinically to facilitate the diagnosis of suspected brain dysfunction (Friedland, 1989; Workshop Panel, 1990). The method has been clinically useful, particularly in cases where the diagnosis is unclear, such as in the early stages of Alzheimer's disease (AD). The purpose of this chapter is to review the use of PET imaging in dementia. Within this chapter we explore the uses of PET imaging in Alzheimer's disease and other dementias of late life. The text of this chapter is divided into six sections. The first section provides an overview of the fundamental principles of PET neuroimaging. The next section discusses the contribution of PET imaging to the understanding of normal aging

KATHLEEN A. WELSH-BOHMER Joseph & Kathleen Bryan Alzheimer's Disease Research Center, and Department of Psychiatry, Duke University Medical Center, Durham, North Carolina 27713.
JOHN M. HOFFMAN Department of Radiology and Medicine, Division of Neurology, Duke University Medical Center, Durham, North Carolina 27713.

Neuroimaging II: Clinical Applications, edited by Erin D. Bigler. Plenum Press, New York, 1996.

and Alzheimer's disease, the most common cause of dementia in the elderly. The functional imaging changes observed in other common forms of dementia are discussed in the next three sections. We start by considering the vascular dementias, move to discussion of the subcortical dementias (Huntington's disease, Parkinson's disease, and progressive supranuclear palsy), and conclude with a description of the PET findings in less common cortical dementias, namely, Pick's disease and Creutzfeldt-Jakob disease. In the final section, the limitations of the PET method alone in the diagnosis of dementia are discussed and the future directions of PET imaging in the clinical arena are considered. Within this last section of the text, we entertain the relative utility of combining methodologies, neuropsychological evaluation, and functional brain imaging–PET as clinical predictors of Alzheimer's disease and other dementias.

Positron Emission Tomography Methods

Positron emission tomography (PET) is one of the newly evolving neuroimaging techniques that may aid in the diagnosis of Alzheimer's disease and other dementing disorders (Cutler, 1988; Friedland, 1989; Pawlik & Heiss, 1989; Rapoport, 1991; Riege & Metter, 1988, for reviews). PET differs from the morphologic imaging techniques of MRI and CT in that it is a *functional* imaging technique, providing quantitative estimates of brain physiology (e.g., cerebral blood flow, glucose metabolism, oxygen metabolism) and biochemistry (e.g., receptor binding kinetics, neurotransmitter distribution). PET methodology takes advantage of the local tissue concentration of biologically important compounds labeled with "positron" emitting isotopes of oxygen, fluorine, nitrogen, carbon, and other elements. When these isotopes are incorporated into glucose, water, oxygen, brain amines, or drug compounds it is possible to monitor the distribution of these compounds in the brain without disturbing their biological properties and physiological behavior. This is accomplished by injecting the compounds of interest intravenously and taking advantage of the unique properties of positron decay to create quantifiable images of brain function. As the injected compound undergoes radioactive decay in the nucleus, a positron is ejected, travels a short distance, and collides with an electron producing two 511-keV photons which are oppositely directed at approximately 180 degrees. This phenomenon permits detection of the two coincident 511-keV photons and is the basis for PET imaging techniques. The positron tomograph is designed such that two detectors are connected in coincidence and will record only those events which occur simultaneously in both detector systems. By using complex mathematical reconstruction algorithms and electronic techniques, the location of the annihilation can be accurately determined. To eventually develop an image, PET requires numerous components including the appropriately labeled compound, a tomograph to measure the distribution of the activity, tracer kinetic models, and complex reconstruction algorithms to convert the data into cross-sectional images. A complete description of this methodology is found in Raichle (1979) and more recently elaborated by Links (1989) and Dannals (1989).

Resting State PET

To explore changes in functional brain activity in dementia, a frequently used approach involves the acquisition of *resting state* images which are then used

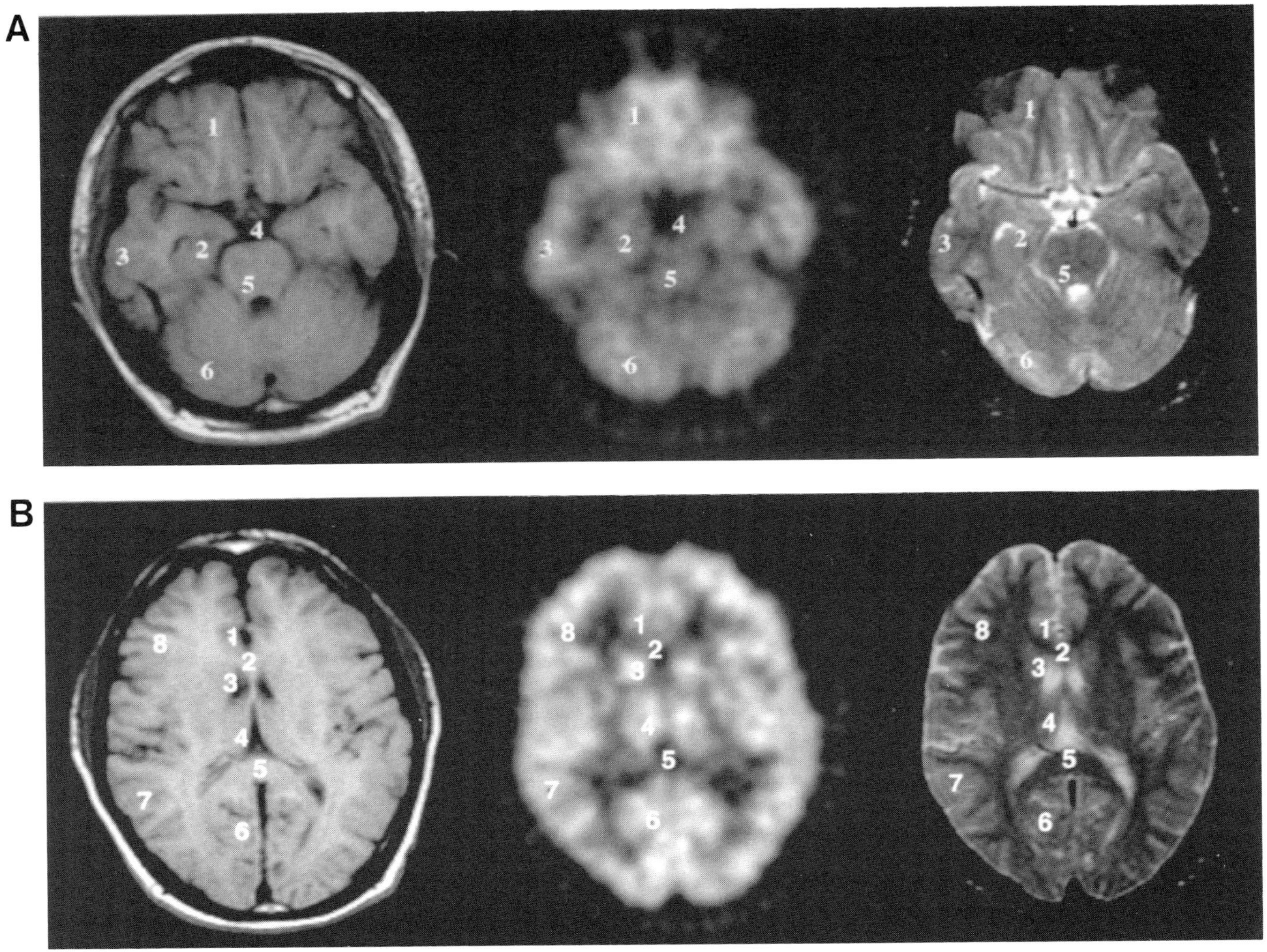

Figure 1. Registered FDG-PET images of a healthy 32-year-old male. Two image sets are shown. The inferior images (A) are through the inferior frontal lobes, the temporal lobes, and cerebellum. In (A): 1, frontal lobe; 2, mesiotemporal lobe (hippocampus complex); 3, lateral temporal lobe; 4, suprasellar cistern; 5, pons; 6, cerebellum. The second image set (B) is at the level of the superior aspect of the ventricles and includes the frontal and parietal regions. In (B): 1, cingulate gyrus; 2, genu of the corpus callosum; 3, body of the caudate nucleus; 4, thalamus; 5, splenium of the corpus callosum; 6, calcarine cortex (visual cortex); 7, parietal lobe; 8, frontal lobe. For both image sets, the left image is a T1-weighted MRI scan (TR = 600 ms, TE = 20 ms), the center image is from an FDG-PET study obtained at the same brain level (on the GE 4096 Plus tomograph), and the right image is from a T2-weighted MRI study (TR = 2500 ms, TE = 80 ms).

to compare differences between patients and controls in the regional magnitude and distribution of glucose utilization (FDG-PET). Applying the method of Sokoloff and coworkers (1977) adapted to PET imaging in humans, the distribution of the ^{18}F-labeled deoxyglucose (FDG), a competitive substrate analog of glucose, is monitored throughout the brain. Because the isotope used in this procedure has a relatively long half-life (110 min), activity over a prolonged interval (30–45 min) can be averaged. This reduces the effects of random variation inherent in the PET method and enhances the signal associated with the variable of interest, namely, glucose metabolism. The degree of statistical precision afforded by this method is an advantage in studies designed to detect reliable differences between subject groups.

The recent advances in the technology permit the acquisition of high-resolution functional images at various levels within the brain and in imaging planes similar to those obtained on MRI. Figure 1 shows an example of an ^{18}FDG-PET scan obtained in the normal resting state and corresponding images with T1 (left panel) and T2 (right panel) weighted MRI images taken at the same brain level. Although the PET images in this scan are presented in shades of gray, often a full color spectrum is used, with low activity represented by black or shades of "cold" colors (blue and green); higher activity is indicated by "warm" colors of yellow, orange, and red or white. Regardless of color convention employed, the intensity of shade or color corresponds to quantitative measures of glucose metabolism with values in milligrams of glucose consumed per minute per 100 grams of brain issue. By superimposing the PET image onto the MRI scan using landmarks on the outer surface of the head or surface of the brain, the PET and MRI scans can be brought into register and quantitative measures of functional activity within specific regions of interest can be obtained (Pelizzari, Chen, Spelbring, Weischselbaum, & Chen, 1989).

Studies conducted in normal subjects during the resting state consistently report a symmetric distribution of cerebral metabolic rate for glucose throughout the brain (Mazziotta, Phelps, Miller, & Kubl, 1981; Mazziotta, Phelps, & Carson, 1982). Regional differences in activity can be seen depending on the "resting" conditions. For example, under conditions of sensory deprivation where the subjects' eyes are closed and ears are occluded, some increased resting metabolism and blood flow has been reported in frontal lobe regions (Greenberg *et al.*, 1981). By contrast, when scanning is conducted with eyes open, increases in blood flow and metabolism are noted in the visual cortex, as one might expect (Phelps, Kuhl, & Mazziotta, 1981).

Resting state PET studies have been used to examine the relationship between neuropsychological test performance and cerebral metabolism. In the typical study design, neuropsychological testing is conducted prior to or following the scanning session and the obtained test scores are then compared to resting state metabolic values. Such studies conducted in healthy normal individuals have failed to reveal significant relationships between resting state metabolism and cognitive test performance (Duara *et al.*, 1984; Haxby *et al.*, 1986). However, in patient groups with brain disease the differences in regional rates of cerebral metabolism are apparent and are related to the severity of illness, type of dementia, behavioral presentation, and resultant patterns of test performance obtained on neuropsychological evaluation (Chase *et al.*, 1984a, 1984b; Friedland *et al.*, 1983; Martin *et al.*, 1986). The latter observations are particularly relevant for brain-behavior study. Through investigation of patients with selective behavioral deficits, the neural substates for these disorders may be specified (Welsh, Hoffman, McDonald, Earl, & Breitner, 1993). Because PET imaging permits investigation of whole brain dynamics, that is, multiple brain systems as they engage or fail to engage in behavior, the method combined with neuropsychological paradigms may permit a level of specificity in understanding the neural organization of behavior and disease not afforded by studies of brain morphometry (e.g., MRI, CT). This is the rationale behind studies of brain activation with PET.

Brain Activation Studies with PET

PET imaging studies conducted *while* the patient is engaged in a cognitive task or under sensory stimulation are referred to as "activation" studies. Activation studies afford unique opportunities to view "on-line" functional brain changes as they occur, a capability not possible with resting state PET. The same principles of resting state PET apply. However, the activation procedure requires isotopes with shorter half-lives in order to allow observation of transient brain changes associated with the experimental procedure. The method typically employed examines cerebral blood flow (CBF) using bolus injections of H_2O or the inhalation of CO_2 labeled with the short-acting radioisotope ^{15}O (half-life = 123 seconds). The CBF is determined over intervals of 40 seconds to 4 minutes with a scanning session of 12–15 minutes (Haxby, Grady, Ungerleider, & Horwitz, 1991). Because of the short half-life of the isotope used in these studies, repeat scans can be obtained within the same individuals within the same imaging session. This capability permits powerful within-subject research designs wherein the various experimental conditions can be presented within the same session and subjects can serve as their own control.

Activation studies conducted in young and elderly normal subjects have been instructive in showing brain metabolic changes related to specific cognitive operations and perceptual tasks. The application of PET in this manner has allowed a more direct investigation of the neural networks supporting very discrete cognitive functions than has been possible with the traditional brain lesion methods of neuropsychology. The potential for PET to generate detailed functional maps of higher cognition has been demonstrated for a variety of functions in language processing (Gur, *et al.*, 1983; Mazziotta, Phelps, & Carson, 1982; Pawlik & Heiss, 1989; Peterson, Fox, Posner, Mintun, & Raichle, 1988, 1989), visual analysis and imagery (Fox, Miezin, Allman, Van Essen, & Raichle, 1987; Fox *et al.*, 1986; Haxby *et al.*, 1991), aspects of attention (Posner & Petersen, 1989), and memory (Fazio *et al.*, 1992).

In studies of dementia, resting state scans are typically employed in order to establish the patterns in brain metabolism characteristic of the disease state. Without some knowledge of the inherent differences in resting states, the results from behavioral activation studies are difficult to interpret. However, with the advances made in the last 10 years toward understanding the changes in cerebral metabolism and blood flow with disease, activation paradigms are beginning to be employed in AD and other dementias. These studies have examined the brain's residual capacity to compensate for injury and the integrity of the brain systems affected by the disease under conditions of behavioral challenge (Duara, Lowenstein, & Barker, 1990; Duara *et al.*, 1992). The following sections review the findings and recent advances in PET imaging in normal aging, Alzheimer's disease, vascular dementia, and some of the subcortical dementias (e.g., Huntington's disease, Parkinson's disease).

PET Studies in Normal Aging and Alzheimer's Disease

Normal Aging

A number of studies have examined cerebral blood flow, oxygen metabolism, and glucose metabolism with normal aging (Duara *et al.*, 1984; Kuhl, Metter, Riege, & Phelps, 1982; Mazziotta, Phelps, Miller, & Kuhl, 1981; Metter, Riege, Kuhl, & Phelps, 1983; Raichle, 1982). Some discrepancies in reported age-related changes have been found. In some studies significant decreases in overall cerebral glucose utilization as a function of age are reported (Hoffman, Guze, Baxter, Mazziotta, & Phelps, 1989; Kuhl *et al.*, 1982); however, in other studies no change or only minimal changes are noted (Frackowiak, Lenzi, Jones, & Heather, 1980). Some methodological differences likely account for these discrepancies (de Leon *et al.*, 1984; Duara *et al.*, 1983, 1984; Yoshii *et al.*, 1988). A major distinction is the use of brain volume corrections. In the studies finding no significant differences between the young and old groups, regional differences were found in the elderly group but disappeared following correction for total brain volume and overall brain atrophy. Although it is imperative to control for apparent reductions in glucose utilization attributable simply to atrophy, the appropriateness of global corrections used in these studies has been argued (Hoffman *et al.*, 1989). This procedure may correct for important regional changes which are not referable to atrophy and therefore not subject to partial volume effects. With enhanced capability to correct for regional brain atrophy, a better appreciation of age-related brain changes will be achieved. Figure 2 presents a normal PET imaging study with arrows pointing to the brain structures typically affected by AD which show, in this case, homogeneous uptake as expected.

Alzheimeir's Disease

Numerous investigations using PET have been conducted in Alzheimer's disease with the hope that by understanding the functional brain deficits associated with the disease, reliable diagnosis of the disease may be achieved. Alzheimer's disease accounts for the vast majority of 2.5 to 4 million cases of dementia among older adults in the United States (Evans *et al.*, 1989). The disease begins insidiously (Crapper-McLachlan, Dalton, Galin, Schlotterer, & Daicar, 1984; Katzman, 1986) and in its early stages may be confused for normal aging (Katzman *et al.*, 1989; Moss & Albert, 1988). Early in the illness, relatively isolated cognitive losses, most commonly memory impairment, are noted (Fuld, Masur, Blau, Crystal, & Aronson, 1990; Knopman & Ryberg, 1989; Masur, Fuld, Blau, Crystal, & Aronson, 1990; Storandt, Botwinick, Danziger, Berg, & Hughes, 1984; Welsh, Butters, Hughes, Mohs, Heyman, 1991). However, as the condition progresses, the losses become more pervasive and pronounced. Deficits in other cognitive and behavioral domains, including general intelligence, orientation, language, visuospatial abilities, and personality, are noted (Moss & Albert, 1988; Rebok, Brandt, & Folstein, 1990; Siegler *et al.*, 1991; Welsh, Butters, Hughes, Mohs, &

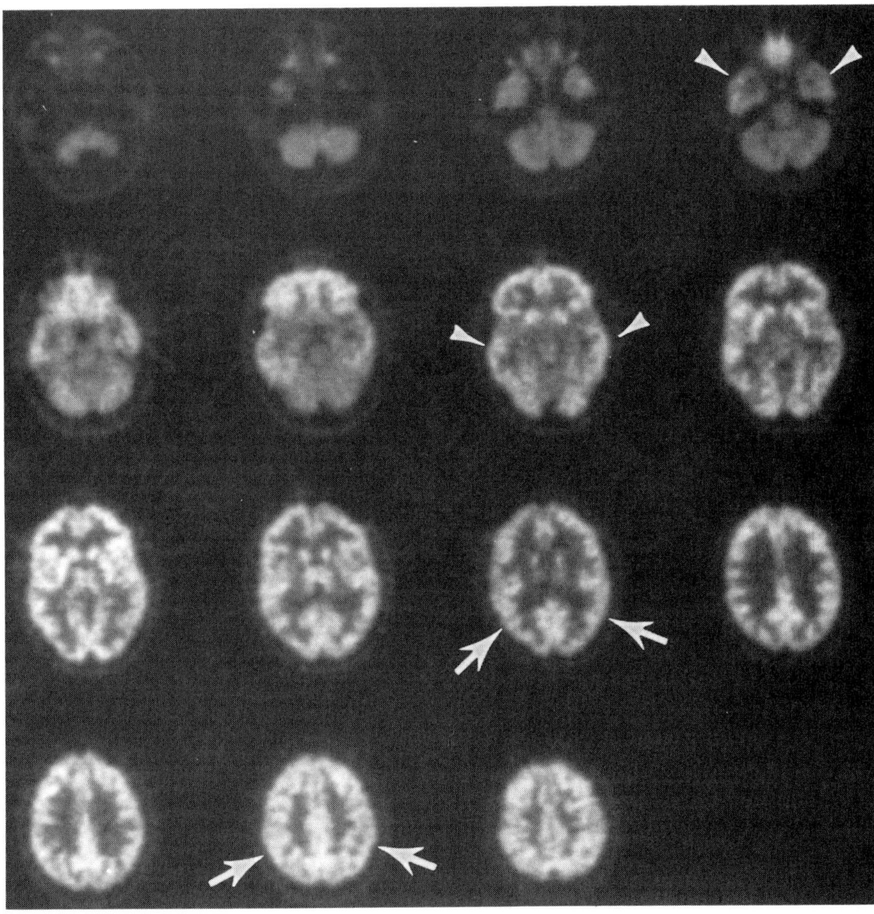

Figure 2. Transverse FDG-PET images covering the entire intracranial contents in a normal individual. Images are oriented such that the top left image is that of the inferior cerebellum with successive images gong superiorly through the brain. There is normal and homogeneous FDG uptake throughout all cortical and subcortical brain structures. Both temporal lobes are shown with small arrows. The parietal region bilaterally is indicated by larger arrows.

Heyman, 1992). Ultimately, the disease leads to progressive incapacitation and death.

In life, Alzheimer's disease is diagnosed as either "possible" or "probable" depending on the weight of the evidence favoring the diagnosis and according to specific criteria established by the National Institute on Communicative Disorders and Stroke and the Alzheimer's Disease and Related Disorders Association (NINCDS-ADRDA) (Khachaturian, 1985; McKhann *et al.*, 1984). The diagnosis of "definite" AD is reserved for pathological verification of the disease upon examination of the brain at autopsy or biopsy. To make the diagnosis, a high density of extracellular senile plaques and neurofibrillary tangles distributed throughout the cortex particularly in the limbic areas (e.g., hippocampus, entorhinal cortex, amygdala, and forebrain) and associated areas of the neocortex are needed. Areas relatively spared by the pathology are the basal ganglia, cerebellum, and primary cortical areas such as the primary visual cortex and the motor strip (Mirra, Hart, & Terry, 1993, for review).

Regional cerebral metabolic rate of glucose use in Alzheimer's disease has most frequently been estimated using resting state PET imaging studies. Through the study of regional brain changes it has been possible to visualize changes in function related to aging and disease. Unlike normal aging, decreased brain glucose metabolism is consistently reported in cases of moderate to severe Alzheimer's disease (Chase *et al.*, 1984b; Chawluk *et al.*, 1990; Duara *et al.*, 1986; Foster *et al.*, 1984; Frackowiak *et al.*, 1981; Friedland *et al.*, 1983; Hoffman *et al.*, 1989; Kuhl, Metter, & Reige, 1985). The pattern of hypometabolism most typically seen in cases of AD is presented in Figure 3. This FDG-PET image, conducted at Duke University, was obtained from a 50-year-old woman with biopsy-proven Alzheimer's disease. In this case, significant reductions in FDG uptake are obvious within the temporal, parietal, and frontal lobes bilaterally.

It should be noted that there is heterogeneity in the metabolic patterns of AD (Grady *et al.*, 1990; Haxby *et al.*, 1988). Nonetheless, bilateral dysfunction in the posterior temporoparietal junction area is the most common finding in AD

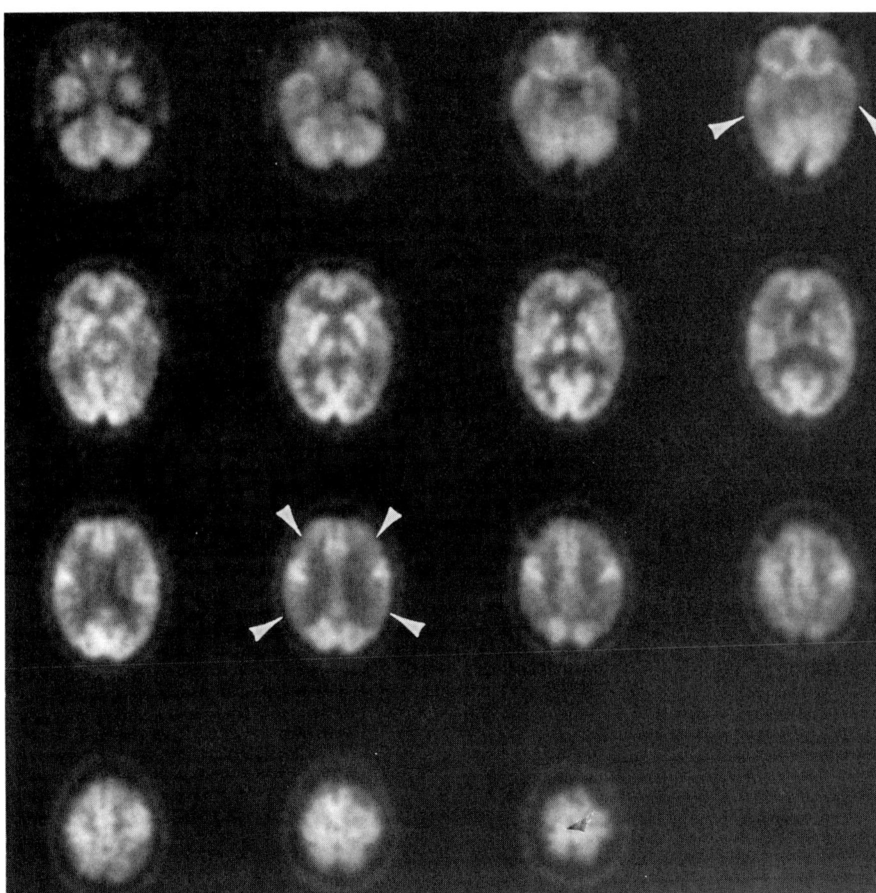

Figure 3. Set of FDG-PET images from a 50-year-old woman with biopsy-proven Alzheimer's disease. Note the significant reduction in FDG uptake bilaterally in the temporal, frontal, and parietal areas (arrows). There is sparing of the visual cortex as well as the motor-sensory areas.

(Grady *et al.*, 1990) so that it is virtually the metabolic signature of the disease (exceptions are discussed later). Some asymmetry in the metabolic rates is also common, usually with the left hemisphere temporoparietal area being more hypometabolic than the same areas on the right (Haxby *et al.*, 1990; Lowenstein *et al.*, 1989). Reductions are also commonly seen in the lateral frontal association cortices and have been reported in some instances within the orbitofrontal cortex, the cingulate cortex, and the posterior medial temporal lobe (Rapoport, 1991). The primary visual cortex, sensorimotor area, cerebellum, brain stem, and subcortical structures are the least affected by the disease and show normal metabolism (Rapoport, 1991).

The resting state findings with FDG-PET suggest specific dysfunction of neocortical metabolism in the temporal lobes and parietal association cortex as well as hypometabolism in some instances within the lateral frontal or orbitofrontal cortices, aspects of the cingulate cortex, and the posterior medial temporal lobe. These findings with FDG-PET are interesting in light of the parallel reported in the regional distribution of AD neuropathology observed at autopsy (Arnold, Hyman, Flory, Damasio, & Van Hoesen, 1991; Hyman, Van Hoesen, Damasio, & Barnes, 1984; Pearson, Esiri, Hjorns, Wilcock, & Powell, 1985). Because the same areas show a preponderance of plaques and tangles, it is possible that the decreased metabolism observed may be reflecting areas of particular neuronal vulnerability or disconnection (Chawluk *et al.*, 1990). For example, the reduced cortical metabolism may be due to decreased neuronal activity secondary to cell loss or perhaps to abnormal cellular function in these specific brain areas. Recent studies exploring PET and postmortem brain changes suggest that the reduced metabolism is primarily a reflection of neuronal loss and gliosis (McGeer, McGeer, Harrop, Akiyama, & Kamo, 1990). However, other laboratories report that cortical neurochemical changes in AD may exceed the actual neuronal cell loss in these areas (Hansen, De Teresa, Davies, & Terry, 1988), suggesting that the cortical hypometabolism observed with PET may reflect dysfunctional neurons as well as neuronal loss. Thus, the source of the reduced brain metabolism is unclear; however, the bulk of the evidence favors a combined effect of cellular loss, malfunctioning of the remaining viable cells, and abnormalities in distant sites due to synaptic disconnection (Smith *et al.*, 1992).

Because the reduction in metabolism within the temporal and parietal lobes is the most consistent observation in AD, this pattern has come to be considered the characteristic "signature" of the disorder in functional imaging studies. This pattern is observed early in the disease course and may precede the appearance of parietal symptoms. Over the course of the illness, the same patterns persist in spite of the advancing dementia. However, there is significant progression in the metabolic asymmetries of the prefrontal, parietal, and lateral temporal cortices (Haxby *et al.*, 1986; Grady *et al.*, 1988).

An important question in the study of brain-behavior relationships is whether the brain changes observed in Alzheimer's disease relate in a meaningful way to cognitive function and behavioral symptomatology. There are numerous studies directed at this question (for review, see Guze, Hoffman, Baxter, Mazziotta, & Phelps, 1991; Hoffman, Hanson, & Coleman, 1993; Rapoport, 1991; Riege & Metter, 1988). A review of these reports indicates considerable heterogeneity across studies in the AD patient groups included. This is due in part to the inherent variability in the clinical presentation of AD and to differences across studies in disease severity. For example, some studies include subjects at very mild stages of disease (Cutler *et al.*, 1985; Haxby, Duara, Grady, Cutter, & Rapoport, 1985) whereas others include subjects across a broader range of severities (de Leon *et al.*, 1983; Haxby *et al.*, 1990; Kiss, Friedland, Ober, & Jagust, 1985). In some instances, individuals with prominent language deficits or apraxia are targeted for study (Martin *et al.*, 1986). Others include diverse patient groups (Grady *et al.*, 1990).

Regardless of the differences in research design across laboratories, consistent and predictable associations in neurobehavioral function are reported. In general, the impairments in verbal abilities associated with AD, such as naming and verbal learning, have been shown to correlate with reduced glucose metabolism in the left hemisphere (Chase, *et al.*, 1984a, 1984b; Martin *et al.*, 1986). Similarly, verbal intelligence (Wechsler Adult Intelligence Scale) shows a relationship with left hemisphere metabolism, particularly with glucose metabolism in the left parasylvian area (Chase *et al.*, 1984a). By contrast, visuoconstructional deficits, apraxia, and impairments in nonverbal intelligence show a high association with right temporal and right parietal hypometabolism, particularly in the right posterior parietal lobe (Chase *et al.*, 1984a; Foster, Chase, Patronas, Gillespie, & Fedio, 1986; Martin *et al.*, 1986).

In studies restricted to patients with prominent aphasias or visuoconstructive defects, clear relationships are seen with respect to regional brain metabolism (Martin *et al.*, 1986; Welsh *et al.*, 1993). In these highly selected patient groups, the visuoconstructive impairments of AD have been related to right parietal metabolism (Chase *et al.*, 1984a, 1984b; Foster *et al.*, 1986; Martin *et al.*, 1986) and right occipital dysfunction (Welsh *et al.*, 1993). These same relationships hold in large group studies where preselection of patients for the cognitive deficits of interest is not conducted (Lowenstein *et al.*, 1989). In one patient series (Welsh, Hoffman, Earl, & Hanson, 1994), a group of 46 patients with probable, possible, or definite AD and 20 patients with subcortical and/or frontal lobe dementias were studied with the neuropsychological battery from the Consortium to Establish a Registry for Alzheimer's Disease (CERAD) and FDG-PET imaging. After adjustment for multiple comparisons, performance on the task of verbal fluency (animal generation) was associated with diminished FDG uptake in the left frontal, temporal, and parietal lobes, whereas visual confrontation naming was associated with reductions in left frontal uptake alone. Constructional praxis (copy of geometric figures) was related to reduced uptake in the right and, to a lesser extent, left parietal lobes, whereas verbal memory, which was consistently defective in all patients, was related to metabolic defects in the left temporal and left parietal lobes. Other studies have shown, interestingly, that the PET changes may not be temporally coincident with observed symptoms but rather that the hypometabolism may actually predate the clinical symptoms and persist thereafter (Haxby *et al.*, 1986). This observation may account for some of the variability seen between studies and certainly needs to be considered when examining potential neural correlates of behavioral symptomatology.

Activation Studies in AD

The intent of activation studies in Alzheimer's disease is to emphasize brain differences associated with disease by directly challenging the system, that is, making the subject perform a cognitive task. In this manner, it may be more readily apparent which brain regions are metabolically affected by the disease at early stages in the degenerative process. Some of the studies conducted in Alzheimer's disease have employed a double injection/double activation ^{18}FDG-PET method while patients engage in various neuropsychological tests (e.g., picture preference task). Use of ^{18}F, which has a long half-life, has an advantage in providing a reliable measure of metabolism but has a disadvantage in that it requires performance of a single task during the entire 30–50 minute scanning session. These studies have been useful in pointing out functional changes occurring in the disease (Duara, Loewenstein, & Barker, 1990; Parks, Loewenstein, & Chang, 1988) as well as pointing out preserved capacities to respond to cognitive demands (Duara *et al.*, 1992). One group of investigators has shown that the patterns of brain activation during tests of memory, verbal fluency, and visual processing is similar in AD patients and elderly controls, even though the patients demonstrate hypometabolism in the resting state (Duara, Lowenstein, & Barker, 1990; Duara *et al.*, 1992). Others have shown a relative shift in temporal lobe activity in AD patients compared to controls when these subjects are imaged while undergoing a memory challenge (Miller *et al.*, 1987). A problem in each of these studies is that there is a large degree of nonspecific brain activation occurring which limits the accuracy in identifying specific regional metabolic changes linked to disease. In part, this is due to task complexity which often involves a variety of brain demands, such as visual attention and analysis, decision making, vocal response, and/or motor movement. Studies using sophisticated behavioral activation paradigms based on cognitive psychology theories show promise in providing techniques which activate specific brain areas and provide control for nonspecific effects (Kessler, Herholz, Grond, & Heiss, 1991). These advancements may allow more precise evaluation of the cognitive processing failures in AD and permit more accurate study of activation differences in very early AD and normal aging.

Vascular Dementia

The terms "vascular dementia" and "multi-infarct dementia" (MID) have often been used interchangeably to describe the dementing syndromes that arise following what are thromboembolic cerebral vascular events, hemorrhagic lesions, or hypoxic-ischemic brain changes. In each of these instances, the dementia observed in the affected patient is believed to arise when a significant amount of brain tissue has been compromised by a series of vascular insults (Tomlinson, Blessed, & Roth, 1970; Tomlinson & Henderson, 1976), although it is now clear that a single, strategically placed lesion may also give rise to pervasive cognitive and functional losses (Kooistra & Heilman, 1988; Mendez, Adams, & Lewandowski, 1989; Scheinberg, 1988; Tatemichi, 1990). As a group, the vascular dementias comprise the second most common form of dementia in the elderly (Roman 1991; Roman *et al.*, 1993). Approximately one half of the dementias in the elderly are believed to be attributable to Alzheimer's disease alone, between 12 and 20% are due to MID, and an additional 16% reflect combined AD and MID (Larsen *et al.*, 1986).

The dementia syndrome resulting from vascular disease has been difficult to define owing to the tremendous variability in neuronal injury (e.g., infarction versus demyelinization), lesion size (e.g., lacunar infarcts versus large vessel occlusions), and the location of the lesions within the brain parenchyma. Attempts to characterize the neuropsychological defects of vascular dementia are further complicated by the existence of AD pathological changes in the brains of many demented individuals with vascular disease (Blessed, Tomlinson, & Roth, 1968; Joachim, Morris, & Selkoe, 1988; Tomlinson, Blessed, & Roth, 1968, 1970) and the presence of vascular pathology or the plaques and tangles of AD in the brains of neurologically normal elderly controls (Arriagada, Marzloff, & Hyman, 1992; Bonne *et al.*, 1992; Mirra, Hart, & Terry, 1993; Roman *et al.*, 1993; Scheinberg, 1988; Tierney *et al.*, 1988).

In spite of the overlap in pathological markers, the differential diagnosis of AD and MID is important given the differences in prognosis, treatment, and genetic risk associated with the two conditions (Schellenberg *et al.*, 1992; St. George-Hyslop *et al.*, 1987; Strittmatter *et al.*, 1993). To assist in the diagnosis of vascular dementia, two different expert panels have recently published criteria for the diagnoses of ischemic vascular dementia (Chui *et al.*, 1992) and the other possible cerebrovascular dementing conditions such as multiple large infarctions (multi-infarct dementia), small vessel disease with dementia (so-called Binswanger's disease (Bennett, Wilson, Gilley, & Fox, 1990), multiple lacunar strokes, hypoperfusion, and hemorrhagic lesions (Roman *et al.*, 1993). In general, the diagnosis of vascular dementia is made if the patient's history is compatible with a vascular cause and the clinical investigation fails to reveal metabolic, infectious, toxic, or structural brain lesions, such as tumor, which can account for the dementing symptoms (Hachinski, Lassen, & Marshall, 1974). Particularly suggestive of a vascular causation is the presence of a known history of neurological symptoms or stroke, CT or MRI evidence of the same, an abrupt onset of the dementia, stepwise deterioration in function, and/or the presence of focal neurological signs on examination. A history of hypertension provides further support for the diagnosis (Kannel, Dawber, Sorlie, & Wolf, 1976; Rosen, Terry, Fuld, Katzman, & Peck, 1980). When these clinical standards are used, the pure forms of AD and MID can be distinguished with 70% to 80% sensitivity and specificity (Chui *et al.*, 1992). Combined syndromes of AD and stroke are more difficult to distinguish regardless of method, thus, even with adherence to standardized criteria the diagnostic accuracy in these situations is less than perfect. A recent prospective study of patients with primary degenerative dementia coming to autopsy indicates that as many as 21% of these cases were erroneously diagnosed as vascular dimentia (Fischer, Jellinger, Gatterbi, & Danielcyzk, 1991).

Given the lack of specificity of the clinical syndrome, it is perhaps not too surprising that the functional brain imaging studies using PET have failed to reveal a specific metabolic pattern associated with vascular dementia. In multi-infarct dementia, multiple focal metabolic deficits in the cortex are commonly seen (see Figure 4). These deficits are frequently asymmetric and variable in their distribution as one might expect given the random nature of lesions in these patients (Benson *et al.*, 1983; Frackowiak *et al.*, 1981). In addition, the degree of hypometabolism often exceeds the structural abnormality on MRI or CT imaging studies, and in some instances a deficit may appear that does not readily correspond to a visualized lesion (Duara *et al.*, 1989; Mazziotta & Phelps, 1986). Hypometabolic foci may appear distant from the area of known infarction in either the ipsilateral or contralateral cerebral hemisphere (Pappata, Tran Dinh, Baron, Cambon, & Syrota, 1987) and may be seen in the contralateral

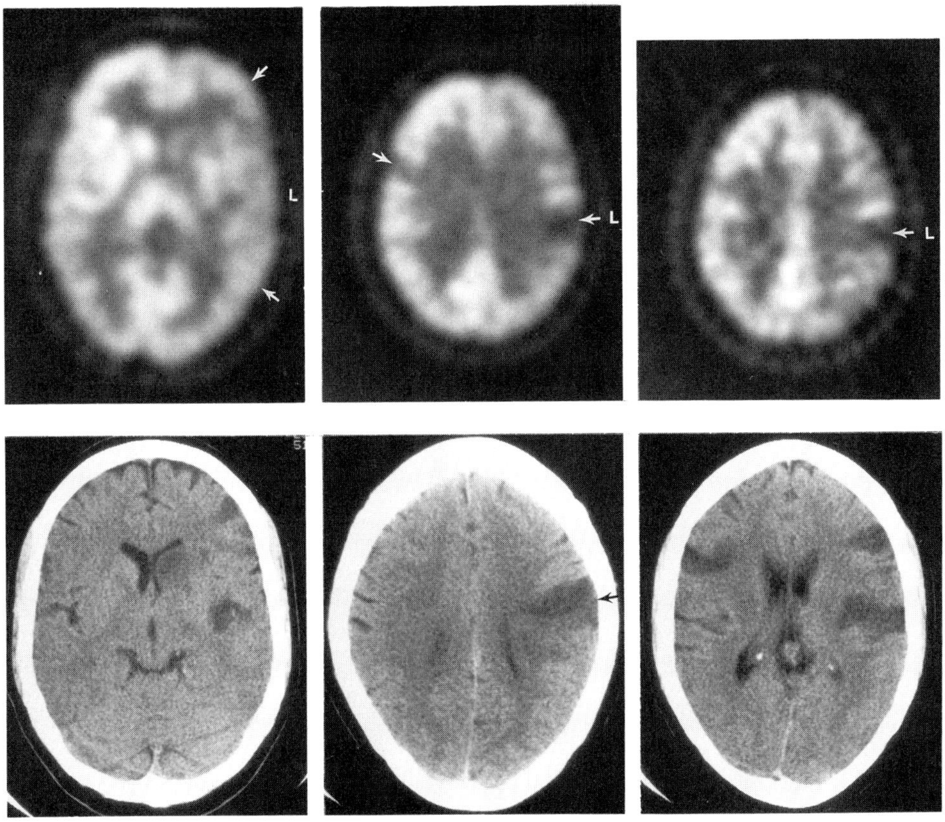

Figure 4. FDG-PET images at three levels (top panels A, B, C) and the corresponding CT scan images (bottom panels D, E, F) in a patient with multi-infarct dementia. In this particular patient there is reduction in FDG uptake throughtout the (A) left frontal and temporal parietal regions (arrows). On image B there is a focal reduction of FDG uptake in the left parietal and right frontal areas (arrows) corresponding to cortical infarctions, visualized on CT images D and E. At a higher image level on the FDG-PET study (C) reduction of FDG uptake in the left parietal area is noted (arrow). Corresponding CT scan (F) shows a hypodense area (arrows).

cerebellar hemisphere (Baron, Bousser, Comar, & Castaigne, 1980). For example, the cerebral cortex may be hypometabolic following an infarction to the ipsilateral thalamus (Baron *et al.*, 1986; Kuhl *et al.*, 1980). The visual cortex has been reported hypometabolic following lesions to the optic radiations (Heiss *et al.*, 1982), and the cerebellar cortex may be hypometabolic contralateral to a supratentorial lesion (Baron *et al.*, 1980; Pappata *et al.*, 1987). This lack of one-to-one correspondence between MRI structural abnormalities and PET functional deficits is believed to reflect dysfunction in anatomically interconnected brain areas (Kuhl *et al.*, 1980; Pappatas *et al.*, 1987). The effects of focal brain injury no doubt include not only local disruption of function but also distant effects through "disconnection" of brain regions interconnected with the injured site (Absher & Benson, 1993; Geschwind 1965; Leuchter *et al.*, 1992). In addition, transsynaptic effects cannot be ruled out in distant sites resulting in structural changes visualizable only at the microscopic level (Pantano *et al.*, 1986).

Comparisons of demented patients with MID and patients with AD have also failed to reveal a distinctive pattern of brain metabolism deficit distinguish-

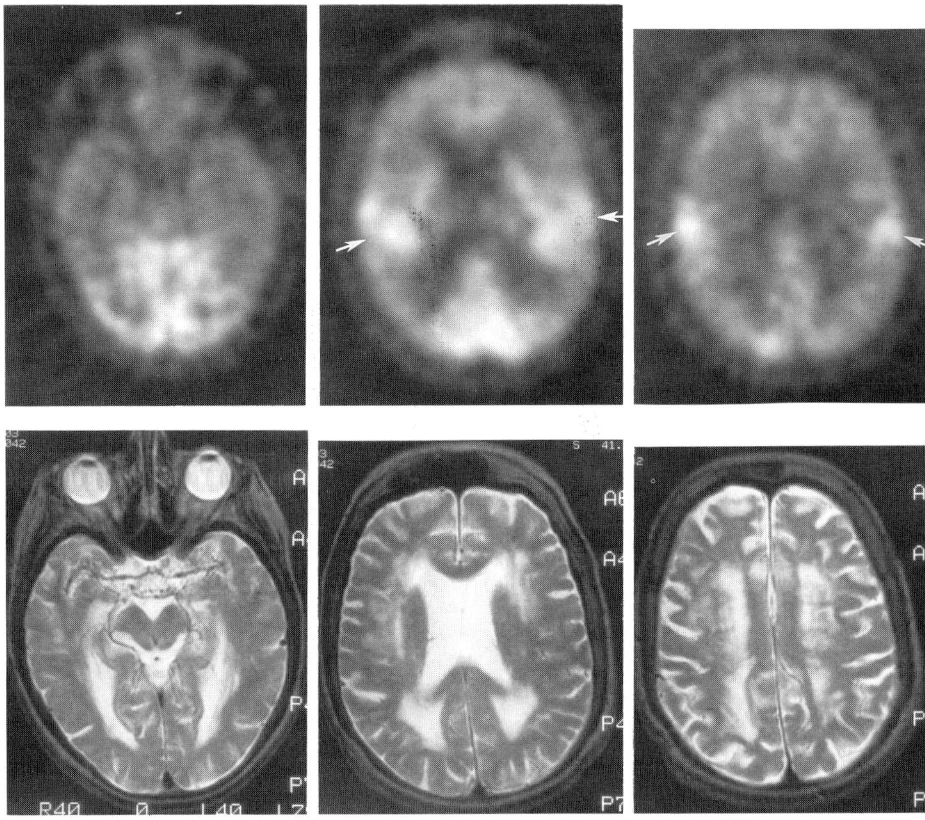

Figure 5. Patient with Binswanger's disease. The top three images are those of the FDG-PET study obtained at the level of the temporal lobe (A), the level of the basal ganglia (B), and in the high parietal area (C). Note significant reduction in FDG uptake in the temporal, frontal, and parietal regions. There is sparing of the sensimotor strip bilaterally (arrows). The pattern of FDG uptake is essentially identical to that obtained in Alzheimer's disease. MRI studies obtained at the same corresponding levels (D, E, F) show extensive high signal abnormalities on the T 2-weighted images corresponding to small vessel changes.

ing these patients groups from one another. In one study, PET imaging proved reasonably sensitive in identifying both Alzheimer's disease (92%) and multi-infarct dementia (87%), however, the specificity was poor (54%). The distinctions between the two disorders was considered "ambiguous at best" (Duara, Barker, Loewenstein, Pascal, & Bowen, 1989). This overlap between the conditions can be appreciated in Figure 5. Presented here are the imaging studies of a patient with Binswanger's disease. The PET imaging study shows reduced glucose metabolism in the temporal, frontal, and parietal lobes bilaterally along with a sparing of the sensorimotor strips bilaterally. The pattern is essentially identical to that of Alzheimer's disease. Yet, MRI studies obtained at the same corresponding levels reveal extensive small vessel changes. Consideration of the PET imaging study alone in this case would lead to a diagnostic conclusion of Alzheimer's disease. However, the history indicated a stepwise course with documented events suggesting strokes, and the MRI supports a vascular component. Thus, although Alzheimer's disease cannot be excluded, the findings suggest that the metabolic abnormality on PET may be secondary to extensive vascular disease.

This case illustrates the importance of considering both the structural and functional brain imaging studies in the diagnosis of dementing disorders.

Subcortical Dementias

In addition to Alzheimer's disease and vascular disease, there are a host of less frequently encountered neurological disorders which result in dementia (Reichman & Cummings, 1990). The so-called subcortical dementias represent the next most frequent group of disorders affecting mentation in middle to late adulthood. Subcortical dementia is not a disease but rather a clinical syndrome characterized by a common pattern of neuropsychological impairments and by primary involvement of brain areas in the rostral brain stem, thalamus, basal ganglia, and their interconnections with the frontal lobe (Cummings, 1990). Typically, patients present with both behavioral and neuropsychological changes including bradyphrenia, depression or other personality change, and deficits in memory, visuospatial skills, sustained attention, and higher-order executive function (e.g., switching response set). These patients rarely suffer from the language disturbance, apraxia, and visual recognition (agnosia) problems characteristic of Alzheimer's disease (Folstein, Brandt, & Folstein, 1990).

Encompassed in the subcortical dementias are a variety of diverse brain disorders, the most common of which are Huntington's disease, Parkinson's disease, and progressive supranuclear palsy. A more recently recognized clinical entity, diffuse Lewy body disease (Yoshimura, 1983), may also be considered a subcortical dementia, although it includes both cortical and subcortical features (Gibb, 1986; Kosaka, Yoshimura, Ikeda, & Budka, 1983) and is often seen within the context of Alzheimer's disease (Rosenblum & Ghatak, 1979). Amyotrophic lateral sclerosis (ALS) may have dementia as an associated feature (Hudson, 1981), however, this occurs rarely and not usually until the late stages of the disorder (Reichman & Cummings, 1990). Other less frequently occurring subcortical conditions include Wilson's disease, idiopathic basal ganglia calcification, thalamic degeneration, and spinocerebellar degeneration syndromes (Cummings, 1990). It should be mentioned that two very common conditions, multiple sclerosis and AIDS-dementia complex, are also considered subcortical dementias (Rao, 1990; Navia, 1990). However, because these two disorders are more prevalent in young-to-middle adulthood rather than middle-to-late adulthood they are not considered here in this review. In general, we have limited the discussion to dementing disorders with average age of onset after age 40. Only Huntington's disease, a disorder with onset in the third or fourth decade of life, is included. Because this condition is well studied from a neuropsychological perspective and often serves as the prototype of the subcortical dementias, it is included as an illustration of the functional imaging changes typical of subcortical disorders.

Huntington's Disease

Huntington's disease (HD) is a genetically transmitted disorder associated with a defect on the distal end of the short arm of chromosome 4 (Gusella *et al.*, 1983; Huntington's Disease Collaborative Research Group, 1993) and is characterized by neurodegenerative changes in the caudate and putamen and their projection areas (Penney & Young, 1988). It affects approximately 5 to 10 individuals per 100,000 (Penney & Young, 1988), although estimates have gone as high as 12 per 100,000 adults (Folstein, Brandt, & Folstein, 1990). The disease follows a progressive course and is invariably fatal. In the early stages, changes in motor, cognitive, and behavioral functions are often subtle (Penney & Young, 1988). However, as the disease advances, personality changes are often reported and the extrapyramidal signs characteristic of the disorder (e.g., involuntary and voluntary motor movements, dysarthria) as well as cognitive deficits emerge.

Numerous neuropsychological studies have been devoted to characterizing the cognitive and behavioral effects of HD (for review, see Folstein, Brandt, & Folstein, 1990). Cognitive deficits have been described in aspects of memory, specifically in the retrieval of new information on demand (Salmon, Kwo-on Yuen, Heindel, Butters, & Thal, 1989); in frontal lobe functions, such as sustained attention, planning, organization, and verbal flexibility (Butters, Sax, Montgomery, & Tarlow, 1978); and in visuomotor integration and visuospatial cognition (Josiassen, Curry, & Mancall, 1983). Spared in Huntington's disease are the apraxia (Brouwers, Cox, Martin, Chase, & Fedio, 1984) and the language problems characteristic of AD (Podoll, Caspary, Lange, & Noth 1988). In addition, the memory problems are unique in the two conditions. Patients with HD do not show the severe encoding and storage deficits of AD nor do they show the loss of semantic representations or intrusional tendencies of AD on tests of memory (Butters, Granholm, Salmon, Grant, & Wolfe, 1987; Butters, Wolfe, Martone, Granholm, & Cenmak, 1985).

In the noncognitive realm, involuntary movements (i.e., choreiform movements; Thompson *et al.,* 1988) and voluntary motor disorders such as bradykinesia, akinesia, slowed motor movements, and/or clumsiness (Bradshaw *et al.,* 1992; Hefter, Hamberg, Lange, & Freud, 1987) are defining features of Huntington's disease. Difficulties in affective regulation and other personality changes are also common (for review, see Folstein, Brandt, & Folstein, 1990). Depression is frequent and often precedes the cognitive and motor problems of the illness (Folstein, Abbott, Chase, Jensen, & Folstein, 1983). Apathetic and inert states may occur later in the illness, and in some instances, explosive irritable behavior and mania are noted.

The neural systems mediating the disease symptoms are also becoming understood. On neuropathological study of affected brain tissue, changes are uniformly observed in the caudate and putamen and are also seen in the lateral and medial globus pallidus, the substantia nigra, the ventral pallidum, areas of the frontal cortex, the thalamus, and the subthalamic nucleus (Von Sattel *et al.,* 1985). Neuroimaging studies (CT and MRI) show atrophy of the caudate nucleus and a loss of definition between the lateral edge of the caudate and the adjacent lateral ventricle (Starkstein *et al.,* 1989). Cortical atrophy is also detectable as the disease progresses, particularly in the frontal lobe (Starkstein *et al.,* 1989). Additional changes have also been noted, albeit inconsistently, in the temporal, parietal, and occipital lobes (for review, see Martin *et al.,* 1992).

Positron emission tomography studies have been shown to be more sensitive than structural imaging studies in identifying the neural deficits of HD. Striatal hypometabolism is characteristic of the disease (Kuhl, Metter, Riege, & Markham, 1984), even in the very early stages (Garnett, Firnau, Nahmias, Carbotte, & Bartolucci, 1984; Grafton *et al.,* 1990, 1992; Hayden, *et al.,* 1987; Mazziotta, *et al.,* 1987). In addition, a series of studies demonstrate cortical abnormalities in the disease, primarily in the frontal cortex and in mesial structures, such as the cingulate cortex (Kuwert *et al.*, 1990; Martin *et al.*, 1992). A typical PET imaging study in the disease is presented in Figure 6. This study was conducted in a 29-year-old male who had a strong family history of Huntington's disease and who was himself symptomatic for 4 years at the time of the imaging study. The study demonstrates reduced metabolism in the heads of the caudate nuclei bilaterally, as well as in the globus pallidus and putamen.

In a recent study the presence of impaired cortical function early in disease was explored. In this study, hypometabolism was indeed seen in the frontal cortex and also in the inferior parietal areas in HD patients with mild symptoms (defined as disease symptoms for less than 5 years) when compared to appropriate age-matched controls (Martin *et al.*, 1992). In clinically more advanced patients, hypometabolic changes were evident throughout the neocortex, except in the temporal lobes. Some studies suggest that these changes, particularly the striatal hypometabolism, may serve as an early indicator of disease. In PET studies conducted in persons at risk of the disease by virtue of their genetic history, a subset of patients have been identified with reduced striatal metabolic rates but no clinical signs of disease at the time of scanning (Mazziotta *et al.*, 1987; Grafton *et al.*, 1990, 1992). In these studies, a few of these at-risk individuals with striatal hypometabolism, but none of those with normal metabolism, have developed the disease over time. The loss of striatal glucose metabolism is particularly notable in individuals who screen positive for genetic markers on chromosome 4, suggesting that neural changes are specific to the disease and

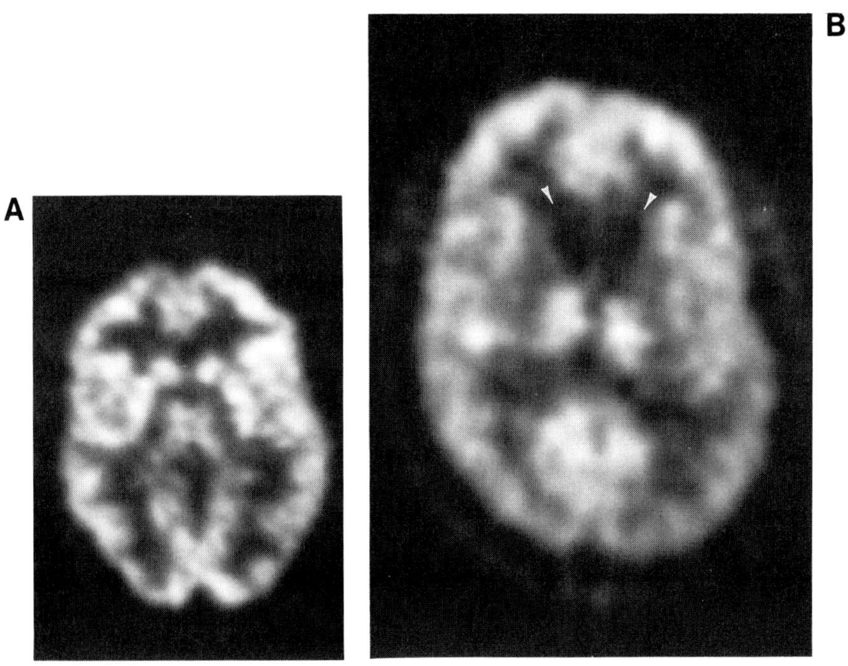

Figure 6. Normal control FDG-PET study (A) at the level of the basal ganglia shows normal FDG uptake in the caudate nuclei bilaterally. (B) FDG-PET study in an individual with Huntington's disease. Note the absence of metabolism in the head of the caudate bilaterally (arrows).

may be occurring for an extended period of time prior to the onset of symptoms (Grafton *et al.*, 1990, 1992).

Thus, the results suggest that PET may be able to detect changes before obvious morphological changes can be visualized on MRI or CT. The findings also suggest that PET methodology may be able to identify people who are harboring the disease but are not yet symptomatic. Although the method appears to be quite sensitive, more longitudinal data is needed to rule out false positives. It should also be noted that the striatal changes reported with FDG-PET are not specific to HD. Reduced striatal metabolism is seen with other basal ganglia disorders including benign heredity chorea (Suchowersky *et al.*, 1986) and Lesch-Nyhan syndrome (Palella *et al.*, 1985). In addition, there are some reports that have failed to demonstrate any striatal hypometabolism in subjects at risk of HD (Young *et al.*, 1987) and another study which reports an individual with a positive PET scan but no DNA marker of HD (Hayden *et al.*, 1987). Thus, PET imaging results alone cannot be considered diagnostic of Huntington's disease or any other striatal disorder for that matter. Since other features of the clinical history and presentation allow easy differentiation between these various movement disorders, consideration of the PET imaging findings in isolation would be foolish. It may be useful, as some have suggested, to consider DNA and PET studies together when evaluating individuals at risk for Huntington's disease (Hayden *et al.*, 1987; Jolles, Chapman, & Alavi, 1989).

Parkinson's Disease

Idiopathic Parkinson's disease (PD) or the "shaking palsy" is an extrapyramidal movement disorder which arises usually between ages 40 and 70 years and affects approximately 1 adult in 1,000 (0.1% of the population) (Kessler, 1978). The cardinal features of the disease are bradykinesia, tremor, and rigidity which result from dysfunction of the nigrostriatal dopamine system (Bernheimer, Birkmeyer, Hornykiewicz, Mellinger, & Seitelberger, 1973). Neuropathological findings reveal a loss of the pigmented cells within the substantia nigra and within other pigmented brain stem nuclei. Lewy bodies, the neuropathologic signature of the disease, are abundant in these brain areas in almost all instances of the disease (Freedman, 1990).

Cognitive impairments are seen in many patients with Parkinson's disease, although the frequency and etiology of "dementia" in the disease have been controversial. Current estimates based on a review of the literature suggest that measurable deficits in memory, visuospatial integration, abstraction, and/or central processing time (i.e., bradyphrenia) occur in approximately 60% of the cases, but estimates vary from 4% to 93% (Mahler & Cummings, 1990). The variability in estimates is due in part to differences in dementia criteria applied across studies and in part to disease heterogeneity. For example, it is now recognized that Parkinson's disease brain changes can occur in isolation, in combination with AD (Cummings & Benson, 1984; Freedman, 1990; Rinne, Rummukainen, Paljarvi, & Rinne, 1989), or with diffuse Lewy bodies throughout the neocortex (Mahler & Cummings, 1990; Xuereb *et al.*, 1990). Because dementia is a part of both AD and diffuse Lewy body disease, the dementia in Parkinson's disease may be referable to the cortical and/or subcortical changes associated with these diseases and not due to the nigrostriatal changes of PD (Hakim & Mathieson, 1979; Perry *et al.*, 1978; Yoshimura, 1983). Although the issue remains under investigation, there is some evidence that the brain-behavior changes associated with Parkinson's disease are neuropsychologically and neuropathologically distinct from those of Alzheimer's disease (Mahler & Cummings, 1990).

Functional brain imaging studies using PET have also shown different patterns of glucose utilization and metabolism in patients with PD and those with AD. In contrast to the diminished metabolism localized to the parietal and temporal lobes in AD (see previous discussion for review), patients with PD show diffusely diminished metabolism throughout the cerebral cortex (Kuhl *et al.*, 1984; Peppard *et al.*, 1992) and in specific regions of the frontal lobe (for review, see Metter, Kuhl, & Riege, 1990). Unlike the subcortical dementia of Huntington's disease, striatal changes are not commonly reported (Kuhl *et al.*, 1984). There are some reports, however, that in nondemented patients with Parkinson's disease, a slight hypermetabolism may occur in the basal ganglia 48 hours after withdrawal of antiparkinsonian medications (Rougemont *et al.*, 1984). In another report, hypermetabolism was reported in cases of hemiparkinsonism contralateral to the symptomatic side (Martin *et al.*, 1984).

It should be noted that none of the changes visualized on PET imaging are an absolute defining feature of PD. As stated above, changes in striatal metabolism are not the common experience (Kuhl *et al.*, 1984). Other studies report that in severe cases of PD with dementia, the pattern of PET metabolism may be indistinguishable from that seen in AD (Rougemont *et al.*, 1984; Peppard *et al.*, 1992). This observation is consonant with our own experience as illustrated in Figure 7. Presented here are FDG-PET images for a patient with idiopathic Parkinson's disease and prominent dementia. In this case there are reductions in cerebral metabolism in the temporal, parietal, and frontal areas in a pattern identical to that of Alzheimer's disease.

The future of PET imaging in Parkinson's disease will likely incorporate presynaptic dopaminergic ligands labeled with positron-emitting radionuclides. The availability of numerous dopaminergic compounds is a distinct advantage for the study of Parkinson's disease. If these compounds can be developed with properties appropriate for PET studies, then their application would allow study of receptor binding and neurochemical changes specific to Parkinson's disease. Studies of potential compounds have been conducted in normal adults and there are a few reports in patients with Parkinson's disease (Hoffman, Hanson, & Coleman, 1993; Perlmutter, 1988). The preliminary evidence is encouraging and suggests decreased dopamine receptor binding in PD patients (Perlmutter, 1988) and decreased dopamine precursor (^{18}F-6-fluorodopa) uptake (Brooks *et al.*, 1992).

Progressive Supranuclear Palsy

Progressive supranuclear palsy (PSP) or Steele, Richardson, Olszewski syndrome, is a disorder often confused with Parkinson's disease because of some shared subcortical features. Like Parkinson's disease, PSP is a chronic, progressive condition with extrapyramidal motor system involvement and dementia. However, the illness includes other features, notably supranuclear opthalmoplegia affecting primarily vertical gaze, pseudobulbar palsy, and axial dystonia (Richardson, Steele, & Olszewski, 1963). Neuropathologically the illness is also different from Parkinson's disease. Neuronal degeneration is seen in the superi-

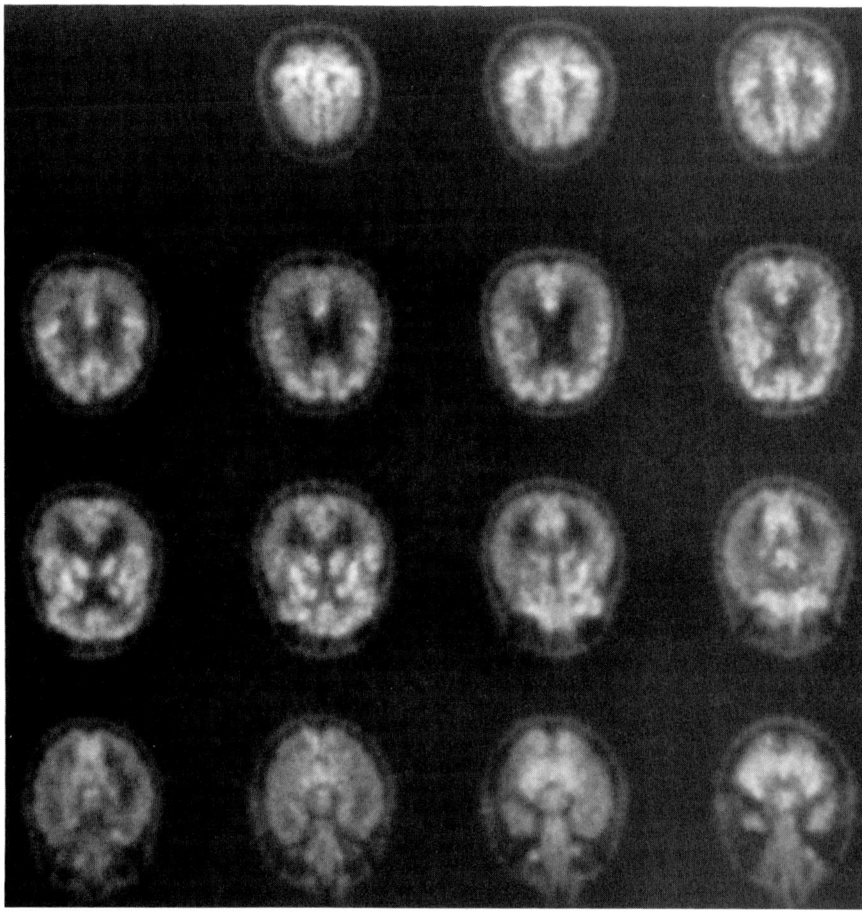

Figure 7. FDG-PET images of the entire intracranial contents obtained in a patient with Parkinson's disease and dementia. Note the similarity between this study and that of Alzheimer's disease. In Parkinson's disease dementia the FDG uptake in the basal ganglia is typically normal. However, there is often a reduction metabolism within the temporal, parietal, and frontal areas, a pattern observed in Alzheimer's disease.

or colliculus, pontine tegmentum, substantia nigra, subthalamic nucleus, and pallidum, and there is a preponderance of argyrophilic neurofibrillary tangles and gliosis within these neural structures. The cerebral and cerebellar cortices are usually spared (Duvoisin, Golbe, & Lepore, 1987). Diagnosis of the disorder requires the presence of the characteristic syndrome and, most importantly, the supranuclear opthalmoplegia which sometimes does not appear until late in the illness or in some rare instances not at all (Davis, Bergeron, & McLachlan, 1985; Jellinger, Riederer, & Tomonga, 1980). In the latter instances, PSP may be suspected but requires autopsy confirmation.

Attempts to facilitate early diagnosis have turned to morphometric analyses of suspected brain structures using CT or MRI. Several studies confirm the expected atrophy in the midbrain tegmentum, superior colliculi, and pons and reveal more generalized cerebral atrophy, particularly in the frontal and temporal lobes. One attempt to characterize the early changes associated with the disease suggests specific atrophy confined to the midbrain when early stage patients are compared to appropriately age-matched controls (Schonfeld, Golbe, Sage, Safer, & Duvoisin, 1987). As the disease progresses, this atrophy becomes more evident and dilatation of the ventricles along with atrophic changes in the temporal lobe appears.

Although PSP is a relatively rare condition, PET imaging studies in the disorder have increased steadily over recent years (Baron *et al.*, 1985; Bhatt, Snow, Martin, Peppard, & Calne, 1991; D'Antona *et al.*, 1985; Goffinet *et al.*, 1989; Karbe *et al.*, 1992). One study using ^{18}FDG-PET revealed a global reduction in cerebral metabolism but particularly prominent hypometabolism in the frontal cortices (D'Antona *et al.*, 1985). Later investigations confirmed this finding and reported selective hypometabolism in the motor and premotor regions of the frontal code. By contrast, the frontal association cortices and the paralimbic regions were relatively less affected (Goffinet *et al.*, 1989). The extensive involvement of the frontal lobe in PSP and the regional pattern of hypometabolism observed differs from the metabolic patterns reported in Parkinson's disease dementia (Metter, Kuhl, & Reige, 1990; Peppard *et al.*, 1992). Other differences between the two disorders have been suggested by studies using labeled dopamine precursors (6-^{18}F-L-dopa) and PET to examine presynaptic dopa uptake (Bhatt *et al.*, 1991). In addition, one study reports a decrease in glucose metabolism in subcortical regions (caudate, putamen, and brain stem) in PSP, whereas no such alteration was seen in PD (Karbe *et al.*, 1992). Together these observations suggest that PET imaging studies may be useful in distinguishing Parkinson's disease from progressive supranuclear palsy in the early stages, in spite of their clinical similarity.

Other Cortical Dementias

Two other dementias, Pick's disease and Creutzfeldt-Jakob disease, are worthy of note. Both are considered cortical dementias, and although they occur relatively infrequently in comparison to Alzheimer's disease, their clinical presentations can be quite similar to AD.

Pick's Disease

Pick's disease is a progressive dementing disorder which is often distinguished from AD by a prominent involvement of the frontal or temporal lobes. Dramatic changes in personality, judgment, and impulse control within the context of variable memory impairment and the absence of visuospatial problems are frequently described in the dementias of the frontal lobe type (Neary, Snowden, Northen, & Goulding, 1988; Knopman *et al.*, 1989). Relatively isolated language deficits have been reported with Pick's disease of the temporal lobe (Holland, McBurney, Moossy, & Reinmuth, 1985; Graff-Radford *et al.*, 1990). The disease tends to have an earlier age of onset than AD, peaking at about age 58 and declining thereafter (Heston & Mastri, 1982). Because of the similarity between the two disorders, the diagnosis of Pick's disease can only be made with any certainty at autopsy. Macroscopically, the disease typically shows circumscribed "lobar" atrophy confined to the frontal and temporal regions (Salmon & Franck, 1989). Neuropathologically, swollen neurons and neurons containing the classic Pick's inclusion bodies are seen throughout these restricted cortical areas and may be seen in other brain areas, often in the basal ganglia and thalamus (Kamo *et al.,* 1987).

There have been several PET studies using FDG in cases of suspected Pick's disease (Chase, Burrows, & Mohr, 1987; Chawluk *et al.,* 1986; Salmon & Franck, 1989) and in rare instances of autopsy-confirmed Pick's disease (Kamo *et al.,* 1987). Although the patients' symptoms varied across these studies, the findings of PET imaging uniformly showed focal hypometabolism in cortical regions corresponding to the patients' respective deficits. In patients presenting with isolated aphasias, pronounced reductions were reported in left hemisphere structures (Chawluk *et al.,* 1986; case 3 in Salmon & Franck, 1989), and in other cases with frontal lobe syndromes sharp decreases in frontal metabolic rates were seen (Chase *et al.,* 1987; Kamo *et al.,* 1987; cases 1 and 2 in Salmon & Franck, 1989). Although it is readily acknowledged that similar dysfunctions can be seen in AD (Haxby *et al.,* 1990; Loewenstein *et al.,* 1989), the important difference between Pick's disease and AD is the sharp restricted nature of the hypometabolism in Pick's disease and the tendency for this dysfunction to be localized within the frontotemporal areas and not within the temporoparietal areas (Kamo *et al.,* 1987). An example of the type of FDG-PET imaging findings seen in Pick's disease is presented in Figure 8. This study was conducted in a patient present-

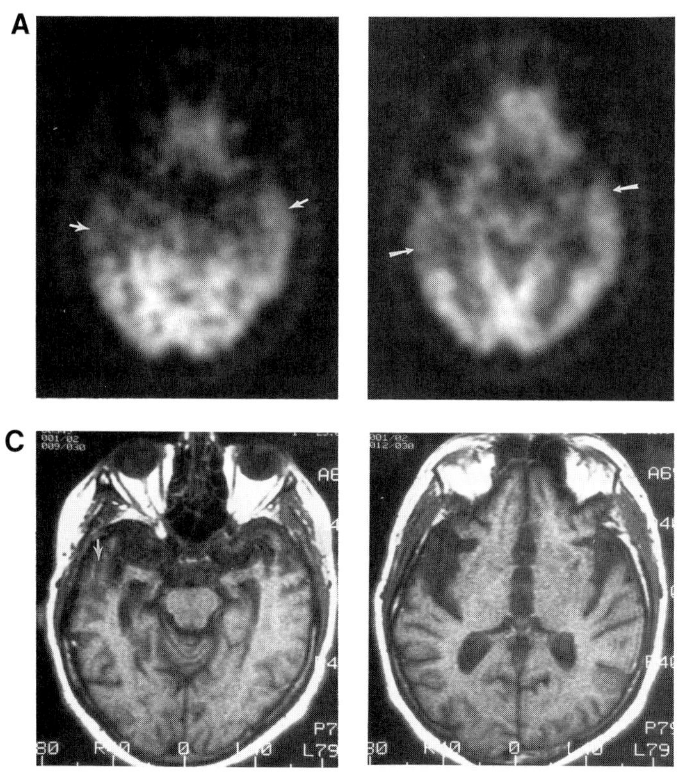

Figure 8. FDG-PET images in a patient with the clinical diagnosis of Pick's disease. The imaging plane at the level of the temporal lobes (A) shows marked reduction in the FDG uptake bilaterally with the right involved more so than the left (arrows). The corresponding MRI study at this level (C) shows marked atrophy on the right arrow). at a slightly higher level (D) there is extensive reduction of FDG uptake throughout the more anterior aspects of the tempooral lobes (arrow). On the corresponding T1 MRI images there is significant atrophy in the emporal lobes (D).

ing with personality change and memory deficits. His condition was diagnosed clinically as Pick's disease. Other conditions such as frontotemporal degeneration or a frontal variant of Alzheimer's disease cannot be excluded without pathological verification of Pick's disease. Regardless, this figure illustrates the marked temporal lobe hypometabolism which can be seen in Pick's disease. The corresponding MRI imaging studies illustrate lobar atrophy in the temporal lobes bilaterally, also typical of the disease.

Creutzfeldt-Jakob Disease

Creutzfeldt-Jakob disease (CJD) is an exceedingly rare dementing disorder which is transmitted through slow infectious agents known as "prions" (Prusiner, 1982). The disease is distinguished from AD and other dementias by its rapid progression, with the typical interval from diagnosis to death usually not exceeding 3 years (Roos, Gajdusek, & Gibbs, 1973; Masters & Richardson, 1978). Interest in the illness has been kindled recently by observations of biological similarities between AD and CJD and the co-occurrence of the condition in some families, suggesting a genetic contribution to both conditions (Brown *et al.*, 1992; Harrison & Roberts, 1991).

PET imaging studies in the disease are few in number. One study reported patchy deficits dissimilar to those seen in AD (Horowitz, Benson, Kuhl, & Cummings, 1982). Another report in a severely demented individual witch CJD reports hypometabolism in the temporal lobes bilaterally, with greater hypometabolism on the left (Friedland, Pruisner, Jagust, Budinger, & Davis, 1984). We have had the opportunity to study one case in the early stage of illness (see Figure 9). This 57-year-old male presented with a pronounced complex visual system disturbance and eventually underwent surgical biopsy because of the aggressive course of his disease. PET imaging studies were completed when this patient was still testable. Revealed were marked reductions in FDG uptake bilaterally in the temporal lobes and parietal association cortices. By contrast, the corresponding MRI imaging study showed only mild atrophy in these areas. In this case, we observed nothing different between these images and those obtained in patients with AD. Thus, although the results in Creutzfeldt-Jakob disease must be considered preliminary until more cases are examined, the findings available to date suggest that PET neuroimaging is not likely to play a diagnostic role in the illness. However, PET imaging may be helpful in confirming the disease and in underscoring the pathophysiology of the condition.

Future Directions of PET in the Diagnosis of Dementia

Limitations of PET in Differential Diagnosis

It would be ideal if the patterns of PET hypometabolism observed in the various diseases were invariant signatures of each, allowing one to differentiate between the various disease entities. Unfortunately, this is not the case. A number of studies have examined the sensitivity and specificity of the PET method in the diagnosis of AD. The data indicate that PET is very sensitive to AD patterns (Duara *et al.*, 1989) and may be able to detect metabolic patterns associated with the disease even when the clinical diagnosis is uncertain (Kuhl *et al.*, 1987). However, the specificity of PET in distinguishing AD from normal aging and other forms of dementia is poor. As already discussed in vascular dementia and

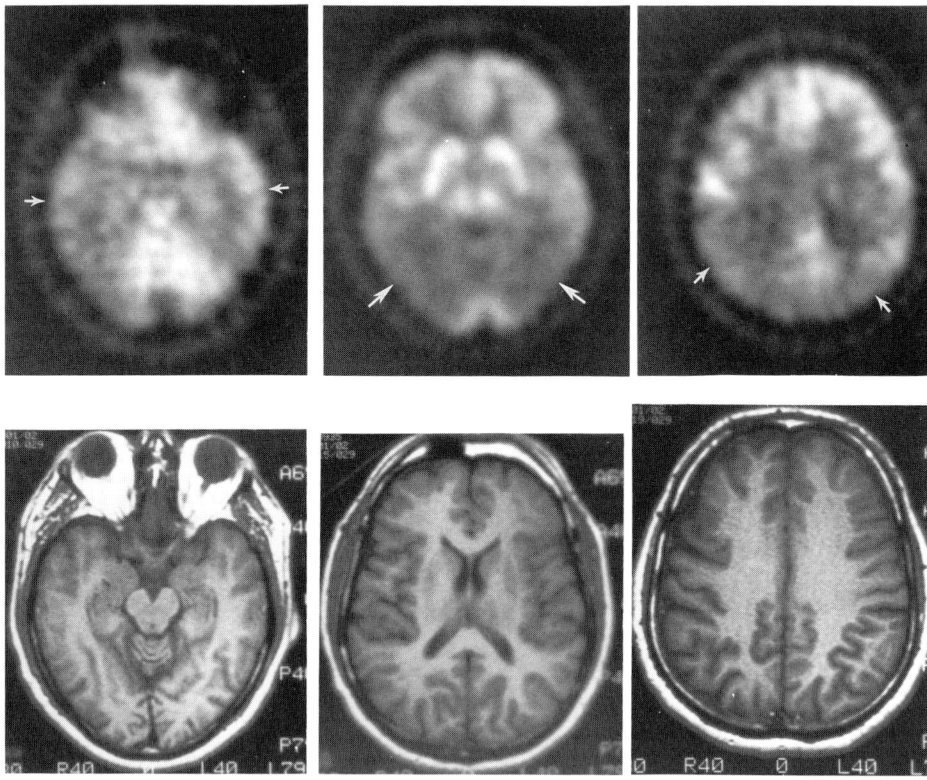

Figure 9. Patient with biopsy-proven Creutzfeldt-Jakob disease. FDT-PET images at three levels (A, B, C) with corresponding T 1 MRI images (D, E, F) are shown. At the level of the temporal lobes (A) there is reduced uptake bilaterally (arrows). At the level of the basal ganglia (B) there is reduced FDG uptake primarily in the parietal association cortex (C, arrows). The corresponding MRI images (D, E, F) show only mild atrophy.

in Parkinson's disease, a pattern resembling that seen in AD can be observed in either of these disorders. The problem of PET specificity is highlighted by the findings of Duara and colleagues (1989) in their evaluation of AD, MID, and normal aging. Although the PET methodology was sensitive in detecting AD and MID, a surprisingly high number of normal elderly individuals studied had abnormal PET findings (54%). A more recent study comparing the diagnostic accuracy of PET in Alzheimer's disease, Parkinson's disease without dementia, and normal aging also indicated that reliance on the classic AD pattern of bilateral temporoparietal hypometabolism for AD diagnosis can dramatically reduce the sensitivity of detection (Powers *et al.*, 1992). Taken together the findings suggest that to the extent that symptoms in disease entities can overlap, PET imaging results will also overlap. Although the method may reliably distinguish between large groups, it may fail in individual cases. As a consequence, reliance on PET imaging data alone as a diagnostic marker test for AD is likely to result in false positive errors and may also miss cases if the so-called classic AD pattern is relied upon too heavily.

The lesson learned with PET imaging as a diagnostic procedure is not a new one. A similar situation has occurred in other areas of medicine. For example, the dexamethasone suppression test (DST) showed promise as a diagnostic test for major depressive disorders (Carroll *et al.*, 1981); however, when put to the clinically challenging test of distinguishing the range of psychiatric conditions (including AD) from depression, the diagnostic accuracy of the test was substantially reduced (Nierenberg & Feinstein, 1988). This observation has led some to suggest that potential diagnostic test procedures, which PET may be considered, should undergo the same rigorous sequential phases of evaluation as investigational drug compounds, thereby assuring efficacy of the test before its widespread clinical use (Nierenberg & Feinstein, 1988). Required is a testing of the method first in typical instances of disease, followed by expanded study of less typical disease cases, including the cohort typical to clinical practice, patients with a host of comorbid illnesses, medications, and prior medical illnesses. The PET methodology has not yet passed this ultimate test.

Future Directions with PET in Clinical Diagnosis

In spite of its limitations, PET continues to be a valuable tool in both research and clinical studies. It provides information about brain function *in vivo* with a degree of sensitivity not possible with other imaging modalities such as SPECT, CT, or conventional MRI scanning (for review, see Jagust & Eberling, 1991). Functional MRI (Kwong *et al.*, 1992; Stehling, Turnera, & Mansfield, 1991) shows an even higher degree of sensitivity but is not yet widely available, and its properties are not as well understood in normal aging and disease states. Thus, studies using PET to better understand the neural mechanisms of behavior and the biology of diseases such as Alzheimer's disease continue with vigor.

In advancing the clinical diagnosis of AD and other dementias, there are several promising avenues of investigation worthy of note. One avenue already discussed is the research using *brain activation* in AD patients. This approach has been likened to the "stress test" used in other areas of medicine such as cardiology (Duara *et al.*, 1992) and attempts to discriminate patients by maximizing the metabolic differences between groups. The studies to date have been few in number and the data are inconclusive. Despite this, the approach has demonstrated utility and has successfully highlighted the brain's surprising reserve capacity to respond to cognitive challenges even when already compromised by AD (Grady *et al.*, 1988; Satz, 1993; Stern, Alexander, Prohovnik, & Mayeux, 1992). With the application of more specific cognitive tasks in the future a higher degree of discrimination between AD and normal aging may be possible. A related area which may advance future diagnosis is the use of *pharmacological challenges* to the human cholinergic system, one of the main neurochemical systems deficient in AD. This approach of developing PET imaging compounds to specific neurochemical systems has been successful in Parkinson's disease (Hagglund *et al.*, 1987; Leenders *et al.*, 1986; Rutgers, Lakke, Panns, Vaalburg, & Korf, 1987; Tedrroff *et al.*, 1988). The compounds developed to date for AD, such as labeled scopolamine, have been tested in normal humans with disappointing results (Frey *et al.*, 1992). Thus, in principle the technique is possible; however, alternative ligands are needed. This is a subject of continued investigation.

Another area with potential diagnostic utility is the use of statistical modeling procedures which take into account patterns of metabolic changes which may predict disease. One such study was able to correctly classify 87% of the mild/moderately demented patients with AD and normal elderly controls (Azari *et al.*, 1993). The model also correctly identified the scan of an individual at risk of AD who had ambiguous symptoms at scanning but who one year later developed clear symptoms of the disease. Other statistical methods which examined changes in isolated regions (and did not take patterns of change into account) misclassified this scan as normal (Pietrini *et al.*, 1991). Additional studies examining disease-specific PET patterns indicate that subjective ratings of brain changes (Powers *et al.*, 1992) are unreliable and that statistical quantitation of change is essential, particularly models which are either multivariate in nature or employ regional covariance to take into account regional interactions (Clark *et al.*, 1985; McCrory & Ford, 1991; Moeller & Strother, 1991). Together these findings suggest that useful diagnostic algorithms might be constructed using quantitative measures of regional changes in FDG-PET metabolism. We have suggested that the incorporation of non-PET variables, such as neuropsychological or MRI volumetric measures, into these statistical models may further increase predictive power (Welsh, Hoffman, & Beam, 1994). We have also suggested that the inclusion of regional MRI volumetric measures, such as those used by Bigler and coworkers (see Naugle & Bigler, 1989, for review), may enhance diagnostic predictions (Welsh *et al.*, 1993). In essence, the procedure suggested mimics the diagnostic process in which the clinician weighs all available information (the history of illness, clinical findings, and MRI and PET imaging results) before arriving at a diagnosis.

With advances in imaging technology and in our understanding of PET changes in dementia, increased diagnostic precision in the evaluation of dementia may be achieved. Until that time, however, reliance on PET imaging results alone is not sufficient for the differentiation of dementia syndromes and is, by today's standards, ill advised. If used judiciously, PET imaging has a place in clinical diagnosis. Whereas in most clinical situations PET imaging is not necessary for the diagnosis of dementia, PET studies can be helpful in diagnostically difficult cases (Hayden *et al.*, 1986; Rapoport, 1991; Hoffman, Welsh, Hanson, Earl, & Coleman, 1992). For example, if a pattern of biparietal deficits is observed and the clinical history indicates cognitive and functional decline, a nonfocal neurological examination, and a normal CT or MRI study, then the results of PET imaging may be considered as supportive of the diagnostic impression of AD. Atypical scans obtained within the same context (e.g., normal metabolism, frontal hypometabolism, striatal hypometabolism, etc.) indicate consideration of other diagnostic possibilities.

Conclusions

The results of resting state FDG-PET imaging studies within the various dementias suggest patterns of metabolism specific to each disease entity. In Alzheimer's disease a fairly typical pattern of hypometabolism in the temporal and parietal association cortex bilaterally is observed. The frontal lobe also is affected later in the illness. In vascular dementias a pattern of patchy hypometabolism occurs, corresponding to structural abnormalities and the disruption of functional interconnections to these lesioned areas. Huntington's disease is characterized by significant hypometabolism in the striatum bilaterally. Also affected are cortical zones in the mesial frontal lobe and inferior parietal lobes bilaterally. By contrast another subcortical dementia, progressive supranuclear palsy, shows prominent frontal lobe hypometabolism and no subcortical changes. Diffuse cortical hypometabolism with a sparing of metabolic function (or hypermetabolism in some cases) in the striatum is typical of Parkinson's disease. Finally, in Pick's disease, focal, asymmetric areas of hypometabolism usually within the frontal or temporal lobes are observed. The only dementia discussed here for which a specific pattern of metabolic change has not been described is Creutzfeldt-Jakob disease. The few cases presented suggest that patchy changes in hypometabolism in the absence of structural lesions are characteristic of the illness.

Objective and accurate indicators of Alzheimer's disease which are sensitive and specific to early illness are required. In this regard, FDG-PET shows promise as an adjunct to the standard clinical examination of dementia. The technique provides unique information about functional brain activity, which when considered with the neuropsychological evaluation and other brain imaging findings, shows potential in advancing diagnostic certainty. To date, many FDG-PET studies have shown parietotemporal hypometabolism in a pattern that is highly characteristic of the diagnosis of Alzheimer's disease. Although the same pattern has been demonstrated in other instances of non-AD dementia (e.g., MID, PD with dementia), many of these false positive instances can be eliminated by consideration of the neuropsychological profile, clinical course, and brain MRI findings. Further development of the technology, including the development of labeled cholinergic ligands and refinement in methods (such as the use of brain activation studies and application of advanced statistical models accounting for regional interactions) may provide even more selective discriminators of AD from other degenerative neurological disorders. As the technology becomes more available and the costs are lowered, PET may be instrumental as a diagnostic tool in AD research and clinical settings. For patients with memory complaints, this may eliminate some of the anxiety associated with uncertainty of diagnosis and may permit the early implementation of treatment at a point in illness when it is likely to be most effective. For research studies, PET may contribute to the identification of the biological factors underlying Alzheimer's disease and other dementias. The technology may also assist in the development of effective drug treatments and provide evaluation of the brain response to the same treatments.

References

Absher, J. R., & Benson, D. F. (1993). Disconnection syndromes: An overview of Geschwind's contributions. *Neurology, 43,* 862–867.

Arnold, S. E., Hyman, B. T., Flory, J., Damasio, A. R., & Van Hoesen, G. W. (1991). The topographical and neuroanatomical distribution of neurofibrillary tangles and neuritic plaques in the cerebral cortex of patients with Alzheimer's disease. *Cerebral Cortex, 1,* 103–116.

Arriagada, P. V., Marzloff, K., & Hyman, B. T. (1992). Distribution of Alzheimer-type pathologic changes in nondemented elderly individuals matches the pattern in Alzheimer's disease. *Neurology, 42,* 1681–1688.

Azari, N. P., Pettigrew, K. D., Schapiro, M. B., Haxby, J. V., Grady, C. L., Pietrini, P., Salerno, J. A., Heston, L. L., Rapoport, S. I., & Horwitz, B. (1993). Early detection of Alzheimer's disease: A statistical approach using positron emission tomographic data. *Journal of Cerebral Blood Flow and Metabolism, 13,* 438–447.

Baron, J. C., Bousser, M. G., Comar, D., & Castaigne, P. (1980). Crossed cerebellar diaschisis in human supratentorial brain infarction. *Annals of Neurology, 8,* 128.

Baron, J. C., D'Antona, R., Pantano, P., Serdaru, M., Samson, Y., & Bousser, M. G. (1986). Effects of thalamic stroke on energy metabolism of the cerebral cortex: A positron emission tomography study in man. *Brain, 109,* 1243–1259.

Baron, J. C., Maziere, B., Lo'ch, C ., Sgouropoulos, P., Bonnet, A. W., & Agid, Y. (1985). Progressive supranuclear palsy: Loss of striatal dopamine receptors demonstrated by positron tomography. *Lancet, 1,* 1163–1164.

Bennett, D. A., Wilson, R. S., Gilley, D. W., & Fox, J. H. (1990). Clinical diagnosis of Binswanger's disease. *Journal of Neurology, Neurosurgery, & Psychiatry, 53,* 961–965.

Benson, D. F., Kuhl, D. E., Hawkins, R. A., Phelps, M. E., Cummings, J. L., & Tsai, S. Y. (1983). The fluorodeoxyglucose 18F scan in Alzheimer's disease and multi-infarct dementia. *Archives of Neurology, 40,* 711–714.

Bernheimer, H., Birkmeyer, W., Hornykiewicz, O., Mellinger, K., & Seitelberger, F. (1973). Brain dopamine and the syndromes of Parkinson and Huntington. *Journal of the Neurological Sciences, 20,* 415–445.

Bhatt, M. H., Snow, B. J., Martin, W. R. W., Peppard, R., & Calne, D. B. (1991). Positron emission tomography in progressive supranuclear palsy. *Archives of Neurology, 48,* 389–391.

Blessed, D. G., Tomlinson, B. E., & Roth, M. (1968). The association between quantitative measures of dementia and of senile change in the cerebral gray matter of elderly subjects. *British Journal of Psychiatry, 114,* 797–811.

Boone, K. B., Miller, B. L., Lesser, I. M., Mehringer, M., Hill-Gutierrez, E., Goldberg, M. A., & Berman, N. G. (1992). Neuropsychological correlates of white-matter lesions in healthy elderly subjects: A threshold effect. *Archives of Neurology, 49,* 549–554.

Bradshaw, J. L., Phillips, J. G., Dennis, C., Mattingley, J. B., Andrewes, D., Chiu, E., Pierson, J. M., & Bradshaw, J. A. (1992). Initiation and execution of movement sequences in those suffering from and at risk of developing Huntington's disease. *Journal of Clinical and Experimental Neuropsychology, 14,* 179–192.

Brooks, D. J., Playford, E. D., Ibanez, V., Sawle, G. V., Thompson, P. D., Findley, L. J., & Marsden, C. D. (1992). Isolated tremor and disruption of the nigrostriatal dopaminergic system: An ^{18}F-dopa PET study. *Neurology, 42,* 1554–1560.

Brouwers, P., Cox, C., Martin, A., Chase, T., & Fedio, P. (1984). Differential perceptual-spatial impairment in Huntington's and Alzheimer's dementias. *Archives of Neurology, 41,* 1073–1076.

Brown, P., Goldfarb, L. G., McCombie, W. R., Nieto, A., Squillacote, D., Sheremata, W., Little, B. W., Godec, M. S., Gibbs, C. J., & Gajdusek, D. C. (1992). Atypical Creutzfeldt-Jakob disease in an American family with an insert mutation in the PRNP amyloid precursor gene. *Neurology, 42,* 422–427.

Butters, N., Granholm, E., Salmon, D. P., Grant, I., & Wolfe, J. (1987). Episodic and semantic memory: A comparison of amnesic and demented patients. *Journal of Clinical and Experimental Neuropsychology, 9,* 479–497.

Butters, N., Sax, D., Montgomery, K., & Tarlow, S. (1978). Comparison of the neuropsychological deficits associated with early and advanced Huntington's disease. *Archives of Neurology, 35,* 585–589.

Butters, N., Wolfe, J., Martone, M., Granholm, E., & Cermak, L. S. (1985). Memory disorders associated with Huntington's disease: Verbal recall, verbal recognition, and procedural memory. *Neuropsychologia, 23,* 729–743.

Carroll, B. J., Feinberg, M., Greden, J. F., Tarika, J., Albala, A. A., Haskett, R. F., James, N. M., Kronfol, Z., Lohr, N., Steiner, M., deVigne, J. P., & Young, E. (1981). A specific laboratory test for the diagnosis of melancholia: Standardization, validation, and clinical utility. *Archives of General Psychiatry, 38,* 15–22.

Chase, T. N., Burrows, H., & Mohr, E. (1987). Cortical glucose utilization patterns in primary degenerative dementias of the anterior and posterior type. *Archives of Gerontology and Geriatrics, 6,* 289–297.

Chase, T. N., Fedio, P., Foster, N. L., Brooks, R., Di Chiro, G., & Mansi, L. (1984a). Wechsler Adult Intelligence Scale performance: Cortical localization by fluorodeoxyglucose F18-positron emission tomography. *Archives of Neurology, 41,* 1244–1247.

Chase, T. N., Foster, N. L., Fedio, P., Brooks, R., Mansi, L., & Di Chiro, G. (1984b). Regional cortical dysfunction in Alzheimer's disease as determined by positron emission tomography. *Annals of Neurology, 15*(Suppl.), S170–174.

Chawluk, J. B., Dann, R., Alavi, A., Hurtig, H. I., Gur, R. E., Resnick, S., Zimmersman, R. A., & Reivick, M. (1990). The effect of focal cerebral atrophy in position emission tomography studies of aging and dementia. *Nuclear Medicine and Biology, 17,* 797–804.

Chawluk, J. B., Mesulam, M. M., Hurtig, H., Kushner, M., Weintraub, S., Saykin, A., Rubin, N., Alavi, A., & Reivich, M. (1986). Slowly progressive aphasia without generalized dementia: Studies with positron emission tomography. *Annals of Neurology, 19,* 68–74.

Chui, H. C., Victoroff, J. I., Margolin, D., Jagust, W., Shankle, R., & Katzman, R. (1992). Criteria for the diagnosis of ischemic vascular dementia proposed by the State of California Alzheimer's Disease Diagnostic and Treatment Centers. *Neurology, 42,* 473–480.

Clark, C. M., Carson, R., Kesseler, R., Margolin, R., Buchsbaum, M., Delisi, L., King, C., & Cohen, R. (1985). Alternative statistical models for the examination of clinical positron emission tomography/fluorodeoxyglucose data. *Journal of Cerebral Blood Flow and Metabolism, 5,* 142–150.

Crapper-McLachlan, D. R., Dalton, A. J., Galin, H., Schlotterer, G., & Daicar, E. (1984). Alzheimer's disease: Clinical course and cognitive disturbances. *Acta Neurologica Scandinavica, 99,* 83–89.

Cummings, J. L. (1990). *Subcortical dementia.* New York: Oxford University Press.

Cutler, N. (1988). Utility of biological markers in the evaluation and diagnosis of Alzheimer's disease. *Brain Dysfunction, 1,* 12–31.

Cutler, N. R., Haxby, J. V., Duara, R., Grady, C. L., Moore, A. M., Parisi, J. E., White, J., Heston, L., Margolin, R. M., & Rapoport, S. I. (1985). Brain metabolism as measured with positron emission tomography: Serial assessment in a patient with familial Alzheimer's disease. *Neurology, 35,* 1556–1561.

Dannals, R. F. (1989). Synthesis of radiotracers. *Journal of Neuropsychiatry and Clinical Neurosciences, 1*(Suppl.), S14–S18.

D'Antona, R., Baron, J. C., Samson, Y., Serdaru, M., Viader, F., Agid, Y., & Cambier, J. (1985). Subcortical dementia. Frontal cortex hypometabolism detected by positron emission tomography in patients with progressive supranuclear palsy. *Brain, 108,* 785–799.

Davis, P. H., Bergeron, C., & McLachlan, D. R. (1985). Atypical presentation of progressive supranuclear palsy. *Annals of Neurology, 17,* 373–343.

de Leon, M. J., Ferris, S. H., George, A. E., Reisberg, B., Christman, D. R., Kricheff, I. I., & Wolf, A. P. (1983). Computed tomography and positron emission transaxial tomography evaluations of normal aging and Alzheimer's disease. *Journal of Cerebral Blood Flow and Metabolism, 3,* 391–394.

de Leon, M. J., George, A. E., Ferris, S. H., Christman, D. R., Fowler, J. S., Gentes, C. I., Brodie, J., Reisberg, B., & Wolf, A. P. (1984). Positron emission tomography and computed tomography assessments of the aging human brain. *Journal of Computer Assisted Tomography, 8,* 88–94.

Duara, R., Barker, W. W., Chang, J., Yoshii, F., Loewenstein, D. A., & Pascal, S. (1992). Viability of neocortical function shown in behavioral activation state PET studies in Alzheimer's disease. *Journal of Cerebral Blood Flow and Metabolism, 12,* 927–934.

Duara, R., Barker, W., Loewenstein, D., Pascal, S., & Bowen, B. (1989). Sensitivity and specificity of positron emission tomography and magnetic resonance imaging studies in Alzheimer's disease and multi-infarct dementia. *European Neurology, 29,* 9–15.

Duara, R., Grady, C., Haxby, J., Ingvar, D., Sokoloff, L., Margolin, R., Manning, R. G., Cutler, N. R., & Rapoport, S. I. (1984). Human brain glucose utilization and cognitive function in relation to age. *Annals of Neurology, 16,* 702–713.

Duara, R., Grady, C., Haxby, J., Sundaram, M., Cutler, N. R., Heston, L., Moore, A., Schlageter, N., Larson, S., & Rapoport, S. I. (1986). Positron emission tomography in Alzheimer's disease. *Neurology, 36,* 879–887.

Duara, R., Loewenstein, D. A., & Barker, W. W. (1990). Utilization of behavioral activation paradigms for positron emission tomography studies in normal young and elderly subjects and in dementia. In R. Duara (Ed.), *Positron emission tomography in dementia. Frontiers of clinical neuroscience: Vol. 10* (pp. 131–148). New York: Wiley-Liss.

Duara, R., Margolin, R. A., Roberson-Tchabo, E. A., London, E. D., Schwartz, M., Renfrew, J. W., Koziarz, B. J., Sundaram, M., Grady, C., Moore, A. M., Ingvar, D. H., Sokoloff, L., Weingartner, H., Kessler, R. M., Manning, R. G., Channing, M. A., Cutler, N. R., & Rapoport, S. I. (1983). Cerebral glucose utilization as measured with positron emission tomography in 21 resting healthy men between the ages of 21 and 83 years. *Brain, 106,* 761–775.

Cummings, J. L., & Benson, D. F. (1984). Subcortical dementia: Review of an emerging concept. *Archives of Neurology, 41,* 874–879.

Duvoisin, R. C., Golbe, L. I., & Lepore, F. E. (1987). Progressive supranuclear palsy. *Canadian Journal of Neurological Sciences, 14,* 547–554.

Evans, D. A., Funkenstein, H. H., Albert, M. S., Scherr, P. A., Cook, N. R., Chown, M. J., Hebert, L. E., Hennekens, C. H., & Taylor, J. O. (1989). Prevalence of Alzheimer's disease in a community sample of older persons: Higher than previously reported. *Journal of the American Medical Association, 262,* 2251–2556.

Fazio, F., Perani, D., Gilardi, M. C., Colombo, F., Cappa, S. F., Vallar, G., Bettinardi, V., Paulesu, E., Alberoni, M., Bressi, S., Franceschi, M., & Lenzi, G. L. (1992). Metabolic impairment in human amnesia: A PET study of memory networks. *Journal of Cerebral Blood Flow and Metabolism, 12,* 353–358.

Fischer, P., Jellinger, K., Gatterbi, G., & Danielcyzk, W. (1991). Prospective neuropathological validation of Hachinski's ischaemic score in dementias. *Journal of Neurology, Neurosurgery, & Psychiatry, 54,* 580–583.

Folstein, S. E., Abbott, M. H., Chase, G. A., Jensen, B. A., & Folstein, M. F. (1983). The association of affective disorder with Huntington's disease in a case series and in families. *Psychological Medicine, 13,* 537–542.

Folstein, S. E., Brandt, J., & Folstein, M. R. (1990). Huntington's disease. In J. Cummings (Ed.), *Subcortical dementia* (pp. 87–107). New York: Oxford University Press.

Foster, N. L., Chase, T. N., Mansi, L., Brooks, R., Fedio, P., Patronas, N. J., & DiChiro,, G. (1984). Cortical abnormalities in Alzheimer's disease. *Annals of Neurology, 16,* 649–654.

Foster, N. L., Chase, T. N., Patronas, N. J., Gillespie, M. M., & Fedio, P. (1986). Cerebral mapping of apraxia in Alzheimer's disease by positron emission tomography. *Annals of Neurology, 19,* 139–143.

Fox, P. T., Miezin, F. M., Allman, J. M., Van Essen, D. C., & Raichle, M. E. (1987). Retinotopic organization of the human visual cortex mapped with positron emission tomography. *Journal of Neuroscience, 7,* 913–922.

Fox, P. T., Mintun, M. A., Raichle, M. E., Miezin, R. M., Allman, J. M., & Van Essen, D. C. (1986). Mapping human visual cortex with positron emission tomography. *Nature, 323,* 806–809.

Frackowiak, R. S. J., Lenzi, G. L., Jones, T., & Heather, J. D. (1980). Quantitative measurement of regional cerebral blood flow and oxygen metabolism in man using ^{15}O and positron emission tomography: Theory, procedure, and normal values. *Journal of Computer Assisted Tomography, 4,* 727–736.

Frackowiak, R. S. J., Pozzili, C., Legg, N. N. J., Du Boulay, G. H., Marshall, J., Lenzi, G. L., & Jones, T. (1981). Regional cerebral ozygen supply and utilization in dementia: A clinical and physiological study with oxygen-15 and positron emission tomography. *Brain, 104,* 753–778.

Freedman, M. (1990). Parkinson's disease. In J. Cummings (Ed.), *Subcortical dementia* (pp. 108–122). New York: Oxford University Press.

Frey, K. A., Koeppe, R. A., Mulholland, G. K., Jewett, D., Hichwa, R., Ehrenkaufer, R. L. E., Carey, J. E., Wieland, D. M., Kuhl, D. E., & Agranoff, B. W. (1992). In vivo muscarinic cholinergic receptor imaging in human brain with [11C] scopolamine and positron emission tomography. *Journal of Cerebral Blood Flow and Metabolism, 12,* 147–154.

Friedland, R. P. (1989). Positron emission tomography in dementia. *Seminars in Neurology, 150,* 735–739.

Friedland, R. P., Budinger, T. F., Ganz, E., Yano, Y., Maathid, C. A., Koss, E., Ober, B. A., Huesman, R. H., & Derenzo, S. E. (1983). Regional cerebral metabolic alterations in dementia of the Alzheimer type: Positron emission tomography with [18F]-fluoro-deoxyglucose. *Journal of Computer Assisted Tomography, 7,* 590–598.

Friedland, R. P., Pruisner, S. B., Jagust, W. J., Budinger, T. F., & Davis, R. L. (1984). Bitemporal hypometabolism in Creutzfeldt-Jakob disease measured by positron emission tomography with [18F]-2-fluorodeoxyglucose. *Journal of Computer Assisted Tomography, 8,* 978–981.

Fuld, P. A., Masur, D. M., Blau, A. D., Crystal, H., & Aronson, M. K. (1990). Object-memory evaluation for prospective detection of dementia in normal functioning elderly: Predictive and normative data. *Journal of Clinical and Experimental Neuropsychology, 12,* 539–548.

Garnett, E. S., Firnau, G., Nahmias, C., Carbotte, R., & Bartolucci, G. (1984). Reduced striatal glucose consumption and prolonged reaction time are early features in Huntington's disease. *Journal of the Neurological Sciences, 65,* 231–237.

Geschwind, N. (1965). Disconnexion syndromes in animals and man, Parts I and II. *Brain, 88,* 237–294, 585–644.

Gibb, W. R. G. (1986). Idiopathic Parkinson's disease and the Lewy body disorders. *Neuropathology and Applied Neurobiology, 12,* 223–234.

Goffinet, A. M., De Volder, A. G., Gullain, C., Rectem, D., Bol, A., Michel, C., Cogneau, M., Labar, D., & Laterre, C. (1989). Positron tomography demonstrated frontal lobe hypometabolism in progressive supranuclear palsy. *Annals of Neurology, 25,* 131–139.

Grady, C. L., Haxby, J. V., Horwitz, B., Sundaram, M., Berg, G., Schapiro, M., Friedland, R. P., & Rapoport, S. I. (1988). Longitudinal study of the early neuropsychological and cerebral metabolic changes in dementia of the Alzheimer type. *Journal of Clinical and Experimental Neuropsychology, 10,* 579–596.

Grady, C. L., Haxby, J. V., Schapiro, M. B., Gonzalez-Aviles, A., Kumar, A., Ball, M. J., Heston, L., & Rapoport, S. I. (1990). Subgroups in dementia of the Alzheimer type identified using positron emission tomography. *Journal of Neuropsychiatry and Clinical Neurosciences, 2,* 373–384.

Graff-Radford, N. R., Damasio, A. R., Hyman, B. T., Hart, M. N., Tranel, D., Damasio, H., Van Hoesen, G. W., & Rezai, K. (1990). Progressive aphasia in a patient with Pick's disease: A neuropsychological, radiologic, and anatomic study. *Neurology, 40,* 620–626.

Grafton, S. T., Mazziotta, J. C., Pahl, J. J., St.George-Hyslop, P., Haines, J. L., Gusella, J., Hoffman, J. M., Baxter, L. R., & Phelps, M. E. (1990). A comparison of neurological, metabolic structural, and genetic evaluations in persons at risk for Huntington's disease. *Annals of Neurology, 28,* 614–621.

Grafton, S. T., Mazziotta, J. C., Pahl, J. J., St.George-Hyslop, P., Haines, J. L., Gusella, J., Hoffman, J. M., Baxter, L. R., & Phelps, M. E. (1992). Serial changes of cerebral glucose metabolism and caudate size in persons at risk for Huntington's disease. *Archives of Neurology, 49,* 1161–1167.

Greenberg, J. H., Reivich, M., Alavi, A., Hand, P., Rosenquist, A., Rintelmann, W., Stein, A., Tusa, R., Dann, R., Christman, D., Fowler, J., MacGregor, B., & Wolf, A. (1981). Metabolic mapping of functional activity in human subjects the [^{18}F]fluorodeoxyglucose technique. *Science, 212,* 678–680.

Gur, R. C., Gur, R. E., Rosen, A. D., Warach, S., Alavi, A., Greenberg, J., Reivich, M. (1983). A cognitive-motor network demonstrated by positron emission tomography. *Neurospychologia, 21,* 601–606.

Gusella, J. F., Wexler, N. S., Conneally, P. M., Nayler, S. L., Anderson, M. A., Tanki, R. E., Watkins, P. C., Ottina, Wallace, M. R., Sakaguchi, A. Y., Young, A. B., Shoulson, I., Bonilla, E., & Martin, J. B. (1983). A polymorphic DNA marker genetically linked to Huntington's disease. *Nature, 306,* 234–238.

Guze, B. H., Hoffman, J. M., Baxter Jr., L. R., Mazziota, J. C., & Phelps, M. E. (1991). Functional brain imaging in Alzheimer-type dementia. *Alzheimer Disease and Associated Disorders, 5,* 215–230.

Hachinski, V. C., Lassen, N. A., & Marshall, J. (1974). Multi-infarct dementia. A cause of mental deterioration in the elderly. *Lancet, 2,* 207–210.

Hagglund, J., Aquilonius, S.-M., Eckernas, S.-A., Hartvig, P., Lundquist, H., Gullberg, P., & Langstrom, B. (1987). Dopamine receptor properties in Parkinson's disease and Huntington's chorea evaluated by positron emission tomography using ^{11}C-N-methyl-spiperone. *Acta Neurologia Scandinavica, 75,* 87–94.

Hakim, A. M., & Mathieson, G. (1979). Dementia in Parkinson's disease: A neuropathologic study. *Neurology, 29,* 1209–1214.

Hansen, L. A., DeTeresa, R., Davies, P., & Terry, R. D. (1988). Neocortical morphometry, lesion counts, and choline acetyltransferase levels in the age spectrum of Alzheimer's disease. *Neurology, 38,* 48–54.

Harrison, P. J., & Roberts, G. W. (1991). "Life Jim, but not as we know it?" Transmissible dementias and the prion protein. *British Journal of Psychiatry, 158,* 457–470.

Haxby, J. V., Duara, R., Grady, C. L., Cutler, N. R., & Rapoport, S. I. (1985). Relations between neuropsychological and cerebral metabolism asymmetries in early Alzheimer's disease. *Journal of Cerebral Blood Flow Metabolism, 5,* 193–200.

Haxby, J. V., Grady, C. L., Duara, R., Schlageter, N., Berg, G., & Rapoport, S. I. (1986). Neocortical metabolic abnormalities precede nonmemory cognitive deficits in early Alzheimer's type dementia. *Archives of Neurology, 43,* 882–885.

Haxby, J. V., Grady, C. L., Koss, E., Horowitz, B., Heston, L., Schapiro, M., Friedland, R. P., & Rapoport, S. I. (1990). Longitudinal study of cerebral metabolic asymmetries and associated neuropsychological patterns in early dementia of the Alzheimer type. *Archives of Neurology, 47,* 753–760.

Haxby, J. V., Grady, C. L., Koss, E., Horowitz, B., Schapiro, M., Friedland, R. P., & Rapoport, S. I. (1988). Heterogenous anterior-posterior metabolic patterns in dementia of the Alzheimer type. *Neurology, 38,* 1853–1863.

Haxby, J. V., Grady, C. L., Ungerleider, L. G., & Horwitz, B. (1991). Mapping the functional neuroanatomy of the intact human brain with brain work imaging. *Neuropsychologia, 29,* 539–555.

Hayden, M. R., Hewitt, J., Stoessl, A. J., Clark, C., Ammann, W., & Martin, W. R. W. (1987). The combined use of positron emission tomography and DNA polymorphisms for preclinical detection of Huntington's disease. *Neurology, 37,* 1441–1447.

Hayden, M. R., Martin, W. R. W., Stoessl, A. J., Clark, C., Hollenberg, S., Adam, M. J., Ammann, W., Harrop, R., Rogers, J., Ruth, T., Sayre, C., & Pate, B. D. (1986). Positron emission tomography in the early diagnosis of Alzheimer's disease. *Neurology, 36,* 888–894.

Hefter, J., Hamberg, V., Lange, H. W., & Freud, H.-J. (1987). Impairment of rapid movement in Huntington's disease. *Brain, 110,* 585–612.

Heiss, W.-D., Vyska, K., Kloster, G., Traupe, H., Freundlieb, C., Hoeck, A., Feinendegen, L. E., & Stoecklin, G. (1982). Demonstration of decreasaed functional activity of visual cortex by ^{11}C-methylglucose and positron emission tomography. *Neuroradiology, 23,* 45–47.

Heston, L. L., & Mastri, A. R. (1982). Age at onset of Pick's and Alzheimer's dementia: Implications for diagnosis and research. *Journal of Gerontology, 4,* 422–424.

Hoffman, J. M., Guze, B. H., Baxter, L. R., Mazziotta, J. C., & Phelps, M. E. (1989). [18F]-fluorodeoxyglucose (FDG) and positron emission tomography (PET) in aging and dementia: A decade of studies. *European Neurology, 29* (Suppl. 3), 16–24.

Hoffman, J. M., Hanson, M. W., & Coleman, R. E. (1993). Clinical positron emission tomography imaging. *Radiological Clinics of North America, 31,* 935–959.

Hoffman, J. M., Welsh, K. A., Hanson, M. W., Earl, N., & Coleman, R. E. (1992). FDG-PET is useful in early detection and confirmation of Alzheimer's disease (AD). *Neurology, 42* (Suppl. 3), 315.

Holland, A. J., McBurney, D. H., Moossy, J., & Reinmuth, O. M. (1985). The dissolution of language in Pick's disease with neurofibrillary tangles: A case study. *Brain and Language, 24,* 36–58.

Horowitz, S., Benson, D. F., Kuhl, D. E., & Cummings, J. L. (1992). FDG scan to confirm Creutzfeldt-Jakob diagnosis (abstract). *Neurology, 32,* A167.

Hudson, A. J. (1981). Amyotrophic lateral sclerosis and its association with dementia, parkinsonism, and other neurological disorders: A review. *Brain, 104,* 217–247.

Huntington's Disease Collaborative Research Group. (1993). A novel gene containing a trinucleotide repeat that is expanded and unstable on Huntington's disease chromosome. *Cell, 72,* 971–983.

Hyman, B. T., Van Hoesen, G. W., Damaisio, A. R., & Barnes, C. L. (1984). Alzheimer's disease: Cell-specific pathology isolates the hippocampal formation. *Science, 225,* 1168–1170.

Jagust, W. J., & Eberling, J. L. (1991). MRI, CT, SPECT, PET: Their use in diagnosing dementia. *Geriatrics, 46,* 28–35.

Jellinger, K., Riederer, P., & Tomonga, M. (1980). Progressive supranuclear palsy: Clinicopathological and biochemical studies. *Journal of Neural Transactions Supplement, 16,* 111–128.

Joachim, C. L., Morris, J. H., & Selkoe, D. J. (1988). Clinically diagnosed Alzheimer's disease: Autopsy rersults in 150 cases. *Annals of Neurology, 24,* 50–56.

Jolles, P. R., Chapman, P. R., & Alavi, A. (1989). PET, CT, and MRI in the evaluation of neuropsychiatric disorders: Current applications. *Journal of Nuclear Medicine, 30,* 1589–1606.

Josiassen, R. C., Curry, L. M., & Mancall, E. L. (1983). Development of neuropsychological deficits in Huntington's disease. *Archives of Neurology, 40,* 791–796.

Kamo, H., McGeer, P. L., Harrop, R., McGeer, E. G., Calne, D. B., Martin, W. R. W., & Pate, B. D. (1987). Positron emission tomography and histopathology in Pick's disease. *Neurology, 37,* 439–445.

Kannel, W. B., Dawber, T. R., Sorlie, P., & Wolf, P. A. (1976). Component blood pressure and risk of atherothrombotic brain infarct: The Framingham study. *Stroke, 7,* 327–331.

Karbe, H., Holthoff, V., Huber, H., Herholz, K., Wienhard, K., Wagner, R., & Heiss, W.-D. (1992). Positron emission tomography in degenerative disorders of the dopaminergic system. *Journal of Neural Transmission, 4,* 121–130.

Katzman, R. (1986). Alzheimer's disease. *New England Journal of Medicine, 314,* 217–218.

Katzman, R., Aronson, M., Fuld, P. A., Kawas, C., Brown, T., Morgenstern, H., Frishman, W., Gidez, L., Eder, H., & Ooi, W. L. (1989). Development of dementing illness in an 80-year-old volunteer cohort. *Annals of Neurology, 25,* 317–324.

Kessler, I. I. (1978). Parkinson's disease in epidemiological perspective. In B. S. Schoenberg (Ed.), *Advances in neurology: Vol. 19* (pp. 355–384). New York: Raven Press.

Kessler, J., Herholz, K., Grond, M., & Heiss, W.-D. (1991). Impaired metabolic activation in Alzheimer's disease: A PET study during continuous visual recognition. *Neuropsychologia, 29,* 229–243.

Khachaturian, Z. S. (1985). Diagnosis of Alzheimer's disease. *Archives of Neurology, 42,* 1097–1105.

Knopman, D. S., Christensen, K. J., Schut, L. J., Harbaugh, R. E., Reeder, T., Ngo, T., & Frey, W. (1989). The spectrum of imaging and neuropsychological findings in Pick's disease. *Neurology, 39,* 362–368.

Knopman, D. S., & Ryberg, S. (1989). A verbal memory test with high predictive accuracy for dementia of the Alzheimer type. *Archives of Neurology, 46,* 141–145.

Kooistra, C. A., & Heilman, K. M. (1988). Memory loss from a subcortical white matter infarct. *Journal of Neurology, Neurosurgery, & Psychiatry, 51,* 866–869.

Kosaka, K., Yoshimira, M., Ikeda, K., & Budka, H. (1984). Diffuse type of Lewy body disease: Progressive dementia with abundant cortical Lewy bodies and senile changes of varying degree: A new disease? *Clinical Neuropathology, 3,* 185–192.

Koss, E., Freidland, R. P., Ober, B. A., & Jagust, W. J. (1985). Differences in lateral hemispheric asymmetries of glucose utilization between early and late onset Alzheimer-type dementia. *American Journal of Psychiatry, 142,* 638–640.

Kuhl, D. E., Metter, E. J., & Riege, W. H. (1985). Patterns of cerebral glucose utilization in depression, multiple infarct dementia, and Alzheimer's disease. In L. Sokoloff (Ed.), *Brain imaging and brain function: Vol. 63* (pp. 211–226). New York: Raven Press.

Kuhl, D. E., Metter, E. J., Riege, W. H., & Markham, C. H. (1984). Patterns of cerebral glucose utilization in Parkinson's disease and Huntington's disease. *Annals of Neurology, 15*(Suppl.), S119–125.

Kuhl, D. E., Metter, E. J., Riege, W. H., & Phelps, M. E. (1982). Effects of human aging on patterns of local cerebral glucose utilization determined by (18F) fluorodeoxyglucose method. *Journal of Cerebral Blood Flow Metabolism, 2,* 163–171.

Kuhl, D. E., Phelps, M. E., Kowell, A. P., Metter, E. J., Selin, C., & Winter, J. (1980). Effect of stroke on local cerebral metabolism and perfusion: Mapping by emission computed tomography of ^{18}FDG and $^{13}NH_3$. *Annals of Neurology, 8,* 47–60.

Kuhl, D. E., Small, G. W., Riege, W. H., Fujikawa, D. G., Metter, E. J., Benson, D. F., Ashford, J. W., Mazziotta, J. C., Maltese, A., & Dorsey, D. A. (1987). Cerebral metabolic patterns before the diagnosis of probable Alzheimer's disease. *Journal of Cerebral Blood Flow and Metabolism, 7,* S406.

Kuwert, T., Lange, H. W., Langen, K.-J., Herzog, H., Aulick, A., & Feinendegen, L. E. (1990). Cortical and subcortical glucose consumption measured by PET in patients with Huntington's disease. *Brain, 113,* 1405–1423.

Kwong, K. K., Belliveau, J. W., Chester, D. A., Goldberg, I. E., Weissfoff, R. M., Poncelet, B. P., Kennedy, D. N., Hoppel, B. E., Cohen, M. S., Turner, R., Cheng, H.-M., Brady, T. J., & Rosen, B. R. (1992). Dyanmic magnetic resonance imaging of human brain activity during primary sensory stimulation. *Proceedings of the National Academy of Science, 89,* 5675–5679.

Larson, D. B., Lyons, J. S., Bareta, J. C., Burns, B. J., Blazer, D. G., & Goldstrom, I. D. (1988). The construct validity of the ischemic score of Hachinski for the detection of dementias. *Journal of Neuropsychiatry, 1,* 181–187.

Leenders, K. L., Palmer, A. J., Quinn, N., Clark, J. C., Firnau, G., Garnett, E. S., Nahmias, C., Jones, T., & Marsden, C. D. (1986). Brain dopamine metabolism in patients with Parkinson's diseasae measured with positron emission tomography. *Journal of Neurology, Neurosurgery, and Psychiatry, 49,* 853–860.

Leuchter, A. F., Newton, T. F., Cook, I. A., Walter, D. O., Rosenberg-Thompson, S., & Lachenbruch, P. A. (1992). Changes in brain functional connectivity in Alzheimer type and multi-infarct dementia. *Brain, 115,* 1543–1561.

Links, J. M. (1989). Data acquisition, display, and quantification. *Journal of Neuropsychiatry and Clinical Neurosciences, 1* (Suppl.), S7–S13.

Loewenstein, D. A., Barker, W. W., Chang, J.-Y., Apicella, A., Yoshii, F., Kothari, P., Levin, B., & Duara, R. (1989). Predominant left hemisphere metabolic dysfunction in dementia. *Archives of Neurology, 46,* 146–142.

Mahler, M. E., & Cummings, J. L. (1990). Alzheimer disease and the dementia of Parkinson's disease: Comparative investigations. *Alzheimer Disease and Associated Disorders, 14,* 133–149.

Martin, A., Brouwers, P., Lalonde, F., Cox, C., Teleska, P., Fedio, P., Foster, N. L., & Chase, T. N. (1986). Towards a behavioral typology of Alzheimer's disease. *Journal of Clinical and Experimental Neuropsychology, 8,* 594–610.

Martin, E. M., Wilson, R. S., Penn, R. D., Foss, J. H., Clasen, R. A., & Jaroy, S. M. (1987). Cortical biopsy results in Alzheimer's disease: Correlation with cognitive deficits. *Neurology, 37,* 1201–1204.

Martin, W. R. W., Beckman, J. H., Calne, D. B., Adam, M. J., Harrop, R., Rogers, J. G., Ruth, T. J., Sayre, C. I., & Pate, B. D. (1984). Cerebral glucose metabolism in Parkinson's disease. *Canadian Journal of Neurological Science, 11,* 169–173.

Martin, W. R. W., Clark, C., Ammann, W., Stoessl, A. J., Shtybel, W., & Hayden, M. R. (1992). Cortical glucose metabolism in Huntington's disease. *Neurology, 42,* 223–229.

Masters, C. L., & Richardson, E. P. (1978). Subacute spongiform encephalopathy (Creutzfeldt-Jakob Disease). *Brain, 101,* 333–344.

Masur, D. M., Fuld, P. A., Blau, A. D., Crystal, H., & Aronson, M. K. (1990). Predicting development of dementia in the elderly with the selective reminding test. *Journal of Clinical and Experimental Neuropsychology, 12,* 529–538.

Mazziotta, J., & Phelps, M. (1986). Positron emission tomography studies of the brain. In M. Phelps, J. Mazziotta, & H. R. Schelbert (Eds.), *Positron emission tomography and autoradiography: Principles and applications for the brain and heart* (pp. 493–579). New York: Raven Press.

Mazziotta, J. C,. Phelps, M. E., & Carson, R. E. (1982). Topographic mapping of human cerebral metabolism: Sensory deprivation. *Annals of Neurology, 12,* 435–444.

Mazziotta, J. C., Phelps, M. E., Miller, J., & Kuhl, D. E. (1981). Tomographic mapping of human cerebral metabolism: Normal unstimulated state. *Neurology, 31,* 503–516.

Mazziotta, J. C., Phelps, M. E., Pahl, J. J., Huang, S.-C., Baxter, L. R., Riege, W. H., Hoffman, J. M., Kuhl, D. E., Lanto, A. B., Wapenski, J. A., & Markham, C. H. (1987). Reduced cerebral glucose metabolism in asymptomatic subjects at risk for Huntington's disease. *New England Journal of Medicine, 316,* 357–362.

McCrory, S. J., & Ford, I. (1991). Multivariate analysis of SPECT images with illustrations in Alzheimer's diseases. *Statistics in Medicine, 10,* 1711–1718.

McGeer, E. G., McGeer, P. L., Harrop, R., Akiyama, H., & Kamo, H. (1990). Correlates with regional postmortem enzyme activities with premortem local glucose metabolic rates in Alzheimer's disease. *Journal of Neuroscience Research, 27,* 612–619.

McKhann, G., Drachman, D., Folstein, M., Katzman, R., Price, D., & Stadlan, E. M. (1984). Clinical diagnosis of Alzheimer's disease: Report of the NINCDS-ADRDA Work Group under the auspices of the Department of Health and Human Services Task Force on Alzheimer's Disease. *Neurology, 34,* 939–944.

Mendez, M. F., Adams, N. L., & Lewandowski, S. K. (1989). Neurobehavioral changes associated with caudate lesions. *Neurology, 39,* 349–354.

Metter, E. J., Kuhl, D. E., & Riege, W. H. (1990). Brain glucose metabolism in Parkinson's disease. In M. B. Streifler, A. D. Korczyn, E. Melamed, & M. B. H. Youdim (Eds.), *Advances in neurology: Vol. 53. Anatomy, pathology, and therapy* (pp. 135–139). New York: Raven Press.

Metter, E. J., Riege, W. H., Kuhl, D. E., & Phelps, M. E. (1983). Differences in regional glucose metabolic intercorrelations with aging. *Journal of Cerebral Blood Flow and Metabolism, 4,* 500–506.

Miller, J. D., de Leon, M. J., Ferris, S. H., Kluger, A., George, A. E., Reisberg, B., Sachs, H. J., & Wolf, A. P. (1987). Abnormal temporal lobe response in Alzheimer's diseasae during cognitive processing as measured by ^{11}C-2-deoxy-D-glucose and PET. *Journal of Cerebral Blood Flow and Metabolism, 7,* 248–251.

Mirra, S. S., Hart, M. N., & Terry, R. D. (1993). Making the diagnosis of Alzheimer's disease: A primer for practicing pathologists. *Archives of Pathology and Laboratory Medicine, 117,* 132–144.

Moeller, J. R., & Strother, S. C. (1991). A regional covariance approach to the analysis of functional patterns in positron emission tomographic data. *Journal of Cerebral Blood Flow and Metabolism, 11,* A121–A135.

Moss, M. B., & Albert, M. S. (1988). Alzheimer's disease and other dementing disorders. In M. S. Albert & M. B. Moss (Eds.), *Geriatric neuropsychology.* New York: Guilford Press.

Naugle, R. I., & Bigler, E. D. (1989). Brain imaging and neuropsychological identification of dementia of the Alzheimer's type. In E. D. Bigler, R. A. Yeo, & E. Turkheimer (Eds.), *Neuropsychological function and brain imaging* (pp. 185–218). New York: Plenum Press.

Navia, B. A. (1990). The AIDS dementia complex. In J. Cummings (Ed.), *Subcortical dementia* (pp. 181–198). New York: Oxford University Press.

Neary, D., Snowden, J. S., Northen, B., & Goulding, P. (1988). Dementia of the frontal lobe type. *Journal of Neurology, Neurosurgery, and Psychiatry, 51,* 353–361.

Nierenberg, A. A., & Feinstein, A. R. (1988). How to evaluate a diagnostic marker test: Lessons from the rise and fall of the dexamethasone suppression test. *Journal of the American Medical Association, 259,* 1699–1702.

Palella, T. D., Hichwa, R. D., Ehrenkaufer, R. L., Rothley, J. M., McQuillan, M. A., Young, A. B., & Kelley, W. N. (1985). 18F-fluorodeoxyglucose PET scanning in HPRT deficiency. *American Journal of Human Genetics, 37,* A70.

Pantano, P., Baron, J. C., Samson, Y., Bousser, M. G., Derouesne, C., & Comar, D. (1986). Crossed cerebellar diaschisis: Further studies. *Brain, 109,* 677–694.

Pappata, S., Tran Dinh, S., Baron, J. C., Cambon, H., & Syrota, A. (1987). Remote metabolic effects of cerebrovascular lesions: Magnetic resonance and positron tomography imaging. *Neuroradiology, 29,* 1–6.

Parks, R. W., Loewenstein, D. A., & Chang, J. Y. (1988). Brain imaging: Positron emission tomography and cognitive functioning. In J. M. Williams, & C. J. Long (Eds.), *Cognitive approaches to neuropsychology* (pp. 189–210). New York: Plenum Press.

Pawlik, G., & Heiss, W. D. (1989). Positron emission tomography and neuropsychological function. In E. D. Bigler, R. A. Yeo, & E. Turkheimer (Eds.), *Neuropsychological function and brain imaging* (pp. 65–138). New York, London: Plenum Press.

Pearson, R. C. A., Esiri, M. M., Hjorns, R. W., Wilcock, G. K., & Powell, T. P. S. (1985). Anatomical correlates of the distribution of pathological changes in the neocortex in Alzheimer's disease. *Proceedings of the National Academy Sciences, 82,* 4531–4534.

Pelizzari, C. A., Chen, G. T., Spelbring, D. R., Weischselbaum, R. R., & Chen, C. T. (1989). Accurate three-dimensional registration of CT, PET, and/or MRI images of the brain. *Journal of Computer Assisted Tomography, 13,* 20–26.

Penney, J. B., & Young, A. B. (1988). Huntington's diseasae. In J. Jankovic & E. Toloso (Eds.), *Parkinson's disease and movement disorders.* (pp. 167–178). Baltimore: Urban & Schwarzenberg Press.

Peppard, R. F., Martin, W. R. W., Carr, G. D., Grochowski, E., Schulzer, M., Guttman, M., McGeer, P. L., Phillips, A. G., Tsui, J. K. C., & Calne, D. B. (1992). Cerebral glucose metabolism in Parkinson's disease with and without dementia. *Archives of Neurology, 49,* 1262–1268.

Perlmutter, J. S. (1988). New insights into the pathophysiology of Parkinson's disease: The challenge of positron emission tomography. *Trends in Neurosciences, 11,* 203–207.

Perry, E. K., Tomlinson, B. E., Blessed, G., Bergmann, K., Gibson, P. H., & Perry, R. H. (1978). Correlation of cholinergic abnormalities with senile plaques and mental test scores in senile dementia. *British Medical Journal, 2,* 1457–1459.

Peterson, S. E., Fox, P. T., Posner, M. I., Mintun, M., & Raichle, M. E. (1988). Positron emission tomographic studies of the cortical anatomy of single-word processing. *Nature, 331,* 585–589.

Peterson, S. E., Fox, P. T., Posner, M. I., Mintun, M., & Raichle, M. E. (1989). Positron emission tomographic studies of the cortical anatomy of single words. *Journal of Cognitive Neuroscience, 1,* 153–170.

Phelps, M. E., Kuhl, D. E., & Mazziotta, J. C. (1981). Metabolic mapping of the brain's response to visual stimulation: Studies in man. *Science, 211,* 1445–1448.

Pietrini, P., Grady, C. L., Haxby, J. V., Heston, L., Salerno, J. A., Gonzales-Aviles, A., & Shapiro, M. B. (1991). Resting cerebral glucose metabolism (CMRglc) does not identify subjects "at risk" for familial Alzheimer disease. *Annals of Neurology, 30,* 287–288.

Podoll, K., Caspary, P., Lange, H. W., & Noth, J. (1988). Language functions in Huntington's disease. *Brain, 111,* 1475–1503.

Posner, M. I., & Petersen, S. E. (1989). The attention system of the human brain. *Annual Review of Neurosciences, 13,* 25–42.

Powers, W. J., Perlmutter, J. S., Videen, T. O., Herscovitch, P., Griffeth, L. K., Royal, H. D., Siegel, B. A., Morris, J. C., & Berg, L. (1992). Blinded clinical evaluation of positron emission tomography for diagnosis of probable Alzheimer's disease. *Neurology, 42,* 765–770.

Prusiner, S. B. (1982). Novel proteinaceous infectious particles cause scrapie. *Science, 216,* 136–144.

Raichle, M. E. (1979). Quantitative *in vivo* autoradiography with positron emission tomography. *Brain Research Reviews, 1,* 47–68.

Raichle, M. E. (1982). Positron emission tomography. In R. A. Thompson & J. R. Green (Eds.), *New perspectives in cerebral localization.* New York: Raven Press.

Rao, S. M. (1990). Multiple sclerosis. In J. Cummings (Ed.), *Subcortical dementia* (pp. 164–180). New York: Oxford University Press.

Rapoport, S. I. (1991). Positron emission tomography in Alzheimer's disease in relation to disease pathogenesis: A critical review. *Cerebrovascular and Brain Metablism Reviews, 3,* 297–335.

Rebok, G., Brandt, J., & Folstein, M. (1990). Longitudinal cognitive decline in patients with Alzheimer's disease. *Journal of Geriatric Psychiatry and Neurology, 3,* 91–97.

Reichman, W. E., & Cummings, J. L. (1990). Diagnosis of rate dementia syndromes: An algorithmic approach. *Journal of Geriatric Psychiatry and Neurology, 3,* 73–84.

Richardson, J. C., Steele, J., & Olszewski, J. (1963). Supranuclear ophthalmoplegia, pseudobulbar palsy, nuchal dystonia and dementia: A clinical report on eight cases of "heterogenous system degeneration." *Transactions of the American Neurological Association, 88,* 25–29.

Riege, W. H., & Metter, E. J. (1988). Cognitive and brain imaging measures of Alzheimer's disease: Review. *Neurobiology of Aging, 9,* 69–86.

Rinne, J. O., Rummukainen, J., Paljarvi, L., & Rinne, U. K. (1989). Dementia in Parkinson's disease is related to neuronal loss in the medial substantia nigra. *Annals of Neurology, 26,* 47–50.

Roman, G. C. (1991). The epidemiology of vascular dementia. In A. Hartmann, W. Kuschinsky, & S. Hoyer (Eds.), *Cerebral ischemia and dementia* (pp. 9–15). Berlin: Springer-Verlag.

Roman, G. C., Tatemichi, T. K., Erkinjuntti, T., Cummings, J. L., Masdeu, J. C., Garcia, J. H., Amaducci, L., Orgogozo, J.-M., Brun, A., Hofman, A., Moody, D. M., Obrien, M. D., Yamaguchi, T., Grafman, J., Drayer, B. P., Bennett, D. A., Fisher, M., Ogata, J., Kokmen, E., Bermego, F., Wolf, P. A., Gorelick, P. B., Bick, K. L., Pajeau, A. D., Bell, M. A., DeCarli, C., Culebras, A., Korczyn, A. D., Bogousslavsky, J., Hartmann, A., & Scheinberg, P. (1993). Vascular dementia: Diagnostic criteria for research studies. Report of the NINDS-AIREN International Workshop. *Neurology, 43,* 250–260.

Roos, R., Gajdusek, D. C., & Gibbs, C. J. (1973). The clinical characteristics of transmissible Creutzfeldt-Jakob disease. *Brain, 96,* 1–20.

Rosen, W. G., Terry, R. D., Fuld, P. A., Katzman, R., & Peck, A. (1980). Pathological verficiation of ischemic score in differentiation of dementias. *Annals of Neurology, 7,* 486–488.

Rosenblum, W. I., & Ghatak, N. R. (1979). Lewy bodies in the presence of Alzheimer's disease. *Archives of Neurology, 36,* 170–171.

Rougemont, D., Baron, J. C., Collard, P., Bustany, P., Comar, D., & Agid, Y. (1984). Local cerebral glucose utilization in treated and untreated patients with Parkinson's disease. *Journal of Neurology, Neurosurgery, and Psychiatry, 47,* 824–830.

Rutgers, A. W. F., Lakke, J. P. W. F., Panns, A. M. J., Vaalburg, W., & Korf, J. (1987). Tracing of dopamine receptors in hemiparkinsonism with positron emission tomography (PET). *Journal of the Neurological Sciences, 80,* 237–248.

Salmon, D. P., Kwo-on Yuen, P. R., Heindel, W. C., Butters, N., & Thal, L. J. (1989). Differentiation of Alzheimer's disease and Huntington's disease with the dementia rating scale. *Archives of Neurology, 46,* 1204–1208.

Salmon, E., & Franck, G. (1989). Positron emission tomographic study in Alzheimer's diseasae and Pick's disease. *Archives of Gerontology and Geriatrics, 1* (Suppl.), 241–247.

Satz, P. (1993). Brain reserve capacity on symptom onset after brain injury: A formulation and review of evidence for threshold theory. *Neuropsychology, 7,* 273–295.

Scheinberg, P. (1988). Dementia due to vascular disease- A multifactorial disorder. *Stroke, 19,* 1291–1299.

Schellenberg, G. D., Bird, T. D., Wijsman, E. M., Orr, H. T., Anderson, L., Nemens, E., White, J. A., Bonnycastel, L., Weber, J. L., Alonso, M. E., Potter, H., Heston, L. L., & Martin, G. M. (1992). Genetic linkage evidence for a familial Alzheimer's disease locus on chromosome 14. *Science, 258,* 668–671.

Schonfeld, S. M., Golbe, L. I., Sage, J. I., Safer, J. N., & Duvoisin, R. C. (1987). Computed tomographic findings in progressive supranuclear palsy: Correlation with clinical grade. *Movement Disorders, 2,* 263–278.

Siegler, I. C., Welsh, K. A., Dawson, D. V., Fillenbaum, G. G., Earl, N. L., Kaplan, E. B., & Clark, C. M. (1991). Ratings of personality change in patients being evaluated for memory disorders. *Alzheimer's Disease and Associated Disorders, 3,* 240–250.

Smith, G. S., de Leon, M. J., George, A. E., Kluger, A., Volkow, N. D., McGrae, T., Golomb, J., Ferris, S. H., Reisberg, B., Ciaravino, J., & La Regina, M. E. (1992). Topography of cross-sectional and longitudinal glucose metabolic deficits in Alzheimer's disease: Pathopohysiologic implications. *Archives of Neurology, 49,* 1142–1150.

Sokoloff, L., Reivich, M., Kennedy, C., Des Rosiers, M. H., Patlak, C. S., Pettigrew, K. D., Sakurada, O., & Shinohara, M. (1977). The [^{14}C]deoxyglucose method for the measurement of local cerebral glucose utilization: Theory, procedure, and normal values in the conscious and anesthetized albino rat. *Journal of Neurochemistry, 28,* 897–916.

Starkstein, S. E., Folstein, S. E., Brandt, J., Pearlson, G. D., McDonnell, A., & Folstein, M. (1989). Brain atrophy in Huntington's disease. A CT-scan study. *Neuroradiology, 31,* 156–159.

Stehling, M. K., Turnera, R., & Mansfield, P. (1991). Echo-planar imaging: Magnetic resonance imaging in a fraction of a second. *Science, 254,* 43–50.

Stern, Y., Alexander, G. E., Prohovnik, I., & Mayeux, R. (1992). Inverse relationship between education and parietotemporal perfusion deficit in Alzheimer's disease. *Annals of Neurology, 32,* 371–373.
St. George-Hyslop, G., Tanzi, R., Polinsky, R., Haines, J. L., Nee, L., Watkins, P. C., Myers, R. H., Feldman, R. G., Pilleri, A., Drachman, D., Growdon, J., Bruni, A., Foncin, J. F., Salvion, D., Frannelt, P., Amaducci, L., Sorbi, S., Piacentino, S., Steward, G. D., Hobbs, W., Conneally, P. M., & Gusella, J. F. (1987). The genetic defect causing familial Alzheimer's disease maps on chromosome 21. *Science, 235,* 885–889.
Storandt, M., Botwinick, R. J., Danziger, W. L., Berg, L., & Hughes, C. P. (1984). Psychometric differentiation of mild senile dementia of the Alzheimer type. *Archives of Neurology, 41,* 497–499.
Strittmatter, W. J., Saunders, A. M., Schmechel, D., Pericak-Vance, M., Enghild, J., Salvesen, G. S., & Roses, A. D. (1993). Apolipoprotein E: High affinity binding to beta amyloid and increased frequency of type 4 allele in late onset familial Alzheimer's. *Proceedings of the National Academy of Sciences, 90,* 1977–1981.
Suchowersky, O., Hayden, M. R., Martin, M. R. W., Stoessl, A. J., Hildebrand, A. M., & Pate, B. D. (1986). Cerebral metabolism of glucose in benign hereidtary chorea. *Movement Disorders, 1,* 33–46.
Tatemichi, T. K. (1990). How acute brain failure becomes chronic: A view of the mechanisms of dementia related to stroke. *Neurology, 40,* 1652–1659.
Tedroff, J., Aquilonius, S.-M., Hartvig, P., Lundqvist, H., Gee, A. G., Uhlin, J., & Långström, B. (1988). Monoamine re-uptake sites in the human brain evaluated by means of 11C-nomifensine and positron emission tomography: The effects of age and Parkinson's disease. *Acta Neurologica Scandinavica, 77,* 192–201.
Thompson, P. D., Berardelli, A., Rothwell, J. C., Day, B. L., Dick, J. P. R., Benecke, R., & Marsden, C. D. (1988). The coexistence of bradykinesia and chorea in Huntington's disease and its implications for theories of basal ganglia control of movement. *Brain, 111,* 223–244.
Tierney, M. C., Fisher, R. H., Lewis, A. J., Zorzitto, M. L., Snow, W. G., Reid, D. W., & Nieuwstraten, P. (1988). The NINCDS-ADRDA work group criteria for the clinical diagnosis of probable AD: A clinicopathologic study of 57 cases. *Neurology, 38,* 359–364.
Tomlinson, B. E., Blessed, G., & Roth, M. (1968). Observations on the brains of nondemented old people. *Journal of the Neurological Sciences, 7,* 331–356.
Tomlinson, B. E., Blessed, G., & Roth, M. (1970). Observations on the brains of demented old people. *Journal of the Neurological Sciences, 11,* 205–242.
Tomlinson, B. E., & Henderson, G. (1976). Some quantitative cerebral findings in normal and demented old people. In R. D. Terry & S. Gershon (Eds.), *Neurobiology of aging* (pp. 183–204). New York: Raven Press.
Von Sattel, J. P., Myers, R. H., Stevens, T. J., Ferrante, R. J., Bird, E. D., & Richardson, E. P. (1985). Neuropathological classification of Huntington's disease. *Journal of Neuropathology and Experimental Neurology, 44,* 559–557.
Welsh, K. A., Butters, N., Hughes, J., Mohs, R. C., & Heyman, A. (1991). Detection of abnormal memory decline in mild cases of Alzheimer's disease using CERAD neuropsychological measures. *Archives of Neurology, 48,* 278–281.
Welsh, K. A., Butters, N., Hughes, J., Mohs, R. C., & Heyman, A. (1992). Detection and staging of dementia in Alzheimer's disease: Use of the neuropsychological measures developed for the Consortium to Establish a Registry for Alzheimer's Disease. *Archives of Neurology, 49,* 448–452.
Welsh, K. A., Hoffman, J. M., & Beam, C. (1994). Positron emission tomography and neuropsychological assessment in the detection of Alzheimer's disease. *Neuropsychological Society: Program & Abstracts, 1,* 42.
Welsh, K. A., Hoffman, J. M., Earl, N. L., & Hanson, M. W. (1994). Neural correlates of dementia: Regional brain metabolism and the CERAD neuropsychological battery. *Archives of Clinical Neuropsychology, 9,* 395–409.
Welsh, K. A., Hoffman, J. M., McDonald, W. M., Earl, N. L., & Breitner, J. C. S. (1993). Concordant but different: Cognitive function, cerebral anatomy, and metabolism in monozygotic twins with Alzheimer's disease. *Neuropsychology, 7,* 158–171.
Workshop Panel. (1990). National Cancer Institute workshop statement: Advances in clinical imaging using positron emission tomography. *Archives of Internal Medicine, 150,* 735–739.
Xuereb, J. H., Tomlinson, B. E., Irving, D., Perry, R. H., Blessed, G., & Perry, E. K. (1990). Cortical and subcortical pathology in Parkinson's disease: Relationship to parkinsonian dementia. In M. B. Streifler, A. D. Korczyn, E. Melamed, & M. B. H. Youdin (Eds.), *Advances in neurobiology: Vol. 53. Parkinson's disease: Anatomy, pathology, and therapy* (pp. 35–40). New York: Raven Press.

Yoshii, F., Barker, W. W., Chang, J. Y., Loewenstein, D., Apicella, A., Smith, D., Boothe, T., Ginsberg, M. D., Pascal, S., & Duara, R. (1988). Sensitivity of glucose metabolism to age, gender, brain volume, brain atrophy, and cerebrovascular risk factors. *Journal of Cerebral Blood Flow and Metabolism, 8,* 654–661.

Yoshimura, M. (1983). Cortical changes in the parkinsonian brain: A contribution to the delineation of "diffuse Lewy body disease." *Journal of Neurology, 229,* 17–32.

Young, A. B., Penney, J. B., Starosta-Rubinstein, S., Markel, D., Berent, S., Rothley, J., Betley, A., & Hichwa, R. (1987). Normal caudate glucose metabolism in persons at risk for Huntington's disease. *Archives of Neurology, 44,* 254–257.

8

Functional and Structural Measures for Determining Risk of Memory Change following Epilepsy Surgery

RICHARD I. NAUGLE, GORDON J. CHELUNE, AND ERIC B. GELLER

INTRODUCTION

Modern surgical intervention for the relief of medically intractable seizures was introduced by Horsley in 1886. Since that time, surgical resection of epileptogenic brain tissue has become a widely accepted medical treatment option for patients with chronic, focal seizure disorders that are refractory to conventional anticonvulsant medications (Anderman, 1987; Engel, 1987). With advanced preoperative screening procedures and methods that are available at specialized epilepsy surgery centers, it is estimated that about 80% of patients with intractable complex partial seizures of temporal origin can become free of seizures or experience marked reductions in seizure tendency following epilepsy surgery (Dreifuss, 1987; Engel, 1987; Ivnik, Sharbrough, & Laws, 1988; Olivier, 1988;

RICHARD I. NAUGLE AND GORDON J. CHELUNE Departments of Psychiatry and Psychology and Neurology, The Cleveland Clinic Foundation, Cleveland, Ohio 44195. ERIC B. GELLER Department of Neurology, The Cleveland Clinic Foundation, Cleveland, Ohio 44195.

Neuroimaging II: Clinical Applications, edited by Erin D. Bigler. Plenum Press, New York, 1996.

Walczak *et al.*, 1990). Although epilepsy surgery has proven efficacy in terms of increasing seizure control, it is not without attendant risks, especially for decrements in neurocognitive function (see reviews by Chelune, 1992; Dodrill, Hermann, Rausch, Chelune, & Oxbury, 1993). As noted by Lüders *et al.* (1987), the goal of epilepsy surgery is to maximally excise the epileptogenic tissue with a minimum of disruption of normal tissue, particularly eloquent tissue that is critical for full functional capacity. Thus, surgical success is defined in terms of a significant reduction in seizure tendency with maximal perseveration or relative improvement in cognitive function and quality of life (National Institutes of Health, 1990).

Since the benefits of increased seizure control must be considered within a risks-to-benefits context, patients contemplating epilepsy surgery need to be advised of their potential chances of experiencing a decrement in functional ability as a result of surgery (Chelune, Naugle, Lüders, & Awad, 1991; Ivnik, Sharbrough, & Laws, 1988). Considerable research has accumulated documenting the typical patterns of cognitive change associated with epilepsy surgery, including possible preoperative variables that appear to moderate such modal patterns of change for the individual patients. While a number of cognitive abilities are at risk, intelligence, language, and memory have been most extensively examined. Of these, memory deficits have been, and remain, the major neurocognitive complication of epilepsy surgery (see reviews by Rausch, 1987; Chelune, 1992; Awad & Chelune, 1993; Naugle, 1992; Saykin *et al.*, 1992). This chapter examines the predictive value of neuropsychological measures of functional ability as well as biomedical tests of structural integrity in identifying patients most at risk for memory decrements following temporal lobe epilepsy surgery. Emphasis is placed on assessing the functional adequacy of the tissue to be resected rather than the functional reserve of the contralateral hemisphere in determining which patients are most at risk for memory deficits. Particular attention is given to the coordinated use of neuropsychological procedures and volumetric magnetic resonance imaging (MRI) in forecasting changes in memory function.

Functional Reserve versus Functional Adequacy

Neuropsychological assessment is typically included as part of a patient's preoperative evaluation for several purposes, including baseline assessment of cognitive function, localization and lateralization, prediction of seizure control, and prediction of neuropsychological outcome (Chelune, 1992, 1994). Disorders of learning and memory have long been associated with temporal lobe dysfunction and have been extensively studied in relation to complex partial seizure disorders (Delaney, Prevey, & Mattson, 1982; Hermann, Wyler, Richey, & Rea, 1987; Loring, Lee, Martin, & Meador, 1988) and the effects of temporal lobe (TL) epilepsy surgery (Chelune *et al.*, 1991; Lee, Loring, & Thompson, 1989; Milner, 1975; Ojemann & Dodrill, 1985; Rausch & Crandall, 1982). While global amnestic syndromes following epilepsy surgery are fortunately rare (Olivier, 1988; Van Buren, 1987; Walczak *et al.*, 1990), milder deficits in material-specific recall are quite common (Naugle, 1992). Nearly half of patients undergoing TL resections of the speech-dominant hemisphere experience a clinically reliable decrement in verbal memory (Chelune, Naugle, Lüders, Sedlak, & Awad, 1993). Because of the frequency of memory deficits following epilepsy surgery, investigators have sought methods of identifying individuals who are most at risk for developing such complications following TL (Ivnik, Sharbrough, & Laws, 1988; Milner, Branch, & Rasmussen, 1962; Ojemann & Dodrill, 1985). While there is virtually universal agreement that the temporal lobes and hippocampi play a central role in memory, there are two schools of thought concerning risk for memory decrements following TL surgery. The traditional model of risk determination suggests that postsurgical memory deficits depend on the capacity or *functional reserve* of the contralateral temporal lobe to support memory functions following surgery. The second model of hippocampal function is less articulated and is still emerging. However, it may be tentatively described as a *functional adequacy* model (Chelune, 1995). That is, postsurgical memory deficits are thought to be dependent on the functional adequacy of the tissue to be resected rather than the memory reserve in the contralateral TL.

The traditional model of hippocampal reserve derives largely from early studies of memory deficit following TL surgery and from investigations involving the intracarotid amobarbital procedure (IAP) or Wada test. Several early researchers presented clinical reports of memory deficit following unilateral TL surgery (e.g., Baldwin, 1956; Walker, 1957), but it was not until Scoville and Milner (1957) reported their study of three patients with persistent anterograde amnesia following bilateral mesial TL surgery that the importance of the mesial temporal cortices and hippocampi for the acquisition and retrieval of new information became widely acknowledged. Penfield and Milner (1958) presented two new cases of postsurgical amnesia, this time following unilateral resections. Because of the similarity between the memory deficits of these patients to those who had undergone bilateral TL resections, Penfield and Milner postulated the presence of an occult lesion in the nonsurgical temporal lobe. This hypothesis was later confirmed in one of the two patients who eventually died of a non-epilepsy-related illness (Penfield & Mathieson, 1994). Since the functional capacity of the contralateral (nonsurgical) TL appeared to moderate the severity of the observed postsurgical memory deficits, Milner and colleagues (Milner, Branch, & Rasmussen, 1962) adapted their use of the IAP to include a memory component to assess the integrity of the contralateral mesial TL and to determine whether it could support memory should the patient undergo surgery.

The IAP was pioneered by Juhn Wada as a means of selectively inducing hemispheric anesthetization to identify language dominance (Wada & Rasmussen, 1960). While the IAP continues to be the principal tool for lateralizing language function among patients who are candidates for epilepsy surgery (Loring, Meador, Lee, & King, 1992), use of the IAP to assess memory function is still controversial (Jones-Gotman *et al.*, 1993). Intracarotid amobarbital injections of the hemisphere ipsilateral to the presumed seizure focus temporarily inactivate the surgical hemisphere and isolate functions in the nonsurgical contralateral hemisphere. Memory testing during this reversible state allows investigators to model the possible effects of the planned temporal lobectomy (Jones-Gotman, 1987; Rausch, Babb, Engel, & Crandall, 1989). The patient who fails memory testing during the ipsilateral IAP is regarded to be at risk for an amnestic event, whereas the patient who passes memory testing is regarded to be at reduced risk for a substantial decrement in memory. The magnitude of decrement in memory is believed to be related to the capacity or *reserve* of the contralateral temporal lobe to support memory after resection of the ipsilateral lobe. This is the *functional reserve* model of hippocampal function.

The value of the IAP in determining the lateralization of the seizure focus has also been clearly demonstrated by numerous investigators (Loring *et al.*, 1990, 1993; Milner *et al.*, 1962; Perrine *et al.*, 1993; Rausch *et al.*, 1989; Sass *et al.*, 1991; Wyllie *et al.*, 1991). However, the IAP has not been found to be very predictive of postsurgical amnesic disorders, perhaps because the occurrence of global amnesia following TL resection is quite rare today, with only six true-positive cases being reported in the experience of 68 advanced surgical centers in a recent survey (Rausch *et al.*, 1993). The value of the ipsilateral IAP has also not been found to be particularly useful in predicting selective memory impairments following TL resection. Loring and colleagues (1990) have observed material-specific memory decrements among some patients who performed poorly on ipsilateral IAP, but nonetheless recommended against "relying solely on amobarbital memory testing to assess the functional ability of the contralateral temporal lobe to sustain global memory prior to temporal lobectomy" (p. 605). Likewise, Wyllie and colleagues (1991) failed to find a significant relationship between the functional reserve of the contralateral temporal lobe as assessed by the IAP and subsequent changes in memory following temporal lobectomy.

Because measures of the functional reserve of the nonsurgical, contralateral temporal lobe cannot be reliably used to predict the commonly observed decrements in memory following TL epilepsy surgery, investigators have begun to explore alternative models. In recent years, there has been a growing body of convergent research using divergent functional and structural measures of biological integrity that suggests that it is the functional adequacy of the tissue to be resected in the ipsilateral TL that determines the nature and extent of memory loss following epilepsy surgery (see Chelune, 1995).

Functional Measures of Adequacy

Several studies have demonstrated an inverse relationship between levels of preoperative memory performance and subsequent changes following TL epilepsy surgery on both traditional neuropsychological tests of memory and memory performance obtained during contralateral IAP injections. In a small series of epilepsy surgery patients, Chelune *et al.* (1991) found that patients with average or above-average baseline functioning showed the largest decrements in performance following surgery, at least among patients undergoing dominant hemisphere TL resections. While the authors suggested that baseline neuropsychological performance may be a useful predictor of subsequent changes in memory, they acknowledged that their findings were based on a small sample that was limited by statistical artifacts (practice effects and regression to the mean) as well as institutional biases and test-specific findings.

To address the limitations noted in the Chelune *et al.* (1991) study, a multicenter replication and extension was conducted at the Cleveland Clinic Foundation in collaboration with the Mayo Clinic, EpiCare Center at Baptist Memorial Hospital, and Long Island Jewish Hospital (Chelune *et al.*, 1992). This study involved 453 TL epilepsy surgery patients and employed several different memory tests. In addition, data from 50 nonoperated seizure control subjects were examined to control for practice effects and regression to the mean. Even with the increased statistical controls, similar findings were obtained across the centers using the Wechsler Memory Scale-Revised (WMS-R; Wechsler, 1987), California Verbal Learning Test (Delis, Kramer, Kaplan, & Ober, 1987), and the Rey Auditory Verbal Learning Test (RAVLT) (Rey, 1964): patients with higher baseline memory functions were at greater risk for memory decrements than those who had lower baseline levels of functioning. This relationship was true for patients whether they underwent dominant or nondominant TL resections. Other centers have also reported an inverse relationship between baseline functioning and subsequent neurocognitive outcome following epilepsy surgery (Bauer *et al.*, 1994).

If one assumes a functional adequacy model of hippocampal functioning, one might expect that assessment of the memory capacity of the surgical hemisphere as assessed via IAP injections of the hemisphere contralateral to the presumed seizure focus would also show an inverse relationship with postsurgical changes in memory similar to that seen with traditional neuropsychological memory tests. Kneebone, Chelune, Dinner, Awad, and Naugle (1992) compared changes in memory functions following TL surgery among 63 patients who had "passed" or "failed" their contralateral IAP memory assessments. As expected, patients having left TL surgery showed significant decrements in verbal memory, but not in visual memory. Furthermore, patients undergoing left TL resection and who had passed the contralateral IAP had significantly larger decrements in memory at follow-up compared to those with failing IAP memory scores. Bauer and colleagues (Bauer *et al.*, 1994) have also reported that patients with the largest decrements in memory following epilepsy surgery tended to have higher IAP memory scores than those who showed little or no change in memory after surgery.

The predictive value of the contralateral IAP in identifying patients at risk for postsurgical memory changes does not appear to be redundant with that provided by baseline neuropsychological tests of memory. In a series of multiple regression analyses, Chelune, Kneebone, Dinner, Awad, and Naugle (1993) found that contralateral IAP memory scores and baseline neuropsychological measures were additive and explained unique sources of variance. Consistent with previous research, both variables were inversely related to subsequent memory loss after epilepsy surgery.

Anatomic Measures of Hippocampal Adequacy

The studies cited above involve measures of functional capacity. Those individuals who demonstrate good functional memory capacity on neuropsychological examination or IAP testing prior to surgery are most at risk for decrements in memory function following TL epilepsy surgery. This is presumably because the hippocampus and adjacent cortex are still functionally adequate. Studies based on anatomical data provide further support that the structural adequacy of the hippocampus in the surgical hemisphere is related to functional memory outcomes.

Early epilepsy surgery research at the Montreal Neurological Institute examining the effects of large versus small resections of the hippocampus suggested that postsurgical changes in memory were proportional to the amount of hippocampus resected (Corkin, 1965; Corsi, 1972; Milner, 1968). However, a more recent study by Leonard (1991) of 278 patients seen at the Montreal Neurological Institute between 1961 and 1980 failed to show any difference in the magnitude of memory changes between patients who had large or small hippocampal resections. Similarly, studies by Loring and colleagues (1991), Wolf and associates (1993), and Hermann, Wyler, and Somes (1993) have failed to find a relationship between extent of resection and subsequent memory loss. While these investigators found changes in material-specific memory, they were unable to relate these changes to extent of resection, possibly due to the moderating effects of hippocampal adequacy. That is, a large resection may produce no more effect on memory than a small resection if the hippocampus is highly dysfunctional.

Histologic studies of hippocampal specimens taken at the time of surgery provide further anatomic support for the hippocampal adequacy model, at least in the speech-dominant hemisphere. Neuronal cell densities in various mesial and lateral temporal structures have been correlated with memory for story passages, simple line drawings, and verbal paired associates (Rausch & Babb, 1987, 1993; Sass *et al.*, 1992a, 1992b, 1990). As might be expected, poor memory performance was correlated with greater extent of hippocampal cell loss, but only among left TL resection patients, in all of these studies. Similarly, hippocampal cell densities have also been related to memory capacity in the surgical hemisphere during IAP injections of the contralateral hemisphere. Rausch and colleagues (1989) observed that those patients with severe neuronal cell loss throughout the epileptogenic hippocampus tended to do poorly on memory testing during IAP injections contralateral to the surgical hemisphere. Sass and colleagues (1991) compared volumetric cell densities in several hippocampal subfields among those patients who passed versus those who failed their IAP memory testing contralateral to the planned side of surgery. Those patients who failed their contralateral IAP memory testing had significantly greater neuronal cell loss in the CA3 region than those who passed with failure rates increasing with increasing degrees of cell loss.

While preoperative memory performance appears to be positively correlated with the degree of anatomic integrity of the hippocampus, postoperative changes in neuropsychological measures of memory following TL resection have been inversely associated with histological indices of hippocampal pathology. Oxbury and Oxbury (1989) reported that left TL surgery patients who did not have histologic evidence of hippocampal sclerosis manifested greater declines in immediate and delayed verbal memory than those whose hippocampi did reveal evidence of extensive pathology. In a prospective study of 40 patients, Hermann and associates (Hermann, Wyler, Somes, Berry, & Dohan, 1992) reported that decrements in memory following temporal lobectomy were inversely related to the degree of hippocampal sclerosis as determined by histologic examination. Memory losses for both verbal and nonverbal material were greater for patients who underwent left TL surgery and who had minimal hippocampal sclerosis. In a larger series of 26 right and 36 left temporal lobectomy patients, Hermann and colleagues (Hermann *et al.*, 1993) demonstrated that the decrements in verbal memory functioning were limited to measures of episodic memory and did not involve semantic memory or verbal memory span.

Some of the most compelling evidence suggesting that hippocampal adequacy of the speech-dominant temporal lobe is a major moderating variable of decrements in postsurgical verbal memory comes from a study by Hermann, Wyler, and Somes (1993). In a randomized surgical trial examining the effects of extent of hippocampal resection on seizure control among 30 left TL surgery patients, these researchers found that decrements in verbal memory were associated with the absence of histologic evidence of hippocampal sclerosis, but not extent of hippocampal resection. Before surgery, patients who were found to have hippocampal sclerosis were clearly inferior in verbal memory to those without hippocampal pathology. However, those patients who had clear hippocampal pathology demonstrated only a minimal decrement in verbal memory following surgery, whereas those patients with relatively healthy hippocampi demonstrated a marked memory loss after surgery and were actually inferior to those with hippocampal sclerosis.

MRI Volumetric Studies of Hippocampal Adequacy

While the relationship between histological data and memory performances both before and after surgery are compelling and complement results based on functional measures of adequacy, they are of limited value in identifying patients at risk for memory loss since histologic analyses are only available after surgical intervention has occurred. Fortunately, advances in magnetic resonance imaging (MRI) techniques, especially thin-slice turboflash sequences, have provided clinicians with a noninvasive means of evaluating hippocampal atrophy before surgery.

Selective atrophy of the anterior temporal lobe and the hippocampus have long been recognized as hallmarks of temporal lobe epilepsy, especially among nonlesional cases (Margerison & Corsellis, 1966). As shown in Figure 1, methods of digitally reconstructing 3-D volumetric images of the hippocampal structures from tracings of coronal or angled T1-weighted thin-slice MRI turboflash im-

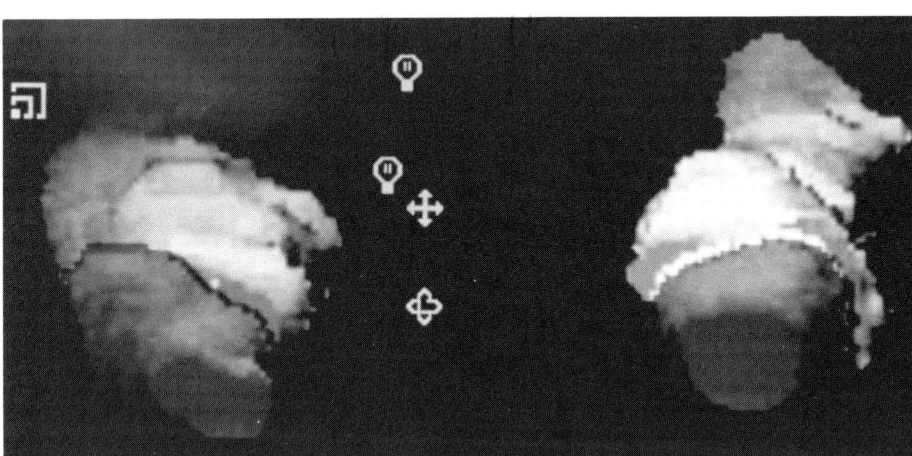

Figure 1. Three-dimensional digital reconstruction of the left and right hippocampal structures from tracings of T1 weighted thin slice MR turboflash images. The darker area of each structure (in the foreground) represents the amygdala. Photograph courtesy of Imad Najm, MD, and Youssef G. Comair, MD, of the Department of Neurological Surgery at the Cleveland Clinic Foundation.

ages have been developed by a number of investigators (Ashtari, Barr, Schaul, & Bogerts, 1991; Cook, Fish, Shorvon, Straughan, & Stevens, 1992; Jack *et al.*, 1989; Watson *et al.*, 1992). These MRI-based hippocampal volume measures have been shown to be highly accurate and reliable (Jack, Bentley, Twomey, & Zinsmeister, 1990; Jack *et al.*, 1988). Right-left asymmetries in MRI hippocampal volumes have been associated with seizure lateralization in a number of studies, with the seizure focus lateralized to the side of the smaller of the two hippocampi (Ashtari *et al.*, 1991; Berkovic *et al.*, 1991; Cendes *et al.*, 1993; Cook *et al.*, 1992; Jack *et al.*, 1990; Loring *et al.*, 1993; Murro *et al.*, 1993; Spencer, McCarthy, & Spencer, 1993). Hippocampal asymmetries have also been associated with anatomic evidence of pathology and degree of neuronal cell loss in the hippocampus (Casino *et al.*, 1991; Cendes *et al.*, 1993; Kuzniecky *et al.*, 1987; Lencz *et al.*, 1992). Jack and colleagues at the May Clinic (Jack *et al.*, 1992) have also found a significant relationship between seizure control and volume in the operated hippocampal formation, with 97% of patients having satisfactory postoperative seizure control when hippocampal asymmetries were concordant with EEG lateralization data.

It is evident that MRI volumetric analysis of the hippocampal formation is highly sensitive to nonlesional pathology, which in turn is associated with seizure lateralization. As depicted in Figure 2, it is also apparent that neuronal cell densities in the hippocampal formation are related to neuropsychological measures of memory both before and after temporal lobectomy. Therefore, assuming that the functional adequacy model accounts for the memory losses often observed after TL surgery, it is reasonable to hypothesize that MRI-based measures of structural adequacy of the hippocampus in the surgical temporal lobe should be related to neuropsychological measures of functional adequacy (i.e., memory). Barr and colleagues (Barr, Ashtari, Schaul, & Bogerts, 1990) reported significant correlations between left hippocampal volumes and verbal memory measures and between right hippocampal volumes and visual memory tests in a sample of 22 temporal lobe epilepsy patients prior to surgery. Lencz and associates (1992) also reported a significant association between volume measurements of the left temporal lobe and presurgical performance on two verbal memory tests among patients with seizure foci in the left temporal lobe. In a study by Loring and colleagues (1993), MRI-based hippocampal volume asymmetries were significantly correlated ($r = .78$) with asymmetries in IAP memory, and

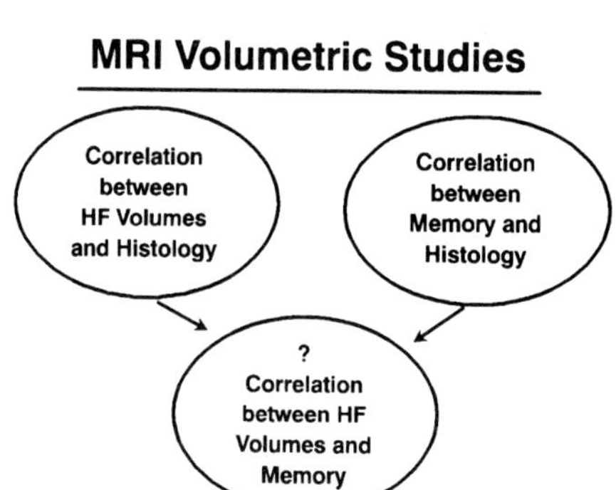

Figure 2. Research has suggested a correlation between neuronal cell density in the hippocampal formation (HF) and neuropsychological measures of memory ability and between memory function and volumetric estimates of the hippocampus. Therefore, it seems reasonable to hypothesize that MR-based mesasures of hippocampal volume may be correlated with performance on memory measures.

when MRI volumetric data was used in combination with the IAP data, 100% of the patients were correctly classified as having either right or left temporal seizure foci. Studies at the Mayo Clinic (Trenerry *et al.*, 1991, 1993) have provided the strongest MRI support to date for the functional adequacy model of hippocampal memory. Investigators found an inverse relationship between relative atrophy of the left hippocampus and verbal memory decline among left TL surgery patients (Trenerry *et al.*, 1991, 1993). Consistent with data from histological studies of hippocampal pathology, patients with large MRI hippocampal volumes in the left temporal lobe showed larger verbal memory decrements following left TL surgery than those with evidence of hippocampal atrophy. Conversely, larger left hippocampal volumes were positively correlated with better verbal memory both before and after right TL surgery (Trenerry *et al.*, 1991).

Our own MRI volumetric research with Drs. Comair and Najm at the Cleveland Clinic also supports the utility of using hippocampal volumes to predict memory changes following temporal lobectomy. While still in progress, our preliminary findings can be briefly summarized here (Chelune, 1995; Chelune, Comair, Najm, *et al.*, 1995). Using normative volume data from 24 controls, we have established statistical criteria for classifying seizure patients as showing either left or right hippocampal atrophy. Comparing these individuals to unoperated seizure controls and neurologically normal controls, right and left TL surgery patients with hippocampal atrophy in the resected temporal lobe show little change in verbal or visual memory following surgery. However, among patients without hippocampal atrophy, there is a clear risk for material-specific memory loss following temporal lobectomy, depending on the side of surgery. That is, patients without evidence of hippocampal atrophy in the surgical left temporal lobe are at increased risk for a decrement in verbal memory, whereas the absence of hippocampal atrophy among right TL surgery patients places them at increased risk for a visual memory loss.

Case Examples

The following cases exemplify the trends that have been described in the preceding review of the research on the prediction of risk for neuropsychological changes following temporal lobectomy. Specifically, the presence of hippocampal atrophy was found to be associated with lower baseline memory scores and less postoperative change in the first case. The second patient, whose MRI did not reveal hippocampal atrophy, showed higher preoperative memory ability than the first patient and substantially greater declines after surgery. Both of the patients had left temporal onset of their seizures and underwent left temporal lobectomies. Both patients were seizure-free at the time of their follow-up examinations approximately 6 months after surgery. They are comparable on the basis of age and gender. However, the two cases differ significantly with regard to the extent of hippocampal atrophy noted prior to surgery and their postsurgical neuropsychological outcomes.

Patient 1. Patient 1 is a 26-year-old male with a high school education who began to have complex partial seizures with frequent secondary generalization at the age of 16 months. Seizures occurred every time he developed a febrile illness; his first nonfebrile seizure occurred at the age of 3 years. He described his auras as beginning with the sensation that he was gasping for air with associated feelings of severe anxiety, after which he lost consciousness. He was observed to then have smacking of the lips and swallowing movements and at times to have uttered loud screams. Postictally, he showed periods of confusion and increased somnolence. At the time of his preoperative examination, he was having approximately two to three, but as many as six, complex partial seizures a week, with two to three seizures becoming secondarily generalized each month. He had up to two seizures a day, but had gone as long as 2 months without an episode. He had tried several anticonvulsants, including Tegretol, Mysoline, Dilantin, Depakote, and Felbamate, alone or in various combinations without success. At the time of his preoperative examination, he was on a combination of Dilantin and Depakote.

His interictal EEG revealed that the majority of spikes (85%) arose from the left anteromesial temporal lobe, maximum at the sphenoidal electrode. There were occasional (5%) generalized spike-wave complexes and rare spikes at the right sphenoidal, P7, O1, and FP2 electrodes. Six seizures were recorded during prolonged preoperative video EEG monitoring. Ictal EEG onset was obscured by muscle artifact, but in five of the six seizures there was a later appearance (>10 seconds) of a rhythmic theta pattern over the left temporal region.

Computerized tomography (CT) scanning in 1985 was unremarkable and his MRI in 1994 was initially considered to be unremarkable. However, further analysis (see Figure 3A) revealed an area of increased T2 signal in the left hippocampus, strongly suggestive of mesial temporal sclerosis (Jack, 1994; Jackson, 1994; Kuzniecky *et al.*, 1987). As seen in Figure 3B, the T1-weighted, thin-section MRI did not show clear hippocampal atrophy. It has been suggested that in such cases, quantitative volume measurements may be useful in demonstrating atrophy (Jack, 1994). In fact, volumetric assessment suggested that the left hippocampus was approximately 71% (3.194 cm^3) the size of the right (4.449 cm^3). Hypometabolism on positron emission tomography using ^{18}F-fluorodeoxyglucose (FDG-PET) may be localized in 60–90% of patients with temporal lobe epilepsy and may demonstrate focal abnormalities even in those instances when EEG is not well localized (Abou-Khalil *et al.*, 1987; Engel *et al.*, 1990, Henry, Mazziotta, & Engel, 1993; Theodore *et al.*, 1983, 1992). As depicted in Figure 4, FDG-PET showed decreased regional cerebral glucose metabolism in the left temporal lobe.

The patient's IAP results suggested left hemisphere dominance for speech. He was able to correctly recognize 10 of 16 stimuli presented after left injection (i.e., using the right hemisphere) and 13 of 16 after right injection (i.e., using the left hemisphere), suggesting bilateral representation of memory. His most recent seizure had occurred 3 days prior to the initiation of his neuropsychological examination, the relevant results of which are presented in Table 1.

Intellectual functioning ranged from borderline to low average. Language studies revealed poor confrontation naming but intact comprehension/execution of commands. The patient's Controlled Oral Word Association was interrupted by a seizure and not completed thereafter. His memory functioning is of particular interest. Note that whereas his Attention/Concentration Index fell within the average range, his Verbal and Visual Memory Indices fell within the borderline range. After a delay, he freely recalled only 50% of the content of the Logical Memory passages that he had initially recalled. His verbal rote learning on the RAVLT was inefficient, and his delayed free recall of that material was impoverished. In contrast, his recognition was rather well pre-

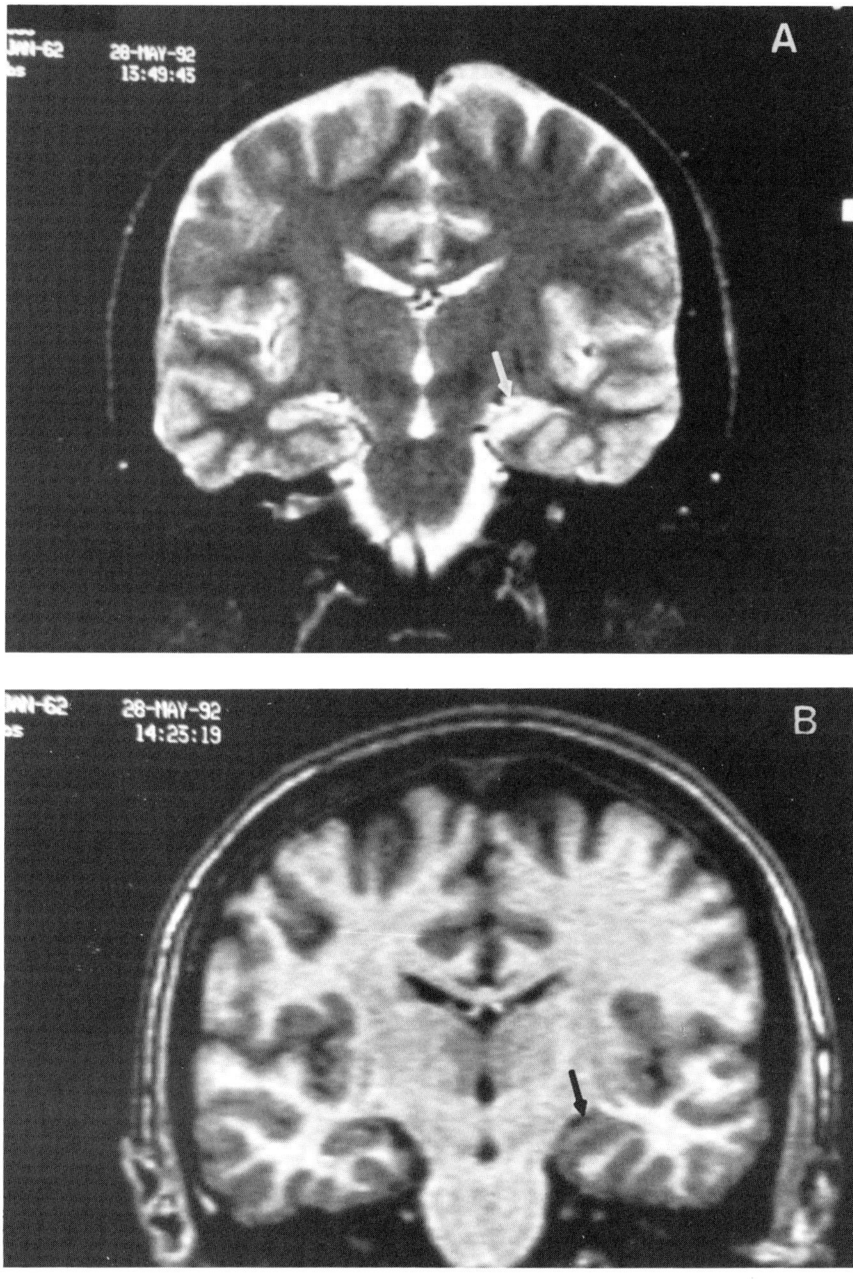

Figure 3. (A) Coronal T2-wighted MRI of a 30-year-old male with left temporal lobe epilepsy (TR = 2500 ms, TE = 90 ms, slice thickness = 6 mm). The patient's left is on the viewer's right. Note the area of increased T2 signal (seen as increased brightness) in the left hippocampus. Increased hippocampal T2 signal is associated wtih mesial temporal sclerosis (see text). (B) The T1-weighted, thin-section (2 mm) slice at approximately the same level does not show clear hippocampal atrophy on visual inspection (TR = 10 ms, TE = 2 ms).

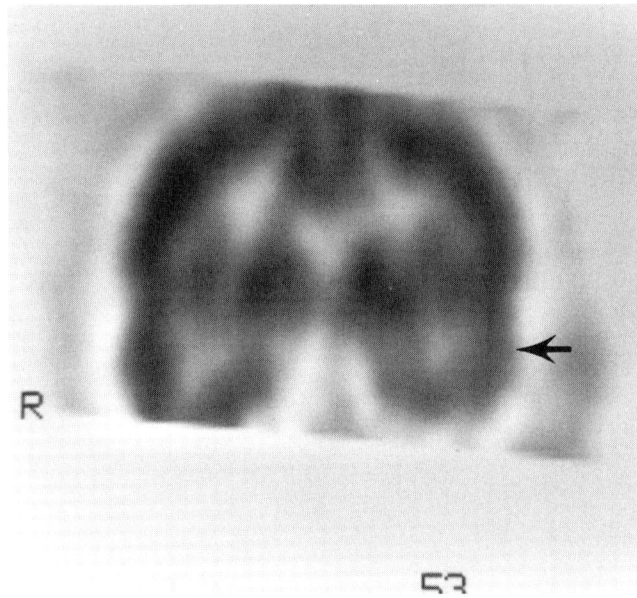

Figure 4. FDG-PET on the same patient as in Figure 2. This is a coronal image at approximately the same level as the MRI images shown, again with the patient's left on the viewer's right. The left temporal lobe cortex is not as dark as on the right, representing decreased FDG uptake in this region. Regional glucose hypometabolism in a temporal lobe is supportive in localizing an epileptogenic focus (see text).

served. Recognition of words and faces, as assessed by the Warrington Recognition Memory Test (RMT), fell within the high average and average ranges, respectively.

The patient underwent a left anterior temporal lobectomy and was seizure-free at the time of his postoperative neuropsychological examination 6 months after surgery. Relevant results of that examination are also presented in Table 1. At that time, he was maintained on Dilantin monotherapy. He reported some memory difficulties, which he characterized as improving, and showed word finding difficulties in his speech.

After surgery, his Attention/Concentration Index again fell within the average range, whereas other indices from the WMS-R generally fell within the low average range. Note that the Verbal Memory Index increased from 73 to 82, whereas the Visual Memory Index increased from 78 to 80. Using test-retest data derived from a sample of unoperated control subjects with intractable epilepsy, a patient with a preoperative Verbal Memory Index of 73 would be expected to have a Verbal Memory Index of 83 on retesting due to practice effect; a posttest Visual Memory Index score of 82 would be anticipated on the basis of his pretest score of 78 (McSweeny, Naugle, Chelune, & Lüders, 1993). In light of these predicted scores, this patient's obtained Verbal and Visual Memory Standard scores on retesting are consistent with expectations even if surgery had not occurred, suggesting that the preoperative to postoperative change was consistent with what would be expected solely on the basis of having been exposed to the test materials on two occasions.

Although data are not yet available to predict the extent of practice effects that might be expected on Logical Memory and Visual Reproduction subtests,

TABLE 1. PREOPERATIVE AND POSTOPERATIVE NEUROPSYCHOLOGICAL EXAMINATION RESULTS: CASE 1

	Preoperative examination	Postoperative examination
Intellectual functioning		
Wechsler Adult Intelligence Scale—Revised		
Verbal IQ	83	96
Performance IQ	72	87
Full Scale IQ	76	91
Speech and language		
Boston Naming Test (Raw score/60)	29	27
FAS	discontinued	25–29 percentile
Multilingual Aphasia Examination		
Token Test	82 percentile	67 percentile
Memory function		
Wechsler Memory Scale—Revised		
Attention/Concentration Index (SS)[a]	95	97
Verbal Memory Index (SS)	73	82
Visual Memory Index (SS)	78	80
Delayed Memory Index (SS)	78	86
Logical Memory I (Raw score/Percentile)	16/13 percentile	20/24 percentile
Logical Memory II[b]	8/7 percentile (50%)	18/31 percentile (90%)
Visual Reproduction I	26/12 percentile	25/11 percentile
Visual Reproduction II[b]	22/13 percentile (85%)	26/21 percentile (100%)
Rey Auditory Verbal Learning Test	5/7/8/8/9	3/5/8/7/8
Total D (Raw score)	5	4
Trial VI (Raw score)	5	1
Recognition (Raw score)	13 (+0 false positive)	9 (+1 false positive)
Warrington Recognition Memory Test		
Words (Raw score/Scaled score)	49/13	44/7
Faces (Raw score/Scaled score)	43/10	46/12
Psychopathology		
Beck Depression Inventory (Total score)	17	13

[a]SS, standard score.
[b]Percent retention scores are given in parentheses.

the patient's postoperative scores clearly demonstrated strong improvements in performance. His delayed free recall of verbal stimuli improved from the 7th to the 31st percentile, whereas his delayed free recall of visuospatial material increased from the 13th to the 21st percentile. Postoperative retention scores were 90% for Logical Memory and 100% for Visual Reproduction. However, he recalled only 3 of the 15 stimuli of the RAVLT during the initial trial and recited only up to 8 of the 15 stimuli over the course of five trials. He freely recalled only one item after the interference trial and identified nine words on the recognition trial while making a single false-positive error. Scores from the Warrington RMT revealed low average recognition of words and high average recognition of faces. Naming to confrontation continued to be impoverished, whereas controlled oral word association and comprehension of commands fell within the average range.

PATIENT 2. Patient 2 showed dramatic changes in memory function following resection that included a hippocampus that did not appear to be atrophied on MRI. Patient 2 is a 42-year-old right-handed male who also had a high school education. He developed seizures at the age of 14, perhaps as a consequence of an episode of encephalitis at the age of 6 months. His seizures occurred approximately one to two times monthly, with a maximum of five seizures monthly. The longest period of time during which he was seizure-free in the year prior to surgery was 2.5 months. He denied having auras and his seizures were characterized by staring and mouth and hand automatisms. On occasion, his seizures progressed to clonic twitching of the right face, version to the right, dystonic posturing of the right hand, and secondary generalization. Postictally, he had prominent dysnomia and expressive aphasia. He had been tried on Dilantin, Phenobarbital, Tegretol, Depakote, and Primodone. His neurological examination was unremarkable.

His initial video EEG evaluation showed that 98% of interictal sharp waves involved the left inferior region (maximum at FC5, also involving FT7, F7, and F5). Only 2% of the interictal sharp waves were maximum at the left sphenoidal electrode (SP1). Ictal EEG was not clearly localizing. Of 14 recorded seizures, only two showed a regional onset in the left inferior frontal/anterior temporal region (FC5 and FT7 maximum). While five seizures were lateralized to the left hemisphere, six were nonlateralized and one had no EEG changes.

In order to better localize the patient's ictal onset zone, further video EEG monitoring was performed using epidural peg and bilateral foramen ovale (FO) electrodes that were inserted surgically, with their multiple contacts lying in the subarachnoid space beneath the mesial temporal lobe (Wyllie & Awad, 1993). The epidural peg electrodes (E) were inserted into the skull through small twist-drill holes and allowed widespread sampling of the cortical surface with a reduction in muscle artifact and EEG attenuation by the skull and scalp found in extracranial recordings (Wyllie & Awad, 1993). The peg electrodes were inserted at approximately the same locations as the scalp electrodes in the 10-20 or 10-10 systems (e.g., ETP7 at the TP7 point).

Interictal sharp waves from this evaluation were located predominantly at FO1 and ETP7 (left mesial and lateral temporal regions), with less frequent sharp waves at ECP1 and ECP5 (left centroparietal region). There were often sharp waves with a dipole (negative at one electrode and positive at the other) at FO1 and ETP7, suggesting that the tissue generating the sharp wave was between the two electrodes. Seven habitual seizures were recorded, with ictal EEG changes initially seen at FO1 (left mesial temporal region) from 40 to 14 seconds prior to clinical onset. Diffuse involvement of the peg electrodes occurred 12 to 43 seconds after seizure onset. This evidence supported localization of the epileptogenic region to the left temporal lobe rather than to the frontal lobe.

As seen in Figure 5, the patient's MRI did not reveal hippocampal asymmetry or increased T2 signal. MRI volumetric assessment revealed that the left hippocampus was comparable in size (4.5 cm^3) to the right (4.4 cm^3). His FDG-PET showed a slight but nonsignificant symmetric hypometabolism in both frontotemporal regions and was considered nondiagnostic. The patient's IAP revealed left hemisphere dominance for speech. Memory was thought to be bilaterally represented, with the right hemisphere appearing to be more functional than the left.

At the time of his presurgical neuropsychological examination, he was on Tegretol monotherapy; his most recent seizure had occurred 3 days prior to the initiation of his evaluation. His preoperative neuropsychological baseline examination results are presented in Table 2. Intellectual function fell within the average range. Controlled oral word association, confrontation naming, and

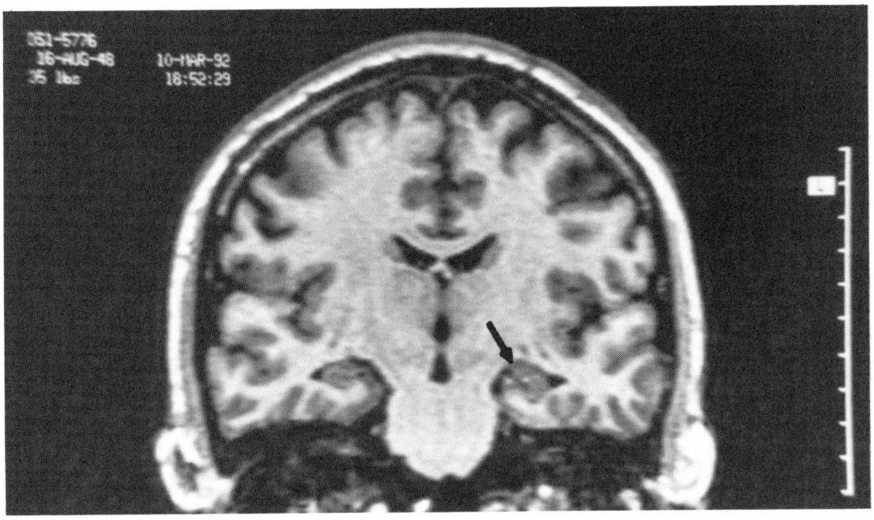

Figure 5. Coronal MRI of a 42-year-old male with intractable left temporal lobe epilepsy. MRI was performed using a T1-weighted (TR = 10, TE = 4), thin-section (2 mm) sequence. The patient's left is on the viewer's right. The gyral anatomy of the neocortex is clearly delineated. The hippocampi are visable as ovoid structures in the mesial temporal lobe. The arrow shows the left hippocampus. There is no apparent hippocampal atrophy. PET of this patient showed no significant hypometabolism (not shown).

execution of commands were preserved. Memory studies revealed that Attention/Concentration Index fell at the high end of the average range, and his Verbal Memory Index was comparable. The Visual Memory Index fell at the high end of the low average range. Verbal rote learning on the RAVLT was less efficient than expected in light of his other scores, and delayed recall of that information was impoverished. However, recognition of those stimuli was preserved. Recognition of words as assessed by the RMT fell within the average range, whereas he had some difficulty with the facial stimuli.

Patient 2 underwent a left frontotemporal craniotomy for temporal lobectomy with microdissection of the mesial structures. In light of the fact that the patient was left hemisphere dominant for speech, the resection was limited to 4 cm from the temporal pole. The hippocampus was removed to approximately 1.5 to 2 cm by subpial dissection and was delivered in one piece. Analysis of the resected specimen revealed focal meningeal fibrosis and mild astrocytosis of the parahippocampus. Following surgery, he was seizure-free and maintained on Tegretol monotherapy. He complained of some difficulty remembering names of persons and objects.

Results of his postoperative neuropsychological examination 6 months following surgery are detailed in Table 2. Note that the patient showed a 30-point drop in the WMS-R Verbal Memory Index. Using the test-retest data derived from the unoperated epilepsy control subjects, a patient with a preoperative Verbal Memory Index of 106 would be expected to have a Verbal Memory Index of 113 on retesting due to practice effects (McSweeny, Chelune, Naugle, & Lüders, 1993). Consequently, the patient's obtained score of 76 is more accurately regarded to be a 47-point drop from the given expected retest score corrected for practice. Because of the failure to demonstrate the expected prac-

TABLE 2. PREOPERATIVE AND POSTOPERATIVE NEUROPSYCHOLOGICAL EXAMINATION RESULTS: CASE 2

	Preoperative examination	Postoperative examination
Intellectual functioning		
Wechsler Adult Intelligence Scale—Revised		
Verbal IQ	104	99
Performance IQ	93	102
Full Scale IQ	99	100
Speech and language		
Boston Naming Test (Raw score/60)	56	51
FAS	30–34 percentile	65–69 percentile
Multilingual Aphasia Examination		
Token Test	82 percentile	82 percentile
Memory function		
Wechsler Memory Scale—Revised		
Attention/Concentration Index (SS)[a]	109	93
Verbal Memory Index (SS)	106	76
Visual Memory Index (SS)	89	80
Delayed Memory Index (SS)	86	71
Logical Memory I (Raw score/Percentile)	31/78 percentile	16/14 percentile
Logical Memory II[b]	15/25 percentile (48%)	4/2 percentile (25%)
Visual Reproduction I	34/59 percentile	29/20 percentile
Visual Reproduction II[b]	22/16 percentile (65%)	20/13 percentile (69%)
Rey Auditory Verbal Learning Test	6/7/10/11/11	5/5/5/3/5
Total D (Raw score)	6	4
Trial VI (Raw score)	7	0
Recognition (Raw score)	15 (+0 false positive)	9 (+1 false positive)
Warrington Recognition Memory Test		
Words (Raw score/Scaled score)	49/14	42/7
Faces (Raw score/Scaled score)	31/<4	40/7
Psychopathology		
Beck Depression Inventory (Total score)	5	9

[a]SS, standard score.
[b]Percent retention scores are given in parentheses.

tice effect as well as the considerable decline from baseline, the patient's obtained standard score of 76 on retesting falls well below expectation (z-score = −3.38) relative to the distribution of retest scores. His Visual Memory Index also dropped, albeit less dramatically, from 89 to 80. Due to his baseline score, a retest score of 91 could be expected on the basis of practice effects. His performance on the other aspects of the memory battery, the RAVLT and RMT, also fell well below his preoperative baseline level of ability. In contrast, speech and language functions as demonstrated at baseline were essentially unchanged.

CONCLUSION

The above cases were carefully selected to show trends that have been detected in research regarding the prediction of preoperative/postoperative neuropsychological change. While they are instructive for that purpose, the reader is reminded that there is considerable variability among temporal lobectomy patients with regard to neuropsychological outcome. The determination of risk for deterioration of cognitive functions is multifactorial and not as clear-cut as might be surmised by these case presentations. Nevertheless, our current understanding of risk determination suggests that the integration of imaging and neuropsychological data holds the most promise for ascertaining the likelihood that a given patient will show noteworthy decrements in memory function following resection of the temporal structures.

References

Abou-Khalil, B. W., Siegel, G. J., Sackellares, J. C., Gilman, S., Hichna, R., & Marshall, R. (1987). Positron emission tomography studies of cerebral glucose metabolism in chronic partial epilepsy. *Annals of Neurology, 22,* 480–486.

Anderman, F. (1987). Identification of candidates for surgical treatment of epilepsy. In J. Engel, Jr. (Ed.), *Surgical treatment of the epilepsies* (pp. 51–69). New York: Raven Press.

Ashtari, M., Barr, W. B., Schaul, N., & Bogerts, B. (1991). Three-dimensional fast low angle shot imaging and computerized volume measurement of the hippocampus in patients with chronic epilepsy of the temporal lobe. *American Journal of Neuroradiology, 12,* 941–947.

Awad, I. A., & Chelune, G. J. (1993). Temporal lobectomy: Outcome and complications. In E. Wyllie (Ed.), *The treatment of epilepsy: Principles and practice* (pp. 1084–1091). Philadelphia: Lea & Febiger.

Baldwin, M. (1956). Modification psychiques survenant aprés lobectomie temporale subtotale. *Neurochirurgie, 2,* 152–167.

Barr, W. B., Ashtari, M., Schaul, N., & Bogerts, B. (1990). Relations between hippocampal volume and memory in patients with intractable seizures (abstract). *Journal of Clinical and Experimental Neuropsychology, 12,* 86–87.

Bauer, R. M., Breier, J., Crosson, B., Gilmore, R., Fennell, E. B., & Roper, S. (1994). Neuropsychological functioning before and after unilateral temporal lobectomy for intractable epilepsy (abstract). *Program and Abstracts of the International Neuropsychological Society,* 60.

Berkovic, S. F., Andermann, F., Olivier, A., Ethier, R., Melanson, D., Robitaille, Y., Kuzniecky, R., Peters, T., & Feindel, W. (1991). Hippocampal sclerosis in temporal lobe epilepsy demonstrated by magnetic resonance imaging. *Annals of Neurology, 29,* 175–182.

Casino, G. D., Jack, C. R., Parisi, J. E., Sharbrough, F. W., Hirschorn, K. A., Meyer, F. B., Marsh, W. R., & O'Brien, P. C. (1991). Magnetic resonance imaging-based volume studies in temporal lobe epilepsy: Pathological correlations. *Annals of Neurology, 29,* 31–36.

Cendes, F., Leproux, F., Melanson, D., Ethier, R., Evans, A., Peters, T., & Andermann, F. (1993). MRI of amygdala and hippocampus in temporal lobe epilepsy. *Journal of Computer Assisted Tomography, 17,* 206–210.

Chelune, G. J. (1992). Using neuropsychological data to forecast postsurgical cognitive outcome. In H. O. Lüders (Ed.), *Epilepsy surgery* (pp. 477–485). New York: Raven Press.

Chelune, G. J. (1995). Hippocampal adequacy versus functional reserve: Predicting memory functions following temporal lobectomy. *Archives of Clinical Neuropsychology, 10,* 413–432.

Chelune, G. J., Comair, Y. G., Najm, I. M., Ruggieri, P., & Lüders, H. O. (1995). MRI volumetrics and memory function in patients with intractable temporal lobe epilepsy. *Epilepsia, 36* (Suppl. 4), 147.

Chelune, G. J., Kneebone, A. C., Dinner, D. S., Awad, I. A., & Naugle, R. I. (1993). Contributions of the IAP and baseline neuropsychological testing in predicting memory outcome following epilepsy surgery: Further support for the "functional adequacy" model of hippocampal functioning (abstract). *Journal of Clinical and Experimental Neuropsychology, 15,* 24.

Chelune, G. J., Naugle, R. I., Lüders, H., & Awad, I. A. (1991). Prediction of cognitive change as a functioni of preoperative ability level among temporal lobectomy patients at six months follow-up. *Neurology, 41,* 399–404.

Chelune, G. J., Naugle, R. I., Lüders, H., Sedlak, J., & Awad, I. A. (1993). Individual change after epilepsy surgery: Practice effects and base rate information. *Neuropsychology, 1,* 41–52.

Chelune, G. J., Trenerry, M. R., Hermann, B. P., Barr, W., Naugle, R., Ivnik, R. J., & Seidenberg, M. (1992). Can we trust our own epilepsy surgery data: An exercise in multicenter replication, generalization, and statistical control (abstract). *The Clinical Neuropsychologist, 6,* 350.

Cook, M. J., Fish, D. R., Shorvon, S. D., Straughan, K., & Stevens, J. M. (1992). Hippocampal volumetric and morphometric studies in frontal and temporal lobe epilepsy. *Brain, 115,* 1001–1005.

Corkin, S. (1965). Tactually guided maze learning in man: Effects of unilateral cortical excisions and bilateral hippocampal lesions. *Neuropsychologia, 3,* 339–352.

Corsi, P. (1972). *Human memory and the medial temporal region of the brain.* Unpublished doctoral dissertation, McGill University, Montreral, Quebec, Canada.

Delaney, R. C., Prevey, M. L., & Mattson, R. H. (1982). Short-term retention with laterealized temporal lobe epilepsy. *Cortex, 22,* 591–600.

Delis, D. C., Kramer, J. H., Kaplan, E., & Ober, B. A. (1987). *California Verbal Learning Test: Adult version.* San Antonio, TX: The Psychological Corporation.

Dodrill, C. B., Hermann, B. P., Rausch, R., Chelune, G. J., & Oxbury, S. (1993). Neuropsychological testing for assessing prognosis following surgery for epilepsy. In J. Engel, Jr. (Ed.), *Surgical treatment of the epilepsies* (2nd ed., pp. 263–271). New York: Raven Press.

Dreifuss, F. E. (1987). Goals of surgery for epilepsy. In J. Engel, Jr. (Ed.), *Surgical treatment of the epilepsies* (pp. 31–49). New York: Raven Press.

Engel, Jr., J. (1987) Approaches to localization of the epileptogenic lesion. In J. Engel, Jr. (Ed.), *Surgical treatment of the epilepsies* (pp. 75–95). New York: Raven Press.

Engel Jr., J., Henry, T. R., Risinger, M. W., Mazziotta, J. C., Sutherling, W. W., Levesque, M. F., & Phelps, M. E. (1990). Presurgical evaluation for partial epilepsy: Relative contribution of chronic depth-electrode recordings versus FDG-PET and scalp-sphenoidal ictal EEG. *Neurology, 40,* 1670–1677.

Henry, T. R., Mazziotta, J. C., & Engel Jr., J. (1993). Interictal metabolic anatomy of mesial temporal lobe epilepsy. Archives of Neurology, 50, 582–589.

Hermann, B. P., Wyler, A. R., Richey, E. T., & Rea, J. M. (1987). Memory function and verbal learning ability in patients with complex partial seizures. *Epilepsia, 28,* 547–554.

Hermann, B. P., Wyler, A. R., & Somes, G. (1993). Memory loss following left anterior temporal lobectomy is associated with hippocampal pathology and not extent of hippocampal resection (abstract). *Journal of Clinical and Experimental Neuropsychology, 6,* 350.

Hermann, B. P., Wyler, A. R., Somes, G., Berry, A. D., & Dohan, F. C. (1992). Pathological status of the mesial temporal lobe predicts memory outcome from left anterior temporal lobectomy. *Neurosurgery, 32,* 652–657.

Hermann, B. P., Wyler, A. R., Somes, G., Dohan, F. C., Berry, A. D., & Clement, L. (1993). Declarative memory following anterior temporal lobectomy in humans. *Behavioral Neuroscience, 108,* 3–10.

Horsley, Sir V. (1886). Brain surgery. *British Medical Journal, 2,* 670–674.

Ivnik, R. J., Sharbrough, F. W., & Laws, E. R. (1988). Anterior temporal lobectomy for the control of partial complex seizures: Information for counseling patients. *Mayo Clinic Proceedings, 63,* 783–793.

Jack, Jr., C. R. (1994). MRI-based hippocampal volume measurements in epilepsy. *Epilepsia, 35* (Suppl. 6), S21–S29.

Jack, C. R., Bentley, M. D., Twomey, C. K., & Zinsmeister, A. R. (1990). MR imaging-based volume measurements of the hippocampus and anterior temporal lobe: Validation studies. *Radiology, 176,* 205–209.

Jack, C. R., Gehring, D. G., Sharbrough, F. W., Felmlee, J. P., Forbes, G., Hench, V. S., & Zinsmeister, A. R. (1988). Temporal lobe volume measurement from MR images: Accuracy and left-right asymmetry in normal persons. *Journal of Computer Assisted Tomography, 12,* 21–29.

Jack, C. R., Sharbrough, F. W., Cascino, G. D., Hirschorn, K. A., O'Brien, P. C., & Marsh, W. R. (1992). Magnetic resonance image-based hippocampal volumetry: Correlation with outcome after temporal lobectomy. *Annals of Neurology, 31,* 138–146.

Jack, C. R., Sharbrough, F. W., Twomey, C. K., Cascino, G. D., Hirschorn, K. A., Marsh, W. R., Zinsmeister, A. R., & Scheithauer, B. (1990). Temporal lobe seizures: Lateralization with MR volume measurements of the hippocampal formation. *Radiology, 175,* 423–429.

Jack, C. R., Twomey, C. K., Zinsmeister, A. R., Sharbrough, F. W., Peterson, R. C., & Cascino, G. D. (1989). Anterior temporal lobes and hippocampal formations: Normative volumetric measurements from MR images in young adults. *Radiology, 172,* 549–554.

Jackson, G. D. (1994). New techniques in magnetic resonance and epilepsy. *Epilepsia, 35* (Suppl. 6), S2–S13.

Jones-Gotman, M. (1987). Commentary: Psychological evaluation—testing hippocampal function. In J. Engel, Jr. (Ed.), *Surgical treatment of the epilepsies* (pp. 203–211). New York: Raven Press.

Jones-Gotman, M., Barr, W. B., Dodrill, C. B., Gotman, J., Meador, K. J., Rausch, R., Sass, K. J.,

Sharbrough, F. W., Silfvenius, H., & Wieser, H. (1993). Controversies concerning the use of intraarterial amobarbital procedures. In J. Engel, Jr. (Ed.), *Surgical treatment of the epilepsies* (2nd ed., pp. 445–449). New York: Raven Press.

Kløve, H., Grabow, J. D., & Trites, R. L. (1969). Evaluation of memory functions with intracarotid sodium amytal. *Transactions of the American Neurological Association, 94,* 76–80.

Kneebone, A. C., Chelune, G. J., Dinner, D. S., Naugle, R. I., & Awad, I. A. (1995). Use of the contralateral intracarotid amobarbital procedure to predict material specific memory change following anterior temporal lobectomy. *EpIlepsia, 36,* 857–865.

Kuzniecky, R., de la Sayette, V., Ethier, R., Melanson, D., Andermann, F., Berkovic, S., Robitaille, Y., Olivier, A., Peters, T., & Feindel, W. (1987). Magnetic resonance imaging in temporal lobe epilepsy: Pathological correlations. *Annals of Neurology, 22,* 341–347.

Lee, G. P., Loring, D. W., & Thompason, J. L. (1989). Construct validity of material-specific memory measures following unilateral temporal lobe ablations. *Psychological Assessment: Journal of Consulting and Clinical Psychology, 3,* 192–197.

Lencz, T., McCarthy, G., Bronen, R. A., Scott, T. M., Inserni, J. A., Sass, K. J., Novelly, R. A., Kim, J. H., & Spencer, D. D. (1992). Quantitative magnetic resonance imaging in temporal lobe epilepsy: Relationship to neuropathology and neuropsychological function. *Annals of Neurology, 31,* 629–637.

Leonard, G. (1991). Temporal lobe surgery for epilepsy: Neuropsychological variables related to surgical outcome. *Canadian Journal of Neurological Science, 18,* 593–597.

Loring, D. W., Lee, G. P., Martin, R. C., & Meador, K. J. (1988). Material specific learning in patients with partial complex seizures of temporal origion: Convergent validation of memory constructs. *Journal of Epilepsy, 1,* 53–59.

Loring, D. W., Lee, G. P., Meador, K. J., Flanigin, H. F., Smith, R. J., Figueroa, R. E., & Martin, R. C. (1990). The intracarotid sodium amobarbital procedure as a predictor of memory failure following unilateral temporal lobectomy. *Neurology, 40,* 605–610.

Loring, D. W., Lee, G. P., Meador, K. J., Smith, J. R., Martin, R. C., Ackell, A. B., & Flanigin, H. F. (1991). Hippocampal contribution to verbal recent memory following dominant hemisphere temporal lobectomy. *Journal of Clinical and Experimental Neuropsychology, 13,* 575–586.

Loring, D. W., Meador, K. J., Lee, G. P., & King, D. W. (1992). *Amobarbital effects and lateralized brain function.* New York: Springer-Verlag.

Loring, D. W., Murro, A. M., Meador, K. J., Lee, G. P., Gratton, C. A., Nichols, M. E., Gallagher, B. B., King, D. W., & Smith, J. R. (1993). Wada memory testing and hippocampal volume measurements in the evaluation for temporal lobectomy. *Neurology, 43,* 1789–1793.

Lüders, H. O., Lesser, R. P., Dinner, D. S., Morris, H. H., Hahn, J. F., Friedman, L., Skipper, G., Wyllie, E., & Friedman, D. (1987). Commentary: Chronic intracranial recording and stimulation with subdural electrods. In J. Engel, Jr. (Ed.), *Surgical treatment of the epilepsies* (pp. 297–321). New York: Raven Press.

Margerison, J. H., & Corsellis, J. A. (1966). Epilepsy and the temporal lobes: A clinical electroencephalographic and neuropathological study of the brain in epilepsy, with particular reference to the temporal lobes. *Brain, 89,* 499–530.

McSweeny, A. J., Chelune, G. J., Naugle, R. I., & Lüders, H. (1993). "T-scores for change:" An illustration of a regression approach to depicting change in clinical neuropsychology. *The Clinical Neuropsychhologist, 7,* 300–312.

Milner, B. (1968). Visual recognition and recall after right temporal-lobe excision in man. *Neuropsychologia, 6,* 191–209.

Milner, B. (1972). Disorders of learning and memory after temporal lobe lesions in man. *Clinical Neurosurgery, 19,* 421–446.

Milner, B. (1975). Psychological aspects of focal epilepsy and its neurosurgical management. In D. Purpura, J. Penry, & J. Walter (Eds.), *Advances in neurology: Vol. 8* (pp. 299–321). New York: Plenum Press.

Milner, B., Branch, C., & Rasmussen, T. (1962). A study of short-term memory after intracarotid injection of sodium amytal. *Transactions of the American Neurological Association, 87,* 244–266.

Mishkin, M., & Appenzeller, T. (1987). The anatomy of memory. *Scientific American, 256,* 80–89.

Murro, A. M., Park, Y. D., King, D. W., Gallagher, B. B., Smith, J. R., Yaghmai, F., Toro, V., Figueroa, T. V., Loring, D. W., & Littleton, W. (1993). Seizure lateralization in temporal lobe epilepsy: A comparison of scalp-sphenoidal EEG and volumetric MRI. *Neurology, 43,* 2531–2533.

National Institutes of Health. (1990). NIH consensus conference: Surgery for epilepsy. *Journal of the American Medical Association, 264,* 729–733.

Naugle, R. I. (1992). Neuropsychological outcome following surgical resection for epilepsy. In H. O. Lüders (Ed.), *Epilepsy surgery* (pp. 637–645). New York: Raven Press.

Ojemann, G., & Dodrill, C. B. (1985). Verbal memory deficits after left temporal lobectomy for epilepsy. *Journal of Neurosurgery, 62,* 101–107.

Olivier, A. (1988). Risk and benefit in the surgery of epilepsy: Complications and positive results on seizure tendency and intellectual function. *Acta Neuroligica Scandinivica, 78* (Suppl. 117), 114–121.

Oxbury, J., & Oxbury, S. (1989). Neuropsychology: Memory and hippocampal pathology. In E. H. Reynolds & M. R. Trimble (Eds.), *The bridge between neurology and psychiatry* (pp. 136–150). Edinburgh: Churchill Livingstone.

Penfield, W., & Mathieson, G. (1974). An autopsy and discussion of the role of the hippocampus in experiential recall. *Archives of Neurology, 34,* 145–154.

Penfield, W., & Milner, B. (1958). Memory deficit produced by bilateral lesions in the hippocampal zone. *Archives of Neurology and Psychiatry, 79,* 475–497.

Perrine, K., Gershengorn, J., Brown, E. R., Choi, I. S., Luciano, D. J., & Devinsky, O. (1993). Material-specific memory in the intracarotid amobarbital procedure. *Neurology, 43,* 706–711.

Rausch, R. (1987). Psychological evaluation. In J. Engel, Jr. (Ed.), *Surgical treatment of the epilepsies* (1st ed., pp. 181–197). New York: Raven Press.

Rausch, R., & Babb, T. L. (1987). Evidence for memory specialization within the mesial temporal lobe in man. In J. Engel, Jr. (Ed.), *Fundamental mechanisms of human brain function* (pp. 103–109). New York: Raven Press.

Rausch, R., & Babb, T. L. (1993). Hippocampal neuron loss and memory scores before and after temporal lobe surgery for epilepsy. *Archives of Neurology, 50,* 812–817.

Rausch, R., Babb, T. L., Engel, J., & Crandall, P. H. (1989). Memory following intracarotid amobarbital injection contralateral to hippocampal damage. *Archives of Neurology, 46,* 783–788.

Rausch, R., & Crandall, P. H. (1982). Psychological status related to surgical control of temporal lobe seizures. *Epilepsia, 23,* 191–202.

Rausch, R., Silfvenius, H., Wieser, H., Dodrill, C. B., Meador, K. J., & Jones-Gotman, M. (1993). Intraarterial amobarbital procedures. In J. Engel, Jr. (Ed.), *Surgical treatment of the epilepsies* (2nd ed., pp. 341–357). New York: Raven Press.

Rey, A. (1964). *L'examen clinque en psychologie.* Paris: Presses Universitaires de France.

Sass, K. J., Lencz, T., Westerveld, M., Novelly, R. A., Spencer, D. D., & Kim, J. H. (1991). The neural substrate of memory impairment demonstrated by the intracarotid amytal procedure. *Archives of Neurology, 48,* 48–52.

Sass, K. J., Sass, A., Westerveld, M., Lencz, T., Novelly, R. A., Kim, J. H., & Spencer, D. D. (1992a). Specificity in the correlation of verbal memory and hippocampal neuron loss: Dissociation of memory, language, and verbal intellectual abaility. *Journal of Clinical and Experimental Neuropsychology, 14,* 662–672.

Sass, K. J., Sass, A., Westerveld, M., Lencz, T., Rosewater, K. M., Novelly, R. A., Kim, J. H., & Spencer, D. D. (1992b). Russell's adaptation of the Wechsler Memory Scale as an index of hippocampal pathology. *Journal of Epilepsy, 5,* 24–30.

Sass, K. J., Spencer, D. D., Kim, J. H., Westerveld, M., Novelly, R. A., & Lencz, T. (1990). Verbal memory impairment correlates with hippocampal pyramidal cell density. *Neurology, 40,* 1694–1697.

Saykin, A. J., Robinson, L. J., Stafiniak, P., Kester, D. B., Gur, R. C., O'Connor, M. J., & Sperling, M. R. (1992). Neuropsychological changes after anterior temporal lobectomy: Acute effects on memory, language, and music. In T. Bennett (Ed.), *The neuropsychology of epilepsy* (pp. 263–290). New York: Plenum Press.

Scoville, W. B., & Milner, B. (1957). Loss of recent memory after bilateral hippocampal lesions. *Journal of Neurology, Neurosurgery, and Psychiatry, 20,* 11–21.

Spencer, S. S., McCarthy, G., & Spencer, D. D. (1993). Diagnosis of medial temporal lobe seizure onset: Relative specificity and sensitivity of quantitative MRI. *Neurology, 43,* 2117–2124.

Theodore, W. H., Newmark, M. E., Sato, S., Brooks, R., Patronas, N., De La Paz, R., Di Chiro, G., Kessler, R. U., Margolhr, R., Manning, R. G., Channing, U., & Porter, R. J. (1983). [^{18}F]Fluorodeoxyglucose positron emission tomography in refractory complex partial seizures. *Annals of Neurology, 4,* 429–437.

Theodore, W. H., Sato, S., Kufta, C., Baush, M. B., Bromfield, E. B., Leiderman, D. B. (1992). Temporal lobectomy for uncontrolled seizures: The role of positron emission tomography. *Annals of Neurology, 32,* 789–794.

Trenerrry, M. R., Jack, C. R., Ivnik, R. J., Sharbrough, F. W., Cascino, G. D., Hirschorn, K. A., Marsh, W. R., Kelly, P. J., & Meyer, F. B. (1991). Memory is correlated with presurgical magnetic resonance imaging hippocampal volumes before and after temporal lobectomy for intractable epilepsy. *Epilepsia, 31* (Suppl. 3), S73.

Trenerry, M. R., Jack, C. R., Ivnik, R. J., Sharbrough, F. W., Cascino, G. D., Hirschorn, K. A., Marsh, W. R., Kelly, P. J., & Meyer, F. B. (1993). MRI hippocampal volumes and memory function before and after temporal lobectomy. *Neurology, 43*, 1800–1805.

Van Buren, J. M. (1987). Complications of surgical procedures in the diagnosis and treatment of epilepsy. In J. Engel, Jr. (Ed.), *Surgical treatment of the epilepsies* (pp. 465–475). New York: Raven Press.

Wada, J., & Rasmussen, T. (1960). Intracarotid injection of sodium amytal for the lateralization of cerebral speech dominance: Experimental and clinical observations. *Journal of Neurosurgery, 17*, 266–282.

Walczak, T., Radtke, R., McNamara, J., Lewis, D. V., Luther, J. S., Thompson, E., Wilson, W. P., Friedman, A. H., & Nashold, B. S. (1990). Anterior temporal lobectomy for complex partial seizures: Evaluation, results, and long-term follow up in 100 cases. *Neurology, 40*, 413–418.

Walker, A. E. (1957). Recent memory impairment in unilateral temporal lesions. *Archives of Neurology and Psychiatry, 78*, 5543–5552.

Watson, C., Andermann, F., Gloor, P., Jones-Gotman, M., Peters, T., Evans, A., Olivier, A., Melanson, D., & Leuroux, G. (1992). Anatomic basis of amygdaloid and hippocampal volume measurement by magnetic resonance imaging. *Neurology, 42*, 1743–1750.

Wechsler, D. (1987). *Wechsler Memory Scale-Revised: Manual.* New York: The Psychological Corporation.

Wolf, R. L., Ivnik, R. J., Hirschorn, K. A., Sharbrough, F. W., Cascino, G. D., & Marsh, W. R. (1993). Neurocognitive efficiency following left temporal lobectomy: Standard versus limited resection. *Journal of Neurosurgery, 79*, 76–83.

Wyllie, E., & Awad, I. A. (1993). Intracranial EEG and localization studies. In E. Wyllie (Ed.), *The treatment of epilepsy: Principles and practices* (pp. 1023–1028). Philadelphia: Lea & Febiger.

Wyllie, E., Naugle, R., Awad, I., Chelune, G., Lüders, H., Dinner, D., Skibinski, C., & Ahl, J. (1991). Intracarotid amobarbital procedure: I. Prediction of decreased modality-specificity memory scores after temporal lobectomy. *Epilepsia, 32*, 857–864.

Wyllie, E., Naugle, R., Chelune, G., Lüders, H., Morris, H., & Skibinski, C. (1991). Intracarotid amobarbital procedure: II. Lateralizing value in evaluation for temporal lobectomy. *Epilepsia, 32*, 865–869.

9

Imaging Functional Reorganization after Brain Injury

CORNELIUS WEILLER, FRANCOIS CHOLLET, AND RICHARD S. J. FRACKOWIAK

INTRODUCTION

Stroke is one of the most common afflictions of patients with neurological symptoms. About 2,400 per 1 million people suffer from a stroke annually. Stroke is a severe disabling disorder and a major socioeconomic health problem. Rehabilitation of stroke patients is a difficult task and requires large interdisciplinary cooperation.

Recovery of function after a stroke is often observed, although the morphological lesion that originally caused the deficit persists. It must be concluded then that brain function is highly adaptable. The mechanisms responsible for recovery were, until recently, largely unknown. However, information on the changes in functional reorganization of the brain is important for the rehabilitation of stroke patients on a rational basis and because, with growing pressure on rehabilitation services, difficult decisions may have to be taken as to which patients should receive such therapy.

Noninvasive functional imaging methods have dramatically improved during the last years, allowing the study of reorganizational changes of the human

CORNELIUS WEILLER Klinik für Neurologie, Friedrich-Schiller-Universität Jena, 07740 Jena, Germany. FRANCOIS CHOLLET Service de Neurologie B, Hôpital Purpan, Toulose 31059, France. RICHARD S. J. FRACKOWIAK Wellcome Department of Cognitive Neurology, Queen Square, London WCI N3B6, England.

Neuroimaging II: Clinical Applications, edited by Erin D. Bigler. Plenum Press, New York, 1996.

brain *in vivo*. Positron emission tomography (PET) can be used to detect significant changes in regional cerebral blood flow (rCBF), which is an indirect measure of local neuronal activity (Raichle, 1987). Focal rCBF changes can be elicited by behavioral, physiological, or other stimuli and can be detected *in vivo* from the whole brain simultaneously. By subtracting images elicited during different conditions, the site of focal CBF change gives information on the cerebral structures associated with or underlying the activity under study. Now, changes in intensity during defined tasks can be studied by MRI. The changes reflect differences in cerebral blood volume, contrast media enhancement, or the rate of deoxyhemoglobin formation and seem related to neuronal activity.

This chapter describes PET work on the functional anatomy of recovery of function after stroke, mainly in the motor system. There are three main PET studies on recovery of motor function, the results of which will be reported here. All three deal with patients who have suffered from a severe hemiplegia due to a first ischemic brain infarct. All patients were studied after recovery of function had occurred, though the corresponding infarction was still present. The recovered function tested was a sequential finger-to-thumb opposition task.

In the first study (Chollet *et al.*, 1991), the contralateral, "unaffected" hand served as control for the recovered hand. In the second study (Weiller, Chollet, Friston, Wise, & Frackowiak, 1992), the pattern of activation elicited by movement of the recovered as well as the contralateral hand was compared with ten normal subjects. In the third study (Weiller, Ramsay, Wise, Friston, & Frackowiak, 1993), a method was developed whereby rCBF changes in a single patient could be compared to the mean reference values from a group of healthy volunteers.

Patients

In all three studies patients who were selected had a dense hemiparesis, mainly of the arm, face, and leg type typically called "pyramidal," and lasting for at least 2 days. The paresis was always due to a first insult corresponding to a single, appropriately sited ischemic infarction on structural imaging, mainly of the internal capsule. The internal capsule, a densely packed bundle of axons carrying information to and from the cortex, includes the fibers of the major descending motor pathways from the cortex.

The neurological deficit affected fractionated finger movements, a task for which the pyramidal tract is thought to be a critical pathway. It was concluded that the lesions did disrupt pyramidal pathways. Indeed, anterograde and retrograde degeneration of corticospinal fibers was demonstrated directly in the MRI of one patient. Despite this persistent disruption of the major efferent motor pathways, recovery of motor function had occurred at the time of PET scanning. The paradigm tested was a sequential finger-to-thumb opposition task, driven by the rhythm of a metronome (three oppositions every 2 seconds). The movements were not forceful but brisk, with large amplitude and precision. Thus the task tested fractionated finger movements that were prominently affected by the former paresis. In the first study, six patients (two females, four males; age grange: 25–71 years) were included (Table 1). One had a hemorrhage, one had a cortical infarct, two had deep white matter infarcts, one had a putaminal infarct, and one had a lacuna affecting the internal capsule. The second study included ten right-handed stroke patients (seven males, three females; age range: 21–62 years, mean: 41 years) all with striatocapsular infarcts (Weiller, Ringelstein, Re-

TABLE 1. PATIENTS AND PARADIGMS STUDIED IN THREE PET STUDIES ON RECOVERY FROM STROKE

Study	Number of patients	Number of controls	Infarctions	Comparisons
Chollet *et al.*, 1991	6	—	Subcortical corticosubcortical	Recovered hand vs. "unaffected" hand
Weiller *et al.*, 1992	10	10	Striatocapsular	Both sides vs. controls and rCBF at rest
Weiller *et al.*, 1993	8	10	Striatocapsular and lacunae (internal capsule)	Recovered side vs. mean of controls

ich, Thron, & Buell, 1990) and ten healthy right-handed volunteers as controls (six males, four females; age range: 28–69 years, mean: 47 years). Eight patients (three females, five males; age range: 21–67 years, mean: 43 years), five with striatocapsular infarcts and three with lacunae located in the internal capsule, were investigated in the third study and compared with the ten controls from the second study. All patients were investigated between 7 weeks and 6 years after the stroke. All patients from the last study, five from the second, and three from the first study had completely recovered. Associated movements of the unaffected hand during movement of the recovered hand were found in one patient from the first study and four patients from the second and third study.

MAPPING OF rCBF CHANGES

The rCBF was measured with a dynamic/integral technique using $C^{15}O_2$ as flow tracer (Lammertsma *et al.*, 1989), which was administered by inhalation for 2 minutes. Twenty-one successive PET images were collected for 0.5 minutes before tracer delivery (background frame) and for 3 minutes after the beginning of tracer administration. Regional cerebral blood flow was measured quantitatively using an arterial input curve for correction of delay and dispersion. Each subject underwent six measurements of rCBF, with three conditions replicated each in a balanced order (ABCCBA). Conditions included "rest" (i.e., eyes closed, "empty your mind," no movement, metronome on) and finger-to-thumb opposition of the right and left hand. Images of all patients and normal subjects acquired during the six different conditions were reoriented parallel to the intercommissural line and then normalized using a linear rescaling and plastic resampling algorithm into the stereotactic space of the anatomical atlas of Talairach and Tournoux (1988) using statistical parametric mapping (SPM; MRC Cyclotron Unit, London, UK; Friston *et al.*, 1989, 1990). This method depends upon a normal distribution of regional blood flow that is not present in the scans of stroke patients. In such scans, low flow picture elements (voxels) in infarcted areas may produce artifacts that interfere with the computational procedure used to resize and reshape the images into the standard anatomical space. Hence template images were constructed for each patient, composed of the mean of the original scan and its mirror image. These were used to compute the geometric translation parameters that were then applied to the original CBF images to effect the necessary spatial transformations. The normalized images (a set of 26 planes 4 mm apart with voxel dimensions 2 × 2 mm in the x and y dimensions) were then smoothed with a Gaussian filter of 10 pixels (at full width, half maximum) to minimize the effects of individual variations in gyral and functional anatomy. Smoothing also spreads the spatial extent of individual distributed foci of activation, increasing the probability that they will interact additively during averaging. A site of additive interaction represents an overlap of foci in which a reliable change in rCBF is seen from scan to scan. The blood flow estimates in each voxel of the smoothed images correspond to a weighted mean rCBF centered on a spherical domain 20 mm in diameter. As the study was designed to examine relative regional changes of cerebral activity under different motor conditions, the influence of global flow differences between the scans of each subject were eliminated by normalizing voxel by voxel to a mean of 50 ml/dl min by an analysis of covariance with measured global flow as the covariate. In the first two studies the scans of each group, patients or normal subjects, in the third study, only the scans of the ten normal subjects were averaged voxel by voxel to give group blood flow maps for each of the three conditions, in which each vowel has an adjusted mean rCBF value and an associated error variance (Friston, Frith, Liddle, & Frackowiak, 1991). Comparisons of the appropriate differences in condition means were made by the t-statistic using the adjusted pixel error variances for each condition estimated from the analysis of covariance. Planned comparisons *within* each group (normal subjects and patients) were made between CBF maps derived during finger opposition of the right hand (i.e., the recovered hand) or the left hand and those derived from the rest states. In the second study activation was compared between patients and the control subjects. To identify areas in which the tasks evoked greater or lesser activation in the patients than in normal subjects, the changes in rCBF between baseline and task were compared between groups to identify pixels at which these changes were significantly different in normal subjects and patients. The error variance for these comparisons was pooled from both groups. In the third study individual patients were compared with the group of ten normals to determine rCBF increases that were significantly greater than predicted by the normal group data. The differences between normal group and individual patient rCBF increases in each image voxel were divided by the standard deviation that was expected by chance (derived from the ANCOVA error variance) to generate a map of z-scores. The z-scores were thresholded according to the normal distribution. For each patient, any voxel was recorded that had a significantly greater change of rCBF from rest than in the normal group during the performance of the motor task, and the location in Talairach space was used to correlate the functional change with anatomy. To assess remote functional effects caused by the lesion which are reflected in rCBF changes that are not necessarily predictable from the site of the infarct, the rest states of the patients were compared to those of the normal subjects in the second study. The resulting set of t values of the different comparisons constituted an SPM. The omnibus significance of the SPMs was assessed by comparing the observed and expected number of pixels above $p < 0.001$ with reference to the Poisson distribution.

Significant pixels from the SPM in this omnibus sense were displayed on coronal, sagittal, and transverse views of the brain.

Functional Disconnection

The effect of decreased rCBF in areas that are not infarcted but which possess functional connections with the infarcted area (diaschisis) has been attributed to hypometabolism of these areas by disruption of afferent (or in rare cases efferent) impulses. To assess these functional disconnections that may influence the potential for recovery, the cortical and subcortical distributions of rCBF were compared in a group of ten patients with striatocapsular infarctions at rest with the distribution of blood flow in a group of age-matched normal subjects (Weiller *et al.*, 1992). In patients at rest, there was a significantly decreased regional CBF in the left striatum and internal capsule (−4 to +12 mm relative to the AC-PC line), corresponding to the site of the lesion (Table 2). In addition, there were significant decreases in the left insula (4 to 16 mm above the AC-PC line), left primary sensorimotor (16 to 28 mm above the AC-PC line) and dorsolateral prefrontal cortices (areas 9 and 10; 16 to 44 mm above the AC-PC line), left thalamus (0 to 12 mm above the AC-PC line), left midbrain/cerebral peduncle, and the right cerebellum (8 to 16 mm below the AC-PC line) (Figure 1). There was a trend toward relative hypoperfusion in the anterior cingulate cortex (areas 24); none of this area was infarcted on structural imaging. In some areas immediately surrounding the infarcts, as in the insula, the hypoperfusion may be caused by selective neuronal loss. However, this cannot be the case for the more distant areas (e.g., dorsolateral prefrontal cortex) and those lying outside the territory of the feeding artery (e.g., thalamus). These areas are components of striato-thalamo-cortical circuits (Alexander & Crutcher, 1990). Most likely the decreased CBF in these areas is due to disruption of these circuits at the level of the basal ganglia. The concept introduced by von Monakow (1914) under the term "diaschisis" supposes that the lesion disrupts part of a neuronal network, of which both the lesion and the deactivated areas are constituent parts. Significant cerebellar hypometabolism has been observed repeatedly in the presence of contralateral basal ganglia and internal capsule lesions. The hypoperfusion in the midbrain/peduncular region can be interpreted as a result of degeneration of the pyramidal tract, the major constituent of this anatomical region. There also were areas in which the rCBF at rest was higher in patients than in normal subjects. These included the right caudate nucleus (12 to 16 mm above the AC-PC line), right angular gyrus (area 39), right premotor cortex (32 to 48 mm above the AC-PC line), and the left posterior cingulate cortex (area 23;

TABLE 2. AREAS WITH SIGNIFICANT DIFFERENT CBF AT REST IN TEN PATIENTS WITH STRIATOCAPSULAR INFARCTS[a]

Region	Talairach coordinates x	y	z	rCBF normals	patients	Z-score
Left striatum	−32	−6	8	71	60	5.1
Left insula	−36	0	12	69	56	5.4
Left thalamus	−10	−26	4	64	57	4.6
Left peduncle/midbrain	−8	−2	−12	62	56	4.5
Left area 1–4	−52	−16	24	51	46	3.7
Left dorsolateral prefrontal cortex	−30	44	24	51	45	3.8
Right cerebellum	14	−62	−16	72	68	3.7

[a]Adjusted group mean rCBF (ml $dl^{-1}min^{-1}$) at rest in the areas indicated on the SPMs to show significant difference in patients (n = 10) and normals (n = 10) (at $p < 0.001$). The Talairach coordinates are in mm and correspond to the stereotactic conventions of the atlas of Talaraich and Tournoux (1988). Coordinates refer to the location of maximum significance with the rCBF. The table shows areas with decreased rCBF in patients: striatum corresponds to the lesion site; insula, prefrontal and primary sensorimotor cortices, thalamus, cerebral peduncle/midbrain, and contralateral cerebellum constitute deactivated areas. (Modified from Weiller *et al.*, 1992).

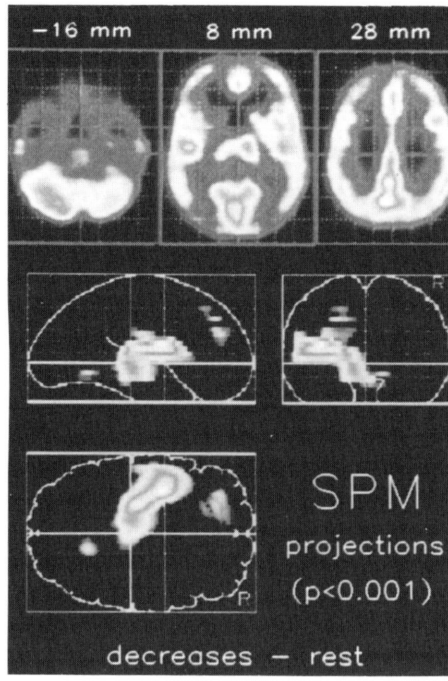

Figure 1. Comparison of rCBF at rest of ten patients with striatocapsular infarcts and ten normal subjects. Sagittal, coronal, and transverse projections of the statistical parametric maps (SPM, in the stereotactic space, obtained from a group-to-group comparison between patients and normal subjects are shown. The grid is the standard, proportional, stereotactic grid of Talairach and Tournoux (1988) that defines the 3-D space into which all the subjects' and patients' brain scans have been normalized. The line drawings are of the horizontal brain contours at the level of the anterior commissure, with the sagittal outline representing the midsagittal plane. Only pixels that are significantly different between both groups at $p < 0.001$ are displayed. There were significant decreases in patients in left striatum and internal capsule, insula, lateral prefrontal and primary sensorimotor cortices, thalamus, cerebral peduncle/midbrain, and right cerebellum. (Modified with permission from Weiller *et al.*, 1992.)

8 to 28 mm above the AC-PC line). Remote effects in the hemisphere contralateral to a lesion have been described and discussed for many years (Kempinsky, 1956), however, they mainly involved a "transhemispheric depression of metabolism." An increase in rCBF contralateral to the lesion was reported for the first time in this study. A functional disinhibition of contralateral premotor cortex and caudate nucleus was hypothesized as forming an integral part of the recovery process.

Activation Studies

The first study compared the pattern of rCBF increases in the brain during movement of the recovered hand with the increases during movement of the contralateral or "unaffected" hand. When the unaffected hand was moved, the pattern of activation was highly lateralized, with rCBF increases in contralateral sensorimotor (SMC) (+24%) and premotor cortices, supplementary motor area (SMA), putamen, insula, parietal cortex [Brodmann area (BA) 40], and ipsilateral cerebellum. There was a nonsignificant increase in ipsilateral SMC (+3.5%) and a small significant increase in ipsilateral BA 40 (+5.3%). During movement of the fingers of the recovered hand, significant increases in rCBF were observed bilaterally in SMC (contralateral: +23%, ipsilateral: +10%) and premotor cortex, insula, area 40, and cerebellum (Figure 2). It was suggested that ipsilateral motor pathways may play a role in recovery of motor function after stroke. This was the first study on this subject. The patient group was rather inhomogeneous,

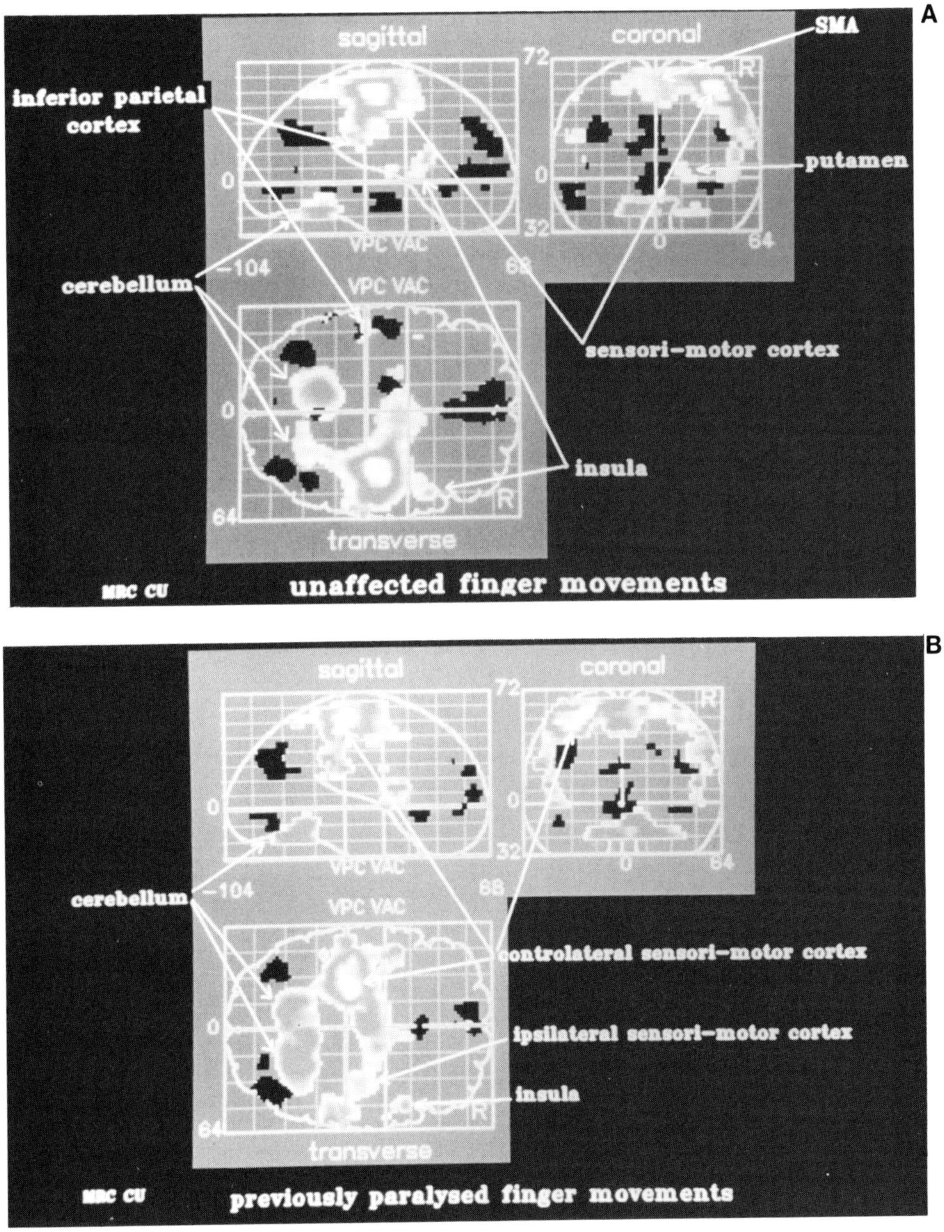

Figure 2. Comparison of rCBF increases in six patients in recovery from motor stroke during movement of "unaffected" hand (A) and the recovered hand (B). Grid and display as in Figure 1. In A, there is clear lateralized activation in right sensorimotor and premotor cortices as well as left cerebellum. There is also activation of the supplementary motor area. In B, there is prominent left sensorimotor cortical activation, but in addition, significant changes are noted ipsilaterally. There is also symmetrical activation of both cerebellar hemispheres, premotor and parietal cortices, and insula. (Modified with permission from Chollet *et al.*, 1991.)

with different types of subcortical and cortical infarcts. It may be argued that the unaffected hand may not represent an ideal control. The occurrence of associated movements of the unaffected hand was not completely assessed and was present in at least one of the patients.

Thus, the second study was restricted to one defined subcortical stroke entity, i.e., striatocapsular infarcts (Weiller *et al.,* 1990). The pattern of activation was compared in the recovered and in the unaffected hand with normal controls. When the patients moved the recovered hand, they activated the contralateral sensorimotor and premotor cortices and the ipsilateral cerebellum to the same extent as normal subjects. However, a bilaterally homologous area comprising middle to anterior aspects of the insula and the most lateral part of the premotor cortex was more active in patients than in normal subjects. In the patients there was also significantly greater activation in bilateral parietal (bilateral area 40), lateral prefrontal, and anterior cingulate cortices; ipsilateral premotor cortex; striatum; and contralateral cerebellum (Figure 3). The pattern of activation was also abnormal when the unaffected hand was moved. Patients activated the insula, striatum, and lateral prefrontal, premotor, and inferior parietal (bilateral area 40) cortices in the undamaged hemisphere more than normal subjects. These results constitute the systematic components of functional reorganization within the motor system after striatocapsular infarction. Be-

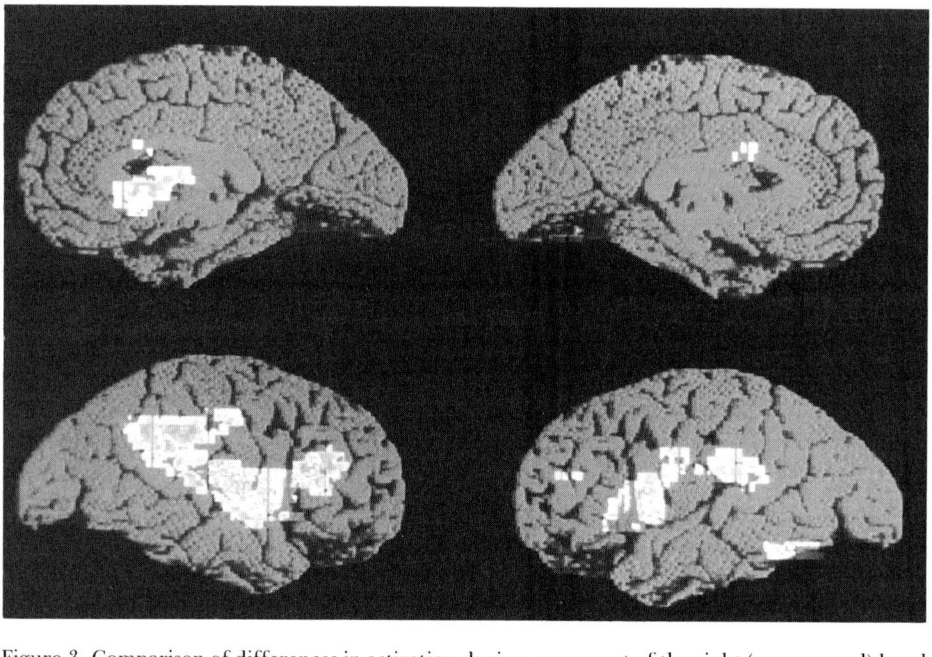

Figure 3. Comparison of differences in activation during movement of the right (= recovered) hand in ten patients with striatocapsular infarcts and ten normal subjects. The grid is as in Figure 1. SPMs are projected onto the lateral and medial surfaces of the brain. The images are presented in groups of four—the top left is the right medial view, the top right is the left medial view, the bottom left is the right lateral surface, and the bottom right is the left lateral surface. Only pixels that showed a significantly greater activation in patients at $p < 0.001$ are shown. Patients activated an area *bilaterally* comprising anterior insula and the most ventral part of the premotor cortex, bilateral area 40, lateral prefrontal and anterior cingulate cortices, ipsilateral premotor cortex, striatum, and contralateral cerebellum more than normal. (Modified with permission from Weiller *et al.,* 1992.)

cause of individual differences in the pattern of recovery shown by stroke patients, and because the final outcome is not easy to predict from the severity of the initial symptoms, it is important to understand how individual patients, each with a specific lesion, are able to recover, that is, how individual patterns of reorganization relate to the specific neural structures damaged.

In the next study a technique was developed whereby rCBF changes in a single patient could be compared to mean reference values from a group of ten healthy volunteers. The first result of this study was that individual patients showed different patterns of redistributed rCBF. Most patients showed increases above normal in the following brain regions in various combinations: in middle to anterior insula together with the most lateral aspects of the premotor cortex in five out of eight; in parietal cortex (area 40) in six out of eight; in the mesial premotor cortex (SMA) in six out of seven (no data were available from this brain region in one patient); in the ipsilateral premotor cortex in eight out of eight; in prefrontal cortex in six out of eight; and in anterior cingulate cortex in three out of eight. (Table 3). In four patients an additional activation above that of the normal group was found in the contralateral SMC along its ventral extent (Figure 4). While in the normal group the contralateral SMC is activated during finger opposition from 44 mm to 60mm above the intercommissural (AC-PC) line with the maximum at 52 mm, the SMC was additionally activated in these patients from as low as 16 mm above the AC-PC line extending up to 52 mm. Thus, the SMC extended more than a centimeter beyond its normal inferior boundary. In these four patients the lesions were located in the posterior limb, the genu, and the anterior part of the posterior limb (Figure 5). The four patients with a lateral

TABLE 3. BRAIN REGIONS ACTIVATED DURING MOVEMENT OF THE RECOVERED HAND AFTER STROKE

	Study[b]		
Brain regions[a]	1	2	3
Contralateral			
SMC	*	*	+[c] (lateral extension)
PMC			
Ipsilateral cerebellum	*	*	*
Ipsilateral			
SMC	+		+ associated movements
PMC	+	+	+
Basal ganglia		+	+
Contralateral cerebellum	+	+	+
Bilateral			
Insula	+ (ipsilateral)	+	+
Parietal cortex		+	+
LPFC		+	+
Mesial structures			
SMA	*	not assessed	+
Cingulate cortex		+	+

[a]Areas that were found in the three studies to show significant increase in rCBF during finger opposition of the recovered hand.
[b]Study 1: Chollet *et al.*, 1991; Study 2: Weiller *et al.*, 1992; Study 3: Weiller *et al.*, 1993.
[c]+ rCBF increases were significantly greater in the recovered hand as compared with the unaffected hand (Chollet *et al.*, 1991) or controls (Weiller *et al.*, 1992, 1993).

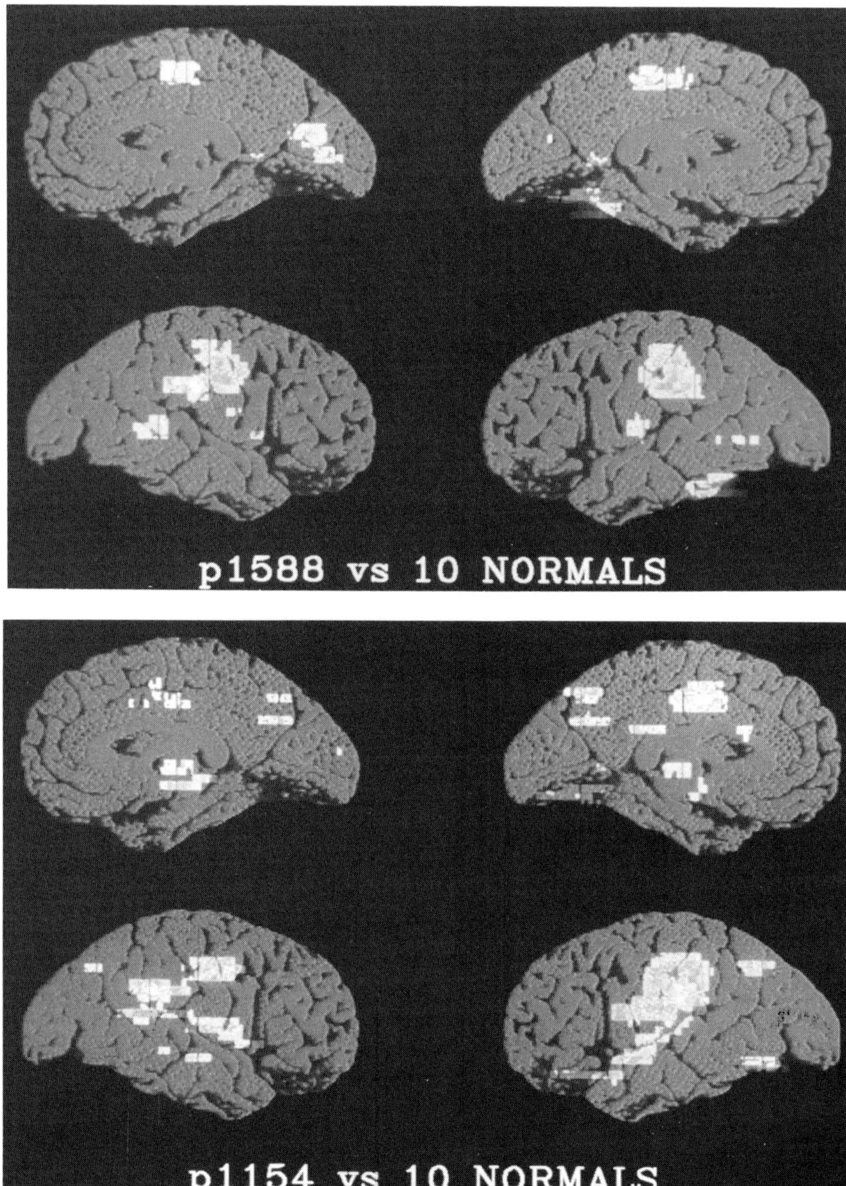

Figure 4. Examples of functional reorganization from two individual patients (A and B) after recovery from motor stroke. The recovered hand was on the right in both cases. Voxels with significantly greater rCBF increases during finger opposition with the recovered hand relative to the group of ten subjects (omnibus $p < 0.001$) constitute the SPMs in the standard, proportional, stereotactic space of Talairach and Tournoux (1988). The dispaly is as in Figure 3. Extension of the hand area into the face area of the contralateral SMC is shown in both patients. Both patients had posterior internal capsular lesions in the territory of the anterior choroidal artery. Activation of ipsilateral SMC wirh associated movements was seen in both patients as well. (Modified with permission from Weiller *et a.*, 1993.)

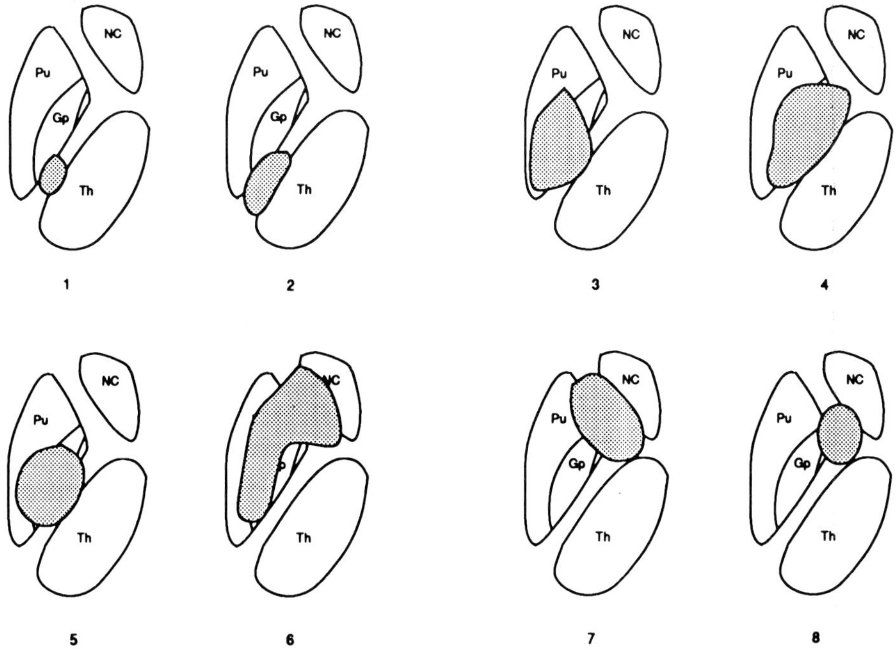

Figure 5. Infarcted areas in the eight patients of the third study. The template is derived from the atlas of Talairach and Tournoux (1988) and represents the basal ganglia at a level 4 mm above the intercommissural line. The lesions were drawn directly from the respective MRI or CT scans and demonstrate their maximal visible extent (NC = caudate nucleus, Pu = putamen, Gp = globus pallidus, Th = thalamus). For ease of comparison all infarcts hve been drawn on the left though patient 4 had a right-sided infarct. (Modified with permission from Weiller *et al.*, 1993.)

extension of the SMC had either no face weakness or very rapid recovery of motor function of the face. The ipsilateral primary sensorimotor cortex was activated more than in normal subjects in only four patients, all of whom exhibited associated movements of the unaffected hand during finger opposition by the recovered hand.

Besides resorption of local edema and necrotic tissue or restitution of reversibly damaged or functional deactivated neurons, several more static or more dynamic theoretical models of functional regeneration have been put forward to explain restoration of brain function in the presence of a persistent defect. These may include redistribution of activity within preexisting neuronal circuits, relearning of originally duplicated function, substitution of function by other areas, the use of alternate pathways run by the original centers, or recruitment of new areas with different pathways. On a cellular level these mechanisms may correlate with synaptic sprouting, increased efficiency of existing neuronal circuits ("unmasking"), or the formation of new polysynaptic connections.

The main conclusion of the three PET studies reported here is that recovery of motor function after disruption of the major efferent motor pathways in the internal capsule is associated with a complex redistribution of rCBF, comprising changes in both the damaged and the undamaged cortex (Table 3); this pattern of redistributed rCBF varies in individual patients. The functional reorganization in association with recovery from motor stroke after capsular infarction comprises plastic changes within the primary sensorimotor cortex; increased activity in nonprimary motor areas, such as SMA, premotor, and cingulate cortices; recruitment of an additional motor system in the parietal cortex and the insula, all within the damaged hemisphere; and a bilateral activation of the motor systems.

Primary Sensorimotor Cortices

Despite disruption of major efferent pathways in the internal capsule an increase in rCBF in the primary sensorimotor cortex was found in the damaged hemisphere in all three studies. In some patients there was even an above-normal increase in rCBF in the lateral part of the contralateral motor cortex, representing a large lateral extension of activity from the hand/arm region of the sensorimotor cortex into the face region. These patients had little or no facial weakness after the stroke and had posterior lesions of the internal capsule that spared the part of the internal capsule thought to carry descending corticobulbar fibers destined for the facial nucleus in the brain stem. Thus a region adjacent to the cortex that has lost all or a part of its output connections with the motor apparatus showed an increased rather than decreased rCBF. This type of reorganization has recently been demonstrated after upper limb deafferentation (Pons *et al.*, 1991) or motor nerve lesion (Donoghue, Suner, & Sanes, 1990). Evidence for reorganization of the motor cortex in man has also been shown after upper limb amputation (Cohen, Bandinelli, Findley, & Hallett, 1991; Hall, Flament, Fraser, & Lemon, 1990). Activation of adjacent motor cortex cells normally destined for facial musculature may be facilitated by lack of inhibition from degenerated intracortical collaterals of motor cortex neurons (DeFelipe, Conley, & Jones, 1986; Gosh & Porter, 1988). Such focal cortical inhibitory activity may be GABA-mediated, as localized pharmacological blockade of cortical inhibition with a GABA antagonist in SMC resulted in the evocation of more widespread movements in response to local stimulation than normal (Jacobs & Donoghue, 1991).

Nonprimary Motor Cortices in the Frontal Lobe

It is also possible that regions of the frontal lobe other than the motor cortex that also contribute substantial numbers of fibers to the corticospinal tract are involved in the recovery process (Dum & Strick, 1991). Such areas include the SMA, the premotor cortex, and cingulate cortex, all of which showed increases in rCBF in most or all of the patients. The motor system in man is conventionally considered to be hierarchically organized, with the primary motor cortex in control of limb movements and executing its influence via direct corticospinal fibers. Premotor cortex (lateral area 6) and SMA are thought to contribute to planning, preparation, or initiation of movement (Wise, 1985). The PET findings suggest that these different motor areas operate in parallel rather than in a hierarchical fashion, and this suggests that they are able to substitute for each other functionally. Experiments in monkeys have convincingly shown that the center motor apparatus consists of multiple cortical fields, each containing a somatotopic representation of muscles (Strick, 1988). Recent experiments suggest that the descending pathways from these multiple cortical motor maps run through different parts of the internal capsule, with fibers descending from the SMA passing through the anterior limb and those from the premotor cortex passing through the anteriomost part of the posterior limb (Fries, Danek, Scheidtmann, & Hamburger, 1993). Thus, small capsular infarcts as studied here may selectively de-efferentiate different motor areas. Depending on the lesion, one or more of the parallel processing motor systems may make up for each other. Other or additional arguments may hold true for the rCBF increases found in the cingulate and dorsolateral prefrontal cortices. The anterior and posterior cingulate cortices appear to form an extensive, reciprocally interconnected system dealing with selective attention or vigilance (Posner & Peterson, 1990). The lateral prefrontal cortex and the angular gyrus (area 39) constitute a brain system concerned with internally generated as opposed to stimulus-driven responses (Frith, Friston, Liddle, & Frackowiak, 1991). This pattern of activation may indicate the need for additional selective attention and intentional mechanisms for the execution of an otherwise simple, repetitive, stimulus-driven motor task, which Brodal (1973) impressively described in his report.

Parietal Association Cortices and Insula

A strong bilateral activation of inferior parietal cortex and the anterior aspects of the insula was found in all three PET studies. This part of the parietal cortex was labeled area 40 by Brodmann (1925). It projects to the inferior part of the premotor cortex (area 6) (Cavada & Goldman-Rakic, 1989). In turn there are projections from inferior area 6 to area 4 (Strick, 1985). The inferior parietal cortex also sends neural projections to the medial parts of the insula. In monkeys the insula itself is related to several somatosensory areas, including SI-SII (Mesulam & Mufson, 1985). Another connection of the insula (at the junction of the granular with the dysgranular cortex) is directed to the ventral portion of area 6 (the precentral operculum), to which the inferior parietal cortex also projects. There is an increase in rCBF in bilateral area 40 and insula in normal subjects as well, although to a lesser degree than in the stroke patients. Hence, the most ventral parts of area 6, insula, and parietal cortex constitute an additional motor system that is recruited above and beyond its normal contribution to the execution of a sequential motor task. Activation of parietal association cortex in the infarcted hemisphere seems to represent a main mechanism in functional reorganization after motor stroke and has recently been reported in a series of stroke patients studied with functional MRI (Rijntjes *et al.*, 1994).

Bilateral Activation of the Motor System

The three PET studies show clearly that the motor system is activated bilaterally. Consistent rCBF increases were found in ipsilateral premotor cortex and basal ganglia and in contralateral cerebellum and have been reported in other studies (Seitz *et al.*, 1991). The role of the ipsilateral primary motor cortex in subserving recovery of function after stroke, however, is not clear. Clinical observations point to a possible role of the ipsilateral cortex, perhaps most remarkably illustrated by the return of function in young patients after hemispherectomy (Glees, 1980). And there is indeed anatomical evidence for ipsilateral corticospinal pathways from the cortex, although this seems to be more pronounced in the control of the proximal muscles (Armand, 1982). Some ipsilateral rCBF increase is found during finger opposition in normal subjects as well (Weiller *et al.*, 1992). However, a recent study with magnetic stimulation in stroke patients did not find any evidence for short latency ipsilateral influences in recovery of function (Palmer, Ashby, & Hajek, 1992). The PET results are not unambiguous. Activation of ipsilateral SMC was found in the first PET study and in some of the patients of the last study, and then related to the occurrence of associated movements of the unaffected hand, while in the second study activation in the ipsilateral motor system was restricted to the premotor cortex. Obviously, the ipsilateral SMC is not recruited in all patients with a motor stroke that is due to capsular infarcts.

Some generalizations can be drawn from these results. There is considerable scope for functional plasticity in the adult human brain. Individual patients do show different patterns of reorganization. There are changes in organization in both the damaged and the undamaged cortex, ipsilateral to the recovered limb. While substitution of function in homotopic brain regions (primary SMC) of the undamaged hemisphere may occur in some cases with motor recovery, the reorganization of parallel-acting multiple motor areas (premotor and cingulate cortices, SMA) or motor related areas (parietal cortex, insula) in the damaged hemisphere may constitute the central mechanism in motor recovery. The site of a subcortical lesion may determine the pattern of reorganization in the cerebral cortex. As shown, a lateral extension of the SMC is only possible if fibers destined for the face are preserved. Similar implications have to be expected for premotor cortex and SMA. It also follows that the more the motor related areas or their efferent pathways are destroyed, the less the potential for reorganization and the poorer the prognosis for the individual patient. The complex pattern of functional reorganization all occurred within areas that are related to motor function under normal conditions as well. Thus, a radical functional reorganization within the frame of a given network rather than substitution of function may be assumed.

What does all this mean to us? Is this complex reorganization that has taken place after stroke really the cause of restitution of function? Is such reorganization actually beneficial to the patients concerned? We hope that these first PET studies on recovery from stroke will encourage the design of new experiments aimed at discovering how such reorganization might contribute to functional recovery. Positron emission tomography has opened in a unique way the potential to study the changes that occur in the human brain after a stroke. Implications for prognosis after stroke and consequences for rehabilitation of stroke patients seem obvious and will be the subject of future studies. The application of PET activation studies in patients recovered from aphasia has recently shown that in language as well, both the damaged and the undamaged hemisphere contribute to recovery of function (Weiller *et al.*, 1993). Magnetic resonance imaging activation studies allow the study of a single subject without the use of radiation. First results from MRI studies of stroke patients are in line with the results and conclusions drawn from the PET studies (Rijntjes *et al.*, 1994).

References

Alexander, G. E., & Crutcher, D. (1990). Functional architecture of basal ganglia circuits: Neural substrates of parallel processing. *Trends in Neurosciences, 13,* 266–271.

Armand, J. (1982). The origin, course and termination of corticospinal fibers in various mammals. *Progress in Brain Research, 57,* 329–360.

Brodal, A. (1973). Self-observations and neuro-anatomical considerations after stroke. *Brain, 96,* 675–694.

Brodmann, K. (1925). *Vergleichende Lokalisationslehre der Großhirnrinde.* Leipzig: Barth.

Cavada, C., & Goldman-Rakic, P. S. (1989). Posterior parietal cortex in rhesus monkeys: Evidence for segregated corticocortical networks linking sensory and limbic areas with the frontal lobe. *Journal of Comparative Neurology, 287,* 422–445.

Chollet, F., DiPiero, V., Wise, R. J. S., Brooks, D. J., Dolan, R. J., & Frackowiak, R. S. J. (1991). The functional anatomy of motor recovery after stroke in humans: A study with positron emission tomography. *Annals of Neurology, 29,* 63–71.

Cohen, L. G., Bandinelli, S., Findley, T. W., & Hallett, M. (1991). Motor reorganization after upper limb amputation in man. *Brain, 114,* 615–627.

DeFelipe, J., Conley, M., & Jones, E. G. (1986). Long-range focal collateralization of axons arising from corticocortical cells in money sensorimotor cortex. *Journal of Neurological Sciences, 6,* 3749–3766.

Donoghue, J. P., Suner, S., & Sanes, J. N. (1990). Dynamic organization of primary motor cortex output to target muscles in adult rats. II. Rapid reorganization following motor nerve lesions. *Experimental Brain Research, 79,* 492–503.

Dum, R. P., & Strick, P. L. (1991). The origin of corticospinal projections from the premotor areas in the frontal lobe. *Journal of Neuroscience, 11,* 667–689.

Fries, W., Danek, A., Scheidtmann, K., & Hamburger, C. (1993). Motor recovery following capsular stroke. *Brain, 116,* 369–382.

Friston, K. J., Frith, C. D., Liddle, P. F., Dolan, R. J., Lammertsma, A. A., & Frackowiak, R. S. J. (1990). The relationship between global and local changes in PET scans. *Journal of Cerebral Blood Flow and Metabolism, 10,* 458–466.

Friston, K. J., Frith, C. D., Liddle, P. F., & Frackowiak, R. S. J. (1991). Comparing functional (PET) images: The assessment of significant change. *Journal of Cerebral Blood Flow and Metabolism, 11,* 690–699.

Friston, K. J., Passingham, R. E., Nutt, J. G., Heather, J. D., Sawle, G. V., & Frackowiak, R. S. (1989). Localisation in PET images: Direct fitting of the intercommissural (AC-PC) line. *Journal of Blood Flow and Metabolism, 9,* 690–695.

Frith, C. D., Friston, K., Liddle, P. F., & Frackowiak, R. S. J. (1991). Willed action and the prefrontal cortex in man: A study with PET. *Proceedings of the Royal Society of London B, 244,* 241–246.

Glees, P. (1980). Functional reorganisation following hemispherectomy in man and after small experimental lesions in iprimates. In P. Bach-y-Rita (Ed.), *Recovery of function: Theoretical considerations for brain injury rehabilitation* (pp. 106–126). Baltimore: University Park Press.

Gosh, S., & Porter, R. (1988). Morphology of pyramidal neurones in monkey motor cortex and the synaptic actions of their intracortical axon collaterals. *Journal of Physiology, 400,* 593–615.

Hall, E. J., Flament, D., Fraser, C., & Lemon, R. N. (1990). Non-invasive brain stimulation reveals reorganised cortical outputs in amputees. *Neuroscience Letters, 116,* 379–386.

Jacobs, K. M., & Donoghue, J. P. (1991). Reshaping the cortical motor map by unmasking latent intracortical connections. *Science, 251,* 944–947.

Kempinsky, W. H. (1956). Spatially remote effects of focal brain injury. Relation to diaschisis. *Transactions of the American Neurological Association, 81,* 79–82.

Lammertsma, A. A., Cunningham, V. J., Heather, J. D., Nutt, J., Frackowiak, R. S. J., & Jones, T. (1989). A combination of dynamic and integral analyses to obtain accurate functional images of rCBF. *Journal of Cerebral Blood Flow and Metabolism, 9* (Suppl.), S582–S583.

Mesulam, M. M., & Mufson, E. J. (1985). The insula of Reil in man and monkey: Architectonics, connectivity, and function. In A. Peters & E. G. Jones (Eds.), *Cerebral cortex: Association and auditory cortices.* New York: Plenum Press.

Palmer, E., Ashby, P., & Hajek, V. E. (1992). Ipsilateral fast corticospinal pathways do not account for recovery in stroke. *Annals of Neurology, 32,* 519–525.

Pons, T., Garraghty, P. E., Ommaya, A. K., Kaas, J. H., Taun, E., & Mishkin, M. (1991). Massive cortical reorganization after sensory deafferentation in adult macaques. *Science, 252,* 1857–1860.

Posner, M. I., & Peterson, S. E. (1990). The attention system of the human brain. *Annual Review of Neuroscience, 13,* 25–42.

Raichle, M. E. (1987). Circulatory and metabolic correlates of brain function in normal humans. In F. Plum (Ed.), *Handbook of physiology: The nervous system: Vol. V* (pp. 643–674). New York: American Physiological Society, Oxford University Press.

Rijntjes, M., Weiller, C., Krams, M., Bauermann, T., Diener, H. C., & Faiss, J. (1994). Functional magnetic resonance imaging in recovery from motor stroke. *Stroke, 25,* 259.

Seitz, R. J., Herzog, H., Steinmetz, H., Roden, W., Benecke, R., & Freund, H. J. (1991). Motor activity after subcortical stroke studied with positron emission tomography (abstract). *Journal of Cerebral Blood Flow and Metabolism, 11* (Suppl. 2), S647.

Strick, P. (1985). How do the basal ganglia and cerebellum gain access to the cortical motor areas? *Behavioral Brain Research, 18,* 107–124.

Strick, P. (1988). Anatomical organization of multiple motor areas in the frontal lobe: Implications for recovery of function. *Advances Neurology, 47,* 293–312.

Talairach, J., & Tournoux, P. (1988). *Coplanar stereotaxic atlas of the human brain.* New York: Georg Thieme Verlag.

von Monakow, C. (1914). *Die Lokalisation im Großhirn und der Abbau der Funktion durch kortikale Herde.* Wiesbaden: Bergmann.

Weiller, C., Chollet, F., Friston, K. J., Wise, R. J. S., & Frackowiak, R. S. J. (1992). Functional reorganization of the brain in recovery from striatocapsular infarction in man. *Annals of Neurology 31,* 463–472.

Weiller, C., Ramsay, S. C., Wise, R. J. S., Friston, K. J., & Frackowiak, R. S. J. (1993). Individual patterns of functional reorganization in the human cerebral cortex after capsular infarction. *Annals of Neurology, 33,* 181–189.

Weiller, C., Rijntjes, M., Isensee, C., Müller, S. P., Kappeler, C., Huber, W., & Diener, H. C. (1993). Recovery from Wernicke's aphasia-a PET study. *Annals of Neurology, 37,* 723–732.

Weiller, C., Ringelstein, E. B., Reiche, W., Thron, A., & Buell, U. (1990). The large striatocapsular infarct. A clinical and pathophysiological entity. *Archives of Neurology, 47,* 1085–1091.

Wise, S. P. (1985). The primate premotor cortex: past, present, and preparatory. *Annual Review of Neuroscience, 8,* 1–19.

10

Neuroimaging and Traumatic Brain Injury

ERIN D. BIGLER

INTRODUCTION

There are two main approaches to imaging the traumatically injured brain: acute diagnostic imaging and determination of long-term structural sequelae (see Gean, 1994; Osborn, 1994; Prayer & Gean, 1994). The actual mechanisms of brain injury have been fully discussed elsewhere (Bigler, 1990; Christman, Grady, Walker, Holloway, & Povlishock, 1994; Newberg & Alavi, this volume; Povlishock, 1992). This chapter will deal exclusively with these two areas of neuroimaging in traumatic brain injury (TBI) and the relationship of acute and chronic findings to neuropsychological outcome.

ACUTE IMAGING

Typically, when a patient sustains a head injury, the first diagnostic imaging procedure is CT scanning (Osborn, 1994). Even though magnetic resonance imaging (MRI) provides much better resolution for imaging brain parenchyma, during the acute phase of injury, CT scanning is preferred for several reasons: a) Images can be acquired quickly; b) CT can handle motion artifact better than MRI; c) medical monitoring or life support devices can coexist within the CT scanning environment and often not with MRI; d) the ability of CT to image acute hemorrhage is excellent; and e) CT can readily detect skull fracture.

ERIN D. BIGLER Department of Psychology, Brigham Young University, Provo, Utah 84602-5372

Neuroimaging II: Clinical Applications, edited by Erin D. Bigler. Plenum Press, New York, 1996.

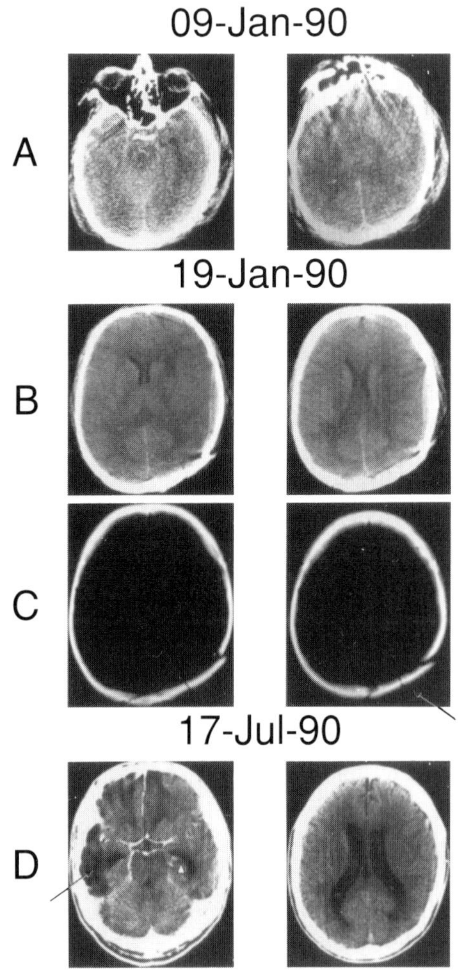

Figure 1. (A) Axial (horizontal) CT images of the brain of the day of injury (January 9, 1990). This patient fell approximately 30 feet, striking his head on a concrete floor. There is frontal bone skull fracture seen in the upper right figure. The brain is very edematous, with obliteration of the ventricular and cisternal systems due to the effects of edema. (B) CT image obtained ten days later on the same patient. The evolution of a subdural hematoma is clearly seen in the right hemisphere just anterior to a major fracture in the posterior right aspect of the patient's skull. There is still midline shift from right to left, but there is better definition of the ventricular system. Using the day-of-injury or the early CT imaging obtained on January 19, 1990 can provide a basis from which to compare this patient (see Figures 2 and 3). (C) Bone window CT film demonstrating the points of fracture. (D) Follow-up CT scan done approximately seven months postinjury. Note the extensive wasting of the inferior frontal and the left temporal region as well as bilateral temporal horn dilation and generalized ventriculomegaly.

Figure 1 depicts acute CT imaging in a patient with significant head injury, including hemorrhage, skull fracture, and cerebral edema, and demonstrates the evolution of trauma-induced degenerative changes over time, based on CT imaging. Figure 2 presents MRI of the same patient over one year postinjury compared to an age-matched noninjured control. (This image should be compared to Figure 1.) Note the extensive wasting of the inferior and anterior aspects of both frontal and temporal lobes (the most common areas injured in

TBI), along with the extensive and generalized enlargement of the ventricular system. Figure 3 depicts a three-dimensional (3-D) reconstruction of acute imaging, the day-of-injury CT (in the patient depicted in Figure 1), and the late effects of injury evidenced in MRI (see Figure 2) by presenting a similar 3-D reconstruction. The day-of-injury scan can be compared to follow-up scans, such as the postinjury MRI scan in Figure 2, to demonstrate the most typical late effects related to brain injury (ventricular dilation and brain wasting; Bigler *et al.*, 1994).

As can be observed from inspection of Figures 1, 2, and 3, initial imaging not only provides the acute pathologic findings at the time of injury along with a baseline from which to monitor acute and subacute pathologic changes (absorption or expansion of hematomas, brain swelling, etc.), but also provides some indication of preinjury brain morphology. Figures 4 and 5 present additional

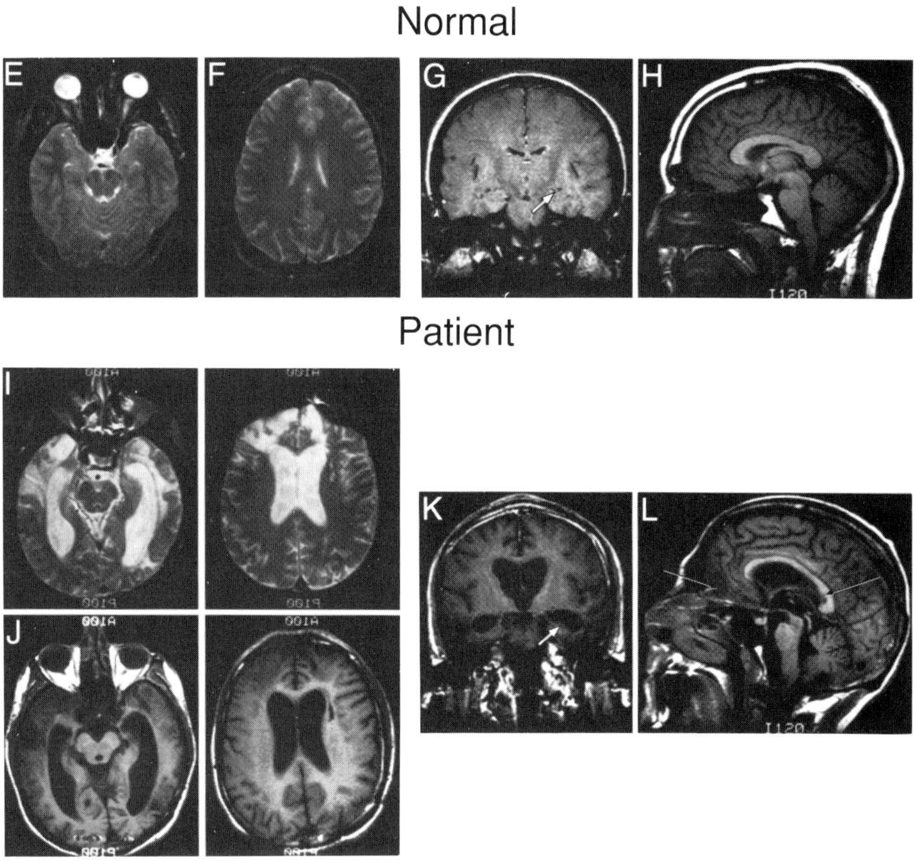

Figure 2. (Top) Normal axial (E and F), coronal (G), and sagittal (H) for comparison with patient (below) over one year postinjury. (I) Axial MR studies with T2 weighting demonstrating extensive wasting of both temporal lobes and marked ventricular dilation (left) and (right) increased lateral ventricular size with encephalomalacic changes in the frontal region bilaterally. (J) Similar level T1-weighted MR studies for comparison. Compare these images to E and F in the normal individual presented above. (K) Coronal image of the brain depicting marked ventricular dilation. The arrow points to the hippocampus which is atrophic. (L) Sagittal view indicating thinning of the corpus callosum, a tear in the corpus callosum (arrow), and anterior and inferior frontal wasting secondary to traumatic brain injury.

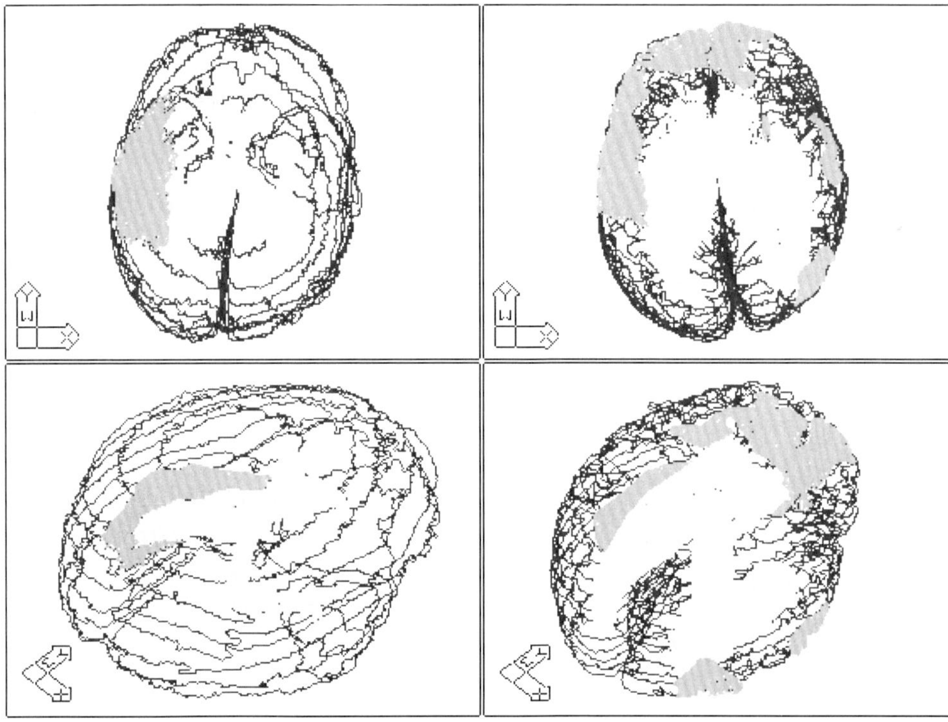

Figure 3. Early CT three-dimensional reconstruction that estimates the preinjury status of this patient's brain and depicts the location of hematoma. This is based on 3-D image reconstruction of the January 9, 1990, scan in this patient (see Figure 1), and the dark area in the left temporal region shows the presence of acute hemorrhagic contusion in that portion of the brain (refer to Figure 1A). (Right) Generalized ventricular dilation presented three-dimensionally (light gray) with extensive wasting in bifrontal and temporal regions (dark gray) (see Bigler, 1994, for description of 3-D imaging technique). This figure presents the 3-D changes in ventricular system observed in the 2-D axial views represented in Figure 2 (I and J).

cases utilizing 3-D reconstruction, demonstrating how the initial or day-of-injury CT scan can be used to demonstrate baseline status and monitor changes over time (Bigler *et al.,* 1994). Although acute imaging analysis reveals many important pathological findings, as demonstrated by these illustrations, in terms of predicting neuropsychological outcome, chronic imaging (i.e., imaging performed at minimum 6 weeks or later, postinjury) provides more robust relationships (see Bigler, 1994; Kurth, Bigler, & Blatter, 1994).

Imaging during the Chronic Phase

TBI-Induced Degenerative Changes. We recently have had several cases wherein preinjury neuroimaging was available (see Bigler, Kurth, Blatter, & Abildskov, 1992; Gale, Johnson, Bigler, & Blatter, 1995). These cases illustrate the ideal circumstance to monitor changes over time, in particular the degenerative changes, and to demonstrate the chronic, lasting effects of trauma on brain morphology. As illustrated in Figures 6 and 7, the most noticeable change is in the ventricular system. It is thought that ventricular changes are *ex vacuo* in nature, related to tissue loss. This occurs because the ventricular system is under

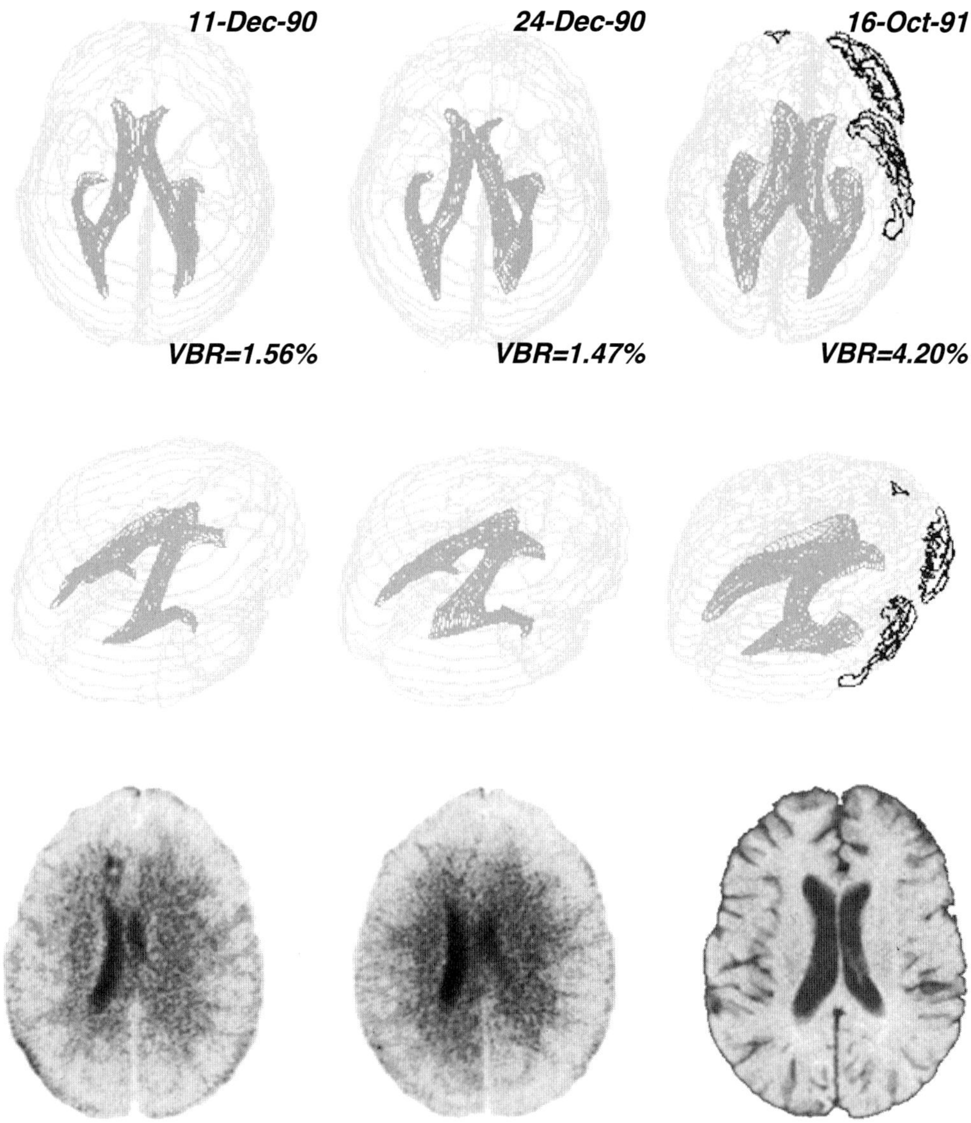

Figure 4. Change in ventricular system over time following trauma. (Left) Day-of-injury CT scan in axial plane at the bottom, with 3-D image reconstruction showing position of ventricular system in the dorsal view (top) and the right posterior oblique (middle). The ventricle-to-brain ratio (VBR) is 1.56 on the day of injury, a score that is within the normal range (see Blatter *et al.*, 1995). (Middle) Corresponding CT 14 days later with similar 3-D image reconstruction above. (Right) MRI studies done 10 months later showing ventriculomegaly with corresponding 3-D image analysis. The VBR has increased dramatically. The dark areas in the frontal and temporal regions represent areas of encephalomalacia.

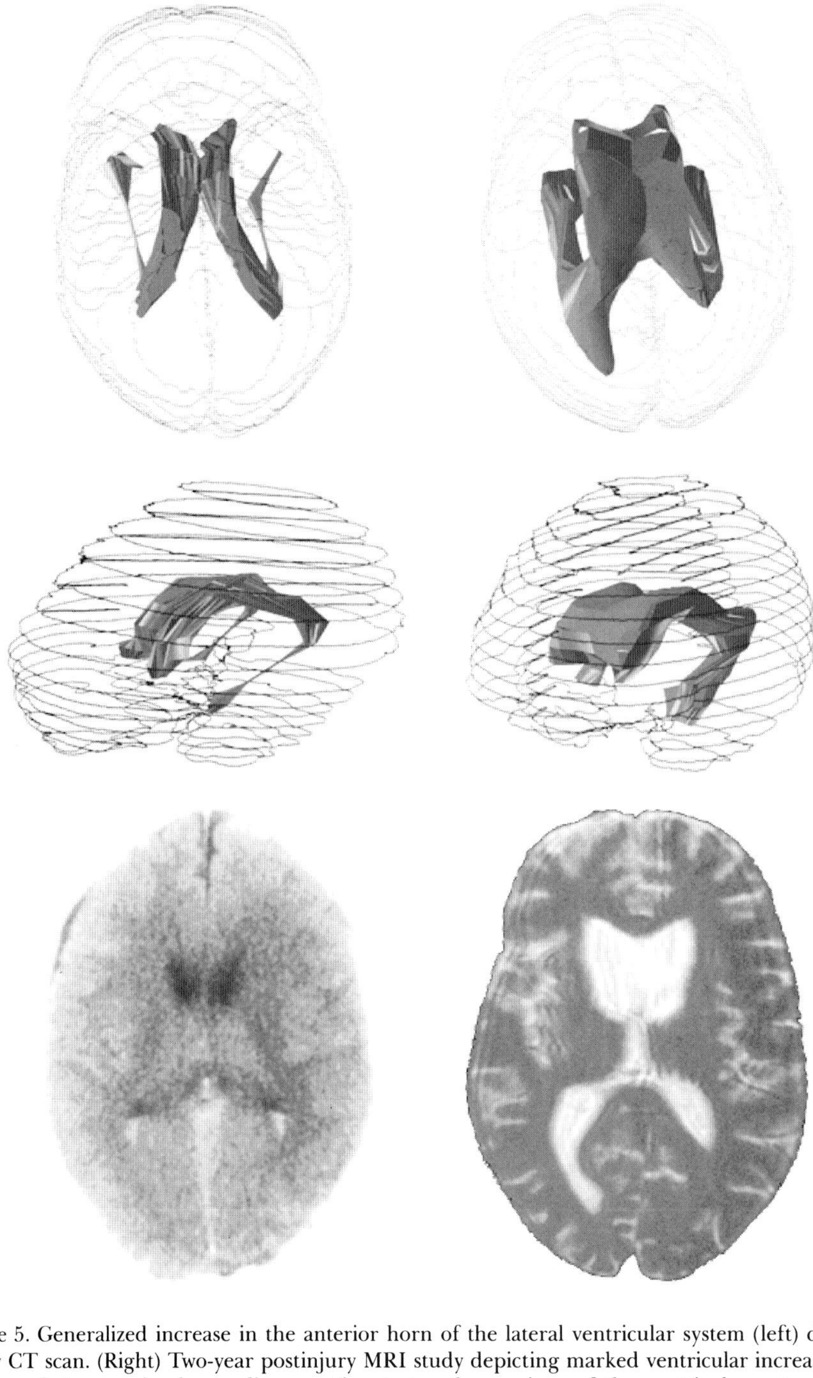

Figure 5. Generalized increase in the anterior horn of the lateral ventricular system (left) day-of-injury CT scan. (Right) Two-year postinjury MRI study depicting marked ventricular increase and dilation of the anterior horns disproportionate to other regions of the ventricular system. It is thought that the ventricular system will dilate corresponding to surrounding wasting of cerebral tissue and that asymmetric enlargement of the ventricular system is an indication of asymmetric parenchymal loss. This patient also displayed classic frontal lobe syndrome in terms of labile affect, inappropriate social behavior, and diminished complex reasoning and problem-solving abilities.

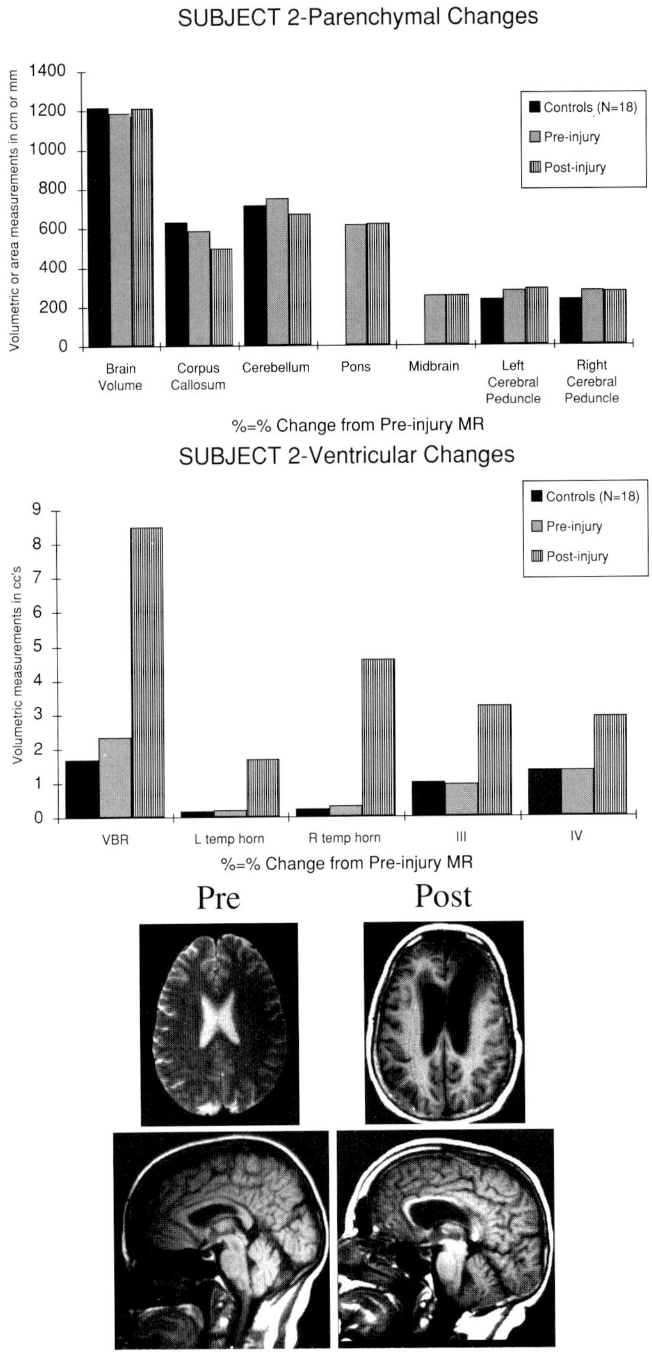

Figure 6. Pre- and post-MRI comparisons. This patient had a preinjury MRI study (bottom left) done to evaluate complaints of dizziness. The scan was interpreted to be within normal limits for her age (69). The postinjury scan (bottom right) is approximately 2 months following a severe traumatic brain injury. The bar graphs presented above depict the morphometric changes in different brain regions that occurred in this patient's brain as a result of trauma in comparison to a group of age-matched, non-brain-injured controls (see Gale *et al.*, 1995b).

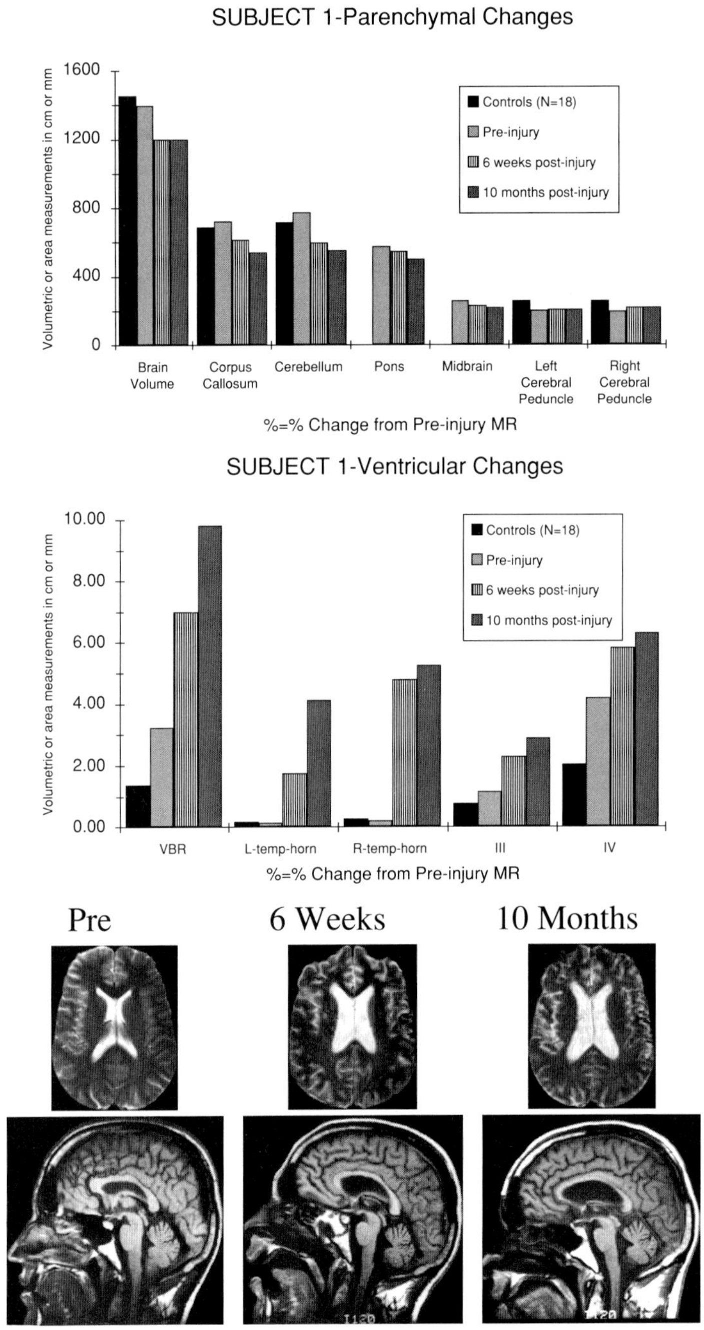

Figure 7. Preinjury, 6-week, and 10-month postinjury comparisons. This patient was a 34-year-old male who had preinjury MRI studies (bottom left) performed to evaluate the origin of persistent headache. No abnormalities were ascertained. The 6-week (bottom middle) and 10-month (bottom right) follow-up MR studies demonstrate significant ventricular enlargement and wasting of the corpus callosum. The bar graphs presented above provide a morphological comparison to a normal control sample of approximately the same age as well as the preinjury comparison (see Gale *et al.*, 1995b).

a constant outward pressure gradient in response to internal cerelarospinal fluid (CSF) production. Accordingly, secondary to trauma-induced parenchymal atrophy, the ventricular system expands. This is probably most impressively observed in the 3-D depiction of the ventricular system postinjury as presented in Figures 4–7.

FRONTAL AND TEMPORAL LOBE FOCUS. Because of the manner in which the frontal and temporal regions of the brain are situated in the anterior and middle fossa of the skull, the frontal and temporal lobes are the most likely regions of the brain to sustain cortical contusion and intraparenchymal hemorrhage (Kurth *et al.*, 1994). Also, it may be that the shear-strain forces reach their maximum in this area of the brain as it is typical to see hemorrhagic petechial lesions in this area as well (see Figure 8). Accordingly, the frontal and temporal regions are the areas most likely to exhibit degenerative changes and this may be

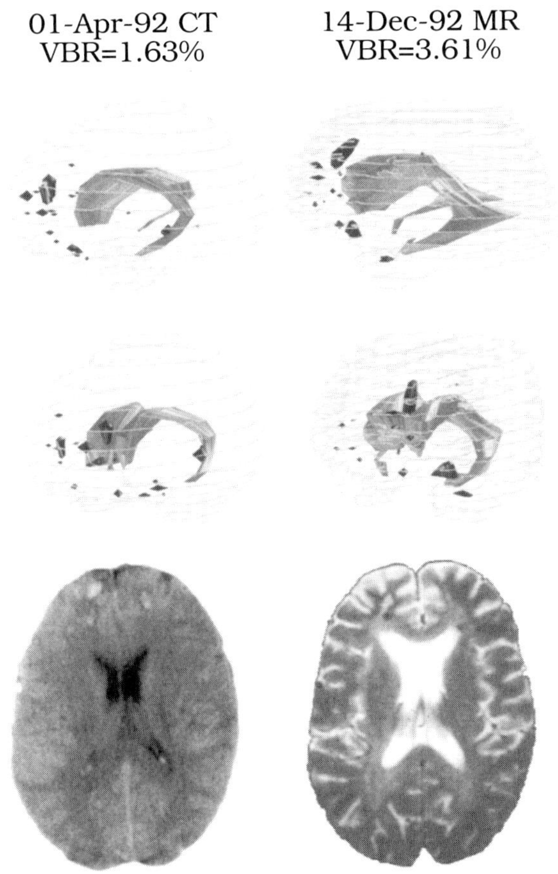

Figure 8. (Left column) Day-of-injury CT scan depicting the predominant locus of frontal and temporal hemorrhagic lesions as a result of a severe traumatic brain injury. The CT scan is presented in axial plane at the bottom left, showing the location of hemorrhages, and these hemorrhages are correspondingly depicted in the 3-D image reconstructions above. (Right column) Follow-up MR scan showing chronic lesion changes in the frontal aspect of the brain with corresponding location of the frontal contusion/hemorrhagic lesion areas (dark gray). In both cases, the ventricular system is presented in light gray for 3-D image analysis (see Kurth *et al.*, 1994).

demonstrated by enlargement of the anterior or frontal and inferior or temporal horns of the lateral ventricular system (recall Figures 2 and 5). Note the temporal horn dilation and the more atrophic anterior aspect of the corpus callosum. These are signs of greater frontal and temporal lobe atrophy in trauma.

WHITE AND GRAY MATTER CHANGES IN TBI. The postmortem studies of TBI victims have demonstrated white matter shearing as a major pathologic factor in many cases of TBI (see reviews by Bigler, 1990; Salazar, 1992). However, there are a multitude of axonal ultrastructure changes associated with trauma (Povlishock, 1992; Christman *et al.*, 1994). Because of the resolution problems of CT and only the recent development of methods for accurate and automated analysis of MR-based morphometry, it is only now that the issue of white matter loss can be addressed via imaging techniques. To this end, Anderson and Bigler (1994, 1995) have approached this problem by comparing day-of-injury CT scans with postinjury (at least 6 weeks) MR scans. A linear measurement of the genu of the corpus callosum (white matter), surface area at the same level of the caudate nucleus, and anterior horn of the lateral ventricle were compared. As previously discussed, it has long been known that the ventricular system enlarges in TBI and that this has been thought to be due to white matter degeneration (see review by Prayer & Gean, 1994). However, there is anterior horn and lateral ventricular system enlargement in Huntington's disease—a disorder of gray matter degeneration (see Hall, Bigler, & Rutledge, 1989; Lenti & Bianchini, 1993). In Huntington's disease, as the caudate degenerates, the anterior horn of the lateral ventricular system expands. To examine this problem, whether anterior horn expansion in traumatic disorder is related to a disproportionate effect on white matter versus gray, Anderson and Bigler (1994, 1995) looked at the relationship between caudate nucleus, anterior horn size of the ventricular system, and white and gray matter structures. If TBI were more selective to injury effects at the level of white versus gray matter, then the corpus callosum measurement should show a decrease while the caudate nucleus measurement remains the same, all in the midst of anterior horn ventricular dilation. This is, in fact, what was observed in the Anderson and Bigler investigation. This has been followed up by a detailed investigation of the corpus callosum by Johnson, Bigler, Burr, and Blatter (1994), with similar results (i.e., ventricular expansion is likely most influenced by white matter loss, in particular the corpus callosum). Figure 9 provides a 3-D perspective of the position of the corpus callosum relative to the lateral ventricular system. Viewing this figure clearly demonstrates the intimate relationship between corpus callosum and the body of the lateral ventricles and why ventricular size increases in response to corpus callosum atrophy. Gale, Johnson, Bigler, and Blattner (1995), in a comparison to normal controls, also demonstrated the differential effect of white matter on brain morphology in traumatic brain injury. In this study, the changes in white matter structures were greater in proportion to gray in TBI victims.

UBIQUITOUS NATURE OF DISSEMINATED DAMAGE IN TBI. In addition to the disproportionate effect of white matter versus gray, Gale (1993) and Gale, Johnson, Bigler, & Blatter (1995) demonstrated the ubiquitous nature of damage to the entire central nervous system in traumatic brain injury. Often, in TBI, it has been assumed that there are more regional or focal injury effects and that outside of the focal or regional damage there may be little structural change. As discussed previously, the frontal and temporal lobe regions are likely more af-

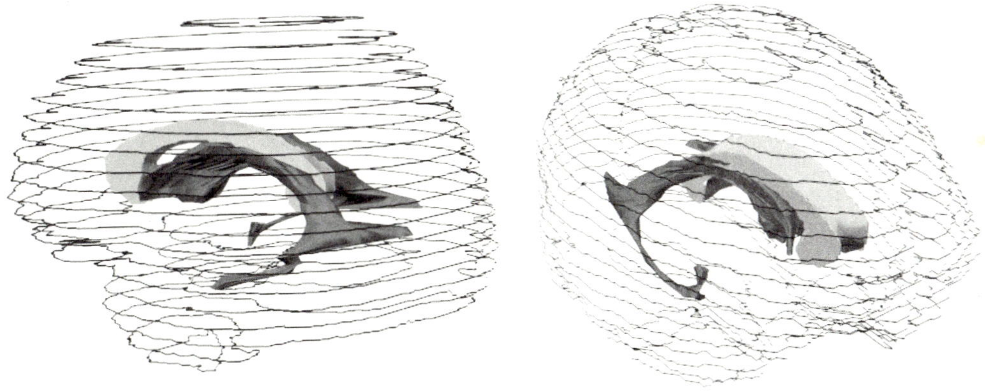

Figure 9. MRI-based 3-D reconstruction of the ventricular system in relationship to the corpus callosum. Each outer rim wireframe outline indicates a different level of the MR scan. This figure is adapted from Johnson *et al.* (1994) and is published with permission from the American Psychological Association. See also Johnson *et al.* (1996).

fected than any other region, yet recent research suggests that the damage is clearly more diffuse, even when there are focal injury effects (see Povlishock, 1992). The research by Gale (1993) and Gale *et al.* (1995) has clearly demonstrated this finding. This research has suggested that damage to the central nervous system may affect the brain at all levels and is related to the severity of injury. In the studies, by Gale *et al.* (Gale, 1993; Gale *et al.*, 1995) detailed morphometric assessment was undertaken of structures at telencephalic, diencephalic, mesencephalic, and rhombencephalic levels, all of which generally demonstrated atrophic change in size in comparison to normal controls. The changes in size increased with the severity of injury. Another study by Anderson and Bigler (1994b) has shown similar findings. Table 1 demonstrates a compari-

TABLE 1. COMPARISON OF SOME MRI-BASED ANATOMIC MEASURES IN TBI PATIENTS WITH NONINJURED CONTROLS[a]

	Control N = 128		TBI - Early[b] N = 78		TBI - Late[c] N = 56	
Brain (C)[d]	1343.7	(35.2)	1340.3	(39.3)	1298.22	(55.5)***
Total intracranial Volume (C)	1463.1	(150.9)	1446.3	(172.9)	1444.26	(148.9)
Ventricle-to-Brain ratio	1.3	(0.46)	1.7	(1.0)	2.73	(1.83)***
Total ventricle (C)	17.4	(5.9)	22.7	(13.7)	34.71	(21.6)***
Lateral ventricle (C)	14.6	(5.7)	19.0	(11.3)	29.8	(19.1)***
Third ventricle (C)	0.73	(0.26)	1.06	(0.61)	1.72	(0.9)***
Fourth ventricle (C)	1.66	(0.53)	1.76	(0.79)	1.71	(0.8)
Left temporal horn	0.21	(0.16)	0.37	(0.50)	0.62	(0.9)***
Right temporal horn	0.23	(0.22)	0.44	(0.77)	0.83	(1.9)***
CSF (C)	103.3	(35.2)	106.6	(39.3)	148.7	(55.6)***
Subarachnoid CSF (C)	85.8	(33.2)	83.1	(33.4)	114.1	(41.8)***
Corpus callosum (surface area)	693.9	(91.1)	—	—	598.6	(117.4)
White matter (C)	651.1	(96.2)	648.5	(126.1)	642.1	(92.1)
Gray matter (C)	701.1	(117.1)	691.9	(125.3)	643.2	(66.3)*

[a]*, 0.05; **,0.01; ***,0.001.
[b]Imaging studies done within 90 days of injury.
[c]Imaging studies done ≥ 90 days postinjury
[d]C, head size correction.

son of several brain structures in TBI victims compared to the normative database presented in Chapter 4 in Volume I of this series (see also Bigler *et al.*, in press; Blatter *et al.*, in press). As can be seen from this table, significant changes in most brain structures occur following TBI.

Predicting Neuropsychological Outcome

The various methods for quantifying brain imaging that have been discussed (see Figures 10 and 11) can also be used to look at the quantitative relationship to neuropsychological test outcome. This is depicted in Figures 12 and 13. This patient sustained a severe traumatic brain injury in a fall while cliff climbing. As can be seen in the figure, the frontal lobe has been extensively damaged. The patient also displays some nonspecific increase in ventricular size and significant wasting of the right frontal area. This neuroimaging data can be

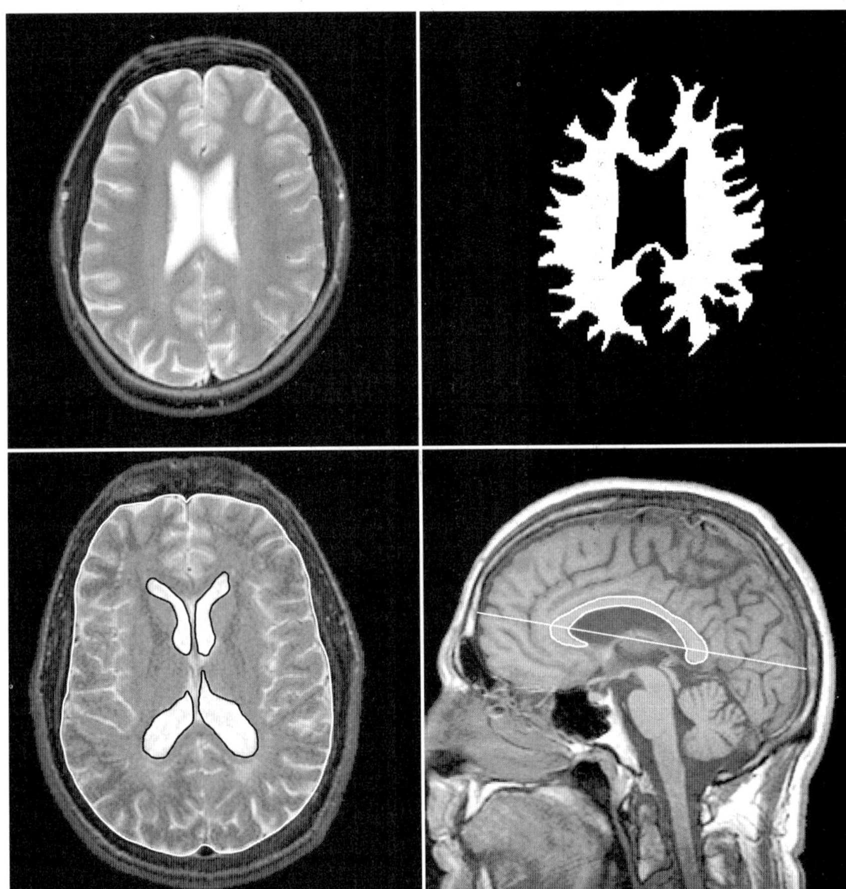

Figure 10. Threshold imaging methods for quantification of surface area change in the brain. The upper left axial MR image is used as the base. Using a thresholding technique based on various gray scale values, CSF field spaces, white matter, and gray matter can be separated. This is depicted in the upper right-hand corner. In the lower left figure, the ventricular system is outlined, as well as the outer rim of cortical surface. In the lower right, the corpus callosum is outlined and an anterior/posterior linear measurement is taken to correct for head size.

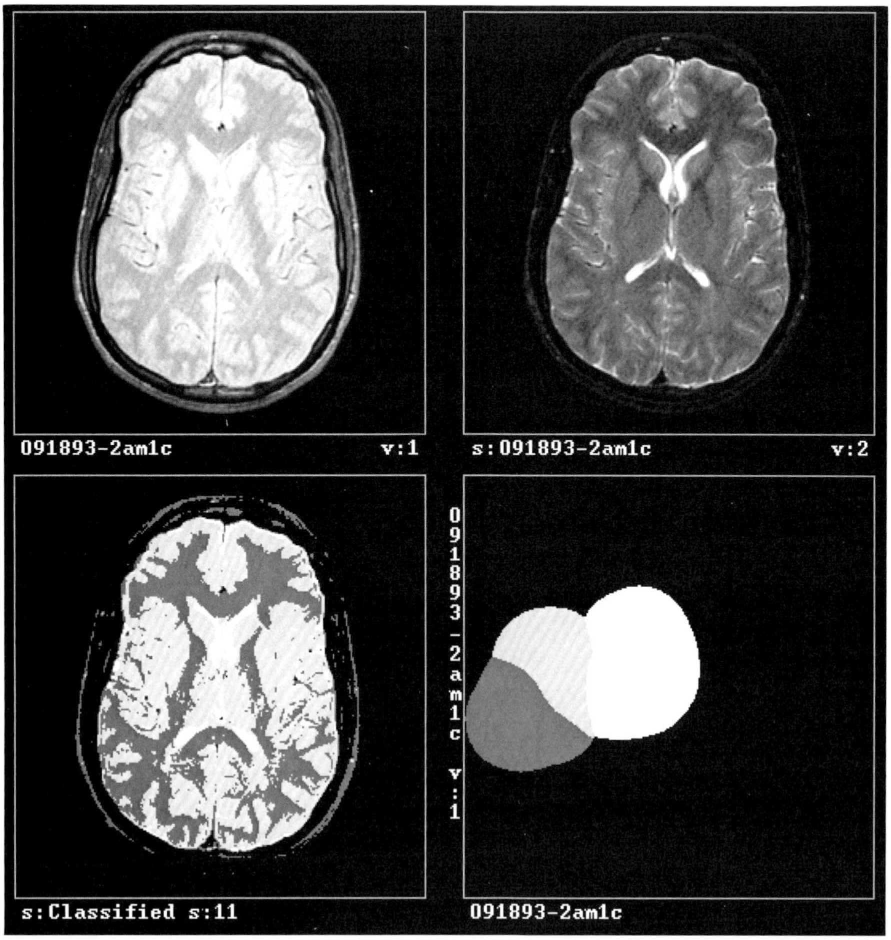

Figure 11. Multispectral image analysis based for quantification of brain structures. (Upper left) Proton weighted image with equally registered T2-weighted image on the right. (Lower right) Feature space map showing the coordinate outline of white matter, gray matter, and CSF space with fully segmented image presented on the left. This segmentation is based on the ANALYZE software program, Mayo Clinic. (See Robb, 1995; also see Figure 3, page 4 of this volume.)

interfaced with the neuropsychological outcome data. Recently, Lisk, Pasteur, Rhoades, Putnam, and Grotta (1994) demonstrated the efficacy of prediction modeling from neuroimaging studies based on early presentation of hemispheric injury due to intracerebral hemorrhage. As this line of research evolves, it is likely that we will eventually be capable of predicting outcome via the quantitative and qualitative analysis of various neuroimaging findings. Currently, our research (Gale *et al.*, 1993, 1995; Johnson *et al.*, 1994) suggests that as ventricular size increases, particularly in the temporal horn region, neuropsychological performance decreases. For example, Table 1 presents quantitative data comparing age-matched normal controls (see Chapter 4, Volume I of this series) with TBI patients scanned within the first 90 days postinjury and those scanned 90 days postinjury. This table demonstrates that with increasing time postinjury, there is greater degenerative change. Our research suggests the expected relationship—that these trauma-induced structural changes relate nega-

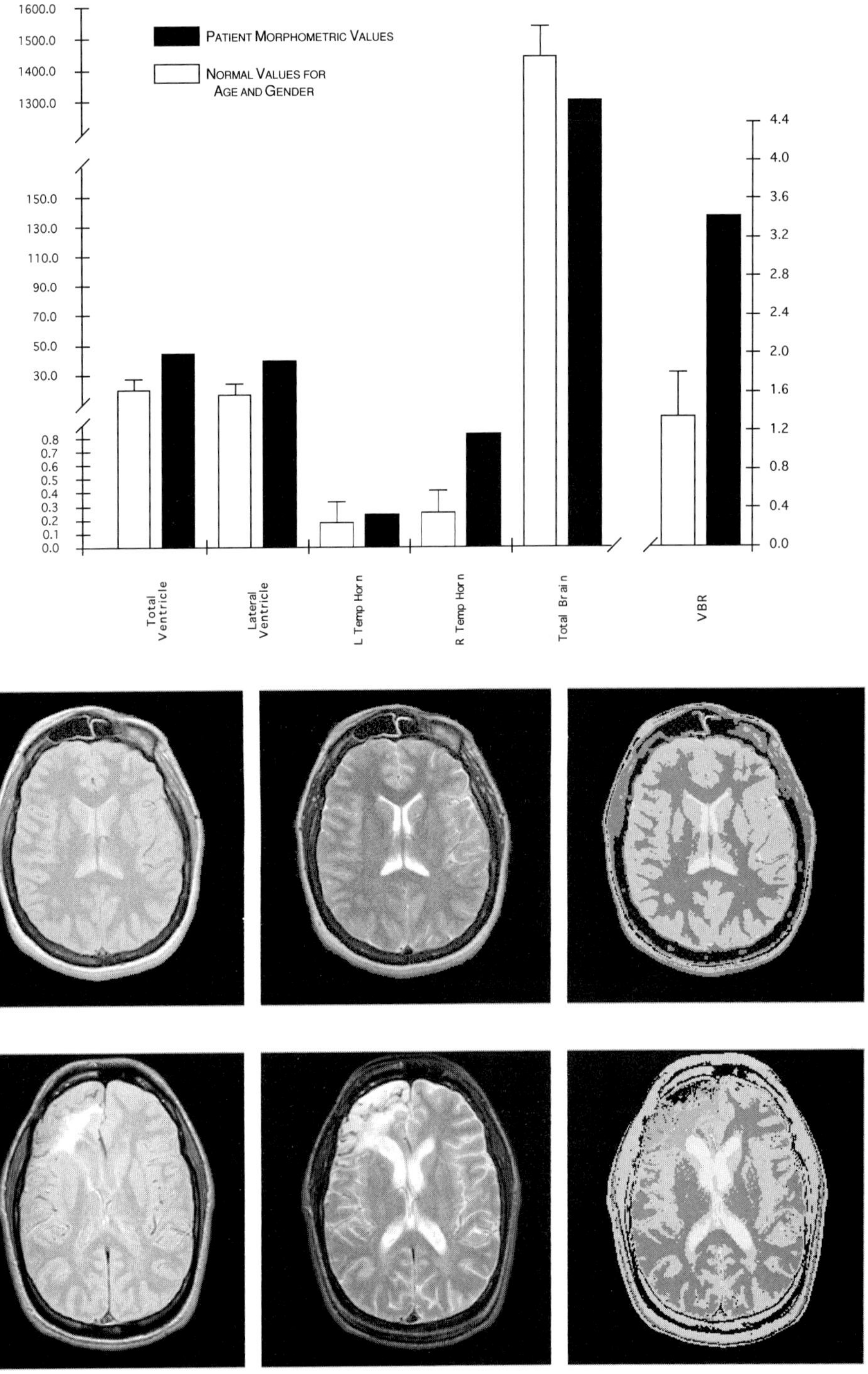

Figure 12. Bar graph depicting aspects of the patient's quantitative neuroimaging findings of the ventricular system based on volumetric estimates from segmented images (see Figures 10 and 11 for details on quantification methods). The top row of MR images depicts a medical control subject with proton density–weighted MR scan on the left, T2-weighted scan in the middle, and the "segmented" image differentiating white matter, gray matter, and CSF space. The bottom row of images is from the TBI patient with similar level MR images and a large right frontal lesion (left is on the reader's right). The patient's neuropsychological data are presented in Figure 13. These quantitative analyses indicate significant ventricular dilation, which in TBI is a sign of white matter loss (see Bigler, 1990).

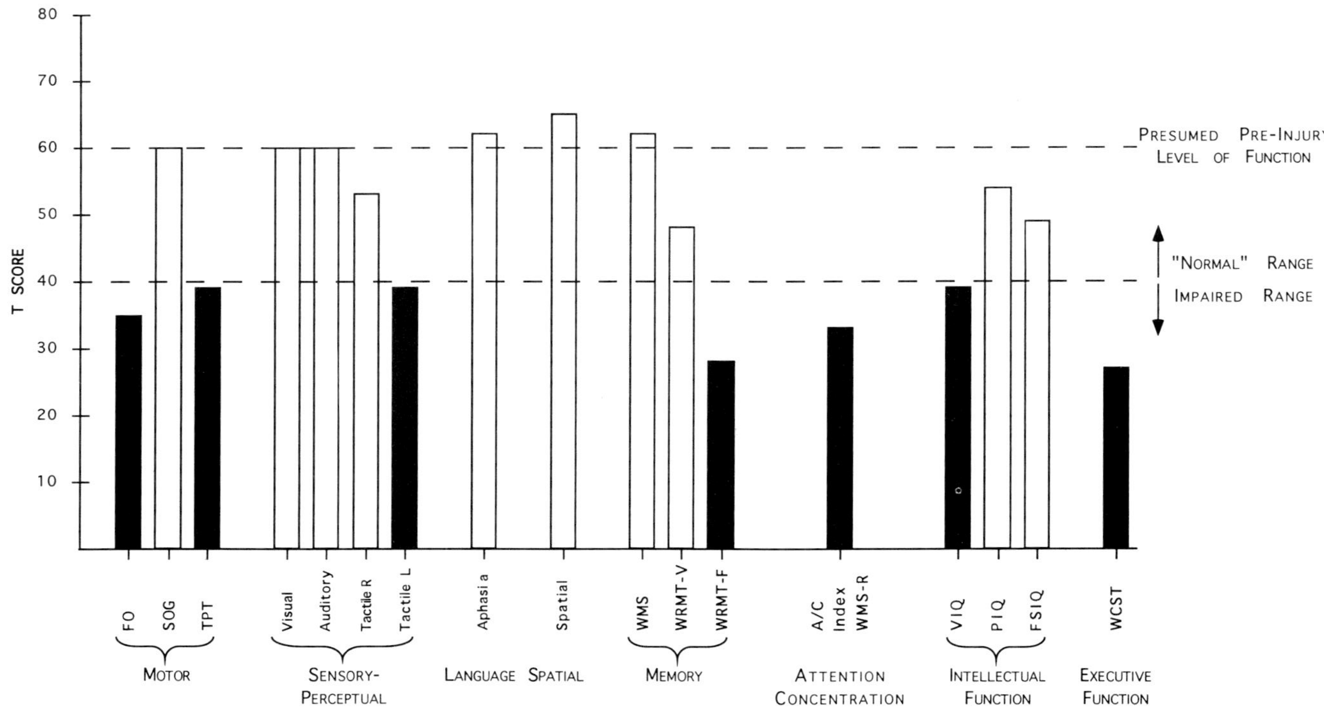

Figure 13. Neuropsychological performance of the patient depicted in Figure 12. The *x*-axis lists various neuropsychological test findings with the scores being converted into a T-score for direct comparison (mean = 50), standard deviation = 10) along the *y*-axis. Abbreviations: FO = finger oscillation; SOG = strength of grip; TPT = tactual performance test; WMS-R = Wechsler Memory Scale-Revised; WRMT-V = Warrington Recognition Memory Test-Verbal; WRMT-F = Warrington Recognition Memory Test-Faces; A/C Index = attention/concentration index; VIQ = verbal intellectual quotient; PIQ = performance intellectual quotient; FSIQ = full scale intellectual quotient based on the results of the Wechsler Adult Intelligence Scale-Revised (1981); WCST = Wisconsin Card Sorting Test. This neuropsychological profile indicates that the patient has no deficit in basic motor strength as reflected by the intact SOG scores. However, fine motor movement (FO) as well as integrative motor control (TPT) is below normal. Sensory-perceptual examination reveals intact visual and auditory processing, bilaterally, but the patient has diminished tactile perceptual processing on the left compared to the right. No deficits are noted in language, spatial, and general memory index score on the Wechsler Memory Scale-Revised. However, he clearly has a deficit in facial recognition memory on the Warrington Recognition Memory Test in comparison to a normal WMS-R general memory index and a normal performance on the Warrington Recognition Memory Test for words (verbal). The patient's attention/concentration index was distinctly impaired on the WMS-R. His intellectual functioning is probably below what would be predicted given his premorbid ability level. He has a distinct deficit in the ability to perform the Wisconsin Card Sorting task, a task which places demands on flexible thinking and cognitive shifting, typically deficit areas for patients with significant frontal lobe damage. This patient was a very successful contractor and builder who sustained a serious traumatic brain injury in a 70-foot fall while rock climbing. He had graduated from high school and had completed 4 years of college but had not graduated. Both parents were college-educated and his father was a university professor. Accordingly, this patient's preinjury intellectual/cognitive status was felt to be in the above-average range, as depicted by the horizontal line. Taken together, these neuropsychological findings suggest a defect in frontal lobe function, with lateralization to the right hemisphere. Accordingly, these neuropsychological findings are consistent with the neuroimaging findings presented in the MR scans in Figure 12.

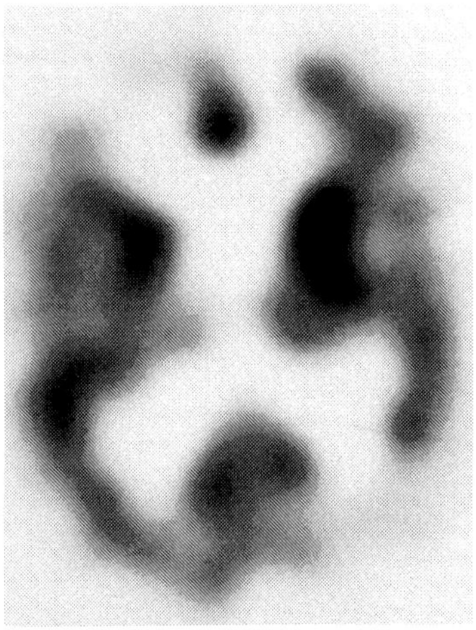

Figure 14. SPECT findings in the patient described in Figures 1, 2, and 3. Note the extensive hypoperfusion in anterior and posterior brain regions. In the normal SPECT scan there should be an even gray scale value that surrounds the outer rim of the brain which is representative of cortical gray matter and metabolic activity. This SPECT scan clearly deviates from the normal pattern, with masked perfusion abnormalities evident.

tively to neuropsychological outcome. In general, the greater the tissue loss the more significant the neuropsychological deficit.

SPECT Findings in TBI

As discussed in Chapters 1 and 3 of this volume, as well as above, there may be changes in brain function that are below the threshold for detection by anatomic imaging. In such cases, blood flow and metabolic imaging procedures have some significant advantage in depicting areas of dysfunction. This is readily seen in Figure 14, which is from the patient presented in Figures 1–3. This shows significant perfusion abnormalities in the brain following trauma that even exceed the boundaries of the well-defined anatomic pathology in the MRI studies of this patient. Accordingly, the combination of structural imaging via CT and/or MR with functional imaging techniques such as single photon emission computed tomography (SPECT) will provide the most comprehensive assessment of the pathophysiologic effects of an injury (Jacobs, Put, Ingels, & Bossuyt, 1994; Ichise, Chung, Wang, Wortzman, Gray, & Franks, 1994).

Conclusions

Contemporary neuroimaging provides excellent visualization of structural damage which occurs as a result of traumatic brain injury. As indicated by several studies, there is a differential effect of brain trauma on white versus gray matter, with white matter being more affected. Accordingly, *ex vacuo* dilation of the ventricular system in trauma is primarily related to white matter loss. The future holds significant promise that a more detailed structure-functional analysis of the brain in correspondence with neuropsychological outcome will be mapped based on regional differences of brain injury. By knowing regional and nonspecific anatomic abnormalities, better predictions about outcome can be made.

Anderson, C. V., & Bigler, E. D. (1994). The role of caudate nucleus and corpus callosum atrophy in trauma-induced anterior horn dilation. *Brain Injury, 8*, 565–569.

Anderson, C. V., & Bigler, E. D. (1995). Ventricular dilation, cortical atrophy, and neuropsychological outcome following traumatic brain injury. *Journal of Neuropsychiatry and Clinical Neuroscience. 7*(1), 42–48.

Bigler, E. D. (1990). *Traumatic brain injury.* Austin, TX: Pro-Ed.

Bigler, E. D. (1994). Neuroimaging and neuropsychological assessment. In C. R. Reynolds (Ed.), *Cognitive assessment* (pp. 1–34). New York: Plenum Press.

Bigler, E. D., Blatter, D. D., Anderson, C. V., Johnson, S. C., Gale, S. D., Hopkins, R. O., & Burnett, B. (in press). Hippocampal volume in normal aging and traumatic brain injury. *American Journal of Neuroradiology.*

Bigler, E. D., Burr, R., Gale, S., Norman, M., Kurth, S., Blatter, D. D., & Abildskov, T. J. (1994). Day-of-injury CT scan as an index to pre-injury brain morphology. *Brain Injury, 8*, 231–238.

Bigler, E. D., Johnson, S. C., Anderson, C. V., Blatter, D. D., Gale, S. D., Russo, A. A., Ryser, D. K., Macnamara, S. E., Bailey, B. J., Hopkins, R. O., & Abildskov, T. J. (1996). Traumatic brain injury and memory: The role of hippocampal atrophy. *Neuropsychology, 10*, 333–342.

Bigler, E. D., Kurth, S., Blatter, D. D., & Abildskov, T. J. (1992). Degenerative changes in traumatic brain injury: Post-injury magnetic resonance identified ventricular expansion compared to pre-injury levels. *Brain Research Bulletin, 28*, 651–653.

Blatter, D. D., Bigler, E. D., Gale, S. D., Johnson, S. C., Anderson, C. V., Burnett, B., Ryser, D. K., Macnamara, S. E., & Bailey, B. J. (in press). MRI based brain and CSF quantification following traumatic brain injury: Correlation with neuropsychological outcome. *American Journal of Neuroradiology.*

Blatter, D. D., Bigler, E. D., Johnson, S. C., Anderson, C., Burnett, B. M., Parker, N., Kurth, S., & Horn, S. (1995). Quantitative volumetric analysis of brain MRI: Normative database spanning five decades (16–65). *American Journal of Neuroradiology, 16*, 241–251.

Christman, C. W., Grady, M. S., Walker, S. A., Holloway, K. L., & Povlishock, J. T. (1994). Ultrastructure studies of diffuse axonal injury in humans. *Journal of Neurotrauma, 11*, 173–186.

Gale, S. D., Burr, R. B., Bigler, E. D., & Blatter, D. D. (1993). Fornix degeneration and memory in traumatic brain injury. *Brain Research Bulletin, 32*, 345–349.

Gale, S. D., Johnson, S. C., Bigler, E. D., & Blatter, D. (1995a). Nonspecific white matter degeneration following traumatic brain injury. *Journal of the International Neuropsychological Society, 1*, 17–28.

Gale, S. D., Johnson, S. C., Bigler, E. D., & Blatter, D. D. (1995b). Traumatic brain injury and temporal horn enlargement: Correlates with tests of intelligence and memory. *Neuropsychiatry, Neuropsychology & Behavioral Neurology, 7*, 160–165.

Gean, A. D. (1994). *Imaging of head trauma.* New York: Raven Press.

Hall, S., Bigler, E. D., & Rutledge, J. N. (1989). Depression preceding choreiform movement in Huntington's disease: A case study. *Archives of Clinical Neuropsychology, 4*, 79–92.

Ischise, M., Chung, D.-A., Wang, P., Wortzman, A., Gray, B. A., & Franks, W. (1994). Techetium-99 HMPAO SPECT, CT and MRI in the evaluation of patients with chronic traumatic brain injury: A correlation with neuropsychological performance. *Journal of Nuclear Medicine, 35*, 217–226.

Jacobs, A., Put, E., Ingels, M., & Bossuyt, A. (1994). Prospective evaluation of technetium-99m-HMPAO SPECT in mild and moderate traumatic brain injury. *Journal of Nuclear Medicine, 35*, 942–948.

Johnson, S., Bigler, E. D., Burr, R., & Blatter, D. (1994). White matter atrophy, ventricular dilation, and intellectual functioning following traumatic brain injury. *Neuropsychology, 8*, 307–315.

Kurth, S., Bigler, E. D., & Blatter, D. (1994). Neuropsychological outcome in quantitative image analysis of acute hemorrhage in traumatic brain injury: Preliminary findings. *Brain Injury, 8*, 489–500.

Lenti, C., & Bianchini, E. (1993). Neuropsychological and neuroradiologic study of a case of early onset Huntington's chorea. *Developmental Medicine and Child Neurology, 35*, 1007–1014.

Lisk, D. R., Pasteur, W., Rhoades, H., Putnam, R. D., & Grotta, J. C. (1994). Early presentation of hemispheric intracerebral hemorrhage: Prediction of outcome and guideline for treatment allocation. *Neurology, 44*, 133–139.

Newberg, A. B., & Alavi, A. (1996). Neuroimaging in neurological disorders. In E. D. Bigler (Ed.), *Handbook of human brain function: Neuroimaging.* New York: Plenum Press.

Osborn, A. (1994). *Neuroradiology*. St. Louis, MO: Mosby Yearbook.

Povlishock, J. T. (1992). Pathobiology of traumatically induced axonal injury in animals and man. *Annals of Emergency Medicine, 22,* 980–986.

Prayer, L. M., & Gean, A. D. (1994). Head trauma. In J. Kucharczyk, M. Moseley, A. J. Barkovich (Eds.), *Magnetic resonance neuroimaging*. Boca Raton, FL: CRC Press.

Salazar, A. M. (1992). Traumatic brain injury: The continuing epidemic. In V. C. Hachinski (Ed.), *Challenges in neurology* (pp. 55–67). Philadelphia: F. A. Davis.

11

Magnetic Resonance Imaging and Memory Disorders

NARINDER KAPUR

INTRODUCTION

Magnetic resonance imaging (MRI) has had two major influences on research in the area of neurologically based memory disorders. First, it has provided us with more reliable anatomical definition of pathological and normal brain tissue. Second, it has allowed better quantification of cerebral lesions and of normal brain structures. In this chapter, I will review both trends and offer examples of each type of advance. The reader is referred to Andreasan (1989) for an introduction to MRI and to the articles edited by Isherwood (1992) and by Perlmutter (1993) for reviews of recent advances in MRI techniques and their clinical applications.

BETTER ANATOMICAL DEFINITION OF CEREBRAL LESIONS

Magnetic resonance imaging (MRI) has made possible the generation of *in vivo* brain images which have a quality and resolution that were previously only available on postmortem examination. There are at least three features of magnetic resonance images that render them particularly valuable for providing improved anatomical definition of cerebral lesions: (a) the high resolution of images, such that structures and lesions as small as 1–2 mm across can be readily

NARINDER KAPUR Wessex Neurological Centre, Southampton General Hospital, Southampton S016 64D, England

Neuroimaging II: Clinical Applications, edited by Erin D. Bigler. Plenum Press, New York, 1996.

visualized; (b) the ability to obtain clear views of structures from coronal, sagittal, and axial views, and at any depth in the brain; and (c) the well-defined demarcation of abnormal signals that reflect underlying brain pathology.

The following are examples of cases where these properties of MRI have been particularly useful in depicting the anatomical basis of memory disorder.

Lesions in Deep Brain Structures

The limbic-diencephalic system, together with other areas such as the basal forebrain, contain many structures that appear to be critical for memory (Mishkin & Appenzeller, 1987). These structures, and lesions to the structures, can be visualized much more clearly with MRI as compared to earlier techniques such as computed tomography (CT).

Fornix. The fornix is the major efferent pathway from the hippocampus, with its primary projection being to the mammillary bodies. It takes a convoluted path from the alveus, arching back and underneath the corpus callosum before forming a further loop where its two columns merge into the mammillary bodies.

Although fornix damage has been reported in patients with cerebral tumors (e.g., Valenstein *et al.*, 1987; Rudge & Warrington, 1991), one of the best examples of discrete fornix lesions is where it is damaged in surgical procedures for the treatment of deep tumors such as colloid cysts. This may result from sectioning of the fornix to gain access to a large colloid cyst or as a result of vascular damage that occurs indirectly during surgery. Hodges and Carpenter (1991) have provided details on two cases where MR imaging clearly showed that there was severance of the fornix, in one case disruption of both anterior limbs of the fornix, and in the other case damage to the left anterior limb of the fornix (Figure 1). Both patients showed significant anterograde memory impairment, with minimal deficits on retrograde memory tasks. It should be remembered that even in cases with discrete lesions such as colloid cysts, the interpretation of memory impairment may sometimes be complicated by the effects of raised intracranial pressure, the effects of hydrocephalus where it is present, and the effects of local displacement pressures of the cyst on neighboring limbic-diencephalic structures. Magnetic resonance imaging will usually show fornix morphology on fine midline sagittal cuts and on coronal views, and it is important to have comparable views from a matched normal control subject so that interruptions in the fornix pathway can be reliably delineated.

Mammillary Bodies. Because of their size, both mammillary bodies are difficult to visualize on routine MR cuts. They are best viewed on fine (1 mm or less) coronal and sagittal sections. Isolated lesions in the mammillary bodies are rare. Hypothalamic tumors such as congenital hamartomas (Soto *et al.*, 1991) or slow growing tumors of recent onset may result in damage to the mammillary bodies, and such tumors have been reported to be associated with memory impairment (Berkovic *et al.*, 1988; Parkin & Hunkin, 1993). However, to date there is no report that provides detailed imaging together with comprehensive neuropsychological assessment in the same set of patients with hypothalamic tumors. We have recently seen a case where there was a lymphoma deep in the brain and where the mammillary bodies were largely obliterated (Figure 2). This patient also had some involvement of the medial thalamus and lower tectum

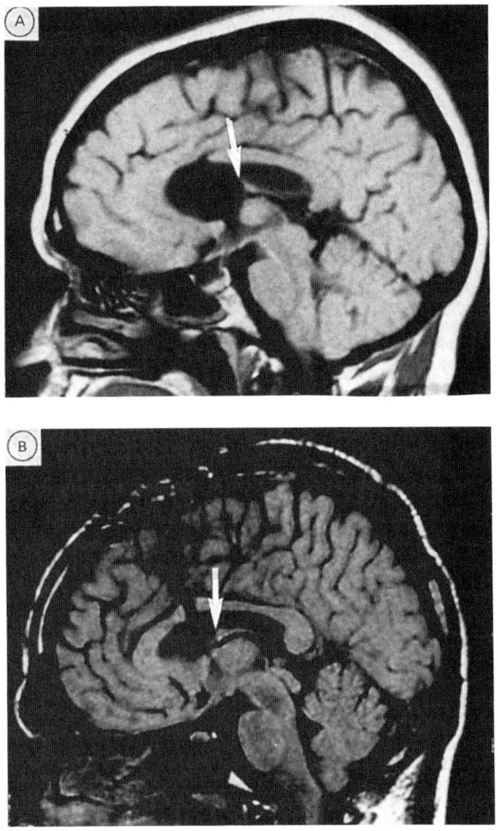

Figure 1. Sagittal midline MR images of two cases showing severance of the fornix. Figure 1A (Case 1) shows interruption of the fornices (arrow) at the level of the formen of Munro directly beneath the surgical corpus callosum defect (Hodges & Carpenter, 1991). In Figure 1B (Case 2) the right fornical column was apparent on contiguous parasagittal cuts, but in the other patient neither column could be demonstrated. (Reproduced with permission of the authors and the *Journal of Neurology, Neurosurgery and Psychiatry*.)

region of the midbrain. He suffered significant anterograde memory impairment, especially on delayed recall tasks, with minimal retrograde memory deficits. Two cases of penetrating head injury that resulted in mammillary body lesions and memory disorder have been reported (Squire, Amaral, Zola-Morgan, Kritchevsky, & Press, 1989; Dusoir, Kapur, Byrnes, McKinstry, & Hoare, 1988). However, in one patient MRI studies also showed damage to the left thalamus, right temporal lobe, and possibly also to the left fornix. In the second patient, a recent PET and quantitative MRI study (Kapur *et al.*, 1994) showed some left hippocampal abnormality.

In alcoholic Korsakoff patients, it has been possible with the aid of MR imaging to identify mammillary body lesions in the form of atrophy of the mammillary bodies which may be obvious on visual inspection or quantifiable (Charness & DeLaPaz, 1987; Bigler, Nelson, & Schmidt, 1989; Squire, Amaral, & Press, 1990). In the case of quantification of such small structures in patients such as alcoholic Korsakoffs, it may be impossible to visualize the mammillary bodies, and allowance for general cerebral atrophy needs to be made by considering correction measures such as intracranial volume.

HIPPOCAMPUS. In human brain pathology, it is rare to find focal lesions that are discrete enough to be restricted to the hippocampus itself. In patients with temporal lobe epilepsy, abnormal signals may not always be seen, and specific

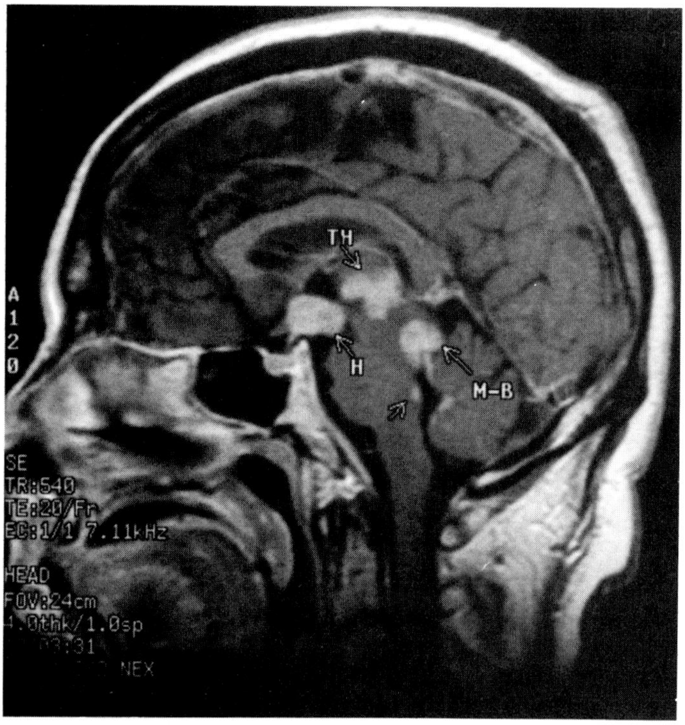

Figure 2. Sagittal midline MR image of patient with multilobular primary cerebral lymphoma showing hypothalamic (H) lesion, medial thalamic (TH) lesion, and midbrain (M-B) lesion.

abnormalities may only appear after quantitative image analysis (Jackson, Connelly, Duncan, Grunewald, & Gadian, 1993; Senzaki *et al.*, 1993). This also applies in cases of cerebral hypoxia (Kritchevsky & Squire, 1993). In contrast to postmortem case reports (e.g., Zola, Morgan, Squire, & Amaral, 1986; Victor & Agamanolis, 1992), *in vivo* MRI studies of focal lesions of the hippocampus have been few in number, and those which have been reported have found selective anterograde memory deficits, with retrograde memory and general cognitive functions appearing to be relatively spared. This has been shown in patients who have suffered paraneoplastic limbic encephalitis (Lacomis, Khoshbin, & Schick, 1990; Kodama, Numaguchi, Gellad, Dwyer, & Kristt, 1991) and in patients with discrete vascular lesions of the hippocampus (Tanabe *et al.*, 1991; Hirasawa, Nakano, Shiohama, Komiya, & Mizuno, 1991). In the latter papers, the authors argued that only region CA1 of the hippocampus was damaged, though such fine anatomical demarcation of lesion boundaries needs to be verified with postmortem correlative evidence before it can be accepted with confidence.

Transient amnesias have traditionally eluded brain imaging correlates, but a few cases have now been described where the occurrence of transient amnesia has been associated with discrete abnormalities on magnetic resonance imaging. Thus, Lee *et al.* (1992) found focal hippocampal abnormality on T_2-weighted images in an epileptic patient with prolonged postictal amnesia. Similarly, Chalmers *et al.* (1991) found discrete bilateral hippocampal abnormality associated with severe amnesia in a diabetic patient who had suffered an attack of hypoglycemia.

Thalamus and Related Pathways. Lesions of the thalamus, especially vascular thrombotic lesions, have usually been easily discriminable on CT scanning (Von Cramon, Hebel, & Schuri, 1985). Magnetic resonance imaging has, however, helped in the more precise anatomical identification of such lesions. Examples of MR-imaged thalamic lesions that have yielded useful anatomical data on the bases of amnesic deficits include those reported by Graff-Radford, Tranel, Van Hoesen, and Brandt (1990) and Hodges and McCarthy (1993) (Figure 3). It would appear that anterior and medial thalamic lesions may be more likely to result in severe memory loss than posterior and lateral lesions and that thalamic lesions which interfere with critical limbic-diencephalic pathways such as the mammillothalamic tract may also be more likely to result in amnesia than those which spare such fibres. In patients with thalamic lesions, retrograde amnesia is often reported to be present in addition to anterograde memory deficits (e.g., Markowitsch, Von Cramon, & Schuri, 1993; see Hodges & McCarthy, 1993; for a review of such cases). The mechanisms of such retrograde memory loss are not clear. Several possibilities come to mind. Firstly, the well-established cortical hypometabolism that is associated with thalamic lesions (e.g., Szelies *et al.*, 1991)

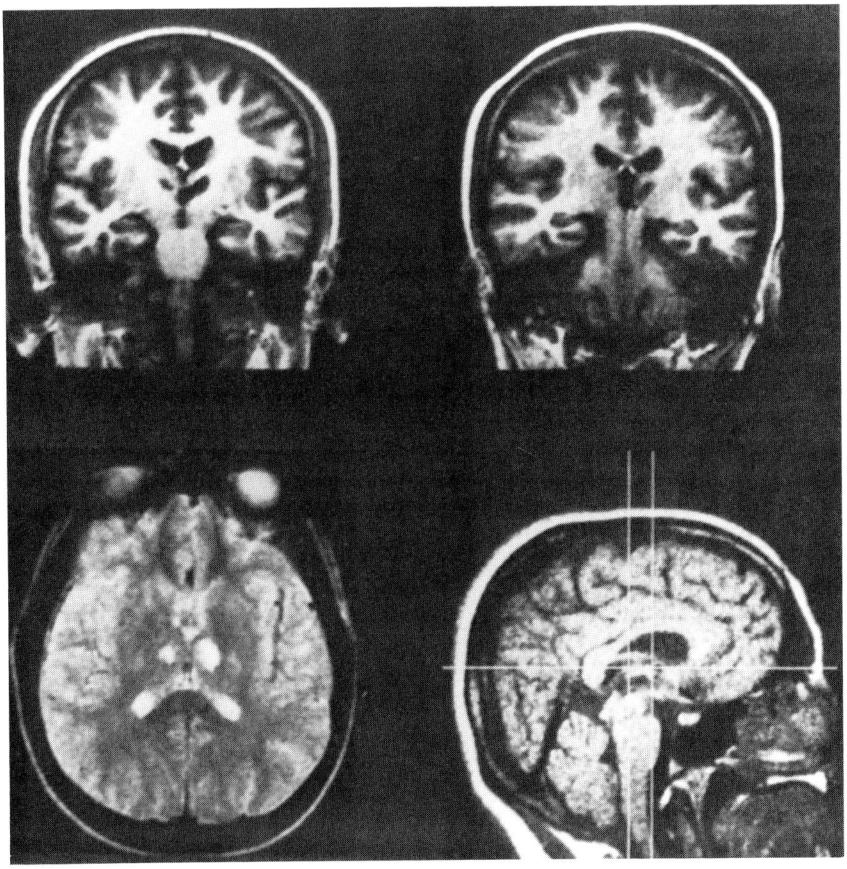

Figure 3. Coronal (top two images) and axial (lower left image) view of a patient with amnesia due to bilateral thalamic infarcts (Graff-Radford, Tranel, Van Hoesen, & Brandt, 1990). The bottom right sagittal image shows the location of the two coronal cuts and the single axial cut. (Reproduced with permission of the authors and *Brain*.)

may result in dysfunction in temporal and frontal cortical areas that are critical to retrograde memory functioning. Secondly, as pointed out by Hodges and McCarthy (1993), a more specific retrieval/executive deficit may be present that arises from damage to thalamic-frontal tracts and leads to a disconnection of anterior from posterior cerebral processing systems. Thirdly, it is possible that since thalamic blood supply usually comes from arteries that are branches of the posterior cerebral artery, the temporal lobe branches of the posterior cerebral artery are also compromised in many patients, thus resulting in a combination of retrograde and anterograde amnesia. Magnetic resonance imaging of thalamic lesions has also helped to discover equally important cases where thalamic lesions are present but without any major memory loss (Kritchevsky, Graff-Radford, & Damasio, 1987). These authors suggested that it was the partial nature of the lesions and/or their dorsomedial locus that may have resulted in the absence of significant memory impairment in their patient.

Lesions in the genu of the internal capsule have been associated with memory loss, and this has been documented in recent MRI studies (e.g., Markowitsch, Von Cramon, Hofmann, Sick, & Kinsler, 1990; Tatemichi *et al.*, 1992). It is possible, as Tatemichi *et al.* have suggested, that the critical neural mechanism underlying such memory loss may be thalamocortical disconnection of white matter tracts, such as the thalamic peduncles, which connect the thalamus with parts of the limbic system.

Basal Forebrain. In recent years, this area of the brain has gained more prominence in the anatomical landscaping of human memory. The basal forebrain, in the posterior inferomedial part of the frontal lobes, is usually considered to comprise the septal nuclei, the nuclei of the diagonal band of Broca, and the nucleus basalis of Meynert. The most common pathology that has been associated with lesions in this area is that resulting from ruptured aneurysms of the anterior communicating artery (Irle, Wowra, Kunert, Hampl, & Kunze, 1992). In patients with marked memory disturbance following such a cerebrovascular lesion, the possible role of additional factors such as infarction in other frontal lobe/limbic system areas or hydrocephalus must be kept in mind. In addition, MRI scanning of such patients may be contraindicated by the presence of ferrous aneurysmal clips. Even where nonferrous clips have been used in the surgical procedure, image artifacts related to the presence of the clip may result in poor anatomical definition of relevant lesions. For these reasons, cases with discrete basal forebrain lesions of other etiologies are of particular interest. One such case has recently been reported (Morris, Bowers, Chatterjee, & Heilman, 1992), with MRI documentation of the basal forebrain lesion. This patient showed both anterograde and retrograde amnesia following resection of a low-grade glioma in the basal forebrain. The patient had a long preamnesia history of poorly controlled complex partial epilepsy, but he was able to hold down a job as a teacher. It remains possible, though difficult to prove, that the basal forebrain lesion may in some way have interacted with latent pathology caused by or related to his epilepsy, and this may have contributed in part to his subsequent amnesia.

Lesions in Neocortical and Adjacent White Matter Structures

Temporal Lobes. Most major lesions in the temporal lobes, such as neoplasms or vascular lesions, have usually been readily discernible on CT scanning.

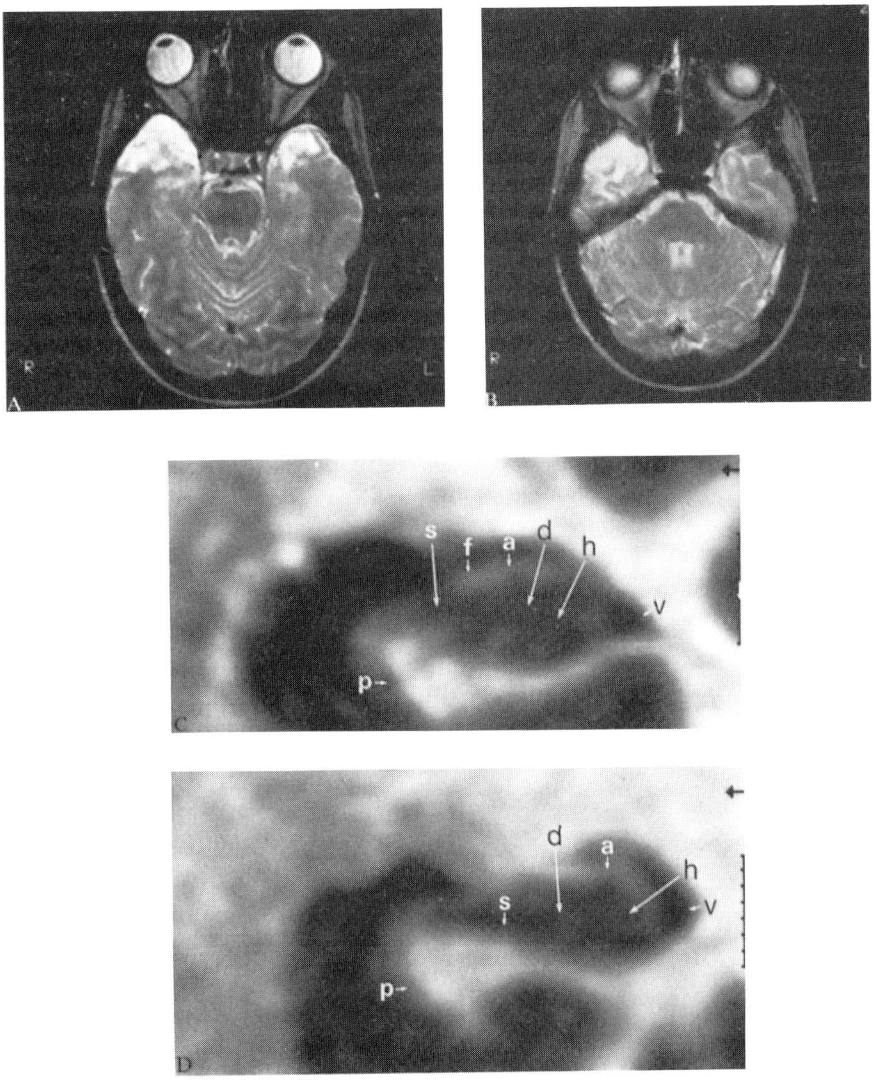

Figure 4. Anterior temporal lobe damage in a patient who suffered a severe closed head injury and was left with a dense, focal retrograde amnesia. (Reproduced with permission of the editor and publishers of *Brain*.)

There are, however, certain lesion pathologies with a predilection for the temporal lobes that are more clearly visible on MR images. Examples of these include patients with blunt head injury (Kapur, Ellison, Smith, McLellan, & Burrows, 1992) (Figure 4), patients who have suffered herpes simplex encephalitis (Figure 5), and patients who have incurred temporal lobe radionecrosis following radiation therapy (Figure 6). Agenesis of the temporal lobe, with a clearly demarcated area of CSF-filled space, also shows up very clearly on MR images (Figure 7), though this condition is usually quite benign and seldom results in a marked memory disorder.

Lesions in Other Neocortical/White Matter Areas. Lesions resulting from demyelination are much more readily visible on MRI than in conventional

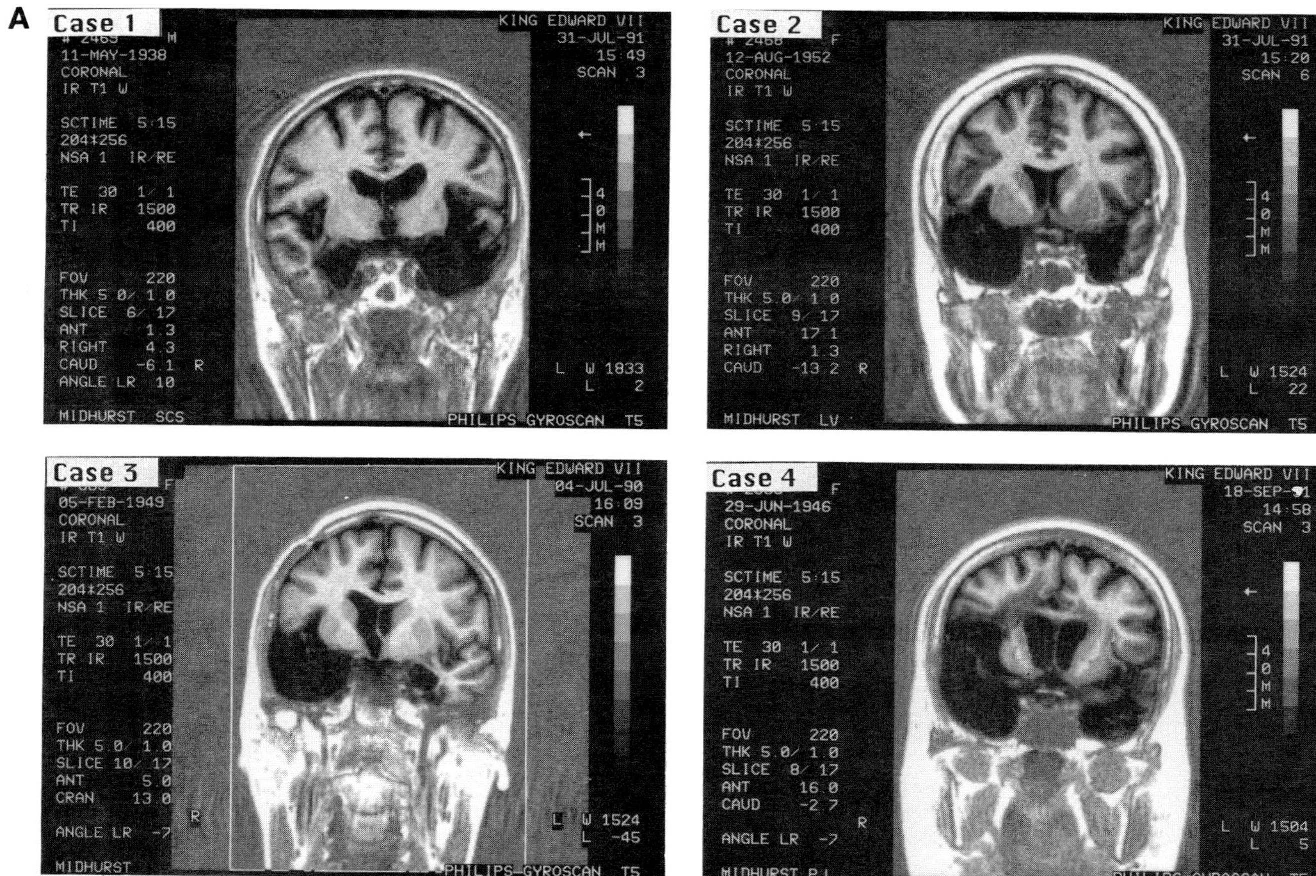

Figure 5. Montage of maximal extent of T1 coronal (A) and T2 axial (B) lesions found in each of ten herpes simplex encephalitis (HSE) patients with residual memory disorder of varying degrees of severity. Case numbers correspond to the numbers allocated to patients in Kapur *et al.*, (1994).

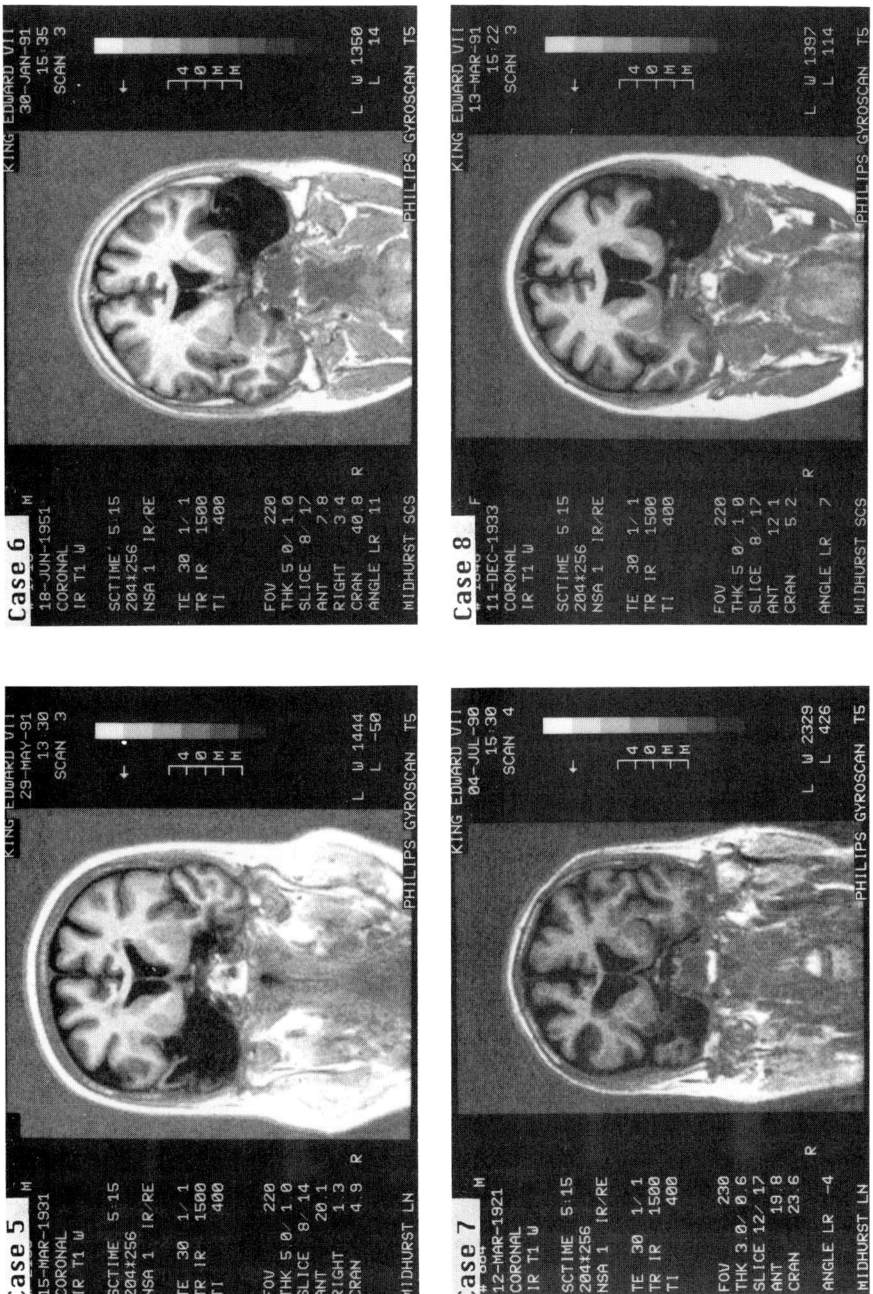

Figure 5. *(Continued)*

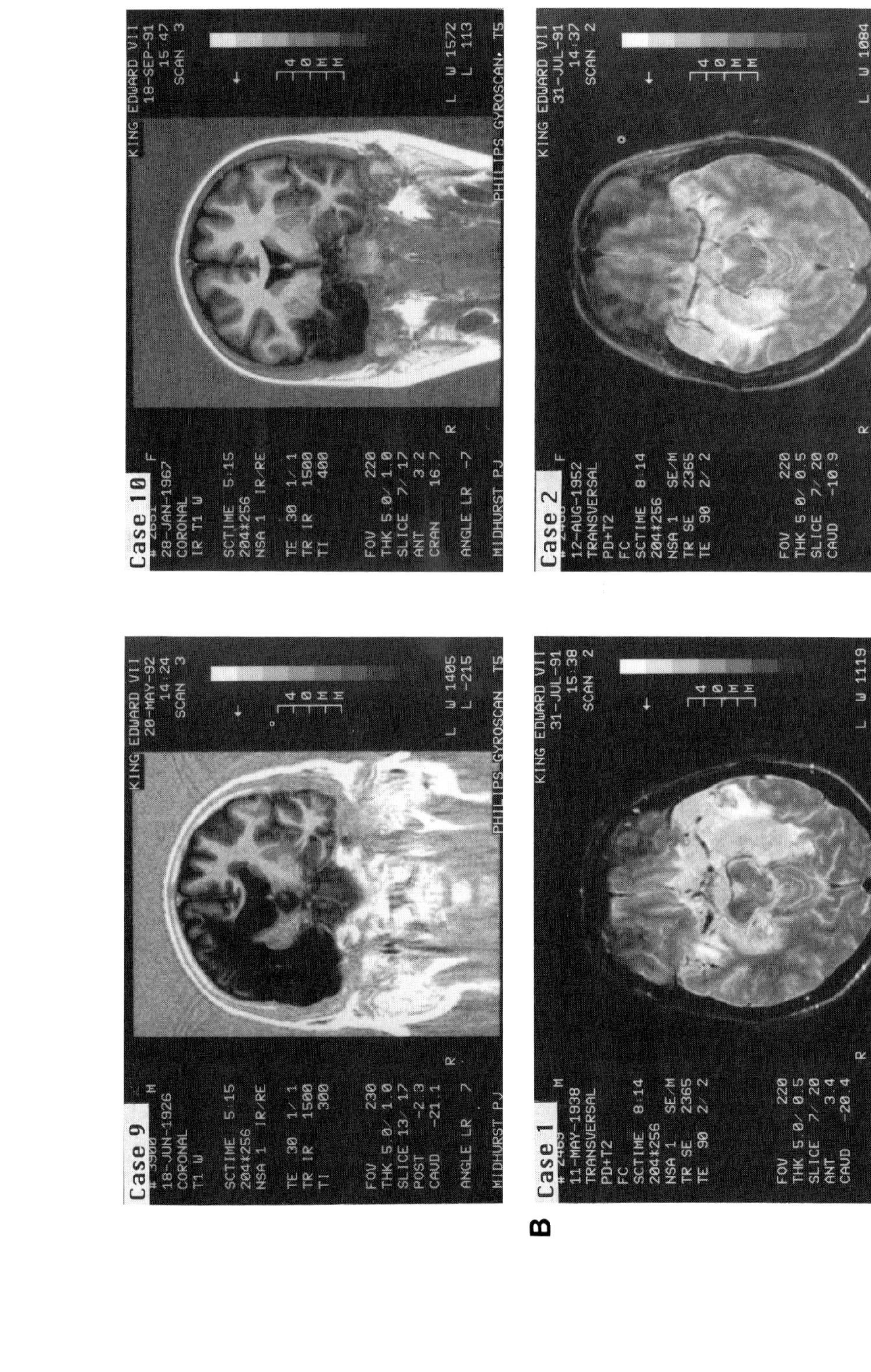

Figure 5. *(Continued)*

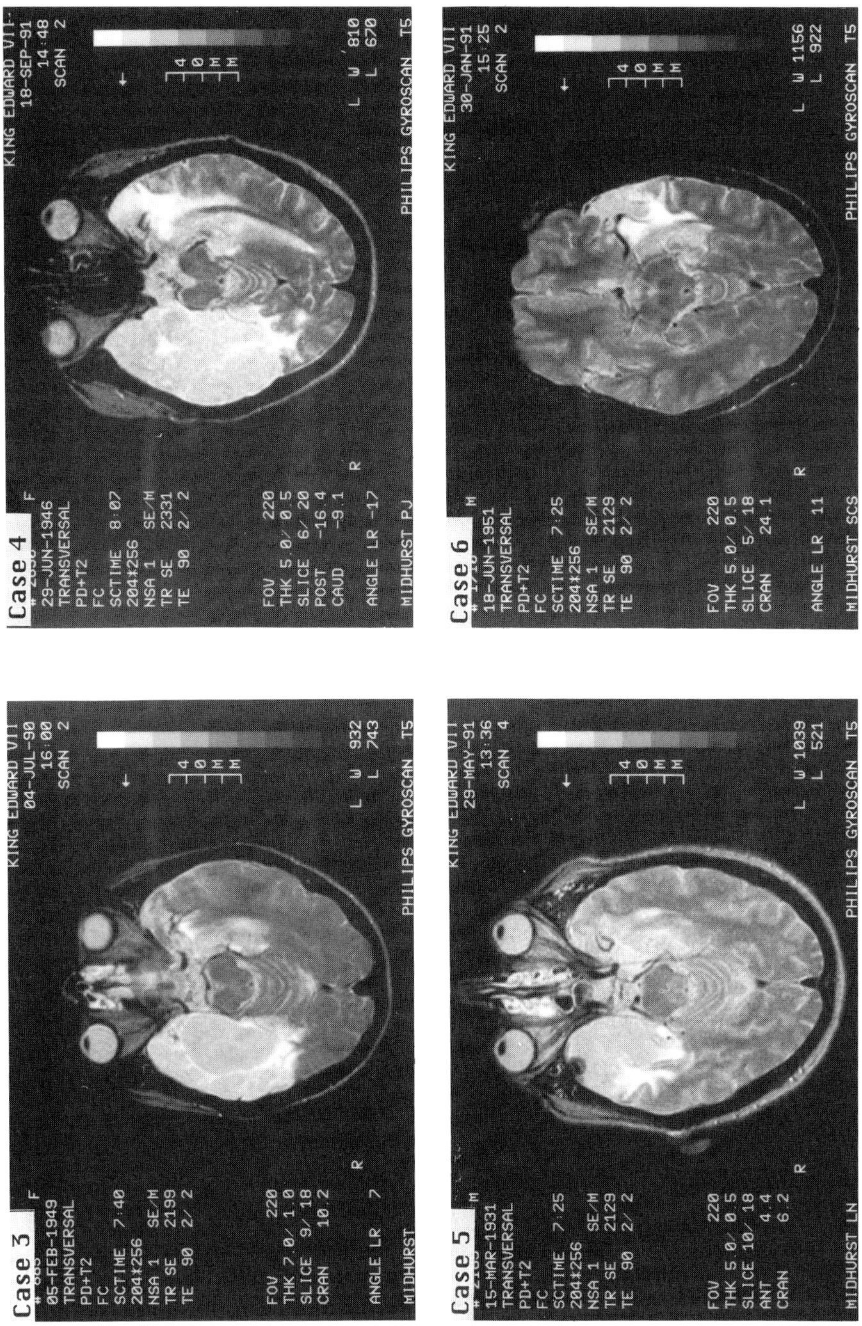

Figure 5. (*Continued*)

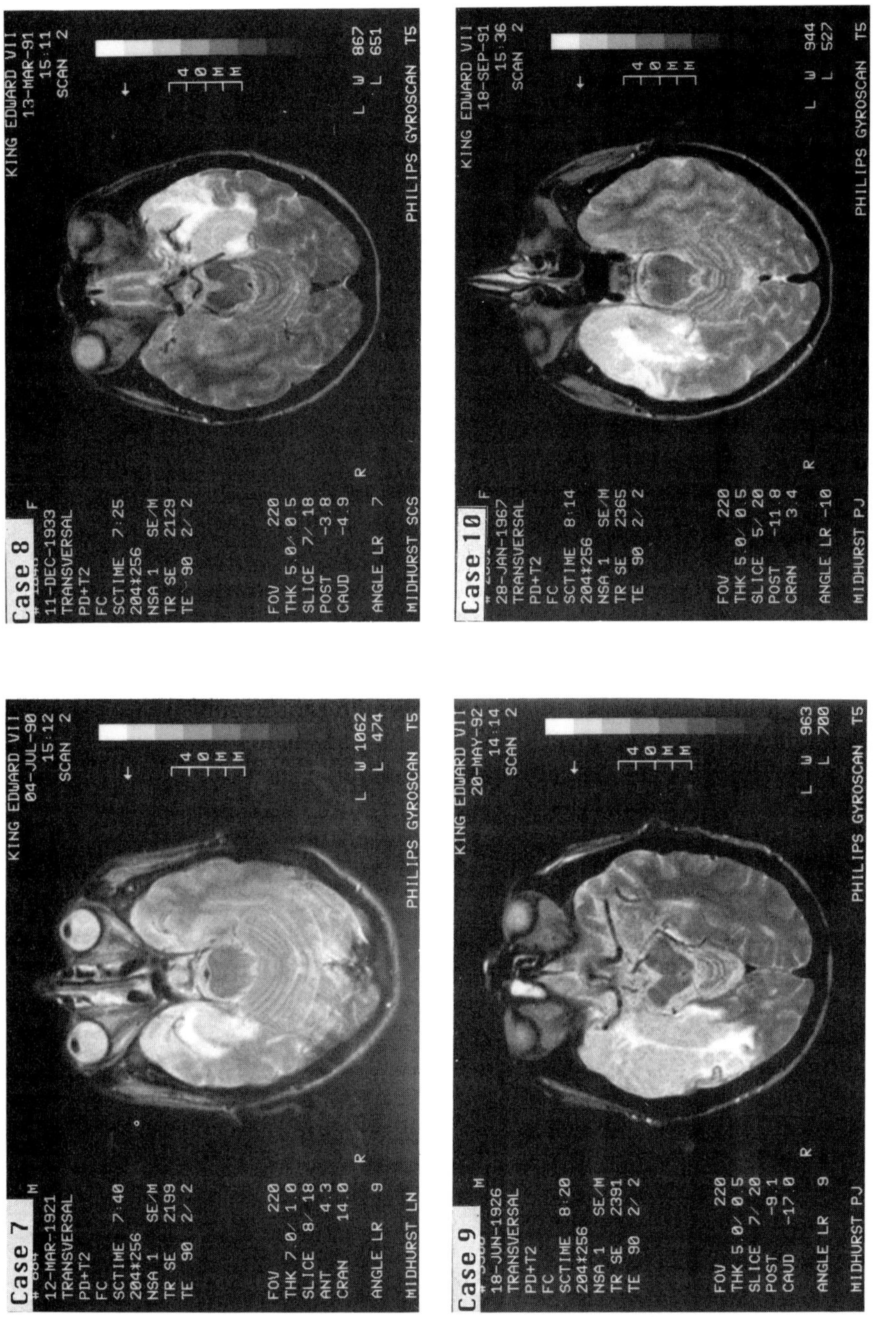

Figure 5. (*Continued*)

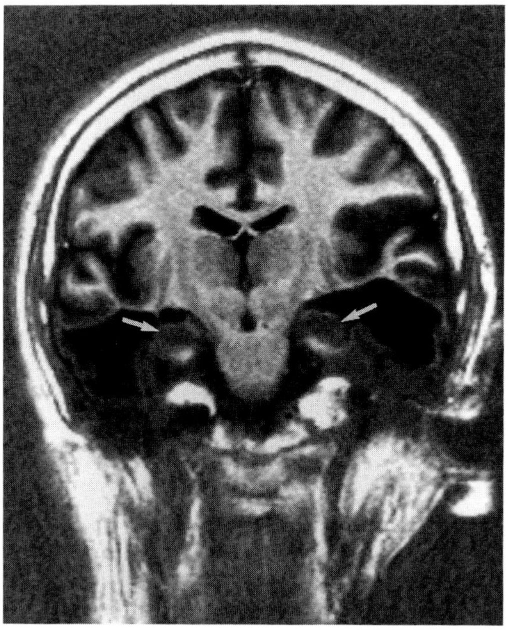

Figure 6. Temporal lobe damage following radionecrosis, with relative sparing of the hippocampus (arrow). (Reproduced with permission of the editor and publishers of *Neuropsychologia*.)

scanning procedures such as CT. Depending on the distribution of such lesions, patients will show a variable degree of memory loss.

Accurate Quantification of Cerebral Lesions and Tissue Integrity

Perhaps more significant than the improved image quality introduced by magnetic resonance procedures is the possibility of producing reliable quantitative indices relating to lesion characteristics and integrity of tissue that may in some cases appear normal but prove to have some subtle abnormalities.

In the following cases, this feature of MRI has been particularly useful in helping us to understand the neural mechanisms underlying memory disorder. Three specific applications will be discussed: quantification of lesion number/size, measures of atrophy in otherwise normal brain structures, and absolute measures of T^1 and T^2 relaxation indices.

Quantification of Lesion Number/Size

It is seldom that MR-derived number/size of lesions has been considered in studies of memory disorder patients. One of the exceptions is in the area of multiple sclerosis (MS), where the presence of multiple discrete lesions has provided researchers with a challenge with respect to the quantification of MR abnormality. Some caution needs to be exercised when considering the range of studies that have examined the relationship between neuropsychological functioning and measures of brain integrity in MS patients. Comi *et al.* (1993) have summarized relevant factors that may explain discrepancies between studies. These include the clinical characteristics of the sample of MS patients (e.g., type

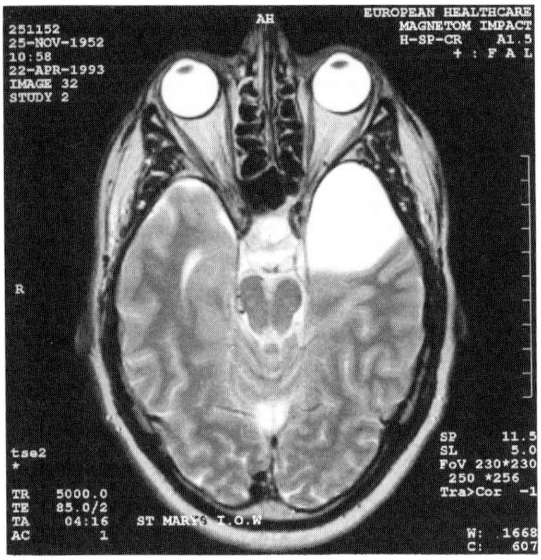

Figure 7. Agenesis of the temporal lobe resulting in a right temporal lobe arachnoidcyst in a 40-year-old patient with mild memory impairment.

of progression of the disease, stage of illness, etc), sample sizes, technical features of the MR scanner and scanning sequences used, type of analysis of MR data/images, type and number of cognitive tests carried out, and type of statistical analysis, that is, whether relevant factors have been covaried in any correlation analyses.

Franklin, Heaton, Nelson, Filley, and Seibert (1988) found that such "whole brain" lesion scores correlated with several measures of retention, including learning and delayed retention of story and visual design materials. However, significant correlations were also found with nonmemory cognitive tasks such as tests of perceptual-motor functioning. Rao, Leo, Haughton, Aubin-Faubert, and Bernardin (1989) reported that total lesion area in their sample of MS patients correlated with performance on a number of memory tasks, including word-list learning, story recall, and spatial memory. Ron, Callanan, and Warrington (1991) found that a total lesion score based on the size and number of lesions correlated with recognition memory for faces and for words, with a rather higher correlation in the case of memory for faces. As in the Franklin *et al.* (1988) study, however, the lesion index also correlated with performance on other cognitive tests, including attention and naming ability. On the basis of quantitative analyses of the MR images of their MS patients, Maurelli *et al.* (1992) failed to find any significant correlation between lesion scores (frontal, periventricular, temporal, or total) and memory test performance. However, when they divided their group into those with extensive periventricular demyelination on the one hand and those with single/multiple discrete lesions in periventricular or extra-periventricular regions, they did find lower performance by the group with extensive periventricular demyelination on verbal paired-associate learning, immediate story recall, and digit span. Comi *et al.* (1993) also gathered multiple MRI indices and found a significant relationship between memory performance, as measured by the Wechsler Memory Scale, and degree of ventricular dilatation, corpus callosum atrophy, cortical atrophy, and total lesion score. The correlation with corpus callosum atrophy was the highest of those that were found.

Izquierdo, Campoy, Jordi, Gonzalez, and Martinez-Parra (1991) reported a significant relationship between periventricular MS lesions and memory performance, and within the group with periventricular lesions those with large lesions fared the worst. Similarly, Anzola *et al.* (1990) found that MS patients with extensive periventricular lesions performed worse on a story recall task than MS patients with multiple, isolated lesions.

A further application of MRI-based analysis of lesion distribution in MS has been in the longitudinal study of such patients. The nature of the course of MS allows serial monitoring of cognitive functions such as memory that may be correlated with changes in lesion profile. In one such study, Mattioli *et al.* (1993) found an absence of any significant relationship between MRI-based lesion changes and changes in memory functioning. Such evidence is nevertheless of interest in highlighting the possibly benign nature of some of the MS lesions visualized using MRI, even though they may appear prominent on routine inspection. Such data also highlight the possibility of pathology in the brains of MS patients, which may not yet be detectable with current MR procedures but may be modulating cognitive performance.

Measures of Structural Change in Brain Morphology

Several neurological conditions, including alcoholic Korsakoff's syndrome, closed head injury, Alzheimer-type dementia, temporal lobe epilepsy, and Cushing's syndrome have given rise to studies where quantitative measures of changes, such as degree of atrophy of anatomical structures, have been correlated with memory performance.

Shear *et al.* (1992), in a study of alcoholic Korsakoff patients, did not find any significant correlation between a measure of cortical atrophy, cortical fluid volume, and performance on a word-list learning task. Likewise, in a similar population of patients, Charness and DeLaPaz (1987) did not find any significant correlation between recall of three objects after 5 minutes and mammillary body volume. In a study of four alcoholic Korsakoff subjects, Squire *et al.* (1990) found some general trends in the relationship between mammillary body volume and degree of amnesia. They noted that one patient, who was the most severely amnesic, was the only one whose mammillary bodies were undetectable in any plane, and that the other three alcoholic Korsakoff patients were difficult to rank in terms of memory impairment. The most amnesic patient, however, also had particularly low thalamic density values in an earlier CT study.

Wilson, Wiedmann, Hadley, and Brooks (1989) found that in patients with closed head injury, performance on visual memory tasks, but not verbal memory tasks, correlated with an MRI-based rating of ventricular enlargement; however, there was an absence of specific correlations between memory measures and lesions in particular hemispheric sites.

In the case of Alzheimer-type dementia, Scheltens *et al.* (1992) examined the relationship between several indices of hippocampal atrophy and several memory test scores. Only a few correlations were significant, namely, between height of the left hippocampus and delayed recall of objects and between width of the left temporal horn and the same delayed recall score. A similar correlation was found between memory performance and subjective ratings of global medial temporal lobe atrophy. Correlations using immediate story recall or digit span did not reach significance with these or other MRI measures. It is of note that a study of healthy elderly subjects (Golomb *et al.*, 1993) reported similar observations—global ratings of hippocampal atrophy showed a stronger relationship with delayed than immediate measures of story recall or verbal paired-associate learning.

Turning to patients with temporal lobe epilepsy, Loring *et al.* (1993) found a significant though modest correlation (r = 0.05) between left hippocampal volume and preoperative verbal memory performance (delayed story recall). There was no corresponding relationship between right hippocampal volume and memory for visual designs. They also noted that differences in picture recognition memory between left and right hemispheres, as indicated during Wada injections that "anesthetized" one or other hemisphere, showed significant correlations with hemispheric differences in hippocampal volume measures. In an earlier study, Lencz *et al.* (1992) found that percent retention of story material from immediate to delayed recall correlated with left hippocampal volume in patients with left temporal lobe epilepsy. There was no corresponding correlations in the case of visual design recall and right hippocampal volumes. Left temporal lobe volumes as a whole (which appeared to include the hippocampus) correlated with word-list recall, with the highest correlation being between left/right temporal lobe ratio and this measure of retention. In contrast to these studies, Trenerry *et al.* (1993) did not find any significant correlation between memory performance across a wide range of anterograde memory tasks and preoperative hippocampal volumes in a group of temporal lobe epilepsy patients undergoing surgery. However, they did find that MRI-based volumetric indices were useful in predicting the amount of decline in verbal memory and visuospatial learning following temporal lobectomy. More specifically, patients with larger right hippocampal volumes showed a greater decline following right temporal lobectomy; in patients undergoing left temporal lobectomy, a greater difference between left and right hippocampal volumes was associated with a larger decline in the percent story recall retained at delayed testing.

In a group study that combined patients with Alzheimer-type dementia, Huntington's disease, alcoholic Korsakoff's syndrome, and two patients with amnesia due to hypoxia, Jernigan and Ostergaard (1993) found that verbal recognition memory correlated with volume of medial temporal lobe structures (taken as the uncus, amygdala, hippocampus, and parahippocampal gyrus). No such correlation was found in the case of more lateral temporo-occipitoparietal neocortex or in the case of the caudate nuclei.

Finally, in a study of patients with Cushing's syndrome, Starkman, Gebarski, Berent, and Schteingart (1992) found that both right and left volumes of the hippocampal formation (taken as dentate gyrus, hippocampus proper, and subiculum) significantly correlated with verbal paired-associate learning and delayed story recall but that there were no similar correlations with memory for visual design material. The authors were able to provide evidence on the specificity of their correlation by showing that there were no similar relationships between their MRI hippocampal indices and performance on a perceptual-motor attention task or between measures of intracranial volume and memory test scores.

In summary, a number of research studies have applied volumetric analytical techniques in studies of memory disorder patients, but the results are rather preliminary and their significance remains to be determined. It is possible that single-subject studies of discrete, identifiable lesions, as reviewed in the first part of this chapter, may provide more definitive data relating to anatomy-memory correlations, though both approaches may eventually prove to have some merit, depending on the questions being addressed.

Absolute Measures of T^1 and T^2 Relaxation Indices

In recent years, absolute measures of T^1 and T^2 relaxation times have been used as sensitive indices of brain abnormality, and they are possibly more sensitive than visually based interpretation or volumetric analyses of images (e.g., Jackson *et al.,* 1993). There are few studies that have related T^1 and T^2 relaxation times with memory performance in neurological patients. Besson *et al.* (1990) examined white matter changes in a group of patients with Alzheimer-type dementia and found correlations between T^1 relaxation times in frontal and temporal regions and both story recall and memory for visual designs. Left temporal lobe abnormality showed a significant correlation with story recall, and right temporal lobe abnormality showed such a correlation with recall of visual designs, but there was no such hemispheric specificity for right and left frontal lobe indices. It is of note that there were no significant correlations between temporal or frontal lobe changes and memory for new faces or memory for famous faces.

Conclusions

It is likely that the full potential of magnetic resonance imaging for advancing our understanding of abnormal and normal memory has yet to be realized. Major developments, such as the use of functional magnetic resonance imaging in normal and memory disorder subjects, may have a significant impact in memory disorders research, but it is too early to say if this will be the case. Other developments should also prove valuable, including faster imaging times, automated volumetric analyses, reconstructive routines for surface and three-dimensional rendering, and the generation of absolute relaxation times that reliably reflect tissue integrity.

Acknowledgments

I am grateful to the Wellcome Trust for financial support (Grant No. 036191). I thank Dr. Simon Barker and Mr. Jason Brice for their comments on this manuscript. I also thank Ms. Lynn Hoggarth, Mr. Keith Scholey, and Dr. Simon Thompson for their assistance.

References

Andreasan, N. C. (1989). Nuclear magnetic resonance imaging. In N. C. Andreasan (Ed.), *Brain Imaging Applications in Psychiatry* (pp. 67–121). Washington, DC: American Psychiatric Press.

Anzola, G. P., Bevilacqua, L., Cappa, S. F., Capra, R., Faglia, L., Farina, E., Frisoni, G., Mariani, C., Pasolini, M. P., & Vignolo, L. A. (1990). Neuropsychological assessment in patients with relapsing-remitting multiple sclerosis and mild functional impairment: Correlation with magnetic resonance imaging. *Journal of Neurology, Neurosurgery and Psychiatry, 53,* 142–145.

Berkovic, S. F., Andermann, F., Melanson, D., Ethier, R. E., Fiendel, W., & Gloor, P. (1988). Hypothalamic hamartomas and ictal laughter: Evolution of a characteristic epileptic syndrome and diagnostic value of magnetic resonance imaging. *Annals of Neurology, 23,* 429–439.

Besson, J. A. O., Crawford, J. R., Parker, D. M., Ebmeier, K. P., Best, P. V., Gemmell, H. G., Sharp, P. F., & Smith, F. W. (1990). Multi-modal imaging in Alzheimer's disease. The relationship between MRI, SPECT, cognitive and pathological changes. *British Journal of Psychiatry, 157,* 216–220.

Bigler, E. D., Nelson, J. E., & Schmidt, R. D. (1989). Mammillary body atrophy identified by magnetic resonance imaging in alcohol amnestic (Korsakoff's) syndrome. *Neuropsychiatry, Neuropsychology, and Behavioural Neurology, 2,* 189–201.

Chalmers, J., Risk, M. T. A., Kean, D. M., Grant, R., Ashworth, B., & Campbell, I. W. (1991). Severe amnesia after hypoglycemia. *Diabetes Care, 14,* 922–925.

Charness, M. E., & DeLaPaz, R. L. (1987). Mammillary body atrophy in Wernicke's encephalopathy: Antemortem identification using magnetic resonance imaging. *Annals of Neurology, 22,* 595–600.

Comi, G., Filippi, M., Martinelli, V., Sirabian, G., Visciani, A., Campi, A., Mammi, S., Rovaris, M., & Canal, N. (1993). Brain magnetic resonance imaging correlates of cognitive impairment in multiple sclerosis. *Journal of the Neurological Sciences, 115,* (Supp.), S66–S73.

Dusoir, H., Kapur, N., Byrnes, D., McKinstry, S., & Hoare, R. D. (1990). The role of diencephalic pathology in human memory disorder: Evidence from a penetrating paranasal brain injury. *Brain, 113,* 1695–1706.

Franklin, G. M., Heaton, R. K., Nelson, L. M., Filley, C. M., & Seibert, C. (1988). Correlation of neuropsychological and MRI findings in chronic/progressive multiple sclerosis. *Neurology, 38,* 1826–1829.

Golomb, J., De Leon, M. J., Kluger, A., George, A. E., Tarshish, C., & Ferris, S. H. (1993). Hippocampal atrophy in normal aging. An association with recent memory impairment. *Archives of Neurology, 55,* 967–973.

Graff-Radford, N. R., Tranel, D., Van Hoesen, G. W., & Brandt, J. P. (1990). Diencephalic amnesia. *Brain, 113,* 1–25.

Hirasawa, M., Nakano, M., Shiohama, N., Komiya, T., & Mizuno, Y. (1991). Amnestic syndrome with bilateral lesions restricted to field CA-1 of both hippocampi. A case report with CT and MRI findings. *Clinical Neurology, 31,* 1147–1150.

Hodges, J. R., & Carpenter, K. (1991). Anterograde amnesia with fornix damage following removal of IIIrd ventricle colloid cyst. *Journal of Neurology, Neurosurgery, and Psychiatry, 54,* 633–638.

Hodges, J. R., & McCarthy, R. A. (1993). Autobiographical amnesia resulting from paramedian thalamic infarction. *Brain, 116,* 921–940.

Irle, E., Wowra, B., Kunert, H. J., Hampl, J., & Kunze, S. (1992). Memory disturbance following anterior communicating artery rupture. *Annals of Neurology, 31,* 473–480.

Isherwood, I. (Ed.). (1992). Neuroimaging. *Current Opinion in Neurology and Neurosurgery, 5,* 841–880.

Izquierdo, G., Campoy, F., Jordi, M., Gonzalez, M., & Martinez-Parra, C. (1991). Memory and learning disturbances in multiple sclerosis. MRI lesions and neuropsychological correlation. *European Journal of Radiology, 13,* 220–224.

Jackson, G. D., Connelly, A., Duncan, D. M., Grunewald, R. A., & Gadian, D. G. (1993). Detection of hippocampal pathology in intractable partial epilepsy: Increased sensitivity with quantitative magnetic resonance T_2 relaxometry. *Neurology, 43,* 1793–1799.

Jernigan, T. L., & Ostergaard, L. (1993). Word priming and recognition memory are both affected by mesial temporal lobe damage. *Neuropsychology, 7,* 14–26.

Kapur, N., Barker, S., Burrows, E., Ellison, D., Brice, J., Illis, L., Scholey, K., Colbourn, C., Wilson, B., & Loates, M. (1994). Herpes simplex encephalitis: Long-term MRI and neuropsychological profile. *Journal of Neurology, Neurosurgery and Psychiatry, 57,* 1334–1342.

Kapur, N., Ellison, D., Smith, M., McLellan, L., & Burrows, E. (1992). Focal retrograde amnesia following bilateral temporal lobe pathology: A neuropsychological and magnetic resonance study. *Brain, 115,* 73–85.

Kapur, N., Scholey, K., Moore, E., Barker, S., Mayes, A., Brice, J., & Fleming, J. (1994). The mammillary bodies revisited: Their role in human memory functioning. In L. Cermak (Ed.), *Neuropsychological Explorations in Memory and Cognition: Essays in Honor of Nelson Butters* (pp. 159–189). New York: Plenum Press.

Kodama, T., Numaguchi, Y., Gellad, F. E., Dwyer, B. A., & Kristt, D. A. (1991). Magnetic resonance imaging of limbic encephalitis. *Neuroradiology, 33,* 520–523.

Kritchevsky, M., Graff-Radford, N. R., & Damasio, A. R. (1987). Normal memory after damage to medial thalamus. *Archives of Neurology, 44,* 959–962.

Kritchevsky, M., & Squire, L. R. (1993). Permanent global amnesia with unknown etiology. *Neurology, 43,* 326–332.

Lacomis, D., Khoshbin, S., & Schick, R. M. (1990). MR imaging of paraneoplastic limbic encephalitis. *Journal of Computer Assisted Tomography, 14,* 115–117.

Lee, B. I., Lee, B. C., Hwang, Y. M., Sohn, Y. H., Jung, J. W., Park, S. C., & Han, M. H. (1992). Prolonged ictal amnesia with transient focal abnormalities on magnetic resonance imaging. *Epilepsia, 33,* 1042–1046.

Lencz, T., McCarthy, G., Bronen, R. A., Scott, T. M., Inserni, J. A., Sass, K. J., Novelly, R. A., Kim, J. H., & Spencer, D. D. (1992). Quantitative magnetic resonance imaging in temporal lobe epilepsy: Relationship to neuropathology and neuropsychological function. *Annals of Neurology, 31,* 629–637.

Loring, D. W., Murro, A. M., Meador, K. J., Lee, G. P., Gratton, C. A., Nichols, M. E., Gallagher, B. B., King, D. W., & Smith, J. R. (1993). Wada memory testing and hippocampal volume measurements in the evaluation for temporal lobectomy. *Neurology, 43,* 1789–1793.

Markowitsch, H. J., Von Cramon, D. Y., Hofmann, E., Sick, C.-D., & Kinsler, P. (1990). Verbal memory deterioration after unilateral infarct of the internal capsule in an adolescent. *Cortex, 26,* 597–609.

Markowitsch, H. J., Von Cramon, D. Y., & Schuri, U. (1993). Mnestic performance profile of a bilateral diencephalic infarct patient with preserved intelligence and severe amnestic disturbances. *Journal of Clinical and Experimental Neuropsychology, 15,* 627–652.

Mattioli, F., Cappa, S. F., Cominelli, C., Capra, R., Marcianoc, N., & Gasparotti, R. (1993). Serial study of neuropsychological performance and gadolinium-enhanced MRI in MS. *Acta Neurologica Scandinavica, 87,* 465–468.

Maurelli, M., Marchioni, E., Cerretano, R., Bosone, D., Bergamaschi, R., Citterio, A., Martelli, A., Sibilla, L., & Savoldi, F. (1992). Neuropsychological assessment in MS: Clinical, neuropsychological and neuroradiological relationships. *Acta Neurologica Scandinavica, 86,* 124–128.

Mishkin, M., & Appenzeller, T. (1987). The anatomy of memory. *Scientific American, 256,* 80–89.

Morris, M. K., Bowers, D., Chatterjee, & Heilman, K. (1992). Amnesia following a discrete basal forebrain lesion. *Brain, 115,* 1827–1847.

Parkin, A. J., & Hunkin, N. M. (1993). Impaired temporal context memory on anterograde but not retrograde tests in the absence of frontal pathology. *Cortex, 29,* 267–280.

Perlmutter, J. S. (Ed.). (1993). Neuroimaging. *Current Opinion in Neurology, 6,* 899–933.

Rao, S. M., Leo, G. J., Haughton, V. M., Aubin-Faubert, P. S. T., & Bernardin, L. (1989). Correlation of magnetic resonance imaging with neuropsychological testing in multiple sclerosis. *Neurology, 39,* 161–166.

Ron, M. A., Callanan, M. M., & Warrington, E. K. (1991). Cognitive abnormalities in multiple sclerosis: A psychometric and an MRI study. *Psychological Medicine, 21,* 59–68.

Rudge, P., & Warrington, E. K. (1991). Selective impairment of memory and visual perception in splenial tumours. *Brain, 114,* 349–360.

Scheltens, P. H., Leys, D., Barkhof, F., Huglo, D., Weinstein, H. C., Vermersch, P., Kuiper, M., Steinling, M., Wolters, E. C., & Valk, J. (1992). Atrophy of medial temporal lobes on MRI in "probable" Alzheimer's disease and normal ageing: Diagnostic value and neuropsychological correlates. *Journal of Neurology, Neurosurgery and Psychiatry, 55,* 967–972.

Senzaki, A., Okubo, Y., Abe, T., Matsuura, M., Moriiwa, M., Asai, K., & Toru, M. (1993). Quantitative magnetic resonance imaging in patients with temporal lobe epilepsy. *Journal of Epilepsy, 6,* 243–249.

Shear, P. K., Butters, N., Jernigan, T. L., Ditraglia, G. M., Irwin, M., Schuckit, M. A., & Cermak, L. S. (1992). Olfactory loss in alcoholics: Correlations with cortical and subcortical MRI indices. *Alcohol, 9,* 247–255.

Soto, A. L., Takahashi, M., Yamashita, Y., Sakamoto, Y., Shinzato, J., & Yoshizumi, K. (1991). MRI findings of hypothalamic hamartoma: Report of five cases and review of the literature. *Computerized Medical Imaging and Graphics, 15,* 415–421.

Squire, L. R., Amaral, D. G., & Press, G. A. (1990). Magnetic resonance imaging of the hippocampal formation and mammillary nuclei distinguish medial temporal lobe and diencephalic amnesia. *Journal of Neuroscience, 10,* 3106–3117.

Squire, L. R., Amaral, D. G., Zola-Morgan, S., Kritchevsky, M., & Press, G. (1989). Description of brain injury in the amnesic patient N. A. based on magnetic resonance imaging. *Experimental Neurology, 105,* 23–35.

Starkman, M. N., Gebarski, S. S., Berent, S., & Schteingart, D. E. (1992). Hippocampal formation volume, memory dysfunction, and cortisol levels in patients with Cushing's syndrome. *Biological Psychiatry, 32,* 756–765.

Szelies, B., Herholz, K., Pawlik, G., Karbe, H., Hebold, I., & Heiss, W.-D. (1991). Widespread functional effects of discrete thalamic infarction. *Archives of Neurology, 48,* 178–182.

Tanabe, H., Hashikawa, K., Nakagawa, Y., Ikeda, M., Yamamato, H., Harada, K., Tsumoto, T., Nishimura, T., Shiraishi, J., & Kimura, K. (1991). Memory loss due to transient hypoperfusion in the medial temporal lobes including hippocampus. *Acta Neurologica Scandinavica, 84,* 22–27, 463.

Tatemichi, T. K., Desmond, D. W., Prohovnik, I., Cross, D. T., Gropen, T. I., Johr, J. P., & Stern, Y. (1992). Confusion and memory loss from capsular genu infarction: A thalamocortical disconnection syndrome? *Neurology, 42,* 1966–1979.

Trenerry, M. R., Jack, C. R., Ivnik, R. J., Sharbrough, F. W., Cascino, G. D., Hirschorn, K. A., Marsh, W. R., Kelly, P. J., & Meyer, F. B. (1993). MRI hippocampal volumes and memory function before and after temporal lobectomy. *Neurology, 43,* 1800–1805.

Valenstein, E., Bowers, D., Verfaellie, M., Heilman, K. M., Day, A., & Watson, R. T. (1987). Retrosplenial amnesia. *Brain, 110,* 1631–1646.

Victor, M., & Agamanolis, D. (1992). Amnesia due to lesions confined to the hippocampus: A clinicopathologic study. *Journal of Cognitive Neuroscience, 2,* 246–257.

Von Cramon, D. Y., Hebel, N., & Schuri, U. (1985). A contribution to the anatomical basis of thalamic amnesia. *Brain, 108,* 993–108.

Wilson, J. T. L., Wiedmann, K. D., Hadley, D. M., & Brooks, D. N. (1989). The relationship between visual memory function and lesions detected by magnetic resonance imaging after closed head injury. *Neuropsychology, 3,* 255–265.

Zola Morgan, S., Squire, L., & Amaral, D. (1986). Human amnesia and the medial temporal lobe region: Enduring memory impairment following a bilateral lesion limited to field CA1 of the hippocampus. *Journal of Neuroscience, 6,* 2950–2967.

12

Functional Magnetic Resonance Imaging

Studies of Cognition

B. J. CASEY, JONATHAN D. COHEN, DOUGLAS C. NOLL, WALTER SCHNEIDER, JAY N. GIEDD, AND JUDITH L. RAPOPORT

INTRODUCTION

Understanding the workings of the human brain is not an easy task. Cognitive functions unique to humans, such as language and its development, ultimately require noninvasive functional measurement of the human brain *in vivo*. With recent advances in neuroimaging techniques, structure/function relations now may be studied in greater detail and in pediatric populations. These technical advances eliminate the invasiveness of conventional neuroimaging techniques in the study of human brain functioning and provide spatiotemporal resolution never before attained.

Hypotheses concerning structure-function relationships in the brain have begun to receive support from a growing number of studies that look at task

B. J. CASEY Western Psychiatric Institute and Clinic, University of Pittsburgh Medical Center, Pittsburgh, Pennsylvania 15213. JONATHAN D. COHEN AND DOUGLAS C. NOLL University of Pittsburgh Medical Center, and Carnegie Mellon University, Pittsburgh, Pennsylvania 15213. WALTER SCHNEIDER University of Pittsburgh, Pittsburgh, Pennsylvania 15213. JAY N. GIEDD AND JUDITH L. RAPOPORT National Institute of Mental Health, Bethesda, Maryland 20892.

Neuroimaging II: Clinical Applications, edited by Erin D. Bigler. Plenum Press, New York, 1996.

performance and brain activation simultaneously. However, conventional neuroimaging techniques, such as positron emission tomography (PET) and single photon emission computed tomography (SPECT), have a number of serious limitations. Besides their reliance on radiation which limits their use with pediatric populations and in extensive longitudinal studies, they are also limited in what information they provide. For example, PET data typically require that images be averaged over a period of 60 seconds or more and summed across subjects to produce sufficient statistical power. This averaging results in poor temporal resolution (60 seconds) and limited spatial resolution (15-mm range).

Recently, a number of scientists (Belliveau *et al.*, 1991; Kwong *et al.*, 1992 Ogawa *et al.*, 1992) have demonstrated the ability to record physiological activity in the brain with 3-mm spatial resolution and a temporal resolution of just a few seconds using functional magnetic resonance imaging (fMRI). Variants of this technique are sensitive to changes in cerebral blood flow, cerebral blood volume, and blood oxygenation (Kwong *et al.*, 1992). These changes may be used to produce functional maps of mental processes.

These techniques are unlike conventional neuroimaging techniques for a number of reasons. First, the resolution of these activation images is close to millimeter spatial resolution, and activated regions can be localized easily by overlaying the activation images on the T1-weighted anatomical images. Second, fMRI is noninvasive and therefore can be used with pediatric populations and in extensive longitudinal studies. Third, since injection of radioactive material is not required, this technique allows for multiple repetitions of several experimental conditions within a single subject. Fourth, activation can be revealed in a single subject, eliminating the need to average across subjects and allowing single-subject case studies. Finally, as stated above, fMRI has temporal resolution within the 2–5 second range, allowing gated acquisitions of images during stimulus and response events.

This chapter discusses the basic underlying principles of functional MRI and applications of this technique to the study of cognitive processes.

Principles of MRI

Conventional anatomical MRI provides images of the brain with high spatial resolution. Magnetic resonance involves application of an energy (radio frequency or RF) pulse across a strong magnetic field. The large magnet used in MR scanners produces a strong ambient magnetic field, which ensures that a number of protons within that field are aligned. To generate an image, the scanner produces a brief RF pulse within the plane to be imaged, which momentarily perturbs or tilts the protons within that plane. It then records the energy that is released as those protons precess back to their original state.

There are two time constants associated with the decay in the MR signal: T1 and T2. The T1 constant corresponds to the decay in the signal that results from the protons gradually returning to their initial state. It is referred to as the longitudinal component, since the protons are "tilting" back up to their aligned or "upright" state. This relaxation is related to the makeup of surrounding tissue, with small molecules (water) relaxing faster than larger lipid molecules. The T2 constant corresponds to the decay in the MR signal that occurs as the protons gradually fall out of phase in their precession. It is referred to as the transverse component, since it is in this plane that the motion of the protons is becoming non-uniform. The T2 relaxation is associated with interactions among nuclear spins and local inhomogeneities in the applied field causing protons to precess at different rates and deviate from the uniform motion of the initial excitation. Localized inhomogeneities in the applied field lead to local differences in the Larmor frequencies, causing a decrease in T2, which is then referred to as T2*. The T2* time constant will be relevant in our discussion of fMRI.

The Basics of Functional MRI

Several research groups (Thulborn, Waterton, Matthews, & Radda, (1982); Rosen, Belliveau, & Chien, 1989; Rosen, Belliveau, Vivea, & Brady, 1990; Ogawa, Lee, Nayak, & Glynn, 1990; Ogawa *et al.*, 1992; Turner, LeBihan, Moonen, Despres, & Frank, 1991; Turner *et al.*, 1992; Detre, Leigh, & Williams, & Koretsky, 1992; Belliveau *et al.*, 1991; Kwong *et al.*, 1992) have all contributed to the development of MRI techniques for recording physiological activity in the brain from localized changes in blood volume or blood flow. All of these techniques take advantage of two basic phenomena: the ability of paramagnetic agents to produce contrast in the MR signal and the fact that regional changes in brain activity are associated with local hemodynamic changes.

Paramagnetism is the ability of an otherwise nonmagnetic material to exhibit magnetic properties in the presence of a magnetic field. The initial efforts to record physiological activity with MRI used paramagnetic contrast agents, such as gadolinium-DTPA, injected into the blood to measure regional blood flow (Rosen *et al.*, 1990). The molecules in the contrast agent introduce local inhomogeneities in the magnetic field, which reduce the coherence of, or "dephase," the signal generated by the protons at that location. The difference in signal due to local dephasing effects can be detected in images acquired using MR parameters that are sensitive to T2* decay. Therefore, paramagnetic contrast agents, in conjunction with T2*-weighted imaging, can be used to produce an image of brain perfusion. By comparing perfusion in activated and nonactivated states, areas of relative brain activity can be identified (Belliveau *et al.*, 1991). One obvious limitation of this approach, however, is the need to use exogenous contrast agents to produce the image.

Recently, several groups have developed methods that do not require the use of injected contrast agents (Ogawa *et al.*, 1990; Turner *et al.*, 1991; Kwong *et al.*, 1992). These techniques are based on the observation that hemoglobin becomes highly paramagnetic in its deoxygenated state. In other words, deoxygenated hemoglobin can be used as a naturally occurring contrast agent, with highly oxygenated areas producing stronger MR signals than less well oxygenated regions. These changes in signal intensity, related to the oxygenation of hemoglobin, therefore, are detectable in T2*-weighted images.

A number of studies (e.g., Fox, Raichle, Mintun, & Dence, 1988; Raichle & Mellinckrodt, 1988) have demonstrated that at least some brain areas increase their blood flow disproportionately to metabolic need when they become active, resulting in a net increase in tissue oxygenation. As a result, the MR signal in those areas increases relative to others, and this contrast can be used to produce MR images of the activated region. This method has already begun to produce provocative results. Bandettini, Wong, Hinks, Tikofsky, and Hyde (1992), Kwong *et al.* (1992), and Ogawa *et al.* (1992) have all shown activity-related changes in occipital cortex in response to visual stimuli comparable to those that have been demonstrated using paramagnetic contrast media to image changes in blood flow directly (Belliveau *et al.*, 1991). However, the percent change in signal is modest at 1.5 to 2%.

All of the initial efforts to image oxygen-related changes directly, without the use of contrast agents, used specialized MRI equipment. Bandettini *et al.* (1992) and Kwong *et al.* (1992) used echo planar imaging (EPI) techniques, which provide a means of acquiring MRI images much faster than ordinary equipment. Customized hardware allows rapid oscillation of gradient magnetic fields that enable the acquisition of a single slice in less than 50 ms. Ogawa *et al.* (1992) and Turner *et al.* (1992) have reported success and increases in signal of 15 to 20% during visual stimulation using high-field (4 T) magnets.

Unfortunately, high-field magnets and the equipment necessary for EPI are not widely available. Recently, however, other laboratories have reported similar results using conventional MRI equipment (Frahm, Bruhn, Merboldt, Hanicke, & Math, 1992; Frahm, Merboldt, & Hanicke, 1993; Gore *et al.*, 1992). Although, the temporal resolution with standard equipment is less (typically 6000 ms to acquire a single slice) than that of EPI, the spatial resolution surpasses that of EPI (roughly 3.125×3.125). Recent developments in spiral scanning, a method for rapidly acquiring (1000 ms/image) high-resolution, T2*-weighted images with standard MR scanners (Noll, Cohen, Meyer, & Schneider, 1995) may be able to bridge the gap between high-resolution but relatively slow conventional scanning methods and the lower resolution but very rapid single-shot EPI methods.

There are constraints on the temporal resolution offered by fMRI, even for specially equipped MR scanners that can acquire images as rapidly as 50 ms. It is constrained by the time course of the physiological processes generating the signal. The MR signal reflects hemodynamic alterations in local levels of deoxyhemoglobin which accompany neuronal activation. Blamire *et al.* (1992) examined the time course of the hemodynamic response during activation of the visual cortex with photic stimulation. Using echo planar imaging on a 2.1 T Bruker Biospec spectrometer, Blamire *et al.* collected data from seven different subjects. Photic stimulation consisted of a flashing checkerboard pattern (Grass stimulation goggles) that was presented during three periods, interleaved with periods of total darkness (baseline). Averaged baseline images (total darkness) were subtracted from images collected during activation (flashing checkerboard). There was a clear delay between stimulus onset and the onset of signal change. Similar to earlier reports (Kwong *et al.*, 1992; DeYoe, Neitz, Bandettini, Wong, & Hyde, 1992; Turner *et al.*, 1992), this delay was roughly 4–8 seconds in duration. Thus, the hemodynamic response apparently takes place over a period of seconds rather than milliseconds. As noted by Kwong *et al.*, this rise time is in reasonable agreement with cerebrovascular transit times based on O-15 PET studies. Blamire *et al.* (1992) also observed similar delays in fall time (i.e., return of signal to baseline following stimulus offset). This time course of changes in deoxyhemoglobin was corroborated by DeYoe *et al.* (1992) in both the visual and motor cortex.

Magnetic resonance imaging poses unique challenges for the empirical scientist in terms of the equipment that is used as well as experimental design. The magnet itself places constraints on the physical characteristics of the hardware that can be used, and the image acquisition process (which is highly sensitive to RF signals) places constraints on the electronics. Ferromagnetic materials cannot be brought near the scanner, and electronic devices must not produce a significant amount of RF noise or must be shielded.

STIMULUS PRESENTATION APPARATUS. We have had success using an active matrix LCD display panel (Sharp 1050A) and a high-luminance overhead projector, placed approximately 5 meters from the subject's head (at the foot of the subject table). Images are projected onto a rear-projection screen, the frame of which is made of plastic or aluminum, and stands at the foot of the subject table. The subject observes the screen through a set of prism glasses. The display panel is connected to a microcomputer which is usually kept in the control room via a standard serial cable. We have found this provides a stable, high-contrast, high-resolution visual image without producing any significant degradation in the MR signal. Other groups have also reported success using specially designed video projector systems, either placed in the scanner room or projecting through the control room window.

The connections between the experimental control computer and the display must be made carefully. The typical MRI scanner room has copper mesh shielding in the walls to minimize radio frequency noise that can arise from radio station transmissions or nearby electronic devices, such as computers. A standard scanner records data at 62.5 MHz. It is critical that any cables entering the room not carry signals of this frequency. High-speed computers and display devices in the 60-MHz range are therefore problematic. Noise produced by such sources will show up as broad lines that obscure the MR image. We have had success running both a 16-MHz IBM-compatible 386 computer with VGA output and a 25-MHz Macintosh IIci with direct video output. Ideally, it is best to use a 1-MHz band pass filter on the video cable, run the cable through a torus, and keep the computer in the control room. However, not all facilities may have a convenient way of running cables from the scanner to the control room. Under such circumstances, we have found that, with a 1.5 T magnet, it is possible to place the computer in the scanner room and use floppy disks, as long as the equipment remains 5 meters or more from the scanner.

It is also possible to present auditory stimuli for feedback and/or auditory studies. This can be done using the pneumatic audio apparatus already equipped in many scanners. However, the noise produced by the scanner during the scanning cycle places serious constraints on the bandwidth and fidelity of the auditory signal that can be presented to the subject. This also precludes the use of ear plugs, which are usually used to reduce the amount of noise that the subject must endure.

For studies in which subject responses must be recorded, we have used a specially constructed, hand-held, fiber-optic response box. This connects, via a fiber-optic cable, to a transducer, which in turn is connected to the microcomputer via a serial input device. This arrangement eliminates the presence of any alternating electrical signal within the scanner that might produce RF interference.

STIMULI. The choice of stimuli, their size, and rate of presentation are other issues to consider. In our studies, we address how one may maximize activation with these stimulus manipulations.

SCANNING PROCEDURE

Most importantly, subjects should be carefully screened for a history of surgery involving metal implants, possible metal fragments in the eyes, a pacemaker, or intrauterine device (IUD). Exclusion criteria would be evidence of medical or psychiatric disease, metal implants, prior claustrophobic history, or weight over 250 lbs.

Because of the high spatial resolution of the MR image, and therefore its sensitivity to movement artifact, it is important to restrain the subject's head during scanning. We have had reasonable success with an inflatable surgical pillow and foam padding, but the bite bar has been most effective in reducing motion.

In our work, each study involves the acquisition of at least two sets of images: activation images and structural images. One of the primary advantages of using MRI is that functional information can be acquired at the same time as, and directly related to, anatomic data, so that areas of activity can be accurately localized. Other techniques, such as PET, require that the functional and anatomic data be acquired at different times, using different apparatus, so that image registration becomes a problem for localizing areas of activation.

We have used T2*-sensitive spiral scan sequences and echo planar sequences to rapidly acquire high-resolution images of activation. The former method consists of applying a multi-slice, spiral-trajectory pulse sequence on a 1.5 T scanner in either a high-resolution (20 interleave, 200 × 200 matrix, 24 cm field of view or FOV) or medium-resolution (10 interleave, 128 × 128 matrix, 24 cm FOV) implementation with a TR of 600 ms, a gradient echo TE of 35 ms, and flip angle of 45 degrees. More recently, we have used a four-spiral pulse sequence to acquire 64 × 64 images at an average rate of three images per second. In the case of our echo planar imaging, we use a gradient echo TE of 40 ms, TR of 6000 ms, flip angle of 90 degrees, 20 cm FOV, and 128 × 64 acquisition matrix. Surface coils in several configurations (single 5-inch or dual 5-inch), head coils, or more recently, brain coils, are applied to improve signal-to-noise ratio of the heavily T2*-weighted (i.e., TE of 35 to 40 ms) scans.

The sequence of events for a typical fMRI study is as follows. First, *T1-weighted images* are acquired using a spoiled gradient sequence (TR = 33 ms, TE = 8 ms, FOV = 24 cm, 256 × 192 matrix) to identify standard Talairach landmarks such as the anterior commissure (AC) vertical and anterior commisure–posterior commisure (AC-PC) line. Second, *T2* images* between 4 and 7 mm thick are acquired covering a specific brain region of interest or the entire brain. *Structural T1-weighted images* are acquired in the same slice locations in order to localize activation. *Magnetic resonance angiograms* (MRA) are obtained at each of these slice locations in order to better distinguish between parenchymal and vascular sources of functional signals. We use a 2-D phase contrast angiography pulse sequence with flow compensation, 25° flip angle, 6 ms TE, 25 ms TR, 192

× 256 matrix, and velocity encoding of 10 cm/sec. Finally, a *volume scan* (TR = 24 ms, TE = 6 ms, flip angle = 45°, 124 slices, 1.5 mm slice thickness, FOV = 24 cm, in-plane resolution = 0.9 × 1.3 mm, SPGR pulse sequence) is acquired for 3-D analyses and, where necessary, cross-session image registration in addition to the purposes of morphometry studies.

Thus, a typical scanning session lasts approximately 2 hours. This time period includes final subject screening and consent (10 min); desensitization to the MR procedures in a simulator (20 min); positioning and entry of the subject in the scanner (10 min); collection of T1 scout images and prescription of desired scanning slice locations (5 min); instructions to the subject, start of behavioral tasks, and functional scan acquisition (30 to 50 min); acquisition of T1 structural scans at the prescribed slice locations (5 min); acquisition of MRA (10 min); and finally the volume scan (10 min).

Data Analysis

Image data are subjected to an extensive series of image processing and data analysis procedures that have been designed to ensure that any patterns of task-related activity that are observed are both valid and reliable. This involves the following steps: (1) mean compensation (correction for gross effects of scanner signal drift); (2) image registration (alignment of all images within a test session to correct for head movement); (3) statistical subtraction (identification of areas of significant task-related activation); (4) MRA filtering (elimination of areas associated with observable large vessels); (5) anatomic localization (identification of areas of activity with specific anatomic structures); and (6) between-subject analyses (determination of group differences). Each of these procedures is described in detail below.

First, we compensate for any drift in scanner sensitivity during the course of the experiment by normalizing the global signal intensity of each image to that of the first acquisition of that slice. Specifically, we determine the ratio of the mean signal intensity for each image (averaged across all voxels) to that of the first image acquired at that location and multiply each voxel by this factor. Following this mean compensation, each subject's images are realigned in 2-D space using Woods' AIR algorithm (Woods, Cherry, & Mazziotta, 1992). This algorithm has been shown to align voxels with an accuracy less than voxel diameter, thus allowing the statistical comparisons (detailed below) to be performed confidently on individual voxels. We use this to quantitatively assess degree of movement within sessions and reject from further analysis data sets that show movement-related artifact.

Areas of significant activation are identified by performing pixelwise *t*-tests between images acquired during two conditions (e.g., visual stimulation and visual fixation conditions). Because an image consists of a large number of pixels (e.g., a 64 × 64 image has 4092 pixels), the probability of a false positive can become quite large, unless some adjustment for multiple comparisons is made. Most frequently, Bonferroni adjustments are made which would be equal to a p value of .00001 ($.05/(64)^2 = .000012$). Instead, we use a split-halves method that produces significance levels comparable to the Bonferroni adjustment ($p <$ $.001^2$ or .000001), but in addition provides us with information regarding the reliability of our data. This method is similar to a cross-validation approach (Tatsuoka, 1988), using each half of the imaging data to predict or validate the other half, as follows. For each slice, the first half of the data acquired in one condition (e.g., a control fixation condition) is compared to the first half of the data acquired in a second condition (e.g., a visual stimulation condition of flashing lights). The same procedure is then applied to the second half of the data. Areas of activation are accepted only if they have significant *t*-values in both comparisons. This split-halves methodology has been widely used in education and psychology for determining test-retest reliability (Snedector & Cochran, 1980) and has been effectively applied to both PET (Squire *et al.*, 1992) and fMRI data (Cohen *et al.*, 1994; Schneider, Casey, & Noll, 1994; Casey, Cohen, *et al.*, 1995).

As described above, typically, statistical subtractions are performed, comparing signal intensity in two conditions (e.g., the activation and control conditions). A well-recognized concern about subtractive methodology (Donders, 1868/1969; Sergent, Zuck, Levesque, & MacDonald, 1992) is whether it is possible to design tasks such that all of the "irrelevant" processes—those that are supposed to be subtracted out—are being performed in the same way or (in the case of neuroimaging studies) by the same anatomic structures in the control and experimental tasks. Failures to satisfy this assumption can produce misleading results. One approach to this problem is to analyze multiple subtraction pairings (e.g., Sergent *et al.*, 1992). For example, if an area shows activation in subtractions with two different controls (e.g., A − B and A − C) it is more likely to reflect processes related to condition A. This design can be further exploited by using a standard ANOVA with voxel signal intensity as the dependent measure and the task conditions as the independent factors. (This approach has also been proposed by Sanders & Orrison, 1993.) This approach is directly analogous to standard factorial designs used in behavioral research.

We have just begun to use a voxelwise multifactorial ANOVA, in place of the standard *t*-test subtraction, to identify areas of significant task-related activation. The use of an ANOVA has at least two advantages over standard subtractions: it allows each task condition to be compared simultaneously against all other conditions of the experiment; and it can be used to control for the effects of nonspecific trends and spurious transients in the MR signal. The ANOVA procedure is performed on a voxelwise basis to construct statistical maps. As with standard subtractions (using *t*-tests), one must establish a significance threshold for the test statistic that takes account of the number of comparisons that are made. Because any image consists of a large number of pixels, the probability of a false positive can become quite large, unless some adjustment for multiple comparisons is made. Most frequently Bonferroni adjustments are made, although this method has been called overly conservative (Friston *et al.*, 1990). Instead, we use a contiguity threshold (Forman *et al.*, 1995) because the probability of detecting a false-positive region of contiguous pixels (or particle) decreases as the number of contiguous pixels increases. Post hoc *t*-tests are performed to determine which conditions differ and whether the observed differences are in increases or decreases in activation.

Even after taking the effects of repeated measures into account, there are still other ways in which significant difference values can arise artifactually. A high *t*-value could be due to one or two spuriously high values or, in simple alternating designs (e.g., ABAB), it could be due to signal drift. A monotonically changing baseline will produce an overall difference in the first condition compared to the second. These concerns can be addressed qualitatively by looking at the raw signal values across time. If significant differences are due to consistent, functionally relevant physiological changes, then signal intensity should alternate regularly and consistently across the behavioral conditions of the task as illustrated in the lower panel of Figure 1. Isolated spikes in one condition as depicted in the upper panel of Figure 1, or baseline drift, suggest artifactual causes for the differences observed.

Areas of significance that coincide with vessels, identified using MRA, are excluded from further analysis. With T2*-weighted fMRI, blood vessels, particularly draining veins, can exhibit activation-related changes in signal (Dyun, Moonen, DeBoer, Iperen, & Luyten, 1993) that may be located away from the actual site of neural activity. Sources of signal, such as draining vessels, can reduce our sensitivity for detecting such dissociations and clearly confound efforts to accurately localize individual areas of activity. We have taken two precautions to guard against vessel-related changes. First, signal changes associated with vascular flow can be mitigated with appropriate choice of pulse sequence parameters, similar to the way that imaging pulse sequences can be sensitized to flow. For example, a single-slice study using a short TR and moderate flip angle will be highly sensitized to flow-related changes (Dyun *et al.*, 1993; Frahm *et al.*, 1993), while longer TR pulse sequences (3 or more seconds) should have little flow sensitization. The pulse sequence parameters we have used in our echo planar studies (6 sec TR and 90 degree flip angle) have very little flow sensitivity. Secondly, we have eliminated any significant regions of activation greater than 6% which have been associated with activation due to large vessels (Menon, Ogawa, Tank, & Ugurbil, 1993). However, the use of MRA filtering provides additional protection.

Regions of significant activation identified by the procedures described above are overlaid on corresponding structural MRI images to permit anatomic localization. Assignment of regions of activity to structures is performed independently by two raters using standard brain atlases as references (Duvernoy, 1991) and individualized surface renderings constructed from the volume scan for each subject. Regions of activity are assigned to coarse structural categories (e.g., corresponding gyri, Brodmann's areas, and/or Talairach coordinates; see Talairach & Tournoux, 1988).

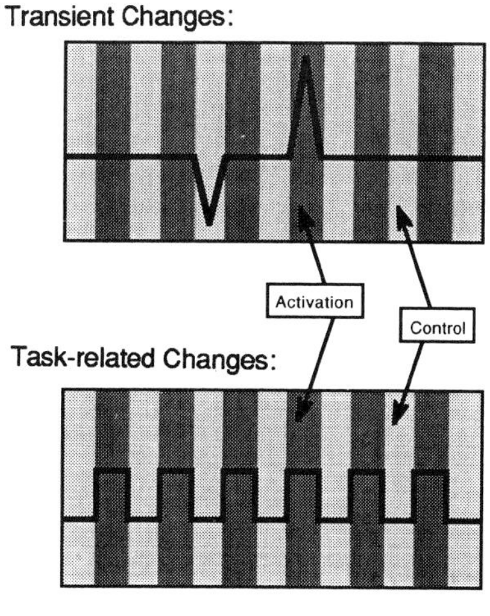

Figure 1. Examples of variations in the MR signal across time due to non-task-related transient events and task-related phasic events.

Hypotheses of group differences are tested by conducting between-subject ANOVA, examining the amount (volume in mm^3) and magnitude (percent change in signal) of activation by location and group. First, we identify areas in the statistical maps for each subject that represent areas of significant effects for each condition of interest. Next, each of these areas is assigned to a specific anatomic structure or area. We measure the total amount of activity and magnitude of activity (i.e., percent differences in signal change) associated with each condition in each structure and subject. These data are analyzed in a between-subject ANOVA, using amount of activity and magnitude of activity as the dependent measures and structure, condition, and group as independent factors.

Cognitive Studies Using Functional MRI

The remainder of this chapter provides an overview of studies using fMRI, with emphasis on the results from noncontrast functional MRI studies of cognition. Only studies of normal adults and children are reviewed. However, it is clear, based on the success of these studies, that fMRI will eventually play an important role in our understanding of brain circuitry in pathological populations as well.

Primary Sensory and Motor Studies

Primary Visual Studies. With the development of fMRI, researchers from multiple labs began to perform visual stimulation and motor tasks to verify its potential. The majority of these initial fMRI studies investigated the functional organization and frequency response of primary visual cortex (V1) and primary motor cortex (M1).

Kwong and colleagues (1992) were the first to obtain completely noninvasive tomographic maps of human brain activity during both visual stimulation and motor tasks. Changes in blood oxygenation were detected using a gradient echo sequence (TR = 3000 ms, TE = 40 ms) sensitive to the paramagnetic state of deoxygenated hemoglobin (i.e., variations in T2*). They also used a T1-sensitive spin echo inversion recovery pulse sequence (TI = 1100 ms, TR = 3500 ms, TE = 42 ms) to measure blood flow change (Kwong *et al.,* 1992). A receive-only RF surface coil was placed behind the occipital pole to improve signal/noise of rest (darkness) and 20 images of visual stimulation (8-Hz diodes emitting flashing red light produced by Grass photic stimulation goggles). Subtraction images (activation minus baseline) revealed a 1.8% increase in signal intensity in the primary visual cortex for seven normal volunteers for both imaging sequences. However, there was a slight difference in average time of activation for the two imaging sequences: 4 seconds for the gradient echo sequence and close to 9 seconds for the inversion recovery sequence.

In an attempt to push the limits of fMRI spatial resolution, Schneider, Noll, and Cohen (1993) examined the retinotopic organization of primary visual cortex. They used a conventional 1.5 T scanner and a T2*-weighted, gradient echo, multiple-slice scanning sequence (TE = 40 ms, TR = 150 ms) with 1.88 × 0.94 mm in-plane resolution. Subjects were presented with an 8-Hz reversing checkerboard pattern presented to partial visual fields (left-right, top-bottom, half rings, and wedges). Signal changes of 1 to 8% were observed which showed contralateral vertically reflected patterns of activation in the visual cortex. For example, left visual field presentation of a stimulus resulted in right occipital cortex activation, while upper visual field presentation resulted in activation of inferior visual cortex.

To compare the spatial topographies across subjects, computer algorithms were used to control for subject-unique folding of cortex, providing a flattened cortical ribbon. Four topographically distinct regions (A–D) were identified. The ribbon analysis of the calcarine fissure area differentiated among the four types of stimulus conditions (left-right, top-bottom, half rings, and wedges), dividing as the stimulus divided the visual field, and is assumed to correspond to V1 (striate cortex). Areas V2 and V3 are divided into two regions superior and inferior to the calcarine fissure. These regions receive input from the lower and upper visual fields, respectively, and match the observed regions B and C. The fourth visual area (D), possibly corresponding to V4 as identified in the monkey, is lateral and inferior to the calcarine fissure, receiving input from the upper and lower fields. This study demonstrates the application of a noninvasive neuroimaging technique with millimeter mapping capability in the study of human cortical function *in vivo*.

Initial functional MRI studies (e.g., Belliveau *et al.*, 1991; Blamire *et al.*, 1992) examining activation of the primary visual cortex used stimulus presentation rates of 8 Hz based on earlier O-15 PET findings (Fox & Raichle, 1985) of maximum cerebral blood flow response at this rate. Using both a MRI cerebral blood flow technique (i.e., spin echo inversion recovery) and MRI blood oxygenation technique (i.e., gradient echo) described above, Belliveau *et al.* (1991) and Kwong *et al.* (1992) reproduced the PET results of Fox and Raichle (1985) that activation of the primary visual cortex peaks at a stimulus presentation rate of 8 Hz. They further illustrated similar curve functions between percent change in PET cerebral blood flow and percent change in MRI signal. A linear increase in MR signal percent change and PET cerebral blood flow percent change was observed with flashing visual stimuli at rates of 1, 4, and 8 Hz, followed by a slow drop-off in signal over rates of 16 and 32 Hz.

Although the visual stimulation frequency appears to peak at 8 Hz, rarely do psychological events occur at this speed. In order to demonstrate the potential of fMRI in the study of higher-level cognitive processes, we must first demonstrate a significant change in signal with meaningful complex stimuli presented at a comprehendible rate. Schneider, Casey, and Noll (1994) recently examined the effects of stimulus type and stimulus presentation rate on cortical activation. Functional MRI was used to record cortical activation across multiple stages in the visual system during single character visual search and reversing checkerboard stimulation at 1, 4, and 8 Hz rates. The T2*-weighted oblique images were acquired using gradient echo sequences with late echo times (TE = 36 ms), a voxel size of 0.94×1.86 mm, and a slice thickness of 4–5 mm (see Figure 3 for slice locations). Character stimuli activated small (average 22 mm^2) reliable particles of gray matter in the calcarine fissure, superior occipital cortex, and fusiform-lingual gyrus. Figure 2 represents the reliable, significant areas of activation observed during the 8-Hz character search condition presented to the right visual field. Activation images were generated by performing pixelwise *t*-tests between images acquired during the control (fixation) and each experimental condition (1, 4, and 8 Hz), separately for each half of the data (split-halves method). Regions of interest revealing significant differences between the

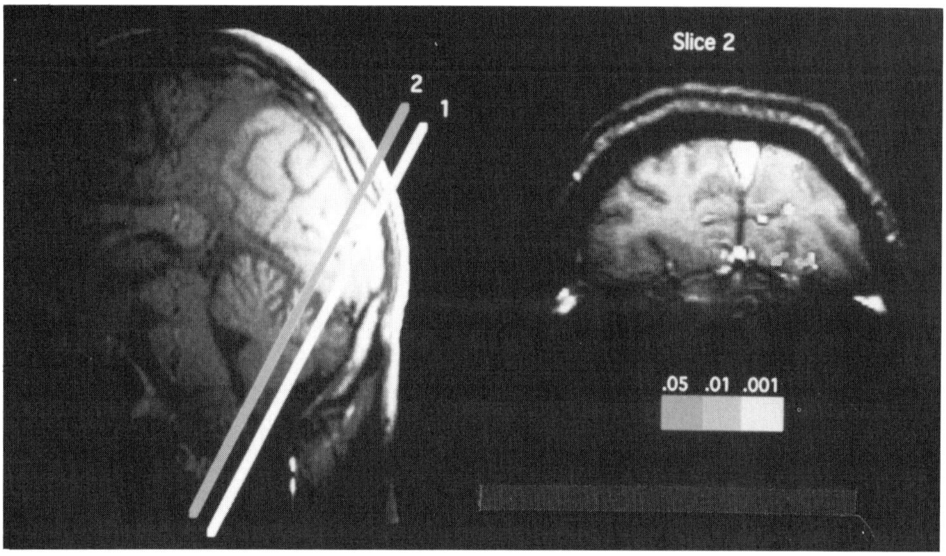

Figure 2. Illustration of slice locations for the activation images (left). Overlay of significant activation on T1-weighted oblique image for slice location 2 (right). (Adapted from Schneider *et al.*, 1994.) For a color representation of this figure, see the color insert facing page 326.

two tasks were identified and overlaid on structural T1-weighted images in order to locate the activation. For the character search, the MRI signal increased from 2.1 to 3.1% as stimulus rate increased from 1 to 8 Hz, independent of the region of processing (see Figure 3). In contrast, the reversing checkerboard stimulus produced higher magnitudes (from 4.5 to 5.2%) of activation for the calcarine fissure and lower magnitudes in the superior occipital and fusiform-

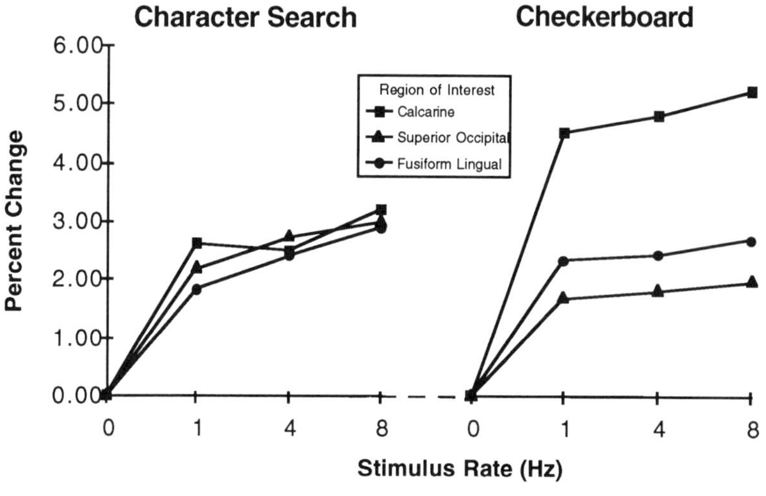

Figure 3. Percent change in MR signal as a function of stimulus presentation rate for each region of interest during the character search (left) and checkerboard search (right).

lingual gyri regions relative to the character search effects. The observed activation to the slowest frequency rate of 1 Hz is important given that complex, meaningful stimuli (e.g., words and sentences) require slower presentation rates.

PRIMARY MOTOR STUDIES. In order to demonstrate that the results observed with visual stimulation are not unique to striate cortex, Kwong *et al.* (1992) investigated the primary motor cortex. An oblique coronal slice through the precentral gyrus was acquired during a repetitive contralateral hand-squeezing task using a gradient echo, echo planar sequence (TR = 2000 ms, TE = 60 ms). Alternating periods of hand squeezing and rest occurred during scanning in a sequence similar to the design of their visual study reported above. The cortical temporal response and percent change in signal in primary motor cortex were similar to that observed within the primary visual cortex.

Figure 4 illustrates significant activation in the motor cortex during a fist-clenching task conducted by Noll *et al.* (1995). This study differs from that of Kwong *et al.* (1992) in that a T2* = sensitive spiral scan sequence (20 interleave, TR = 100 ms, TE = 36 ms, 200 × 200 matrix, 24 cm FOV, 7 mm slice thickness) was used to acquire the images on a conventional 1.5 T scanner without any specialized hardware. A 5-inch surface coils was placed on the contralateral side of the head to improve the signal/noise ratio. Activation images were generated by performing pixelwise *t*-tests between images acquired during rest and fist-clenching conditions. Regions revealing significant differences between the two tasks were identified and overlaid on structural T1-weighted images (Figure 4C). Figure 4 illustrates activation observed in the *t*-map (Figure 4B) which is overlaid on its corresponding structural T1-weighted image (Figure 4A). An increase in signal of 1 to 3% was observed in the gray matter of the motor cortex.

ATTENTION

THE STROOP EFFECT. One of our first cognitive studies using functional MRI utilized the Stroop paradigm (Stroop, 1935), the classic paradigm for examining selective attention. The Stroop task requires subjects to respond to specific attributes of a stimulus while suppressing or inhibiting a competing response. In the conflict condition, a color name is presented in a conflicting color (e.g., RED printed in blue ink). The task is to name the color while suppressing the learned and interfering response of reading the word. In the congruent condition (i.e., facilitation) the word and the color of the word are congruent (e.g., RED printed

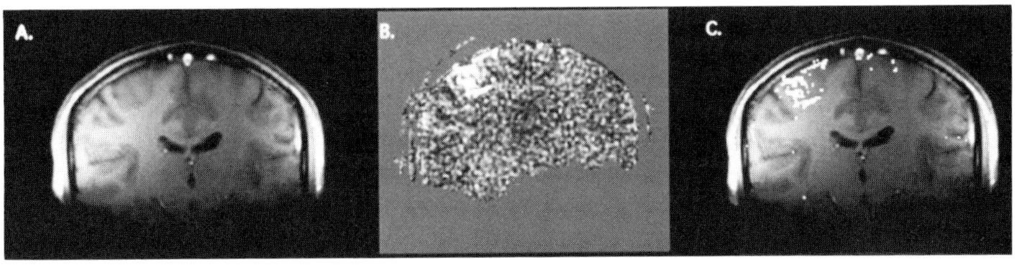

Figure 4. (A) The T1-weighted coronal image of motor cortex. (B) The corresponding *t*-map image generated from the comparison of fist clinching and rest. (C) The overlay of the *t*-map image on the corresponding T1-weighted image. (Adapted from Noll *et al.*, 1995.)

in red ink). Other conditions include the use of a color bar or a neutral (i.e., noncolor) word. The color bar condition consists of the subject naming the color of color bars without the inclusion of word information. The neutral word condition consists of a noncolor word presented in color (e.g., CAR printed in blue ink). The color bar and neutral conditions avoid the inherent confound of word reading in the case of the congruent condition (e.g., RED in red ink).

Although the Stroop paradigm has been used extensively in both experimental and clinical settings (Macleod, 1991), it is unclear whether the interference of the conflict condition occurs during stimulus selection between the word (e.g. RED) and the color of ink (in this case blue) or during response selection between the word RED and the response blue (Taylor, Kornblum, Minoshima, Oliver, & Koeppe, 1994). Based on robust PET activation of the right anterior cingulate during performance of the Stroop conflict condition relative to the congruent condition, Pardo, Pardo, Janer, and Raichle (1990) concluded that the interference observed during the conflict condition was due to interference during response selection and was subserved by the right anterior cingulate.

We examined prefrontal activation during the performance of the Stroop task using gradient echo EPI (TE = 40 ms, TR = 6000 ms) on a 1.5 T scanner. Eight contiguous, 5-mm coronal slices covering the prefrontal cortex were acquired during the Stroop conflict condition and a control condition. In order to control for the possibility of subjects reading the word as opposed to naming the color of the word in the congruent condition (i.e., RED in red ink), we used the more conservative neutral word condition as our control task. As reported by Casey, Trainor, *et al.* (1993), all six subjects we scanned failed to show activation in the prefrontal cortex. Our failure to find prefrontal cortex activation when comparing the Stroop conflict condition with a neutral word condition corroborates the PET data reported by Bench *et al.* (1993). They observed activation in the parietal cortex but no activation in the prefrontal cortex when comparing these conditions.

When we failed to see activation in the anterior cingulate without a stringent control task, we decided to attempt to replicate earlier work by Pardo *et al.* (1990). When replicating the paradigm of Pardo *et al.* exactly using Stroop conflict and congruent conditions, reliable activation of the right anterior cingulate was observed, as depicted in Figure 5. The left two images reveal significant

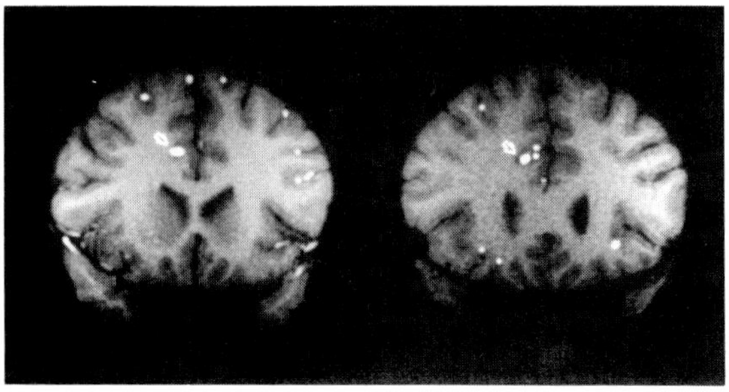
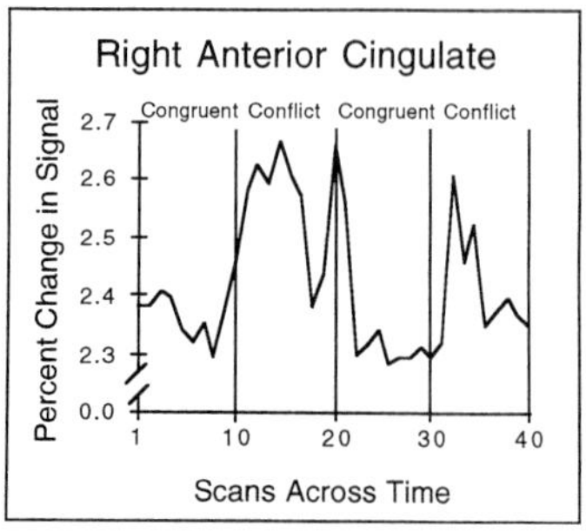

Figure 5. Overlays of significant activation observed during the Stroop conflict paradigm relative to the congruent control condition on two coronal images located 10 to 20 mm anterior to the anterior commissure (left). Time series analysis of change in MR signal in the right anterior cingulate as a function of the experimental manipulation (right).

activation in the right anterior cingulate in coronal slices between 10 and 20 mm anterior to the anterior commissure. The far right panel of Figure 5 reveals that these changes in the MR signal correspond with our experimental manipulation. Conclusions from this subtraction are unclear since the subjects could have been reading the words in the congruent condition relative to the control as opposed to color naming. Other reports from PET studies using the Stroop paradigm (Bench *et al.*, 1993; George *et al.*, 1994) suggest a lack of robustness of this task to activate discrete regions of interest when using well-controlled neutral conditions. Furthermore, Carter, Mintun, and Cohen (in press) reported activation in the right anterior cingulate even when subtracting the neutral condition from the congruent condition. Thus, the robust activation of the right anterior cingulate originally observed by Pardo *et al.* (1990) in the Stroop conflict may be attributed to a difference between color naming and word reading or attentional processes other than response inhibition.

Language

Speech Perception. Speech perception is the ability to process meaningful phonological stimuli. Lesion studies suggest that this process relies heavily on the functioning of the left temporal lobe in the perisylvian region. A recent PET study by Zatorre, Evans, Meyer, and Gjedde (1992) reported bilateral activation of superior temporal gyri during passive word presentation and not during the presentation of noise. Using functional MRI, Rao and colleagues (1992) further examined activation in the superior temporal gyri to see if the spatial and temporal resolution of this technique would reveal lateral hemispheric asymmetries not discerned with PET. Imaging was performed on a General Electric 1.5 T Signa system with a gradient echo EPI pulse sequence (TE = 50 ms, TR = 1000 ms, FOV = 24 cm). A single coronal slice through the superior temporal gyrus was selected and scanned repeatedly throughout the study (128 sequential images). Subjects listened to words being read at a rate of one every 3 seconds in between periods of rest with eyes open. Preliminary results revealed bilateral activation of the superior temporal gyri during listening to words, with greater activation (2.4%) in the left superior temporal gyrus relative to the right superior temporal gyrus (1.5%). These findings are consistent with results from both lesion and PET studies. However, the current study did not use a noise control task necessary for establishing that the observed signal change is specific to speech perception *per se*.

Speech Production. Both the comprehension and the production of spoken language clearly distinguishes human beings from other organisms. Recently, a number of O-15 PET studies (Petersen, Fox, Posner, Mintun, & Raichle, 1989; Petersen, Fox, Snyder, & Raichle, 1990) have identified brain regions associated with speech production and language processing. For example, Petersen and colleagues (1990) demonstrated activation of the left inferior frontal cortex during a verb generation task. Blamire and associates (1992), in an attempt to measure activity in the frontal cortex with echo planar imaging, used a variation of the verb generation task.

Functional images using a EPI spin echo sequence with T2* weighting of 70 ms were acquired of a single axial slice, 10 mm thick, and centered 8 mm below the AC-PC basal brain line during several conditions, including resting baseline, mouth movements without speech or subvocalization, repeating words being read (one per 1.5 s), and generating related verbs to nouns being read (e.g., responding "fly" to the noun "bird"). Additional control conditions included passive listening to words being read, passive listening to letters being read, and covert verb generation. In agreement with PET studies, increased activity in the left inferior frontal cortex during the verb generation condition compared to baseline was observed (roughly 10–15% change in signal). Little or no activation was reported in this region during passive listening, mouth movements, or covert generation, but a slight increase in signal was observed during the word repetition task, but not to the extent that was observed during verb generation.

In a related study by investigators at the National Institutes of Health (Appollonio *et al.,* 1993), functional images were acquired during a covert (silent) word generation task and compared to an automated speech task (recitation of the days of the week). Activation was observed in the left inferior frontal cortex (Brodmann's areas 44 and 45) and in motor and premotor cortex (Brodmann's areas 4 and 6). These data are in disagreement with Blamire and colleagues (1992), who reported little to no activation in the left inferior frontal cortex during their covert word generation task. Possible explanations for these conflicting reports may be due to how the regions of interest were defined or in where the slices were prescribed for the two different studies.

Finally, Hinke, Hu, Stillman, Kim, Merkle, Salmi and Ugurbil (1993) also used a covert word generation task. Nine right-handed, English-speaking subjects were scanned using a 4 T Siemens/SISCo system with a gradient-recalled echo sequence. The task was to silently generate words from the alphabet (e.g., A-apple, B-boat, C-car, etc.). The T2*-weighted images of the perisylvian region were obtained during a sequence of 10 baseline, 10 task, and then 15 recovery images. The baseline and recovery periods consisted of the subjects resting quietly with eyes closed. A percentage intensity change map was generated between images obtained during control and experimental tasks and included only those pixels that passed the student *t*-test at 99.9% confidence level. A greater region of Broca's area was active (53 mm^2) relative to its right-side homologue (16 mm^2) during the word generation task. However, signal intensity change was comparable for the left and right at approximately 4 and 4.5%.

The results of the previous studies suggest that Broca's area may be involved in word generation and not in motor aspects of articulation. However, subtle mouth movements could have been involved in subvocalizations during both silent word generation tasks. There was no mention of control for this potential confound.

Memory

Nonspatial Working Memory. Studies in nonhuman primates have suggested that dorsolateral prefrontal cortex plays an important role in supporting memory representations needed to select task-appropriate actions (Goldman-Rakic, 1987). Several PET studies (Jonides *et al.,* 1993; Weinberger, Berman, & Zec, 1986) have lent support to this hypothesis. In a recent study (Cohen, Noll, & Nystrom, 1995) T2*-sensitive spiral sequence images were acquired in an attempt to provide more detailed anatomic localization of areas associated with working memory. Scanning was performed on a standard 1.5 T General Electric Signa, using two 5-inch surface coils mounted on each side of the head. Seven-millimeter scans were acquired at six contiguous locations (see Figure 6). Six subjects alternated between performance of a control and experimental task (see Figure 7). For the control task, subjects monitoring a sequential display of single letters (one per 1.5 s) for the target "X" responded by pressing a button on a hand-held, fiber-optic response box. In the experimental task, the subjects responded only when the current letter was the same as the letter two before it (e.g., A-F-A, but not A-G-Q-A or A-A). Responses were recorded by computer. Presentation rate, target frequency, and the frequency of repeated letters was identical in the two tasks. Both the control and experimental tasks required that the subjects encode each letter, evaluate its identity, and respond to a target by pressing a button. However, the experimental task had the additional requirement that the subject keep in mind the two previous letters in order to identify targets.

Ten sets of scans were acquired during each of ten 60-second intervals, alternating between control and experimental tasks. Activation images were generated by performing pixelwise *t*-tests between images acquired during the control and experimental tasks, separately for each half of the data (split-halves method). Regions of interest revealing significant differences between the two tasks were identified and overlaid on structural T1-weighted images (refer to Figure 6). Eleven out of 12 subjects showed significant, reliable areas of activation in dorsolateral prefrontal cortex with an average change in signal of 1.2%.

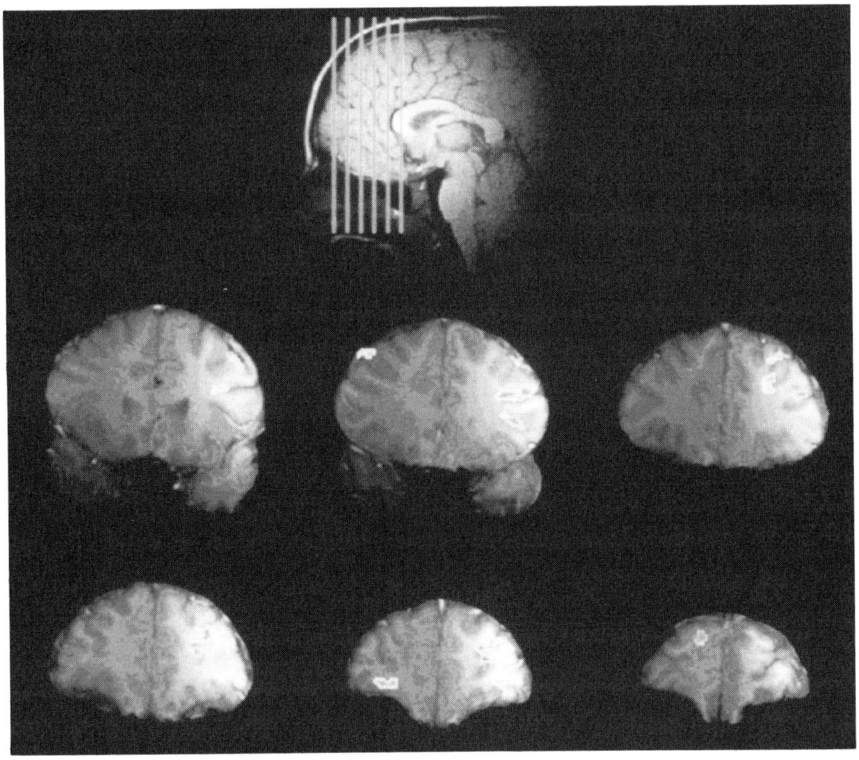

Figure 6. Slice locations and overlay of significant activation for each of six slice locations during the performance of a working memory task. (Adapted from Cohen *et al.*, 1994.) For a color representation of this figure, see the color insert facing page 326.

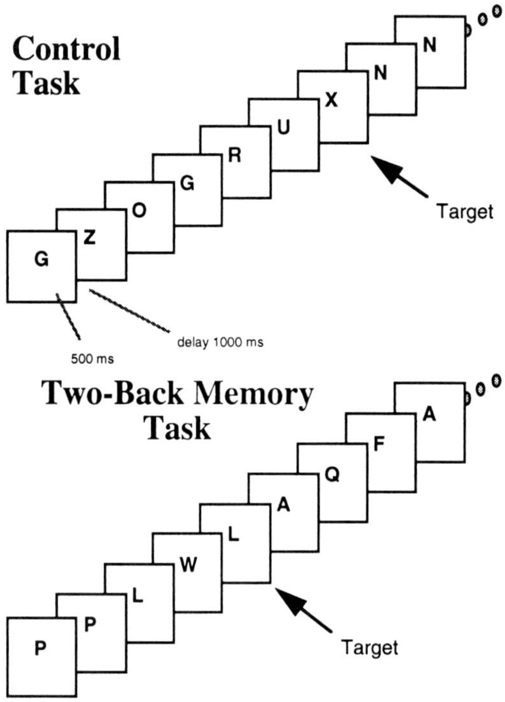

Figure 7. Illustration of the working memory and control tasks.

These results support the hypothesis that the frontal cortex is involved in the maintenance of representation needed to select task-appropriate action.

Imagery

Visual Imagery. When we construct mental images, we experience a sensation in the absence of an external stimulus. An intriguing question of cognitive psychology has been whether imagining an event involves the same processes as perceiving an event. In other words, when we imagine a stimulus, do we activate the same sensory systems as the actual stimulus itself activates? Although mental images can be auditory or tactile or can involve some other modality, cognitive psychologists have typically confined their studies to visual imagery (e.g., Cooper & Shepard, 1973). In a unique study by LeBihan and colleagues at the National Institutes of Health (LeBihan, Turner, Jezzard, Cuenod, & Zeffiro, 1993), functional activation maps of visual cortex were obtained with gradient echo EPI during mental representation of an imagined visual stimulus. Coronal slices intersecting the calcarine fissures (2.5 mm resolution, 10 mm slice thickness, 25 ms acquisition time) were acquired during periods of photic stimulation (flashing red light–emitting diodes produced by Grass stimulation goggles), interleaved with three periods of total darkness. In place of one of the photic stimulation episodes, the subject was instructed to imagine the visual stimulus. The order of conditions was stimulus on, stimulus off, imagine stimulus, stimulus off, stimulus on, stimulus off. During photic stimulation, each of the seven subjects showed a significant increase (2.8%) in signal relative to stimulus baseline in the primary visual cortex along the calcarine fissure. Significant increases of 1.5% were observed in the same regions for five of the seven subjects during the imagined stimulus relative to baseline. It is unclear whether these effects can be attributed to processes involved in imagery or to processes involved in short-term memory of the visual stimulus, since the stimulus was present less than 30 seconds prior to imagining it.

Developmental Studies of Cognition

Working Memory. Perhaps one of the most important aspects of fMRI is its utility in studying human brain development *in vivo*. We have recently completed two studies with pediatric populations. The first study (Casey, Cohen, *et al.*, 1995) examined activation in the prefrontal cortex in children during performance of a nonspatial working memory task. The task was identical to our paradigm described earlier in this chapter, requiring the subject to respond whenever the current letter was the same as two letters back (e.g., A-B-A), but not repeats (e.g., A-A). Again, the control task was the detection of a single letter.

We acquired gradient echo, echo planar images (TE = 40 ms, TR = 6000 ms) using a General Electric 1.5 T Signa scanner in eight 5-mm coronal slice locations covering the entire prefrontal cortex. Forty repetitions of each slice location per condition were collected. Analysis of these echo planar images revealed reliable bilateral activation in the middle and inferior frontal gyri and some activation of the orbital frontal cortex and anterior cingulate (see Figure 8).

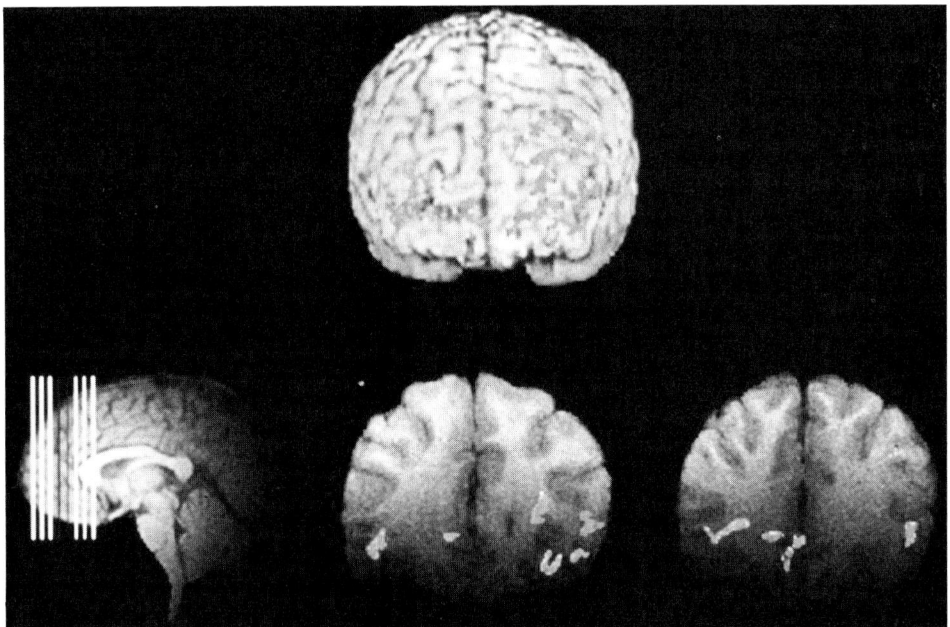

Figure 8. Overlay of prefrontal cortex activation on a surface rendering of an 11-year-old subject's brain during performance of a working memory task (top). Coronal view of prefrontal cortex activation for the same subject in slice locations between 40 and 50 mm anterior to the anterior commisure (bottom). For a color representation of this figure, see the color insert facing page 326.

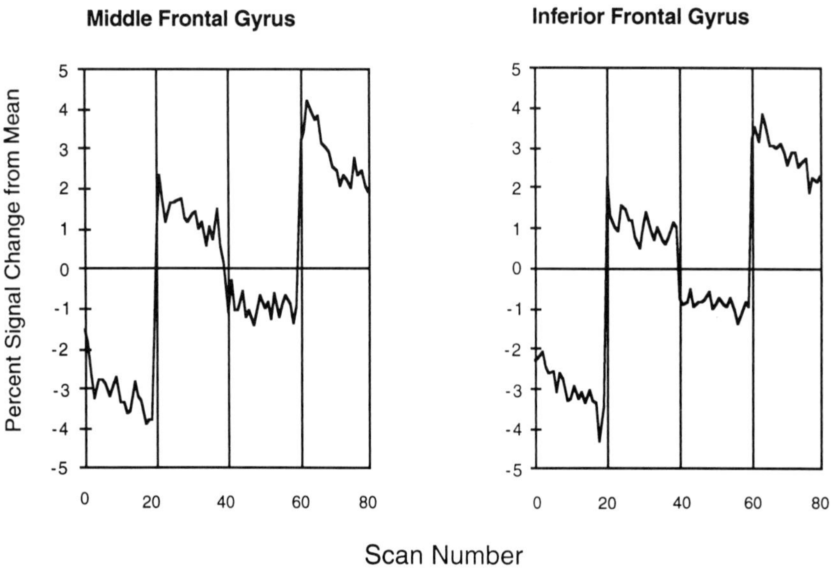

Figure 9. Percent change in MR signal intensity across time for the inferior and middle frontal gyri for all subjects across alternating blocks of control and activation conditions, respectively, with 20 images per block.

Figure 9 depicts the percent difference in signal across scans and averaged across subjects for the middle and inferior frontal gyrus. Both these regions were reliably activated across subjects. Although all experiments were run in an ABBA paradigm, the results have been plotted by alternating condition (ABAB). As demonstrated, in the middle and inferior frontal gyrus across five prepubertal children the change in percent difference corresponds nicely with the experimental manipulation. These results corroborate those of our earlier work using the identical task with adults and a spiral scanning sequence (Cohen *et al.*, 1994).

RESPONSE INHIBITION. In a second study of prefrontal activation in children, we used a version of the classic go-no-go task (Casey, Trainor, *et al.*, 1995). The task requires the subject to respond to any letter by an *X* with 75% of the trials being targets or non *X*s (refer to Figure 10). During the first minute of scanning the trials were target trials (20 of 20 trials), and the next minute of trials consisted of 50% nontarget trials (10 of 20), keeping the probability of a target at 75%. A second control task with 100% targets (10 of 10) and longer interstimulus intervals to control for the fewer number of responses in the 50% nontarget condition was also included. We hypothesized that the go-no-go task would activate target brain regions of the prefrontal and limbic circuits involved in inhibitory processes associated with response execution. Specifically, we expected activation in the medial orbital frontal region and anterior cingulate gyrus during blocks of trials requiring the subject to inhibit a response. This hypothesis is based on animal (Mishkin, 1964) and human (Malloy, Bihrlr, & Duffy, 1993) lesion studies and O-15 PET studies (Bench *et al.*, 1993; Pardo *et al.*, 1990; Taylor *et al.*, 1994) relating the medial orbital frontal and anterior cingulate cortex to response inhibition.

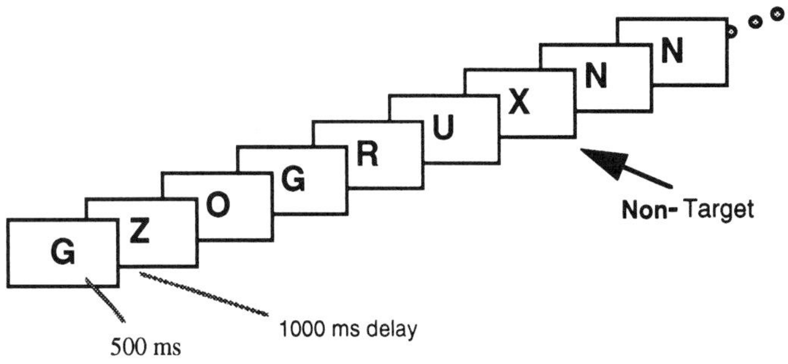

Figure 10. Illustration of the response inhibition paradigm.

We used a combination of a Bonferroni correction ($p < .05/4096$ or .00001) for regions activating less than 50 mm^3 (1 voxel) and a cluster-size threshold ($p < .001$) for larger regions (i.e., two or more contiguous pixels) to detect significant activation while guarding against false positives. The preliminary results from 18 subjects, between the ages of 7 and 24 years (nine adults and nine children), revealed reliable activation in the middle and inferior frontal, anterior cingulate, and orbital frontal gyri for adults and children (see Table 1). The volume (mm^3) of activation was significantly correlated with age ($p <.0009$) and was statistically different between groups ($p < .001$) with larger volumes of activation observed in children (490 mm^3) relative to adults (182 mm^3) (see Figure 11). On the other hand, the magnitude (percent change in MR signal) of activation was correlated with behavioral performance ($p < .009$), whereby poorer performance (i.e., those subjects having the greatest difficulty on the task) correlated with greater MR signal change. Correlational analyses by region of interest revealed that the

TABLE 1. STATISTICS FOR AREAS OF ACTIVATION FOR REGIONS OF INTEREST[a]

Regions of interest	Number of subjects	Brodmann's areas	Mean volume(mm^3)	Mean percent difference	Mean F
Children					
Anterior cingulate	5	24,32	580	1.86	18.73
Inferior frontal	8	45,10,47	533	1.71	18.76
Middle frontal	7	46,10,9,8	509	1.44	19.68
Orbital frontal	6	11	434	1.95	21.86
Superior frontal	3	9,8,6	326	1.72	19.76
Adults					
Anterior cingulate	5	24,32	187	1.60	26.60
Inferior frontal	7	45,47,10	200	1.41	18.62
Middle frontal	6	46,10,9,8	205	1.53	21.35
Orbital frontal	6	11	153	1.49	20.88
Superior frontal	3	9,8,6	122	1.76	21.44

[a]This table reports statistics for areas of activation in the anterior cingulate, inferior frontal gyrus, middle frontal gyrus, orbital gyrus, and superior frontal gyrus. For each region of interest the following data are reported: the number of subjects showing activation in that structure; the corresponding Brodmann's areas for each structure; the mean volume of activation, computed first within subjects, then calculated across subjects; the percentage difference in signal intensity averaged across subjects; mean F values for activated area averaged across subjects. Only eight of nine children and eight of nine adults showed statistically significant activation in the prefrontal cortex. Averages are across only those subjects showing activation in the corresponding structure.

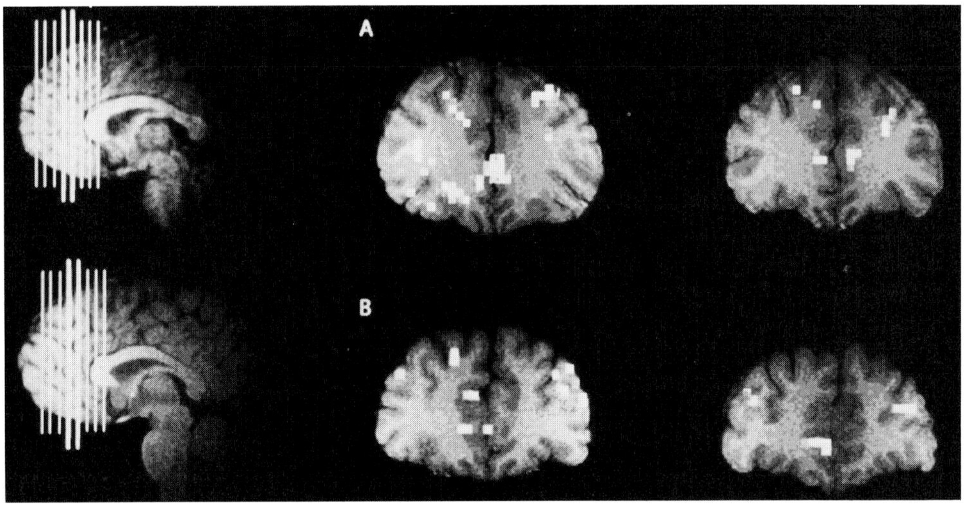

Figure 11. (A) Overlay of activation on T1-weighted coronal images located approximately 40 and 45 mm anterior to the anterior commissure for a right-handed 11-year-old male. (B) Corresponding overlays for right-handed 25-year-old male. The left midsagittal images depict the exact slice locations.

relation between the magnitude of the activation (percent change in MR signal) and false alarm rate was significant for the inferior frontal gyri bilaterally ($p <$ .05) and right anterior cingulate ($p < .001$).

Because the average number of false alarms made by children during performance of the go-no-go task was significantly higher ($p < .0001$) than for adults (20 and 6, respectively), we performed a partial correlational analysis to determine whether or not the previously reported correlations and group differences could be explained by behavioral performance alone. The results revealed significant correlations between the volume of activation and age even after controlling for behavioral performance ($p < .0001$). Likewise, there was a significant correlation between number of false alarms and percent change in MR signal even after controlling for age. These preliminary results suggest that while developmental differences in activation are represented as more diffuse patterns of activity in children relative to more focal activation in adults, individual differences in task performance may be represented in the magnitude of the signal. This latter finding is consistent with our previously published developmental fMRI findings (Casey, Cohen, *et al.*, 1995) of greater percent change in signal in children relative to previously published adult data (Cohen *et al.*, 1994) on an identical working memory task that was more demanding for the children. These results are also consistent with recent work by Braver *et al.* (1995) demonstrating monotonic increases in MR signal as a function of increasing memory load.

To determine whether our developmental differences in volume of activation may be associated with greater movement in our pediatric sample relative to our adult sample, we examined the in-plane motion for each subject. Based on motion correction logfiles of our image registration software (Wood's automated image registration) on our sample of 18 subjects reported above, the in-plane movement in our pediatric sample was almost identical (slightly less, actually) to

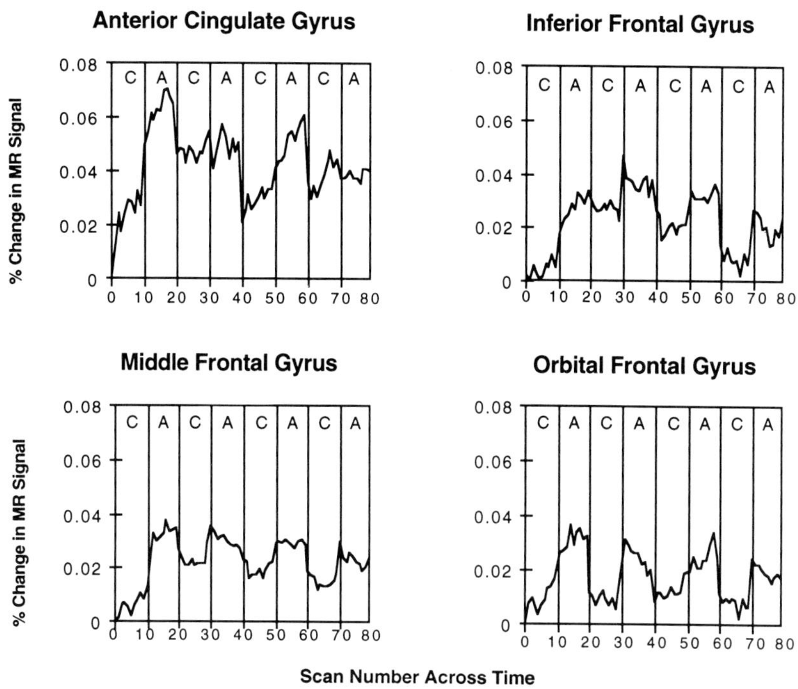

Figure 12. Percent change in MR signal intensity across time for the anterior cingulate, inferior and middle frontal, and orbital gyri for all subjects across alternating blocks of control (C) and activation (A) conditions with 10 images per block.

our adult sample, with mean movement in the *x*-axis of 0.045 mm, in the *y*-axis of 0.41 mm, and 0.05 degrees in rotation.

Finally, to demonstrate the reliability of our data, we plotted time series analyses for activated regions across subjects for the anterior cingulate, inferior and middle frontal, and orbital gyri. These data are averaged across both children and adults since there were no significant developmental differences in the percent change in MR signal (see Figure 12).

Discussion

The work presented in this chapter is just the beginning of a new era in cognitive neuroscience. With the continued advances in fMRI, understanding of uniquely human functions such as language and its development can proceed as never before. Pediatric neuroimaging studies which have been limited due to radiation exposure inherent in traditional techniques can now be done as well. However, just as with PET during its initial use and still today, the quantification of analyses and standard for localizing regions of interest are undergoing development and advancement.

As exciting and important as this new technique is, it also has limitations. We now consider some of the important issues concerning fMRI, as it relates to other methods that are currently available.

Safety and Spatial Resolution

The two primary advantages of fMRI are its safety and its high spatial resolution. As we discussed above, the fact that the procedure is noninvasive and nontoxic means that we can use this technique with pediatric populations and for longitudinal studies. Furthermore, multiple scans can be obtained from a single subject within a single session making it possible and practical to conduct within-subject averaging of images, which is extremely valuable given the degree of between-subject variability that has been observed in cortical anatomy. Most PET and SPECT studies require that images be averaged across subjects in order to obtain significant effects. This is because only a limited number of scans can be performed on a single subject (due to restrictions in the maximum allowable exposure to the radioactive tracer agents that are involved). As we have seen, however, anatomy can vary dramatically between subjects, significantly reducing both the resolution and accuracy of the results that are obtained when data are averaged across subjects. Because there are no exposure restrictions associated with fMRI, image averaging can be performed within a single subject, and areas of activity can then be localized in structural images obtained from the same subject.

Nevertheless, the anatomic diversity observed across subjects poses a serious challenge to the functional neuroimaging enterprise as a whole, especially as it begins to take on higher cognition. The problem is that as we begin to focus on such questions, the relevant anatomy has not been prespecified (as it has for primary sensory and motor function). Furthermore, it seems at least plausible to assume that, as we move into more cognitive domains, the mapping of function onto structure will become increasingly complex and unique to each subject. This is important also in determining how we will be referencing these regions of activation. We have chosen to reference the location of activation to regional landmarks (e.g., distance from the occipital pole or from the calcarine fissure in the case of visual stimulation studies) and in Talairach coordinates (Talairach & Tournoux, 1988) or at least in reference to the AC-PC line which is easily identified with high-resolution T1-weighted MR images. However, with the variability observed from subject to subject in gyral formations, normalization of activation across subjects (e.g., Talairach coordinates; Talairach & Tournoux, 1988) may result in a loss of spatial resolution.

Physiological and Anatomic Limitations

Ultimately, fMRI is limited by the physiological processes it is used to measure. Current methods appear to be detecting changes in blood oxygenation that result from the brain's local regulation of its blood supply in response to local neural activity. Animal studies suggest that this response can begin as rapidly as 500 ms following the onset of neural firing. However, even using high-speed MRI equipment (with temporal resolutions of 250 ms; see below) researchers have not seen changes in the MR signal until about 2–4 seconds after stimulation. As described earlier, these changes do not appear to peak until 4–6 seconds (Blamire *et al.*, 1992; Bandettini *et al.*, 1992; DeYoe *et al.*, 1992). It may be possible to improve on this, with gated acquisition of images (see below) and signal averaging, however such methods have yet to be explored.

Another potential limitation is that some brain areas may be more efficient in regulating their blood supply than the primary sensory and motor areas that have been studied most intensively so far. This is suggested by the growing PET literature. It has been much more difficult to generate detectable levels of activation in association cortex than it has in primary sensory or motor cortex. The use of magnetic contrast agents, which provide additional sensitivity, may be necessary for studying subtle effects in these areas. In any event, both PET and fMRI will benefit by efforts to better our understanding of the basic physiological processes that regulate cerebral blood flow and its relationship to blood oxygenation and metabolism.

Finally, there are anatomic constraints on the use of fMRI. Some brain areas lend themselves better to MR imaging than others. In particular, it is difficult to obtain good images from areas that lie near large cavities such as the frontal sinuses. The differences in magnetic susceptibility that occur along the boundaries of these areas produce a distortion of the ambient magnetic field, which degrades the MR signal. It is possible to compensate for these distortions by introducing appropriate additional fields, referred to as shimming. However, this process is time-consuming and not always fully successful.

Temporal Resolution

Functional MRI provides temporal resolution that is better than PET or SPECT but substantially less than that of event-related potentials (ERP) or magnetoencephalography (MEG). At present, specially equipped MR scanners can acquire images as rapidly as four per second (100 ms acquisition time, 150 ms between images). However, temporal resolution is also constrained by the time course of the physiological processes that are generating the signal. Current methods rely on hemodynamic responses which appear to take place over seconds rather than milliseconds (see below). This means that, at present, behavioral conditions must produce a steady state that lasts several seconds. This is an improvement on PET and SPECT, which require that conditions last a minute or more. However, it places current fMRI techniques out of the range of being able to measure many cognitive processes, which can occur in 100 ms or less. At present, ERP and MEG are the only noninvasive neuroimaging techniques that offer this level of temporal resolution. In the future, it is likely that these techniques will be used to complement one another.

Technical Issues

There are still technical problems associated with fMRI. As we noted earlier, the high resolution of the MR image makes it particularly susceptible to motion artifact. This can manifest in a variety of ways, the most notable of which is the appearance of gray matter-white matter boundaries in difference images which can compromise the fidelity and significance of activation-related changes. These arise as structures of different magnetic susceptibility traverse the field of view of a given set of pixels. If this happens in a gradual or regular way across the conditions of the experiment, it can represent a confounding source of difference in the MR signal, which will show up in the difference image (Hajnal *et al.*, 1994). A number of approaches have been taken to physically immobilize the head. In the long term, however, the most promising approach will be the development of computer algorithms that will permit post hoc correction of images.

A second source of artifact is transient scanner effects. On occasion, the MR scanner can produce an aberrant image. Since activity-related changes are small (< 5%), a single bad image out of 10 or 20, with large discrepant values, can significantly bias analysis results. It is important, therefore, to examine all images and eliminate ones of poor quality.

Blood flow can also produce artifact. In areas where a large volume of blood is moving during the scan, this will produce a spatial frequency echo (see Huk, Gademann, & Friedmann, 1986). The T2*-weighted images may be particularly sensitive to venous blood flow. Another type of artifact has been observed by a number of groups working with standard MRI equipment which may also be attributed to blood flow and or respiration. This appears as large, low spatial frequency variations (bands) in difference images. The source of this artifact has not yet been identified. It appears to be related to transient shifts in the strength of the MR signal during image acquisition. One possible source of this effect may be diffuse changes in cerebral blood flow related to the cardiac cycle. If this or some other transient change in signal intensity is the cause, it should be possible to eliminate this effect by appropriately normalizing the data for each acquisition period. This approach is currently being explored.

Another type of artifact is associated with inhomogeneity in the magnetic field of certain brain regions. For example, the inferior frontal cortex is often a trouble area due to the sinuses since different localized fields result when going from brain tissue to air. The magnetic field within such areas can be made more homogeneous by shimming before scanning, that is, correcting the detected field inhomogeneity using a set of magnetic shim coils.

These sources of artifact require that images be carefully examined and screened before *t*-test images are considered to reflect functional brain activity. These forms of artifact produce patterns in *t*-test images and time series analyses that are distinct from patterns produced by functional activity. Of course, *t*-test images should always be compared with structural images, to verify that putative areas of activation lie within gray matter and that time series data reflect changes in the MR signal that correspond to the experimental manipulation.

Finally, in fMRI studies, as in any other type of study, replication is the best measure of reliability. Replications are best performed on the same subject in the same session. The location of a given anatomic area can vary between subjects by as much as a centimeter or more, even after correction for overall brain size. Two of the current authors in collaboration with statisticians (Drs. William Eddy and Christopher Genovese) at Carnegie Mellon University have developed a statistical analysis for determining the reliability of fMRI data. The method takes the information from reliability maps about the number of times a voxel is active across at least three tests and uses maximum likelihood to estimate two probabilities: the chance that a truly active voxel is classified as active and the chance that a truly inactive voxel is classified as active. These are the formal equivalents, in signal detection terms, of the hit and false alarm rates and provide a well-understood and quantitative measure of the reliability of our signal based on multiple (test-retest) determinations (Cohen *et al.*, 1995). Figure 13 is an illustration of reliability maps of test-retest data using this methodology. These data indicate that there is variability in the spatial configuration of areas of activity that are detected across different test sessions for the same subject.

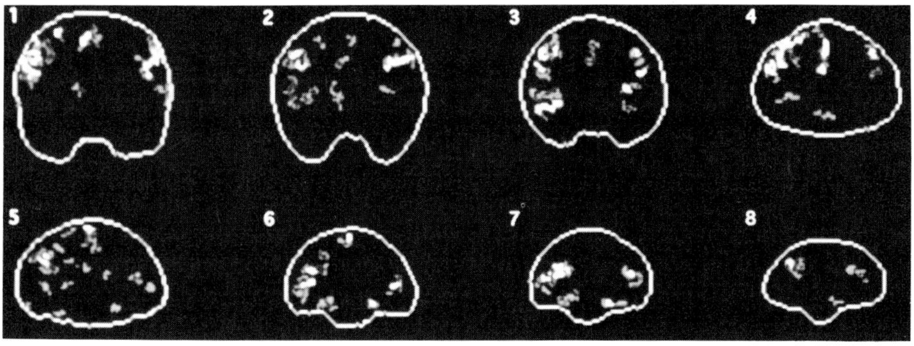

Figure 13. Reliability maps of test–retest data. The number of test sessions in which each voxel was found to be significantly active is indicated. (Adapted from Cohen *et al.*, 1995.) For a color representation of this figure, see the color insert facing page 327.

However, they also demonstrate that the overall distribution of areas is consistent across tests—most areas detected have a reliable center, with a penumbra of lesser reliability. This pattern suggests that fMRI is able to detect areas of reliable activation in cognitive tasks, although there is some degree of variability in the spatial configuration of each area from test to test.

Current and Future Directions

High-Field Magnets. The strength of the MR signal is directly related to the strength of the ambient field used to align the protons in their resting state. Most scanners that are available in clinical settings are 1.5 T. However, commercial scanners are now available at 3 T, and there are a number of 4 T research scanners that are in use. The benefit of increased field strength, in terms of signal-to-noise ratio appears to fall somewhere between a linear and an exponential improvement (Turner *et al.*, 1992; Ogawa *et al.*, 1992). Clearly, improvements in SNR will provide improved spatial and, possibly, temporal resolution. Spectroscopy (for example, the ability to image phosphorus) will also benefit by increased magnetic field strength. However, as field strength increases, so too does the energy of the RF pulse needed to generate a signal. At present, this energy is well below levels that are considered to be safe for human subjects. However, the safety limit, along with the technical problems associated with building larger magnets, are likely to significantly constrain developments in this direction.

High-Speed Imaging. While larger magnets are not likely to become widely available in the immediate future, faster imaging techniques are. There have been a number of recent developments in MR technology aimed at reducing image acquisition time. These include echo planar imaging (e.g., Bandettini *et al.*, 1992; Kwong *et al.*, 1992), spiral scan pulse sequences (Noll *et al.*, 1995), and others. Some of these require additional apparatus, and all rely on special software. The technical details of these developments are beyond the scope of this paper. What they offer, however, is much more rapid imaging times: images can be acquired as quickly as four per second rather than once every 7 seconds. One or more of these techniques are likely to become more widely available within the next few years. These will open up a number of ways for improving functional studies: 1) A given number of images can be acquired over a much shorter period of time, reducing subject fatigue, minimizing motion artifact, etc.; 2) A greater number of images can be acquired for the same amount of time, permitting better averaging, more detailed time course data, etc.; and 3) The duration of each behavioral condition can be reduced, more closely approximating mixed designs. As noted above, the temporal resolution of MR is currently limited by the time course of local hemodynamic responses in the brain. However, as new measurement techniques are developed that more closely measure neural activity the actual temporal resolution of MR may improve.

Gated Acquisition. As the temporal resolution of MRI improves, it will become increasingly valuable to synchronize image acquisition with stimulus presentation (or response generation). This will permit image averaging in the way it is currently done with ERP and MEG. This approach might allow detection of physiological responses closer to their onset than have been observed to date, possibly as early as 500 ms after behavioral stimulation. The primary constraint on this approach is that MR images must be acquired at regular intervals. The time between images determines the degree to which protons have returned to their resting state, which in turn determines the magnitude of the signal recovered in subsequent images. Uneven intervals will produce confounding variations in signal strength.

Use of Multiple Imaging Modalities. The greatest promise for fMRI, as for all of the other approaches to neuroimaging, is its use in conjunction with complementary techniques. Each method has its own strengths and potentials, as well as its areas of weakness. Functional MRI offers tremendous spatial resolution. However, it is does not have the temporal resolution of ERP or MEG. Conversely, it has proven very difficult to localize activity identified using ERP and MEG. By applying these techniques together, however, it may be possible to identify an area of activity with fMRI and then trace the details of is time course using ERP and/or MEG. In turn, PET and SPECT can provide neurochemical specificity not currently possible with other techniques. These can be used to trace the underlying neurotransmitter systems involved in functions whose anatomic location and time course have been identified using a combination of MRI, ERP, and MEG. The next major phase of growth in neuroimaging research, no doubt, will be the development of techniques for coordinating the use of multiple modalities. This has important conceptual significance.

Summary and Conclusion

Functional MRI is a promising new addition to the array of functional neuroimaging techniques that are becoming increasingly available to cognitive neuroscientists and psychologists in general. It provides a noninvasive means for imaging hemodynamic changes that appear to be closely related to functional brain activity. A major advantage of fMRI is that it does not require the injection of contrast agents and does not pose any currently known risk to the subject. As such, it permits within-subject designs, which can be used to dramatically im-

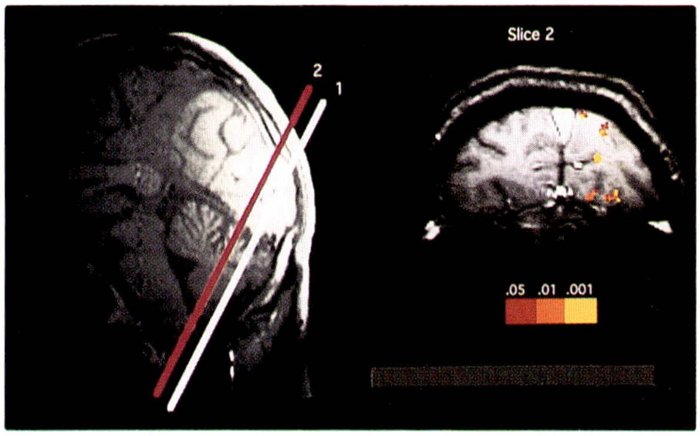

12.2. Illustration of slice locations for the activation images (left). Overlay of significant activation on T1-weighted oblique image for slice location 2 (right). (Adapted from Schneider *et al.*, 1994.)

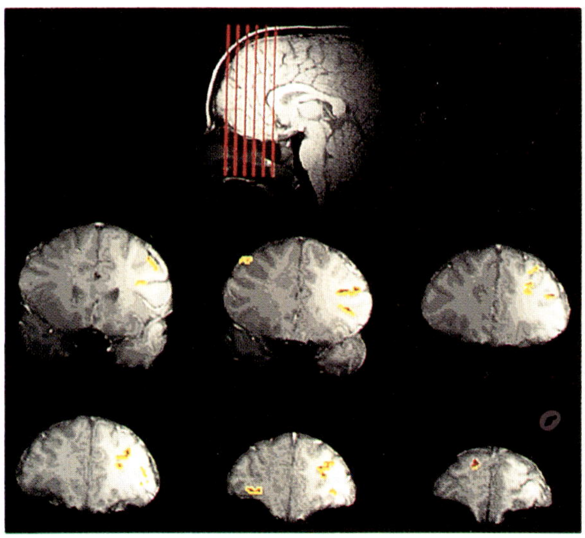

12.6. Slice locations and overlay of significant activation for each of six slice locations during the performance of a working memory task. (Adapted from Cohen *et al.*, 1994.)

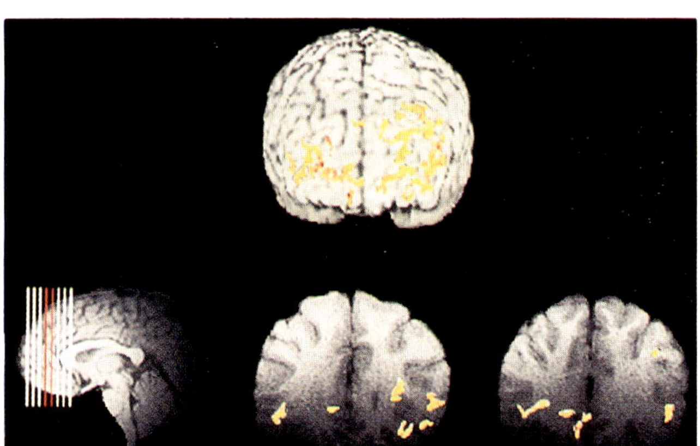

12.8. Overlay of prefrontal cortex activation on a surface rendering of an 11-year-old subject's brain during performance of a working memory task (top). Coronal view of prefrontal cortex activation for the same subject in slice locations between 40 and 50 mm anterior to the anterior commisure (bottom).

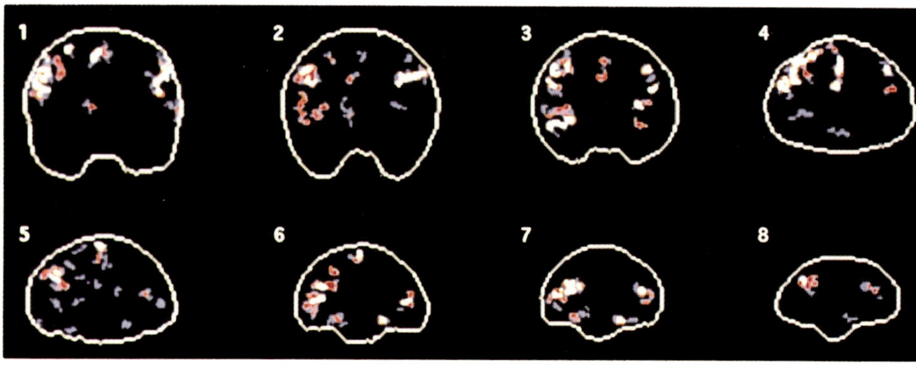

12.13. Reliability maps of test–retest data. Colors indicate the number of test sessions in which each voxel was found to be significantly active (white, all three tests; red, two tests; blue, one test). (Adapted from Cohen *et al.*, 1995.)

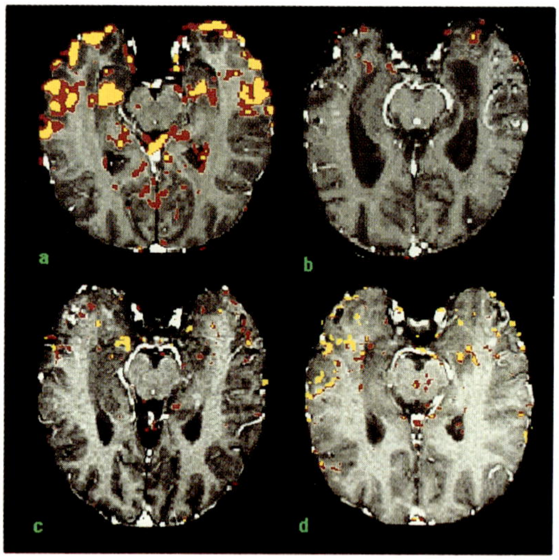

13.3. Auditory-verbal memory activation images. (a) Healthy young adult, (b) patient with early Alzheimer's disease, (c) healthy young adult during cyclical auditory-verbal learning task, (d) patient with schizophrenia during cyclical visual-verbal encoding task.

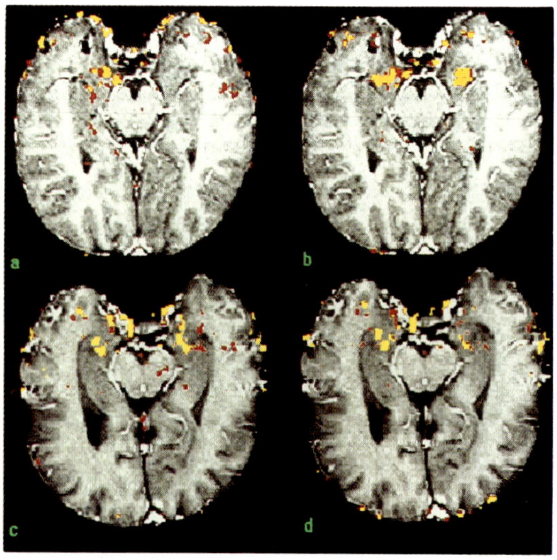

13.5. Activation images for cyclical memory for words and faces. (a) Healthy young adult encoding words, (b) healthy young adult encoding faces, (c) patient with right temporal lobe seizure focus encoding words, (d) patient with right temporal lobe seizure focus encoding faces.

prove spatial resolution and test the replicability of findings. In comparison to other neuroimaging techniques, it offers the best available spatial resolution and temporal resolution that is superior to PET and SPECT but is inferior to ERP and MEG.

Acknowledgments

The authors would like to thank Drs. William Eddy, Christopher Genovese, Peter Jezzard, Manny Kanal, Mark Mintun, Bruce Rosen, and Robert Turner for their encouragement and guidance of work reported in this chapter. Reprint requests should be forwarded to Dr. Casey, Western Psychiatric Institute and Clinic, Room E-519, University of Pittsburgh Medical Center, 3811 O'Hara Street, Pittsburgh, PA 15213.

References

Appollonio, I., Rueckert, L., Partiot, A., Grafman, J., Jezzard, P., Litvan, I., Johnson, R., & Turner, R. (1993). Magnetic resonance functional neuroimaging during covert letter fluency and automatic speech. *Proceedings of the Society for Neuroscience Meeting, USA, 19,* 843.

Bandettini, P. A., Wong, E. C., Hinks, R. S., Tikofsky, R. S., & Hyde, J. S. (1992). Time course of EPI of human brain function during task activation. *Magnetic Resonance in Medicine, 25,* 390–397.

Belliveau, J. W., Kennedy, D. N., McKinstry, R. C., Buchbinder, B. R., Weisskoff, R. M., Cohen, M. S., Vevea, J. M., Brady, T. J., & Rosen, B. R. (1991). Functional mapping of the human visual cortex by magnetic resonance imaging. *Science, 254,* 716–719.

Bench, C. J., Frith, C. D., Grasby, P. M., Friston, K. J., Paulesu, E., Frackowiak, R. S. J., & Dolan, R. J. (1993). Investigations of the functional anatomy of attention using the Stroop test. *Neuropsychologia, 31,* 907–922.

Blamire, A. M., Ogawa, S., Ugurbil, K., Rothman, D., McCarthy, G., Ellerman, J., Hyder, F., Rattner, Z., & Schulman, R. G. (1992). Echo-planar imaging of the activated human visual cortex shows a time delay between stimulus and activation. *Proceedings of the 11th Annual Scientific Meeting of the Society of Magnetic Resonance in Medicine, Germany, 3,* 823.

Braver, T. S., Cohen, J. D., Jonides, J., Smith, E. E., Awh, E., Schumacher, E. L., Lauber, E., & Noll, D. C. (1995, November). *A parametric study of frontal cortex involvement in human working memory using functional MRI.* Paper presented at the meeting of the Society for Neuroscience, San Diego, CA.

Carter, C. S., Mintun, M. A., & Cohen, J. D., (1995). Interference and facilitation effects during selective attention: An $H_2^{15}O$ study of stroop task performance. *Neuroimage, 2,* 264–272.

Casey, B. J., Cohen, J. D., Jezzard, P., Turner, R., Noll, D. C., Tranior, R. J., Giedd, J., Kaysen, D., Hertz-Pannier, L., & Rapoport, J. L. (1995). Activation of prefrontal cortex in children during a nonspatial working memory task with functional MRI. *Neuroimage, 2,* 221–229.

Casey, B. J., Cohen, J. D., Noll, D. C., Forman, S., & Rapoport, J. L. (1993). Activation of the anterior cingulate during the Stroop conflict paradigm using functional MRI. *Proceedings of the Society for Neuroscience Meeting, USA, 19,* 1285.

Casey, B. J., Trainor, R., Orendi, J., Giedd, J., Castellanos, X., Noll, D., Cohen, J. D., Haxby, J., Jezzard, P., & Rapoport, J. L. (1995, November). Activation of anterior cingulate in children during a response inhibition task with functional MRI. *Proceedings of the Society for Neuroscience Meeting, USA, 21*(3), 1924.

Cohen, J. D., Forman, S. D., Braver, T. S., Casey, B. J., Servan-Schreiber, D., & Noll, D. C. (1994). Activation of prefrontal cortex in a non-spatial working memory task with functional MRI. *Human Brain Mapping, 1,* 293–304.

Cohen, J. D., Noll, D. C., & Nystrom, L. (1995). Qualitative and quantitative assessment of test-retest reliability of functional MRI data. *Proceedings of the Society for Magnetic Resonance and the European Society for Magnetic Resonance in Medicine and Biology, France, 1,* 234.

Cooper, L. A., & Shepard, R. N. (1973). Chronometric studies of the rotation of mental images. In W. G. Chase (Ed.), *Visual information processing* (pp. 75–176). New York: Academic Press.

Detre, J. A., Leigh, J. S., Williams, D. S., & Koretsky, A. P. (1992). Perfusion imaging. *Magnetic Resonance in Medicine, 23,* 37–45.
DeYoe, E. A., Neitz, J., Bandettini, P. A., Wong, E. C., & Hyde, J. S. (1992). Time course of event-related MR signal enhancement in visual and motor cortex. *Proceedings of the 11th Annual Scientific Meeting of the Society of Magnetic Resonance in Medicine, Germany, 3,* 1824.
Donders, F. C. (1969). On the speed of mental processes. In W. G. Koster (Ed. and Trans.), *Attention and performance II.* Amsterdam: North-Holland. (Original work published 1868)
Duvernoy, H. M. (1991). *The human brain: Surface, three dimensional sectional anatomy and MRI.* New York: Springer-Verlag Wien.
Dyun, J. F., Moonen, C. T. W., DeBoer, R. W., Iperen, G. W., & Luyten, P. R. (1993). In-flow versus deoxyhemoglobin effects in BOLD fMRI using gradient echo at 1.5 T. *Proceedings of the 12th Annual Scientific Meeting of the Society for Magnetic Resonance in Medicine, USA, 1,* 168.
Forman, S. D., Cohen, J. D., Fitzgerald, M., Eddy, W. F., Mintun, M. A., & Noll, D. C. (1995). Improved assessment of significant activation in functional magnetic resonance imaging (fMRI): Use of a cluster-size threshold. *Magnetic Resonance in Medicine, 33,* 636–647.
Fox, P. T., & Raichle, M. E. (1985). Stimulus rate determines regional brain blood flow in striate cortex. *Annuals of Neurology, 17,* 303–305.
Fox, P. T., Raichle, M. E., Mintun, M. A., & Dence, C. (1988). Nonoxidative glucose consumption during focal physiologic neural activity. *Science, 241,* 1445–1448.
Frahm, J., Bruhn, H., Merboldt, K. D., Hanicke, W., & Math, D. (1992). Dynamic MR imaging of human brain oxygenation during rest and photic stimulation. *Journal of Magnetic Resonance Imaging, 2,* 501–505.
Frahm, J., Merboldt, K. D., & Hanicke, W. (1993). Functional MRI of human brain activation at high spatial resolution. *Magnetic Resonance in Medicine, 29*(1), 139–144.
Friston, K. J., Frith, C. D., Liddle, P. F., Dolan, R. J., Lammertsma, A. A., & Frackowiak, R. S. (1990). The relationship between global and local changes in PET scans. *Journal of Cerebral Blood Flow and Metabolism, 10* 458–466.
George, M. S., Ketter, T. A., Parekh, P. I., Rosinsky, N., Ring, H., Casey, B. J., Trimble, M. R., Horwitz, B., & Herscovitch, P. (1994). Regional brain activity when selecting a response despite interference. *Human Brain Mapping, 1,* 194–209.
Goldman-Rakic, P. S. (1987). Circuitry of primate prefrontal cortex and regulation of behavior by representational memory. *Handbook of Physiology: The Nervous System, 5,* 373–417.
Gore, J. C., McCarthy, G., Constable, R. T., Anderson, A. W., Kennan, R. P., Rattner, Z., & Zhong, J. (1992). Imaging regional brain activation at 1.5 T using conventional imaging techniques. *Proceedings of the 11th Annual Scientific Meeting of the Society of Magnetic Resonance in Medicine, Germany, 3,* 1826.
Hajnal, J. V., Myers, R., Oatridge, A., Schwieso, J. E., Young, I. R., & Bydder, G. M. (1994). Artifacts due to stimulus correlated motion in functional imaging of the brain. *Magnetic Resonance in Medicine, 31,* 283–291.
Hinke, R. M., Hu, X., Stillman, A. E., Kim, S. G., & Ugurbil, K. (1993). MR functional imaging of silent speech. *Journal of Magnetic Resonance Imaging, 3,* 184.
Huk, W. J., Gademann, G., & Friedmann, G. (1986). *Magnetic resonance imaging of central nervous system diseases.* New York, Springer-Verlag.
Jonides, J., Smith, E. E., Koeppe, R. A., Awh, E., Minoshima, S., & Mintun, M. A. (1993). Spatial working memory in humans as revealed by PET. *Nature, 363,* 623–625.
Kwong, K. K., Belliveau, J. W., Chesler, D. A., Goldberg, I. E., Weisskoff, R. M., Poncelet, B. P., Kennedy, D. N., Hoppel, B. E., Cohen, M. S., Turner, R., Cheng, H.-M., Brady, T. J., & Rosen, B. R. (1992). Dynamic magnetic resonance imaging of human brain activity during primary sensory stimulation. *Proceedings of the National Academy of the Sciences, USA, 89,* 5675–5679.
LeBihan, D., Turner, R., Jezzard, P., Cuenod, C. A., & Zeffiro, T. (1993). Activation of human visual cortex by mental representation of visual patterns. *Proceedings of the 12th Annual Meeting of the Society for Magnetic Resonance in Medicine, USA, 1,* 311.
Macleod, C. M. (1991). Half a century of research on the Stroop effect: An integrative review. *Psychological Bulletin, 109,* 163–203.
Malloy, P., Bihrlr, A., & Duffy, J. (1993). The orbitomedial frontal syndrome. *Archives of Clinical Neuropsychology, 8,* 185–201.
Menon, R. S., Ogawa, S., Tank, D. W., & Ugurbil, K. (1993). 4 Tesla gradient recalled echo characteristics of photic stimulation-induced signal changes in the human primary visual cortex. *Magnetic Resonance in Medicine, 30,* 380–386.

Mishkin, M. (1964). Perseveration of central sets after frontal lesions in monkeys. In J. M. Warren and K. Akert (Eds.), *The frontal granular cortex and behavior* (pp. 219–241). New York: McGraw Hill.

Noll, D. C., Cohen, J. D., Meyer, H. C., & Schneider, W. (1995). Spiral K-space MR imaging of cortical activation. *Journal of Magnetic Resonance Imaging, 5,* 49–56.

Ogawa, S., Lee, T. S., Nayak, A. S., & Glynn, P. (1990). Oxygenation-sensitive contrast in magnetic resonance image of rodent brain at high magnetic fields. *Magnetic Resonance in Medicine, 14,* 68–78.

Ogawa, S., Tank, D. W., Memon, R., Ellermann, J. M., Kim, S. G., Merkle, H., & Ugurbil, K. (1992). Intrinsic signal changes accompanying sensory stimulation: Functional brain mapping using MRI. *Proceedings of the National Academy of Sciences, USA, 89,* 5951–5955.

Pardo, J. V., Pardo, P. J., Janer, K. W., & Raichle, M. E. (1990). The anterior cingulate cortex mediates processing selection in the Stroop attentional conflict paradigm. *Proceedings of the National Academy of Sciences, USA, 87,* 256–259.

Petersen, S. E., Fox, P. T., Posner, M. I., Mintun, M., & Raichle, M. E. (1989). Positron emission tomographic studies of the processing of single words. *Journal of Cognitive Neuroscience, 1,* 153–170.

Petersen, S. E., Fox, P. T., Snyder, A. Z., & Raichle, M. E. (1990). Activation of extrastriate and frontal cortical areas by visual words and word-like stimuli. *Science, 249,* 1041–1044.

Raichle, M. E., & Mellinckrodt, E. (1988). In V. B. Mountcastle, F. Plum, & S. R. Geiger (Eds.), *Handbook of physiology: The nervous system.* Bethesda, MD: American Psychological Society.

Rao, S. M., Bandettini, P. A., Wong, E. C., Yetkin, F. Z., Hammeke, T. A., Mueller, W. M., Goldman, S., Morris, G. L., Antuono, P. G., Estkowski, L. D., Haughton, V. M., & Hyde, J. S. (1992). Gradient-echo EPI demonstrates bilateral superior temporal gyrus activation during passive word presentation. *Proceedings of the 11th Annual Meeting of the Society for Magnetic Resonance in Medicine, Germany, 3,* 1827.

Rosen, B. R., Belliveau, J. W., & Chien, D. (1989). Perfusion mapping by nuclear magnetic resonance. *Magnetic Resonance Quarterly, 5,* 263–281.

Rosen, B. R., Belliveau, J. W., Vivea, J. M., & Brady, T. J. (1990). Perfusion imaging with NMR contrast agents. *Magnetic Resonance in Medicine, 14,* 249–165.

Sanders, J. A., & Orrison, W. W. (1993). ANOVA tests for identification of FMRI activation. *Proceedings of the 12th Annual Meeting of the Society of Magnetic Resonance in Medicine, USA, 3,* 1376.

Schneider, W., Casey, B. J., & Noll, D. (1994). Functional MRI mapping of stimulus rate effects across visual processing stages. *Human Brain Mapping, 1,* 117–133.

Schneider, W., Noll, D. C., & Cohen, J. D. (1993). Functional topographic mapping of the cortical ribbon in human vision with conventional MRI scanners. *Nature, 365,* 150–153.

Sergent, J., Zuck, E., Levesque, M., & MacDonald, B. (1992). Positron emission tomography study of letter and object processing: Empirical findings and methodological considerations. *Cerebral Cortex, 1,* 68–80.

Snedecor, G. W., & Cochran, W. G. (1980). Uses of the correlational coefficient. In G. W. Snedecor & W. G. Cochran (Eds.), *Statistical methods* (7th ed.). Ames, IA: Iowa State University Press.

Squire, L. R., Ojemann, J. G., Miezin, F. M., Petersen, S. E., Videen, T. O., & Raichle, M. E. (1992). Activation of the hippocampus in normal humans: A functional anatomical study of memory. *Proceedings of the National Academy of Sciences, USA, 89,* 1837–1841.

Stroop, J. R. (1935). Studies of interference in serial verbal reactions. *Journal of Experimental Psychology, 18,* 643–662.

Talairach, J., & Tournoux, P. (1988). *Co-planar stereotaxic atlas of the human brain.* New York: Thieme.

Tatsuoka, M. M. (1988). *Multivariate analysis: Techniques for educational and psychological research* (2nd ed.). New York: Macmillan.

Taylor, S. F., Kornblum, S., Minoshima, S., Oliver, L. M., & Koeppe, R. A. (1994). Changes in medial cortical blood flow with a stimulus-response compatibility task. *Neuropsychologia, 32,* 249–255.

Thulborn, K. R., Waterton, J. C., Matthews, P. M., & Radda, G. K. (1982). Oxygenation dependence of the transverse relaxation time of water protons in whole blood at high field. *Biochimica Biophysica Acta, 714,* 265–270.

Turner, R., LeBihan, D., Moonen, C. T. W., Despres, D., & Frank, J. (1991). Echo-planar time course MRI of cat brain oxygenation changes. *Magnetic Resonance in Medicine, 22,* 159–166.

Turner, R., Jezzard, P., Wen, H., Kwong, K., LeBihan, D., & Balaban, R. (1992). Functional mapping of the human visual cortex at 4 tesla using deoxygenation contrast EPI. *Proceedings of the 11th Annual Scientific Meeting of the Society of Magnetic Resonance in Medicine, Germany, 1,* 304.

Weinberger, D. R., Berman, K. F., & Zec, R. F. (1986). Physiological dysfunction of dorsolateral prefrontal cortex in schizophrenia: I. Regional cerebral blood flow evidence. *Archives of General Psychiatry, 43,* 114–125.

Woods, R. P., Cherry, S. R., & Mazziotta, J. C. (1992). Rapid automated algorithm for aligning and reslicing PET images. *Journal of Computer Assisted Tomography, 16,* 620–633.

Zatorre, R. J., Evans, A. C., Meyer, E., & Gjedde, A. (1992). Lateralization of phonetic and pitch discrimination in speech processing. *Science, 256,* 846–849.

13

Functional Magnetic Resonance Imaging

Studies of Memory

ANDREW J. SAYKIN, HENRY J. RIORDAN, ROBERT B. BURR, LAURA A. FLASHMAN, ARTHUR C. MAERLENDER, KATHLEEN A. FLANNERY, AND JOHN B. WEAVER

INTRODUCTION

Functional magnetic resonance imaging (fMRI) is an exciting new technique that permits mapping of subtle changes in local blood oxygenation in the brain with excellent spatial and temporal resolution, thus improving on the resolution of other functional imaging modalities, including positron emission tomography (PET), rCBF, and SPECT. Functional MRI is also noninvasive and repeatable, and it does not require radioisotopes. Substantially more accurate mapping of brain structure to function is made possible at the individual subject level because there is intrinsic coregistration of anatomic and functional images with resolution on the order of millimeters.

ANDREW J. SAYKIN Departments of Psychiatry and Radiology, Dartmouth Medical School, Lebanon, New Hampshire 03756; and New Hampshire Hospital, Concord, New Hampshire 03301. HENRY J. RIORDAN AND KATHLEEN A. FLANNERY Department of Psychiatry, Dartmouth Medical School, Lebanon, New Hampshire 03756. ROBERT B. BURR Department of Neurosurgery, University of Utah Medical Center, Salt Lake City, Utah 84112. LAURA A. FLASHMAN AND ARTHUR C. MAERLENDER New Hampshire Hospital, Concord, New Hampshire, and Department of Psychiatry, Dartmouth Medical School, Lebanon, New Hampshire 03756. JOHN B. WEAVER Department of Radiology, Dartmouth Medical School, Lebanon, New Hampshire 03756.

Neuroimaging II: Clinical Applications, edited by Erin D. Bigler. Plenum Press, New York, 1996.

Functional MRI was initially applied to studies of sensory and motor stimulation (Belliveau *et al.*, 1991; Connelly *et al.*, 1993; Kim *et al.*, 1993; Kwong *et al.*, 1992; Ogawa *et al.*, 1992; Rao *et al.*, 1993), but over the last one to two years there has been a progressive increase in application to cognitive domains (e.g., Cohen & Bookheimer, 1994; Binder *et al.*, 1994), including auditory (Binder *et al.*, 1994) and visual processing (see Chapter 12, this volume) and language (Frahm, Bruhn, Merboldt, & Hanicke, 1992). There has been very little work to date on memory. Our group at Dartmouth Medical School has concentrated on applications of fMRI to better our understanding of the neural basis of normal and disordered human memory. Capitalizing on the high resolution and repeatability of activations, we are attempting to clarify brain regions and networks important for subtypes of memory and the differential breakdown of memory function across disorders. In this chapter, we briefly review relevant background concepts on brain activation, fMRI, and types of memory and present some preliminary fMRI studies of memory from our center.

Brain Activation, Functional MRI, and Memory

Mechanisms of Brain Activation

Brain activation involves short-term biochemical changes, primarily at the synapse, in response to an activating stimulus and can be distinguished from slow changes in overall brain states such as general arousal or the sleep-wake cycle. The reader is referred to Roland (1993) for a comprehensive, systems-level review of brain activation.

Neuronal Metabolism and rCBF. As early as 1890, Roy and Sherrington observed that cerebral blood flow (CBF) is regulated by brain activity. It is now established that under most conditions, both at rest and during activation, there is a coupling between neuronal activity, energy metabolism, and regional CBF and cerebral blood volume (rCBF and rCBV). The fundamental mechanism linking rCBF to neuronal activity remains unknown although several hypotheses have been advanced regarding regulation via chemical mechanisms at the precapillary level (Roland, 1993). Biochemical activation processes require oxygen, glucose, amino acids, fatty acids, and vitamins that enter the brain through the capillaries, which are impermeable to most other substances. Glucose is transported across the blood-brain barrier in proportion to local neuronal activity demands. The transport capacity increases as a function of capillary surface that dynamically shifts as needed through frequent opening and closing of capillaries.

Regional Differences during Memory Activation. There are regional differences in activation within the brain, with capacity for activation being proportional to the density of synapses. Regions that have the greatest capacity for activation require the greatest influx of nutrients, particularly glucose and oxygen, and are associated with the highest level of glucose metabolism (regional cerebral metabolic rate for glucose, rCMRGlc). When metabolic demand increases, rCBF and rCBV also increase. In his review based on tracer methods, Roland (1993) suggested a three-compartment model for regional differences in activation potential: (a) Gray matter with high synaptic density, such as primary cortical and association areas, has high baseline metabolism and high capacity to increase; (b) gray matter composed of cell bodies of neurons and astrocytes are intermediate in activity; and (c) white matter including axons and glial cells are lowest in resting activity and have low capacity to increase.

Memory is particularly suited to activation studies because the synapse is the critical site for most processes important for memory. New learning alters the efficiency of synaptic connections (Hebb, 1949; Kandel, Schwartz, & Jessell, 1991, for review). Long-term potentiation (LTP) is a sustained increase in synaptic efficiency following stimulation and has been studied extensively in the hippocampus, the structure most closely associated with long-term memory (Scoville & Milner 1957; Squire, 1987). Long-term depression (LTD), by contrast, is a sustained decrease in synaptic efficiency following stimulation. Long-term depression has been primarily studied in cerebellum and may be important for motor learning. Long-term memory is thought to be subserved by an as yet unknown mechanism of protein synthesis. Hypotheses regarding the molecular mechanism of LTM include (a) increases in number and density of synapses, (b) changes in the shapes of synapses, and (c) increases in the number of synaptic vesicles and consequently amount of neurotransmitters.

Temporal Factors. Thoughts, perceptions, and fine movements typically occur in a time frame of 50 ms to a few seconds. There is a convergence of data to suggest that measurable changes in brain activity occur very shortly after sensory or cognitive stimulation, that is, in the range 200 ms. The time course of the hemodynamic response to stimulation has not yet been fully characterized (Friston, Jezzard, & Turner, 1994; Roland, 1993), but optical imaging studies in the primate suggest that localized oxygen delivery occurs 200–400 ms after neuronal activation begins (Frostig, Lieke, Tso, & Grinvald, 1990). After 300–400 ms local blood volume increases, and of particular relevance for fMRI, at 1000 ms oxyhemoglobin is significantly increased.

Functional MRI

Methods of fMRI are discussed in detail elsewhere (Bandettini, Wong, Hinks, Tikofsky, & Hyde, 1992; Frahm *et al.*, 1992; Frahm, Merboldt, & Hancke, 1989; Binder *et al.*, 1994; see also Chapter 12, this volume). Changes in MRI signal intensity during functional activation are thought to result from a local surplus of oxyhemoglobin relative to metabolic demand, shifting the ratio of oxyhemoglobin to deoxyhemoglobin (Bandettini *et al.*, 1992; Belliveau *et al.*, 1991; Belliveau *et al.*, 1992; Kwong *et al.*, 1992; Ogawa & Lee, 1990; Ogawa *et al.*, 1992). The signal alterations which are typically subtle particularly at conventional MRI field strengths, are due to differential magnetic susceptibility of oxyhemoglobin (diamagnetic) and deoxyhemoglobin (paramagnetic). These local oxygen-related perturbations introduced in the magnetic field essentially play the role of an endogenous contrast agent (Ogawa, Lee, & Barrere, 1993) and are hypothesized to reflect microvascular changes specific to the dynamic activity of local neural circuits and their associated induced local field gradients (Lai *et al.*, 1993). The onset of changes in signal intensity detectable by fMRI occurs maximally at 5 to 8 seconds postactivation, with drop-off approximately 5 to 9 seconds after cessation of stimulation (Bandettini, Jesmanowicz, Wong, & Hyde, 1993). Peak activation to motor or sensory stimulation has occurred in most studies within several seconds, typically reaching a peak by 10 seconds. Signal decreases after termination of stimulation and poststimulation hypoactivation (below baseline) have also been observed, and have been interpreted as a postactivation decrease in oxygenated hemoglobin concentration in capillaries (Stern, Kwong, Belliveau, Baker, & Rosen, 1992; Turner *et al.*, 1992; Binder *et al.*, 1994).

There are numerous technical considerations in fMRI research, including acquisition hardware and software issues, as well as methods of image analysis. Although many of the above studies utilized upgraded (echo planar) scanning equipment, results have been obtained using conventional clinical scanners at 1.5 Tesla (e.g., Cohen *et al.*, 1994; Connelly *et al.*, 1993). Factors related to signal intensity include choice of planes, pulse sequences, magnetic field strength, selection of gradient and radio frequency coils, shimming and tuning of the magnetic field, and acquisition parameters including TE and TR, flip angle, field of view, pixel/voxel size, and slice thickness (Connelly *et al.*, 1993; Binder *et al.*, 1994, for review). Rate of stimulation has also been shown to influence degree of brain activation in PET and fMRI experiments (e.g., Roland 1993). Choice of image processing methodology may substantially influence the ability to detect subtle changes in signal associated with experimental activations. These issues have been reviewed in detail by Bandettini *et al.* (1993) and Friston, Jezzard, & Turner (1994).

Changes in signal may also be artifactually related to inflow from vessels perpendicular to the plane of view and large and medium vessels in plane including draining veins. This is a potential threat to the interpretation of activation as an increase in local blood oxygenation. Therefore, activation must be interpreted with caution, and some groups use MR angiography (MRA) to obtain visualization of large vessels in the vicinity of the fMRI acquisition plane. Slight patient movements or very subtle pulsatile motion of the brain due to respiration, cardiac cycle, or cerebrospinal fluid (CSF) flow can introduce artifact as well. Although ultrafast and echo planar imaging (EPI) systems have the potential to reduce motion-related artifact, several centers using EPI suggest acquiring images at a much slower rate than the potential rapidity of EPI in order to preserve spatial resolution of images. An alternative is software-based motion correction algorithms for microrealignment.

There are special considerations in the statistical analysis of fMRI data. These issues were recently reviewed by Bandettini *et al.* (1993) and Friston *et al.* (1994). Major approaches to analysis of functional imaging data have included intra- and intersubject averaging and subtraction strategies (Fox, Mintun, Reiman, & Raichle, 1988; Fox, 1991; Ford, McColl, McCormack, & McCrory, 1991), local-maximum sampling (Fox *et al.*, 1988), correlational/covariance approaches (Clark, Kessler, Buchbaum, Margolin, & Holcomb, 1984), and statistical parametric mapping (Friston *et al.*, 1990; Friston, Frith, Liddle, & Frackowiak, 1991). Bandettini *et al.* (1993) reviewed and evaluated several methods for fMRI analysis, including subtraction and parametric mapping which have limitations in the fMRI context. In the same paper, these authors introduced a novel method for fMRI analysis using Fourier transformation in the frequency domain and compared it to subtraction methods on experimental motor stimulation data sets. Comparison is made between a reference waveform and the obtained data by a modified correlation coefficient. Bandettini *et al.* (1993) evaluated the temporal course plots of activation, response peaks in the frequency domain, phase information from Fourier-transformed data, cross-correlations using time domain, and rate of activation data. Friston *et al.* (1994) reviewed available methods for analyzing fMRI time course data and pointed out that the hemodynamic response should be incorporated in models of fMRI activation. This is particularly important for ultrafast data acquisition such as in echo-planar studies where time domain sampling can be as rapid as 10–20 ms. For longer sampling periods such as 6–9 s per image, as used in many studies, hemodynamic response may be somewhat less of a factor, particularly if subjects have reached a steady state during performance.

Functional Brain Imaging and Memory

Types of Memory. There are many ways to potentially categorize memory (Schacter & Tulving, 1994; Squire & Butters, 1992). Key memory dimensions and hypothesized brain regions are summarized in Figure 1. This multidimensional view of memory demonstrates that the investigation of specific memory processes (e.g., encoding versus recall) for different types of memory (e.g., explicit versus working) can be studied in terms of the contribution of specific brain regions (e.g., hippocampal formation versus frontal lobe) using functional brain imaging.

PET Studies of Memory. The available data utilizing PET methodology to investigate the neural substrates of explicit memory has produced equivocal results with regard to regional specificity. Squire *et al.* (1992) used visual presentation of words and found that the right hippocampus was activated in memory tasks in the absence of semantic or phonetic analysis. Roland, Gulyas, Seitz, Bohm, and Stone-Elander (1990) examined visual learning during rCBF measurement, but reported widespread activation including the anterior hippocampus. When George *et al.* (1993) used visual presentation of faces in a PET (^{15}O) study, they reported activation in the right and left temporal lobes.

When the modality of input has been auditory, investigators have also reported widespread areas of activation and deactivation. Grasby *et al.* (1993, 1994) used an auditory-verbal memory task and reported that memory load was

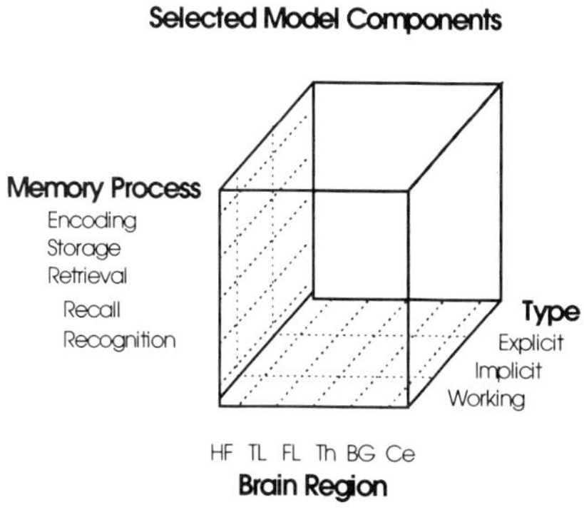

Figure 1. Key memory dimensions and hypothesized brain regions (HF = hippocampal formation, TL = temporal lobe neocortex, FL = prefrontal lobe cortical areas, Th = medial thalamus, BG = basal ganglia, Ce = cerebellum).

positively correlated with rCBF in the cerebellum and thalamus, superior and middle frontal gyri, anterior insular regions bilaterally, anterior cingulate, precuneus, and left and right lateral premotor areas. Andreasen *et al.* (1995) examined short-term and long-term memory for words in a PET (^{15}O) study. They reported that large right frontal areas, biparietal areas, and the left cerebellum were activated in both short-term and long-term memory. However, in the short-term memory condition a relatively large region in the left prefrontal cortex was also activated. They concluded that the neural circuitry for episodic memory has much larger cortical and cerebellar components than previously noted. In a verbal memory study, Tulving *et al.* (1994) using PET (^{15}O) described an "extended strip" of increased blood flow in the right dorsolateral prefrontal cortex, anterior portion of area 6, left anterior cingulate sulcus, and bilaterally in the parietal lobes, and decreased blood flow bilaterally in the temporal lobes. Similar PET findings were obtained by Becker *et al.* (1994).

Raichle *et al.* (1994) in an examination of practice-related changes in human brain functional anatomy during nonmotor learning, concluded that one critical factor in determining neuronal circuitry appears to be the degree to which a task is learned or automatic. Specifically, they had subjects perform a naive and practiced simple verbal response selection task (saying an appropriate verb for a visually presented noun). During naive performance, the areas of the brain most active included the anterior cingulate, left prefrontal, and left posterior temporal cortices, and the right cerebellar hemisphere. These areas were all significantly less active during practiced performance.

fMRI Studies of Working Memory. Although fMRI has not been previously applied to explicit memory, it has been applied to the study of working memory, defined by Baddeley (1992) as "the temporary maintenance and manipulation of information." Working memory differs from explicit memory in that it does not involve long-term storage and retrieval of information or synaptic mechanisms of long-term potentiation. Cohen *et al.* (1994) used fMRI to study activation patterns in the prefrontal cortex while normal controls performed a nonspatial working memory task. They found activation in the middle and inferior frontal gyri. McCarthy *et al.* (1994) also examined spatial working memory and found an increased signal in the middle frontal gyrus, right greater than left. In contrast to the PET studies, the fMRI studies of working memory have shown converging evidence for focal activation in the prefrontal cortex.

Preliminary Results of the fMRI Memory Activation Studies

The objectives of our initial studies were to examine explicit memory processes in normal individuals and in patient groups with disordered memory including epilepsy, Alzheimer's disease, and schizophrenia. Our hypotheses were that we would see focal activation in structures known to be critical for memory in normals. Further, we thought that this activation would be generally consistent with the type of information obtained in lesion studies and from more established functional imaging studies such as PET. We examined the underlying anatomic distribution of memory activation patterns, that is, hippocampal versus neocortical, anterior versus posterior, left versus right hippocampal formation.

Also, we were interested in material-specific functions of memory, component processes (e.g., encoding versus recall versus recognition), and modality of input (auditory, visual, and material specificity). This section presents three experiments designed to examine these important factors in explicit memory. All three experiments followed an established protocol using the same General Electric Signa scanner (1.5 Tesla) at Dartmouth-Hitchcock Medical Center.

Subject Preparation and Task Considerations

Careful attention to subject preparation, rapport, motivation, task demands, and environmental conditions is critical for reproducible brain activation studies (Gur, Erwin, & Gur, 1992). Administration of all experimental procedures must be highly standardized, and all potential sources of subject distraction or discomfort must be minimized. The fMRI context places special constraints on task development, for example, the need for absence of movement, including speech. Subjects were screened for auditory and visual impairments prior to scanning. Subjects were fitted with headphones prior to entry to the MRI chamber. The head was positioned with the laser alignment beams, and a deformable head holder was molded to each subject's head to minimize movement. The room was darkened to reduce activation of visual cortex. Although it is very loud in the scanner, we have achieved satisfactory results placing the headphones tightly in the subjects' ears and surrounding the ears with foam pads.

Careful consideration was given to movement artifact which is a potential problem in clinical studies and we made sure that all subjects understood the importance of lying completely still. It is difficult, but quite possible, to obtain movement-free data on impaired or disturbed patients. Because all images are coregistered to a high-resolution anatomic image in the same plane, we are able to determine precisely whether there is any shift from one functional image to the next. In addition to preliminary reviews by gross inspection, any movement during the study can be quantitated by placement of markers on ROIs to determine whether there are local shifts in the image series. We are also developing an automated image alignment routine to adjust for subtle movement.

MRI Localization Protocol

All three experiments used the same localization protocol (Figure 2) to obtain an oblique axial plane containing the maximal amount of hippocampal gray matter. Briefly, to acquire the localized plane for analysis, we first obtained a coronal image defining the hippocampus (2a). We then resliced the image sagittally (2b), so that the maximal amount of hippocampal gray was observed. We resliced again along the long axis of the hippocampus to show both maximal head and tail of the hippocampus (2c).

Experiment 1: Auditory-Verbal Memory

The first memory activation task employed 20 low-imagery words as stimuli that were adapted from a paper and pencil version (Saykin *et al.*, 1992). These words were presented through headphones at a rate of 2.5 seconds per word, 90 seconds per condition. Conditions included resting baseline, encoding, silent

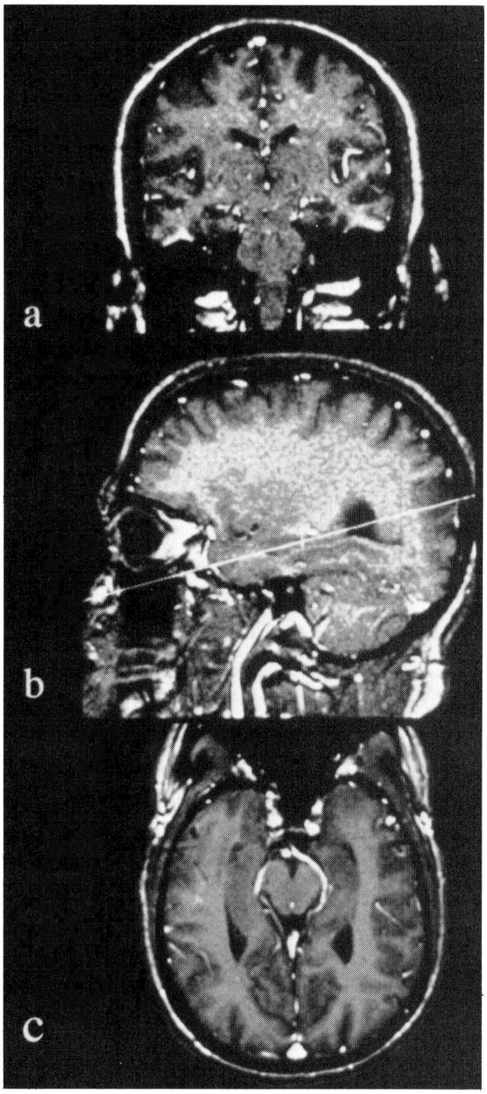

Figure 2. Localization protocol for oblique axial plane through hippocampus.

free recall, recognition, and poststimulation resting baseline. During the recognition condition, subjects responded with bilateral foot movement to signal identification of a target word. This permitted us to obtain information on performance without introducing movement artifact from speech. Each condition consisted of ten images acquired in the same oblique axial plane over 90 seconds. Data were preprocessed by cropping and masking images so that only pixels containing brain were analyzed. A 3 × 3 Gaussian filter was used to smooth the images to reduce noise. Images were averaged within each task condition and the mean initial resting baseline was then subtracted from the mean of each memory condition. These difference maps between initial resting state and memory activation were normalized prior to thresholding. Activated pixels were determined by thresholding the normalized subtraction images at 1 and 2 *SD*. This yielded

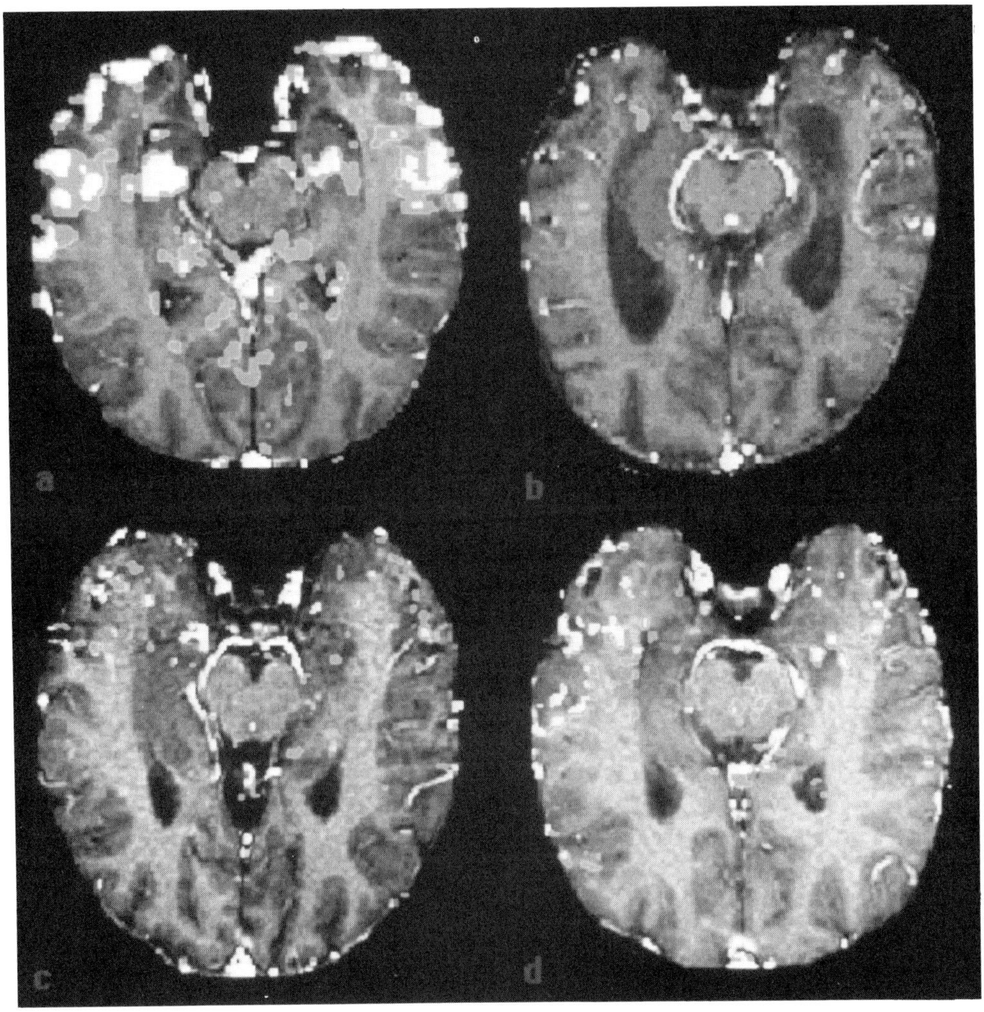

Figure 3. Auditory-verbal memory activation images. (a) Healthy young adult, (b) patient with early Alzheimer's disease, (c) healthy young adult during cyclical auditory-verbal learning task, (d) patient with schizophrenia during cyclical visual verbal encoding task. For a color representation of this figure, see the color insert facing page 327.

widespread temporal neocortical and anterior hippocampal activation, greater on the left than on the right in healthy controls (Saykin, Burr, Riordan, Maerlander, & Weaver, 1995). For example, Figure 3a shows the encoding minus baseline comparison image for a healthy young adult control. The activated pixels are overlaid on the high-resolution anatomic image (left side of the brain appears on the left side of the image). There were individual differences in activation pattern across healthy subjects. These results were generally consistent with PET studies of memory which typically report widespread bilateral activation (Squire *et al.*, 1992; Grasby *et al.*, 1993). We also examined the time course of activation for ROIs in the left anterior and posterior hippocampal formation across memory conditions (Figure 4A). Note the steep increase in signal at the onset of encoding in the left anterior hippocampal ROI and the very gradual decrease after a peak during free recall. Figure 3b shows the activation

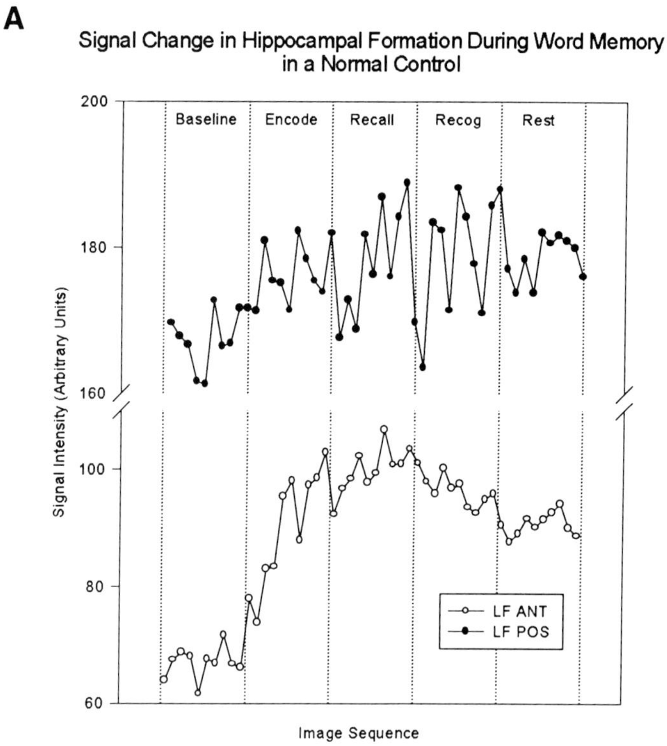

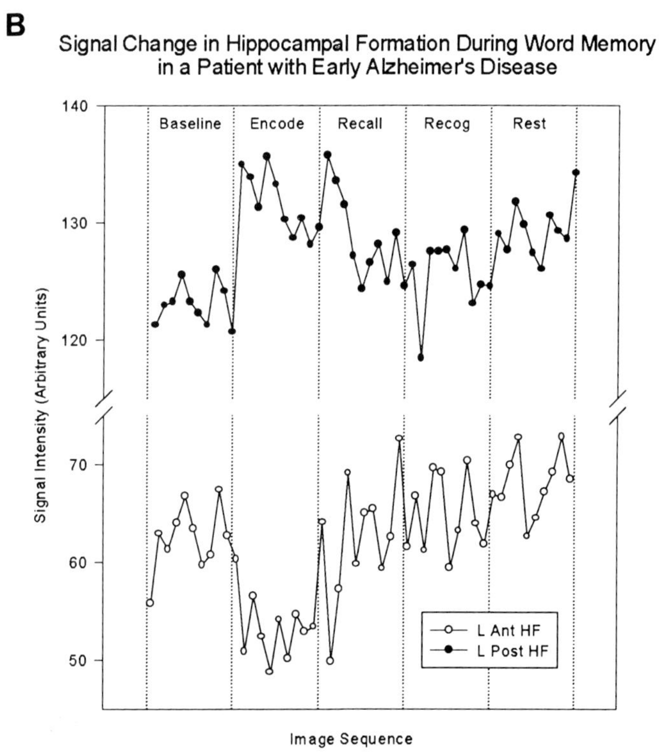

Figure 4. Time course profiles. (A) Healthy young adult, (B) patient with early Alzheimer's disease.

image on the same task for a patient with early Alzheimer's disease who had asymmetric hippocampal atrophy (right > left). Note the relatively small areas of activation that are largely restricted to the anterior temporal regions. Figure 4B shows the time course performance for this patient and indicates a selective decrease during encoding in the left anterior hippocampus, but a corresponding increase in the left posterior hippocampal formation (HF). This left posterior HF activation failed to reach threshold on the activation map. Our initial studies of auditory-verbal memory indicated the feasibility of using fMRI to examine memory processes in healthy and memory disorder subjects. We also observed variable levels and patterns of activation across subjects and task conditions.

Experiment 2: Auditory-Verbal Learning in a Cyclical Design

In Experiment 1 we confirmed that we could obtain activation in the HF. However, the design did not control for language processing. To isolate signal changes due to verbal encoding independent of language processing we moved to a cyclical stimulation paradigm (six 1-minute cycles) that alternated 30-second conditions of perception and encoding. In the encode condition (listening with an explicit instruction to memorize the words), the same list of ten concrete nouns was presented for each trial. All word lists were equated for frequency (Francis & Kucera, 1982), concreteness, and letter length. In the perceive-only condition (listening with no memory demand), a new list was presented for each trial. The cyclical paradigm permitted implementation of specialized image processing strategies designed to optimize detection of focal changes in signal intensity from perceive to encode conditions. In this "constrained subtraction model," we set a stringent threshold criteria for activation on a pixel by pixel basis. During the transition from perceive to encode, a pixel had to activate at least 1 *SD* and increase at least 5% in magnitude, aggregated across the six cycles. Further, from encode to perceive, the pixel had to decrease, on average, 1 *SD*. The obtained images showed considerably more focal activation in the HF, yet individual differences in lateralization of activation were observed (Saykin, Burr, Riordan, Maerlander, & Weaver, 1995). Results for a healthy control are shown in Figure 3c. The activated pixels are overlaid on the high-resolution anatomic image (left side of the brain appears on the left side of the image). Note the more focal activation largely confined to the left anterior HF. These more focal activation patterns have been characteristic of cyclical designs and related image processing strategies.

Experiment 3: Visual Memory for Words and Faces

In the next experiment we shifted to the visual modality and examined material specificity while continuing to utilize a similar cyclical perceive/encode design. Stimuli were back-projected from outside the scanner chamber to a 1 m^2 screen at the foot of the scanner. Words and faces were viewed by the subject through a mirror in the head coil assembly. In the word memory task, we used a series of lists similar to those employed in experiment 2 (Francis & Kucera, 1982). Facial stimuli were adapted from a battery of tests developed at the Brain Behavior Laboratory of the University of Pennsylvania (Gur *et al.*, 1990).

Each task consisted of four cycles of perceive encode conditions with 15 stimuli (words or faces, respectively) in each condition in a counterbalanced design. As in the prior experiment we used the constrained subtraction method for identification of activated pixels. Figures 5a and 5b show encoding for words and faces in a healthy young adult control. On the word encoding task (Figure 5a), activation was largely restricted to the left hippocampus as in the above auditory verbal memory task. For the facial memory task (Figure 5b), more focal bilateral anterior hippocampal activation was observed.

We have studied several patients with known hippocampal abnormalities on these memory activation tasks. Figures 5c and 5d show results for one candidate for anterior temporal lobectomy for medically refractory epilepsy. This patient was left language–dominant by Wada test with worse right hemispheric memory performance. As can be seen in the figure, she showed bilateral activation in the anterior HF during the word task (Figure 5c), but showed left lateralization for the facial task (Figure 5d). This suggested a material-specific activation deficit consistent with her right mesial temporal seizure focus and with her facial memory performance.

We also examined patients with schizophrenia on the visually presented

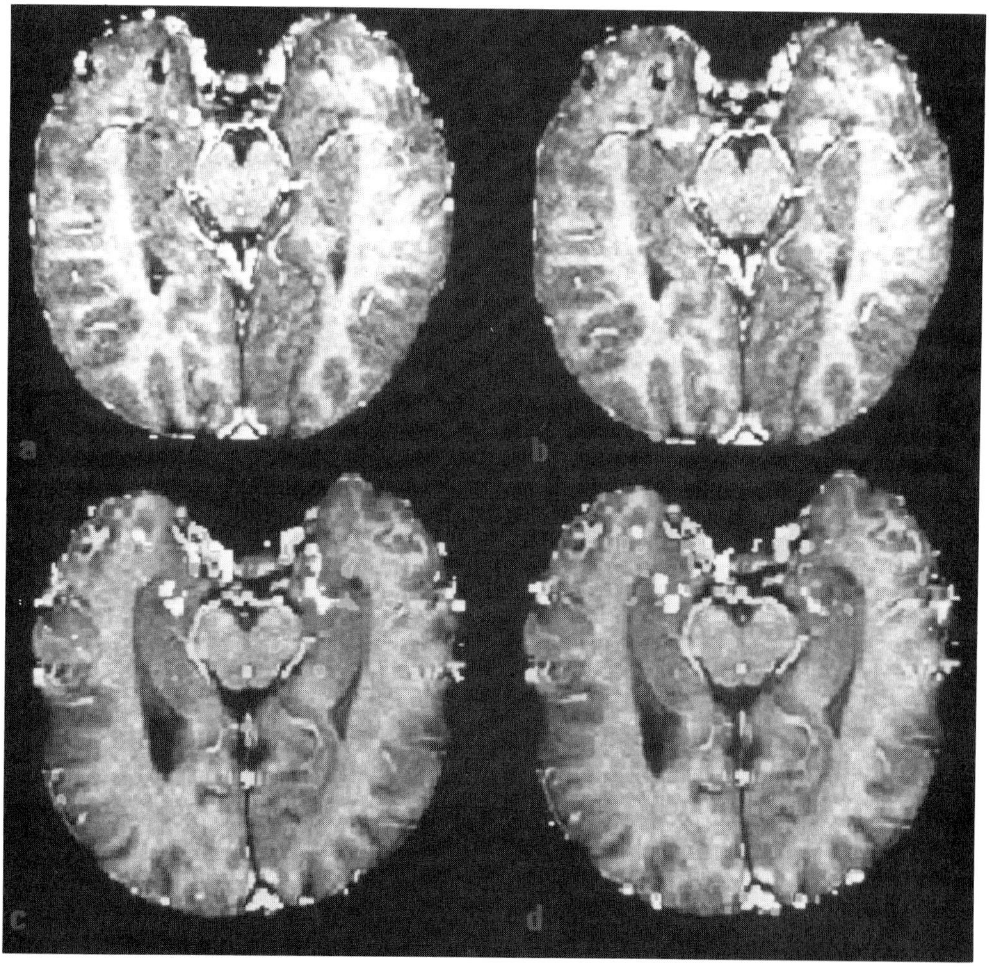

Figure 5. Activation images for cyclical memory for words and faces. (a) Healthy young adult encoding words, (b) healthy young adult encoding faces, (c) patient with right temporal lobe seizure focus encoding words, (d) patient with right temporal lobe seizure focus encoding faces. For a color representation of this figure, see the color insert facing page 327.

word memory task. Patients with schizophrenia frequently exhibit selective deficits in verbal memory against a background of generalized neuropsychological deficits (Saykin *et al.*, 1991). Figure 3d shows activation during word encoding for a patient with schizophrenia. This patient's performance on free recall for the word memory stimuli after scanning was severely impaired. Of interest in his activation image was the relative absence of left hippocampal activation, as well as the predominance of atypical diffuse activation in the left lateral neocortical areas.

A similar pattern of activation (more neocortical than hippocampal) was observed in a group of patients with schizophrenia across both the word and facial memory tasks. A quantitative ROI analysis of signal changes in the neocortical and hippocampal regions suggested an abnormal activation pattern during encoding, with bilateral underactivation in the hippocampal formation and overactivation in adjacent middle temporal lobe neocortical regions ($p < .05$) for medial/lateral × diagnosis interaction; Saykin, Riordan, Weaver, *et al.*, 1995). Within-group correlational analyses for patients indicated significant relationships between fMRI activation and performance and clinical variables. Changes in signal intensity on the word task were positively correlated with memory performance during scanning and clinical neuropsychological tests of verbal learning and inversely related to negative symptoms.

Methodological Considerations

Several methodological issues should be considered when employing fMRI to study memory. Most currently available hardware/software combinations do not permit simultaneous high-resolution acquisition of the entire brain. Therefore choice of imaging plane or planes is critical. Current hypotheses regarding structures critical for long-term memory emphasize the hippocampal formation and associated limbic regions. For a single plane study, use of an oblique axial image through the long axis of the hippocampus permits recording of signal from both the head and tail of the hippocampus. However, because of the curvilinear structure of the hippocampus it is often difficult to record simultaneously from the body of the hippocampus as well. Relatively little attention has been paid to anterior versus posterior differentiation in the hippocampus in humans or animals, particularly with regard to memory, and fMRI is likely to improve our understanding of this anatomic dimension. We have found that with experience, acquisition of the oblique axial plane containing maximal gray matter of the anterior and posterior HF is quite reproducible for longitudinal studies.

The hippocampal formation sits in proximity to the P3 segment of the posterior cerebral artery, the choroidal artery, and many small vessels (see Duvernoy, 1988). The lateral and anterior horns of the ventricular system are also in proximity. Artifactual changes in signal may occur due to inflow from vessels perpendicular to the plane of view or from large and medium vessels within plane. Magnetic resonance angiography (MRA: see Edelman, 1992, for review) is often helpful for visualization of large vessels. Figure 6 shows examples from the MRA series of a recent subject with epilepsy including axial (Figure 6a) and coronal (Figure 6b) sequences indicating the position of major vessels near the longitudinal hippocampal fMRI plane.

Other forms of susceptibility artifact related to changes in tissue can also be

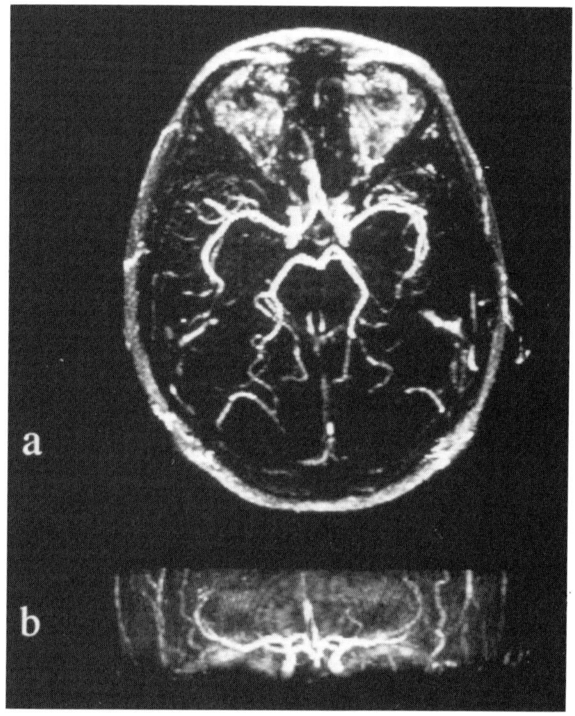

Figure 6. MR angiography. (a) Axial image showing major vessels in close proximity to the functional imaging plane. (b) Coronal image indicating position of vessels in the temporal lobe region.

a problem. Structural MRI data acquired in a volumetric format can be reformatted in planes above and below the functional plane or planes to clarify the location of vessels and brain CSF or other tissue transitions. In our initial studies, examination of the memory activation maps indicated that large vessels and draining veins do not show task-related activation whereas hippocampal parenchyma is frequently activated. We have noted that in cases of extremely low activation during memory, when a within-subject subtraction method is used in the absence of a cyclical constrained subtraction method, large vessels can show pseudoactivation, probably due to physiological pulsatile motion. This is almost always eliminated with a cyclical task design and analytic strategy.

Conclusions

Our initial studies suggest that fMRI performed on a conventional scanner at 1.5 T is a feasible methodology for studying memory activation in healthy normals and patients with disordered memory. High spatial resolution permits examination of selective hippocampal activation, including subcomponents of the hippocampal formation at an individual subject level. There are individual differences in patterns of activation across subjects and for component processes of memory (e.g., encoding, recall, recognition). Functional MRI appears promising for the examination of memory dimensions including input modality (auditory/visual) and material specificity (words/faces). Level and pattern of activation appear to be associated with memory performance during scanning, as well as baseline memory abilities on standardized neuropsychological memory testing.

Our data suggest that fMRI memory activation results are meaningful as evidenced by relationships between signal changes and important clinical variables such as lateralization of seizure focus in temporal lobe epilepsy and the extent of negative symptoms in schizophrenia.

Future Prospects and Applications

Rapid advances in MRI hardware and software will permit the examination of multiple brain regions simultaneously, with high temporal and spatial resolution. In an ideal situation, one would be able to obtain three-dimensional activation data for the entire brain in physiological real time while preserving millimeter spatial resolution. Aspects of this technology such as 3-D capability and higher field strength magnets are already becoming available. In addition to improved instrumentation, refinements in experimental task design and image and signal processing software are also important. Our memory activation data suggest that cyclical task designs maximize the signal of interest while permitting the rejection of artifact and noise unrelated to the stimulation. This has also been shown for other types of activation (e.g., Bandettini *et al.*, 1993).

Functional MRI is already being used clinically in a few centers to assist with preoperative brain mapping of sensorimotor functions in an attempt to spare functionally critical tissue during neurosurgical procedures. There has been considerable recent interest in application of fMRI to presurgical mapping of cortical zones necessary for speech and language capabilities. Preoperative identification of regions subserving memory is critical for patients undergoing temporal lobectomy as well (Saykin, Weaver, Burr, *et al.*, 1995). Functional MRI may have the potential to supplement or ultimately replace some invasive procedures such as the intracarotid amobarbital (Wada) test and intraoperative mapping. However, extensive validation and outcome studies are required to determine this potential, including test-retest studies of the reproducibility of patterns of activation. Finally, of both clinical and theoretical relevance, fMRI may provide an improved method for examination of brain plasticity after injury or anomalous representation of memory functions triggered by an early neurodevelopmental insult.

It may also be possible to optimize treatment of brain disorders affecting memory with fMRI assessment. Through longitudinal studies, responses to neuropharmacological and psychotropic medications could be determined. The extent to which fMRI could predict response to treatment or identify optimal treatment warrants investigation. Identification of brain areas that activate or fail to activate during memory processing might suggest specific rehabilitation strategies tailored to the individual activation profile. The functional consequences of extent and pattern of fMRI memory activation could be examined in relation to measures of activities of daily living.

Functional MRI is an extremely promising noninvasive technology for the study of higher cognitive functioning that has numerous advantages over other available methods. It is rapidly changing the landscape of our understanding of the normal brain and neurological and psychiatric disorders.

Acknowledgments

This research was supported in part by funding from the NINDS (RO1-NS-28813), the Alzheimer's Association, the Hitchcock Foundation, and the Rita

Rudel Foundation, the New Hampshire Hospital, and the Department of Radiology at DHMC. We thank Kevin Carroll, Lauren Schlesinger, and Robert Ferranti for assistance with image acquisition and analysis; Barbara Baker for help with task and manuscript preparation; and Denise Bono for subject recruitment and testing. We also thank Drs. Thomas McAllister, Peter Williamson, Vijay Thadani, David Roberts, Theo Manschreck, E. Michael Kahn, and Thomas Oxman for providing patients and Drs. Alexander Mamourian, Terrance Darcey, Laurence Cromwell, and Peter Spiegel for assistance with imaging protocols.

References

Andreasen, N. C., O'Leary, D. S., Arndt, S., Cizadlo, T., Hurtig, R., Rezai, K., Watkins, G. L., Ponto, L. L. B., & Hichwa, R. D. (1995). Short-term and long-term verbal memory: A positron emission tomography study. *Proceedings of the National Academy of Sciences, 92*(11), 5111–5115.

Baddeley, A. D. (1992). Working memory. *Science, 255,* 556–559.

Bandettini, P. A., Jesmanowicz, A., Wong, E. C., & Hyde, J. S. (1993). Processing strategies for time-course data sets in fMRI of the human brain. *Magnetic Resonance in Medicine, 30*(2), 161–173.

Bandettini, P. A., Wong, E. C., Hinks, R. S., Tikofsky, R. S., & Hyde, J. S. (1992). Time course EPI of human brain function during task activation. *Magnetic Resonance in Medicine, 25,* 390–397.

Becker, J. T., Mintun, M. A., Diehl, D. J., Dobkin, J., Martidis, A., Madoff, D. C., & DeKosky, S. T. (1994). Functional neuroanatomy of verbal free recall: A replication study. *Human Brain Mapping, 1,* 284–292.

Belliveau, J. W., Kennedy, D. N., McKinstry, R. C., Buchbinder, B. R., Weisskoff, R. M., Cohen, M. S., Vevea, J. M., Brady, T. J., & Rosen, B. R. (1991). Functional mapping of the human visual cortex by magnetic resonance imaging. *Science, 254,* 716–719.

Belliveau, J. W., Kwong, K. K., Kennedy, D. N., Baker, J. R., Stern, C. E., Benson, R., Chesler, D. A., Weisskoff, R. M., Cohen, M. S., Tootell, R. B. H., Fox, P. T., Brady, T. J., & Rosen, B. R. (1992). Magnetic resonance imaging mapping of brain function: Human visual cortex. *Investigative Radiology, 27,* S59–S65.

Binder, J. R., Rao, S. M., Hammeke, T. A., Yetkin, F. Z., Jesmanowicz, A., Bandettini, P. A., Wong, E. C., Estkowski, L. D., Goldstein, M. D., & Haughton, V. M. (1994). Functional magnetic resonance imaging of human auditory cortex. *Annals of Neurology, 35,*(6), 662–672.

Clark, C. M., Kessler, R., Buchbaum, M. S., Margolin, R., & Holcomb, H. H. (1984). Correlational methods for determining regional coupling of cerebral glucose metabolism: A pilot study. *Biological Psychology, 19,* 663–678.

Cohen, J. D., Forman, S. D., Braver, T. S., Casey, B. J., Servan-Schreiber, D., & Noll, D. C. (1994). Activation of the prefrontal cortex in a nonspatial working memory task with functional MRI. *Human Brain Mapping, 1,* 293–304.

Cohen, M. S., & Bookheimer, S. Y. (1994). Localization of brain function using magnetic resonance imaging. *Trends in Neurosciences, 17,*(7) 268–277.

Connelly, A., Jackson, G. D., Frackowiak, R. S. J., Belliveau, J. W., Varga-Kladem, F. & Gadian, D. G. (1993). Functional mapping of activated human primary cortex with a clinical magnetic resonance imaging system. *Radiology, 188,* 125.

Duvernoy, H. M. (1988). *The human hippocampus: An atlas of applied anatomy.* New York: Springer-Verlag.

Edelman, R. R. (1992). Basic principles of magnetic resonance angiography. *Cardiovascular Intervention Radiology, 15,* 3–13.

Ford, I., McColl, J. H., McCormack, A. G., & McCrory, S. J. (1991). Statistical issues in the analysis of neuroimages. *Journal of Cerebral Blood Flow and Metabolism, 11,* A89–A95.

Fox, P. T. (1991). Physiological ROI definition by image subtraction. *Journal of Cerebral Blood Flow and Metabolism, 11,* A79–A82.

Fox, P. T., Mintun, M. A., Reiman, E. M., & Raichle, M. E. (1988). Enhanced detection of focal brain responses using intersubject averaging and change-distribution analysis of subtracted PET images. *Journal of Cerebral Blood Flow and Metabolism, 8,* 642–653.

Frahm, J., Bruhn, H., Merboldt, K., & Hanicke, W. (1992). *Journal of Magnetic Resonance Imaging, 2,* 501.

Frahm, J., Merboldt, K. D., & Hancke, W. (1993). Functional MRI of human brain activation at high spatial resolution. *Magnetic Resonance in Medicine, 29,* 139–144.

Francis, W. N., & Kucera, H. (1982). *Frequency analysis of English usage: Lexicon and grammar.* Boston: Houghton Mifflin.

Friston, K. J., Frith, C. D., Liddle, P. F., Dolan, R. J., Lammertsma, A. A., & Frackowiak, R. J. J. (1990). The relationship between global and local changes in PET scans. *Journal of Cerebral Blood Flow and Metabolism, 10,* 458–466.

Friston, K. J., Frith, C. D., Liddle, P. F., & Frackowiak, R. S. J. (1991). Plastic transformations of PET images. *Journal of Computer Assisted Tomography, 15,* 634–639.

Friston, K. J., Jezzard, P., & Turner, R. (1994). The analysis of functional MRI time series. *Human Brain Mapping, 1,* 153–171.

Frostig, R. D., Lieke, E. E., Tso, D. Y., & Grinvald, A. (1990). Cortical functional architecture and local coupling between neuronal activity and the microcirculation revealed by in vivo high-resolution optical imaging of intrinsic signals. *Proceedings of the National Academy of Sciences USA, 87,* 6082–6086.

George, M. S., Ketter, T. A., Gill, D. S., Haxby, J. V., Ungerleider, L. G., Herscovitch, P., & Post, R. M. (1993). Brain regions involved in recognizing facial emotion or identity: An oxygen-15 PET study. *Journal of Neuropsychiatry and Clinical Neurosciences, 5*(4), 384–394.

Grasby, P. M., Frith, C. D., Friston, K. J., Bench, C., Frackowiak, R. S. J., & Dolan, R. J. (1993). Functional mapping in brain areas implicated in auditory-verbal memory function. *Brain, 116,* 116–120.

Grasby, P. M., Frith, C. D., Friston, K. J., Simpson, J., Fletcher, P. C., Frackowiak, R. S. J., & Dolan, R. J. (1994). A graded task approach to the functional mapping of brain areas implicated in auditory-verbal memory. *Brain, 117,* 1271–1282.

Gur, R. C., Erwin, R. J., & Gur, R. E. (1992). Neurobehavioral probes for physiologic neuroimaging studies. *Archives of General Psychiatry, 49,* 409–414.

Gur, R. C., Saykin, A. J., Benton, A., Kaplan, E., Levin, H., Kester, D. B., & Gur, R. E. (1990). Behavioral imaging: III. Inter-rater agreement and reliability of weightings. *Journal of Neuropsychiatry, Neuropsychology and Behavioral Neurology, 3,* 113–124.

Hebb, D. O. (1949). *The organization of behavior.* New York: Wiley.

Kandel, E. R., Schwartz, J. H., & Jessell, T. M. (1991). *Principles of neural science* (3rd ed.). New York: Elsevier.

Kim, S., Ashe, J., Georgopoulos, A. P., Merkle, H., Ellermann, J. M., Menon, R. S., Ogawa, S., *et al.* (1993). Functional imaging of human motor cortex at high magnetic field. *Journal of Neurophysiology, 69*(1), 297–302.

Kwong, K. K., Belliveau, J. W., & Chesler, D. A., Goldberg, I. E., Weisskoff, R. M., Poncelet, B. P., *et al.* (1992). Dynamic magnetic resonance imaging of human brain activity during primary sensory stimulation. *Proceedings of the National Academy of Sciences, USA, 89,* 567.

Lai, S., Hopkins, A. L., Haacke, E. M., Li, D., Wasserman, B. A., Buckley, P., Friedman, L., *et al.* (1993). Identification of vascular structures as a major source of signal contrast in high resolution 2D and 3D functional activation imaging of the motor cortex at 1.5T: Preliminary results. *Magnetic Resonance in Medicine, 30,* 387.

McCarthy, G., Blamire, A. M., Puce, A., Nobre, A. C., Bloch, G., Hyder, F., Goldman-Rakic, P., & Shulman, R. G. (1994). Functional magnetic resonance imaging of human prefrontal cortex activation during a spatial working memory task. *Proceedings of the National Academy of Sciences, USA, 91*(18), 8690–8694.

Ogawa, S., & Lee, T. M. (1990). Magnetic resonance imaging of blood vessels at high fields: In vivo and in vitro measurements and image stimulation. *Magnetic Resonance in Medicine, 16,* 9–18.

Ogawa, S., Lee, T. M., & Barrere, B. (1993). The sensitivity of magnetic resonance image signals of a rat brain to changes in the cerebral venous blood oxygenation. *Magnetic Resonance in Medicine, 29,* 205.

Ogawa, S., Tank, D. W., Menon, R., Ellermann, J. M., Kim, S. A., Merkle, H., *et al.* (1992). Intrinsic signal changes accompanying sensory stimulation: Functional brain mapping with magnetic resonance imaging. *Proceedings of the National Academy of Sciences, USA, 89,* 5951.

Press, G. A., Amaral, D. G., & Squire, L. R. (1989). Hippocampal abnormalities in amnesic patients revealed by high-resolution MRI. *Nature, 341,* 54–57.

Raichle, M. E., Fiez, J. A., Videen, T. O., MacLeod, A. M., Pardo, J. V., Fox, P. T., & Petersen, S. E. (1994). Practice-related changes in human brain functional anatomy during nonmotor learning. *Cerebral Cortex, 4*(1), 8–26.

Rao, S. M., Binder, J. R., Bandettini, P. A., Hammeke, T. A., Yetkin, F. Z., Jesmanowicz, A., Lisk,

L. M., Morris, G. L., Mueller, W. M., & Estkowski, L. D. (1993). Functional magnetic resonance imaging of complex human movements. *Neurology, 43*(11), 2311–2318.

Roland, P. E. (1993). *Brain activation.* New York: Wiley-Liss, Inc.

Roland, E., Gulyas, B., Seitz, R. J., Bohm, C., & Stone-Elander, S. (1990). Functional anatomy of storage, recall, and recognition of a visual pattern in man. *NeuroReport, 1*(1), 53–56.

Roy, C. S., & Sherrington, C. S. (1890). On the regulation of the blood-supply of the brain. *Journal of Physiology, 11,* 85–108.

Saykin, A. J., Burr, R. B., Riordan, H. J., Maerlander, A. C., & Weaver, J. B. (1995). Temporal neocortical and hippocampal memory activation as measured by functional magnetic resonance imaging. *Journal of the International Neuropsychological Society, 1,* 187.

Saykin, A. J., Gur, R. C., Gur, R. E., Mozley, D., Mozley, L. H., Resnick, S. M., Kester, D. B., & Stafiniak, P. (1991). Neuropsychological function in schizophrenia: Selective impairment in memory and learning. *Archives of General Psychiatry, 48,* 618–624.

Saykin, A. J., Riordan, H. J., Weaver, J. B., Manschreck, T., Flashman, L. A., Kahn, E. M., Flannery, K. A., Mamourian, A., & McAllister, T. W. (1995). Memory activation in schizophrenia: Preliminary observations using functional magnetic resonance imaging (fMRI). *Schizophrenia Research, 15,* 97.

Saykin, A. J., Robinson, L. J., Stafiniak, P., Kester, B., Gur, R., O'Connor, M. J., & Sperling, M. (1992). Neuropsychological changes after anterior temporal lobectomy: Acute effects on memory, language, and music. In T. L. Bennett (Ed.), *The neuropsychology of epilepsy.* New York: Plenum Press.

Saykin, A. J., Weaver, J. B., Burr, R. B., Riordan, H. J., Roberts, D. W., Williamson, P. D., Thadani, V. M., Darcey, T. M., & Cromwell, L. D. (1995). Functional magnetic resonance imaging in the evaluation of epilepsy surgery patients: A memory activation study. *Epilepsia, 35,* 86.

Schacter, D. L., & Tulving, E. (1994). *Memory systems 1994.* Cambridge, MA: MIT Press.

Scoville, W. B., & Milner, B. (1957). Loss of recent memory after bilateral hippocampal lesions. *Journal of Neurology, Neurosurgery, and Psychiatry, 20,* 11–21.

Squire, L. R. (1987). *Memory and brain.* New York: Oxford University Press.

Squire, L. R., & Butters, N. (1992). *Neuropsychology of memory* (2nd ed.). New York: Guildford Press.

Squire, L. R., Ojemann, J. G., Miezin, F. M., Petersen, S. E., Videen, T. O., & Raichle, M. E. (1992). Activation of the hippocampus in normal humans: A functional anatomical study of memory. *Proceedings of the National Academy of Sciences, USA, 89*(5), 1837–1841.

Stern, C. E., Kwong, K. K., Belliveau, J. W., Baker, J. R., & Rosen, B. R. (1992). MR tracking of physiological mechanisms underlying brain activity. In *Book of Abstracts, 11th Annual Meeting, Society for Magnetic Resonance in Medicine* (p. 1821). Berlin.

Tulving, E., Kapur, S., Markowitsch, H. J., Craik, F., Habib, R., & Houle, S. (1994). Neuroanatomical correlates of retrieval in episodic memory: Auditory sentence recognition. *Proceedings of the National Academy of Sciences, USA, 91*(6), 2012–2015.

Turner, R., Jezzard, P., Wen, H., Kwong, K., Le Bihan, D., & Balaban, R. (1992). Functional mapping of the human cortex at 4 tesla using deoxygenation contrast EPI. In *Book of Abstracts, 11th Annual Meeting, Society for Magnetic Resonance in Medicine* (p. 304). Berlin.

III

Future Direction and Clinical Application

14

Neuroimaging and Biology

Concluding Remarks

RONALD A. YEO

> The caverns of the mind are obscure, and shadowy, or pervaded with a lustre, beautifully bright indeed, but shining not beyond their portholes.
>
> —Percy Bysshe Shelley, *Speculations on Metaphysics*

Remarkably, the caverns of the mind are indeed shining beyond their portholes, though the terrain revealed sometimes appears new and strange. Being able to see inside the heads of living, thinking people has revolutionized the way human neuroscience is conducted and has captured the imagination of neuroscientists around the world. Our technological tools have developed so rapidly, however, that there is substantial uncertainty as to how they are best utilized. The promise of neuroimaging is unfulfilled, though a respectable downpayment has been made. The chapters in these two volumes provide a superb summary of past achievements and numerous insights regarding technical and experimental innovation. In this brief chapter, I will not attempt to summarize these chapters, but rather, offer some comments designed to place the field of neuroimaging in a larger scientific context, in particular, that of theoretical biology.

The rate of progress across different types of neuroimaging studies has been uneven. Structural neuroimaging studies have had a significant impact on diverse realms of inquiry. These include the study of both clinical applications, such as the diagnostic utility of volumetric hippocampal analysis described by Naugle and colleagues in temporal lobe epilepsy (Chapter 8, Volume II), and the study of theoretical issues, such as the case-oriented analysis of the substrates of

---

RONALD A. YEO Department of Psychology, University of New Mexico, Albuquerque, New Mexico 87131

Neuroimaging II: Clinical Applications, edited by Erin D. Bigler. Plenum Press, New York, 1996.

amnesia described by Kapur (Chapter 11, Volume II). In the area of neuropsychiatric disorders, ventricular enlargement and reduced mesial temporal volumes in schizophrenia are now central to any comprehensive account of the disorder (see Marsh *et al.,* Chapter 4, Volume II). Functional neuroimaging of disease states has also made great strides. The rather distinct "signatures" of the different dementias described by Welsh and Hoffman (Chapter 7, Volume II) and the varied patterns of activation seen with recovery from hemiplegia in stroke patients described by Weiller and colleagues (Chapter 9, Volume II) are excellent examples. The observation of frontal hypometabolism in schizophrenia is so well established that any comprehensive account of schizophrenia must deal with this finding.

The functional imaging of cognition and emotion has probably had relatively less impact on current theory and practice than either structural neuroimaging or functional neuroimaging of disease states. Undoubtedly, cognitive and emotional studies present greater technical difficulties. And as many authors of chapters in this volume have commented, there are complex and uncertain assumptions underlying the process whereby regional activations are associated with cognitive processes (see especially Chapter 5 by Bookheimer, Volume I, and Chapter 12, this volume). Studies using PET and fMRI have provided repeated demonstrations of retinotopic and somatotopic organization but at present have had only minimal impact on cognitive theories, such as theories of language or memory. Indeed, as clearly described by Bookheimer in the first volume of this series, PET studies of hippocampal activation have provided results that are both inconsistent and confusing. This is despite the fact that studies of rodents and nonhuman primates have provided a great deal of insight into the anatomy and function of the hippocampus. Similarly, conditions under which activation is seen in Wernicke's area, another of the "better known" brain regions, have proven remarkably difficult to identify. Yet despite such difficulties, functional imaging of cognition and emotion represents the most exciting application of the new neuroimaging technologies. There is something almost magical about the ability to "see thought," and with continued technical development, and especially the development of various experimental methods (e.g., correlational designs), functional imaging will undoubtedly have a great impact on our theories of cognition and emotion.

Many chapters in this volume provide cogent criticisms of the current state of neuroimaging and suggestions as to how to achieve the scientific promise provided by new imaging technologies. In the same spirit, the next few pages will provide some additional suggestions relevant to issues of experimental design and biological theory, stemming from considerations of the fundamental biological constructs of natural selection and individual variation.

Biology and Psychology

One of the most important developments in psychology over the past two decades has been the systematic attempt to integrate biological theory and psychological explanation. Neuroimaging studies represent only a portion of this trend and, along with focal lesion studies and single-unit recording studies, tackle only one aspect (or level) of a biological approach to explanation—that of "mechanism." They focus on the machinery making the computation, its inputs, outputs, anatomical locus, and conditions of activation. In this sense, the brain is not conceptualized as different in principle than any other type of information processing device. Yet the brain is different and not just because it is composed of neurons rather than silicon chips. It is an evolved organ shaped over millions of years by natural selection. Its growth and development within our lifetime and the manner in which it is influenced by environmental inputs is regulated by a relatively small number of genes (at least in comparison to the total number of neural elements). These considerations provide fundamental constraints to our theories of brain function. The actual mechanisms of brain function are only a subset of theoretically possible mechanisms, constrained by natural selection and phylogenetic variability. Further, the way in which actual information processing mechanisms emerge is constrained by the limited information available in the genome. Only those mechanisms which can be shaped from 40,000 or so genes, with appropriate environmental inputs, will actually be instantiated.

A balanced and complete biological explanation thus entails not only an understanding of mechanism, but also an understanding of evolutionary/genetic underpinnings and ontogeny. Our ability to integrate and develop neuroimaging data at the level of mechanism will be aided by attention to these two other aspects of biological explanation. There are a few time-honored experimental methods that facilitate the development of such insights, and in comparison with the immense technological problems in neuroimaging already solved, they are relatively easy to accomplish.

First, there are the tried and true analytical methods of behavior genetics, such as comparison of concordance rates between monozygotic and dyzygotic twins. Conventional behavior genetic designs allow not only an estimation of heritability (h^2), but also an estimate of the particular classes of environmental factors that are most important. A recent study by Tramo and colleagues (1995) was able to conclude that variation in the anatomy of the left and right hemispheres is differentially affected by genetic influences. There is greater population variability in the morphology of the left hemisphere than the right, and this variability was found to be under genetic rather than environmental influence. This observation is quite interesting in light of the fact that language probably represents a more recent evolutionary adaptation than the spatial skills of the right hemisphere, consistent with Geschwind and Galaburda's (1987) notion of the "conservatism" of right hemisphere development. One can also investigate the manner in which genes influence phenotype–either in an additive fashion or through epistasis, or "emergenesis," as described by Lykken, McGue, Tellegen, and Bouchard (1992). Some traits, such as extroversion and certain types of EEG responses, appear to be under significant genetic influence, but reflect emergent properties of gene configurations rather than additive influences. Emergent genetic influences can be easily determined. When dyzygotic pair correlations are less than one half the difference between the monozygotic pair correlations and zero, emergenesis is likely.

Most behavior genetic designs allow determination of the nature of environmental influences, that is, whether they reflect environmental factors shared within families (e.g., such as economic status, attitudes toward education) or environmental factors unique to an individual in the family (e.g., differential treatment by parents or brain trauma). Understanding the manner in which genetic and environmental factors influence brain anatomy and function is a fundamental aspect of our efforts to reveal the brain's information processing mechanisms. Given the expense of neuroimaging studies, might it not be worthwhile to invest a small amount of additional effort into subject recruitment? One could as easily study mechanisms of memory or language in a sample of twins as in a sample of unrelated subjects and have the bonus of additional information about environmental and genetic influences. In the large population centers where neuroimaging studies are typically conducted, twin samples are readily available.

Brain mechanisms develop over time. The manner in which they develop provides fundamental insights into such issues as the nature of modularity and the nature and timing of environmental determinants of design. There are, of course, several excellent examples of developmental neuroimaging research programs, including the superb anatomical studies from Sowell and Jernigan (Chapter 3, Volume I), Neville's studies of the differential maturation rates and localization of syntactic versus semantic linguistic systems (Neville, Mills, & Lawson, 1992), and Thatcher's (1992) EEG studies of cyclic reorganization in childhood. Functional MRI and MEG studies are probably best suited for these types of investigation, given their repeatability.

Individual Differences

For many researchers individual differences are simply "nuisance" variables that need to be controlled in order to reduce error variance in one's data. There are, of course, important questions concerning the significance of individual differences, issues extending far beyond minimizing error variance. The dimension of individual differences receiving most attention thus far is probably normal human aging, as described by Raz (Chapter 6, this volume). Sex differences have been less frequently studied (however, see Farace & Turkheimer, Chapter 5, this volume). The recent PET study of sex differences in phonological processing PET by Shaywitz and colleagues (1995) should provide impetus for more investigations of this type. Studies of individual differences can provide insights into the major goal of functional neuroimaging, the identification of specific processing mechanisms, through analysis of the covariance of performance levels and anatomic variables. As Raz points out, the covariation of changes in memory ability and brain structure in normal aging provides insights into the general neural mechanisms of memory.

But for many appropriate applications of neuroimaging technology, individual difference variables truly remain nuisances that need to be dealt with. We can deal with such unwanted sources of variance in more sophisticated ways than is typical of neuroimaging studies. To be blunt, there is more to individual differences than handedness. It is commonplace to study only right-handers, because, as we all know, left-handers may have bilateral or right hemisphere language dominance. The logic of controlling for variations in handedness is inescapable—failure to do so increases error variance and makes one's data a muddle. But handedness only weakly predicts individual variability in the lateralization of language processing. It may be desirable to actually assess language lateralization, with conventional dichotic listening or tachistoscopic techniques, rather than simply rely upon right-handedness as a control for all individual differences in lateralization. At the risk of seeming repetitive, neuroimaging studies are expensive. Lateralization studies are cheap. A bit more effort devoted to characterizing subjects, and perhaps studying only those who are modal in terms of lateralization of a given trait, might increase statistical power.

When one does assess handedness, one should probably get a brief measure of relative hand skill. This allows one to distinguish "extreme" right-handers from modal right-handers, something quite difficult to do with conventional questionnaires of hand preference. We have found that extreme right-handers share interesting similarities with left-handers—more minor physical anomalies and greater fluctuating anatomic asymmetry (see Markow, 1994), as well as a higher incidence of left-handed parents, than modal right-handers (Yeo & Gangestad, 1993; Gangestad & Yeo, 1994). We have also found (Yeo *et al.*, under review) that deviation from population-typical functional asymmetries, in either direction from the mean, for both linguistic and nonlinguistic tasks, is predicted by a composite measure of minor physical anomalies and fluctuating anatomic asymmetry, putative markers of developmental instability (defined as the imprecise expression of the genetic plan for development). These observations suggest that selecting subjects that are "moderate" right-handers, as defined by relative hand skill, with fewer minor physical anomalies and fluctuating asymmetry, may be more effective in terms of reducing variability across subjects in lateralization than merely limiting one's study to self-professed right-handers. It is also important to note that lateralization scores from different cognitive tasks, such as language and spatial processing, are typically uncorrelated (Boles, 1991; Bryden, Hecaen, & DeAgostini, 1983). That is, knowing the direction and degree of lateralization on one cognitive task does not allow prediction of the direction and degree of lateralization on a different task. Obviously, one cannot rely a verbal dichotic listening test to provide information on lateralization of visuospatial processing.

Many chapters in these volumes (e.g., Farace & Turkheimer, Chapter 5, this volume; Kertesz, Chapter 6, volume I) comment on the problems attendant with neuroanatomic variability. There is much variability, for example, in brain size. It now appears most probable that there is a correlation between brain size and IQ (e.g., Willerman, Schultz, Rutledge, & Bigler, 1992), albeit a weak one. Further, intelligence is related to the extent of in-task brain activation (Haier, Siegel, Tang, Abel, & Buchsbaum, 1992). Given the ubiquity of the positive manifold phenomenon, individual differences in intelligence are likely correlated with individual differences in performance level on a wide variety of the cognitive tasks favored by neuroimaging investigators. Meaningful differences in patterns of activation may thus be related to overall brain size and IQ. Attention to these variables, either in subject selection or in statistical analysis, may reduce error variance.

Anatomic asymmetries are quite numerous, as are individual differences in anatomic asymmetries. As anatomic asymmetries may well be correlated with patterns of intellectual ability and intrahemispheric organization (Yeo, Turkheimer, Raz, & Bigler, 1987; Burke, Yeo, Delaney, & Conners, 1993), they represent markers of important individual differences. They are also easily measured from MRI scans. A recent study by Karbe *et al.* (1995) provides an example of how attention to individual differences in anatomic asymmetry can facilitate interpretation of functional neuroimaging data. They investigated activation of Brodmann's area 22 in a PET study of single word repetition. The extent of activation in this region of the left hemisphere had a strong inverse correlation with the extent of leftward asymmetry of the planum temporale (r = −.71). Such results clearly draw attention to the manner in which anatomic asymmetries might interact with physiologic activation. Schultz and colleagues (1994) recently reported that detailed, reliable anatomic measures failed to discriminate dyslexics from controls. Combining physiologic measures with anatomic studies may help reveal unique patterns of abnormality in this and other disorders.

Summary

Despite the enormous advantages in human neuroscience made possible by the various neuroimaging technologies, an understanding of the enormous complexity of human brain function remains beyond our grasp. In part, this reflects the rather "piecemeal" fashion in which neuroimaging data is collected. By conducting multimodal investigations, for example, combining the fine spatial resolution of fMRI with the fine temporal resolution of MEG, a more integrated picture of human brain function will undoubtedly emerge. Understanding the factors leading to individual variation, both environmental and genetic, and the specific nature of individual variation will facilitate the application of neuroimaging methodologies to important human problems.

References

Boles, D. B. (1991). Factor analysis and the cerebral hemispheres: Pilot study and parietal functions. *Neuropsychologia, 29,* 59–91.

Bryden, M. P., Hecaen, H., & DeAgostini, M. (1983). Patterns of cerebral organization. *Brain and Language, 20,* 249–262.

Burke, H. L., Yeo, R. A., Delaney, H. C., & Conners, L. (1993). CT scan cerebral asymmetries: Predictors of recovery from aphasia. *Journal of Clinical and Experimental Neuropsychology, 15,* 191–204.

Gangestad, S. W., & Yeo, R. A. (1994). Parental handedness and relative hand skill: A test of the developmental instability hypothesis. *Neuropsychology, 8,* 572–578.

Geschwind, N., & Galaburda, A. M. (1987). *Cerebral lateralization: Biologic mechanisms, associations, and pathology.* Cambridge, MA: MIT Press.

Haier, R. J., Siegel, B., Tang, C., Abel, L., & Buchsbaum, M. (1992). Intelligence and changes in regional cerebral glucose metabolic rate following learning. *Intelligence, 16,* 415–426.

Karbe, H., Wurker, M., Herholz, K., Ghaemi, M., Pietrzyk, U., Kessler, J., & Heiss, W. D. (1995). Planum temporale and Brodmann's area 22: Magnetic resonance imaging and high-resolution positron emission tomography demonstrate functional left-right asymmetry. *Archives of Neurology, 52,* 869–874.

Lykken, D. T., McGue, M., Tellegen, A., & Bouchard, T. J. (1992). Emergenesis. Genetic traits that do not run in families. *American Psychologist, 47,* 1565–1577.

Markow, T. A. (1994). *Developmental instability: Its origins and evolutionary implications.* Dodrecht: Kluwer Academic Publishers.

Neville, H. J., Mills, D. L., & Lawson, D. S. (1992). Fractionating language: Different neural subsystems with different sensitive periods. *Cerebral Cortex, 2,* 244–258.

Schultz, R. T., Cho, N. K., Staib, L. H., Kier, L. E., Fletcher, J. F., Shaywitz, S. E., Shanweiler, D. P., Katz, L., Gore, J. C., Duncan, J. S., & Shaywitz, B. A. (1994). Brain morphology in normal and dyslexic children: The influence of sex and age. *Annals of Neurology, 35,* 732–742.

Shaywitz, B. A., Shaywitz, S. E., Pugh, K. E., Constable, R. T., Skludarski, P., Fulbright, R. K., Bronen, R. A., Fletcher, J. M., Shankweller, D. P., Katz, L., & Gore, J. C. (1995). Sex differences in the functional organization of the brain for language. *Nature, 373,* 607–609.

Thatcher, R. W. (1992). Cyclic cortical reorganization during childhood. *Brain and Cognition, 20,* 24–50.

Tramo, M. J., Loftus, W. C., Thomas, C. E., Green, R. L., Mott, L. A., & Gazzaniga, M. S. (1995).

Surface area of human cerebral cortex and its gross morphological subdivisions: *In vivo* measurements in monozygotic twins suggest differential hemispheric effects of genetic factors. *Journal of Cognitive Neuroscience, 7,* 292–302.

Willerman, L., Schultz, R., Rutledge, J. N., & Bigler, E. (1992). In vivo brain size and intelligence. *Intelligence, 15,* 223–228.

Yeo, R. A., & Gangestad, S. W. (1993). Developmental origins of variation in human hand preference. *Genetica, 89,* 281–296.

Yeo, R. A., Turkheimer, E., Raz, N., & Bigler, E. D. (1987). Volumetric asymmetries of the human brain: Intellectual correlates. *Brain and Cognition, 6,* 15–23.

Index

ISBN 0-306-45229-4
90000
9 780306 452291